T0191804

UNIPA Springer Series

More information about this series at http://www.springer.com/series/13175

Giovanni Falcone

Editor

Lie Groups, Differential Equations, and Geometry

Advances and Surveys

Editor
Giovanni Falcone
Dipartimento di Matematica e Informatica
University of Palermo
Palermo, Italy

ISSN 2366-7516   ISSN 2366-7524   (electronic)
UNIPA Springer Series
ISBN 978-3-319-87249-0   ISBN 978-3-319-62181-4   (eBook)
DOI 10.1007/978-3-319-62181-4

Printed on acid-free paper

This Springer imprint is published by Springer Nature
The registered company is Springer International Publishing AG
The registered company address is: Gewerbestrasse 11, 6330 Cham, Switzerland

Preface

This volume collects articles from a number of authors who, mostly, met in the context of the scientific project Lie Groups, Differential Equations, and Geometry, which was supported by the European Union's Seventh Framework Programme (FP7/2007–2013) under grant agreement no. 317721.

Besides the many secondments, it behooves me to mention the exciting atmosphere of the conferences in Batumi (17–22 June 2013), Palermo (30 June–5 July 2014), and Chongqing (12–18 June 2015). The Georgian and Chinese hospitality was undoubtedly unrivaled, and while the forthcoming conference in Modica (8–14 June 2016) will be a pale shadow of those in Batumi and Chongqing, it will still be an occasion for fruitful discussions.

Last but not least, I would like to pay credit to the contributions of mathematicians of excellent caliber who kindly and ardently participated in this task, even if not supported by the project. It is an honor for me to introduce such a collection and a pleasure to thank the authors for granting me the privilege of their friendship.

Palermo, Italy, February 11, 2016
Giovanni Falcone

A few days before this volume went to the editor, my dearest friend and mentor, Karl Strambach, passed away. I cannot but dedicate to Karl, who was undoubtedly a main actor in the project, the small efforts I made in editing this volume.

Introduction

Lie theory takes its name from the Norwegian mathematician Marius Sophus Lie who, in 1871, at the age of 29, moved to Erlangen and visited Felix Klein to start a fruitful collaboration that unfortunately ended in reciprocal accusations. Today, Lie theory finds a major place in many branches of mathematics, both abstract and applied. It is current not only important in algebra and geometry, but also plays a significant role in some fields of physics, for example quantum theory, and in control theory (cf., e.g., [1]). The reason for the wide presence of Lie theory probably lies in the fact that it links several basic questions in mathematics, that is, linearization and approximation, exponentiation, differentiation, and nonassociativity.

In Chap. 1, we begin from the very first definitions, giving short surveys of basic topics in the theory of Lie algebras and in Representation theory, two essays of which are given in Chaps. 2 and 3. In the former, the author deals with the representation of the Heisenberg Lie algebra in an infinite dimensional Hilbert space, which, in Quantum Mechanics, describes the commutation relation between the position Q and the momentum P. In the latter, the author, led by the most influential Kazhdan–Lusztig conjecture posed in 1979, aims to describe representation theory from a categorical point of view and shows how this approach comes to determine irreducible characters, that is, the traces of the images of an irreducible representation, which are a basic tool that summarizes much information of the representation itself.

In Chap. 1, Sect. 1.1.3, the connection between Lie algebras and Lie groups is sketched. After showing that the exponential map transforms the Heisenberg Lie algebra into the Heisenberg group, we give the first few terms of the Baker–Campbell–Hausdorff formula, which is applied, in Chap. 4, to the problem of the computation of order conditions for exponential splitting schemes, together with an alternative approach for finding the order conditions by inspecting the coefficients of leading Lyndon–Shirshov words in an exponential function of a sum of Lie elements. Lyndon–Shirshov words produce a basis of a free Lie algebra (cf. [2, 3]), and Chibrikov [4] generalized the Shirshov construction of a free basis for a Sabinin algebra. On the other hand, in the words of Sabinin and Mikheev [5], a "more natural context for the Baker–Campbell–Hausdorff formula is the completion of the universal enveloping algebra of a relatively free Sabinin algebra on two generators," thus indicating a connection with the theory of loops. Roughly speaking, a loop is a group where associativity is, at best, replaced by some weaker algebraic law. Extending previous results of Malcev and Kuzmin, Sabinin and Mikheev (see also [6]) showed that the correspondence between Lie algebras and local Lie groups is valid in the nonassociative case, as well, by defining an algebraic structure on the tangent space of any analytic loop, which is called, in fact, Sabinin algebra.

In Chap. 1, Sect. 1.2.1, we give a short recap of the theory of loops and its relation to the foundations of Geometry, a connection almost forgotten today. In Chap. 6, the authors study half-automorphisms of Moufang loops and, more generally, of dissociative 2-loops, that is, such that any two elements generate a group, and introduce the well-known Cayley–Dickson construction of a sequence of algebras of dimension $2n$, whose multiplicative loops in dimensions 8 and 16 are the Moufang loop of octonions and the sedenion loop, respectively. Besides Lie algebras and Cayley (or octonion) algebras, among nonassociative algebras we also find Jordan algebras and Gerstenhaber algebras, which play an important role in Mathematical Physics. Gerstenhaber algebras are a main object of Chap. 5, and in Chap. 1, Sect. 1.2.2, we give their definition.

As among Lie groups one finds all closed subgroups of the Euclidean group of motions, it is not surprising that Lie theory has found applications in control theory. Chapter 7 conducts a survey of classic and recent results on left-invariant affine control systems and the associated Hamilton-Poisson systems, extensively in the cases of the three-dimensional Lie groups, the $(2n + 1)$-dimensional Heisenberg groups, and the six-dimensional orthogonal group. Chapter 8, on the other hand, is devoted to optimal control problems whose behavior is described by quasilinear differential equations of the first order on the plane with nonlocal boundary conditions, proving an existence and uniqueness theorem of a generalized solution, and obtaining an a priori estimate.

Chapters 9 and 10 deal with differential equations, a field which has been motivating for Lie theory since its birth. It is well known that Lie's original aim was to develop a theory for ordinary differential equations, equivalent to Galois theory for algebraic equations, and it must be acknowledged that his approach unifies many solving techniques and also provides new ones which apply to both ordinary and partial differential equations. The book by Olver [7] is a recommended source for all beginners.

Indeed, the composition of differential operators is a nonassociative product; thus they often produce a Lie algebra of operators. In Chap. 1, Sect. 1.1.3, we give the definition of the Weyl algebra of differential operators, showing its connections to the Heisenberg algebra and to the irreducible representations of sl(2;k).

The Lie group symmetry is often also applied to Finsler geometry, where the computations are usually more difficult than in Riemannian geometry. In Chap. 1, Sect. 1.3, we give an account of Finsler geometry. By constructing tangent Lie algebras to the holonomy group of a non-Riemannian Finsler manifold, in Chap. 12 the authors introduce the reader to a method for the investigation of its holonomy properties, giving a unified treatment of their previous results. Some advances in the smooth solutions of Hilbert's fourth problem in the regular case are presented in Chap. 11, where the authors give a classification of some families of projectively flat Finsler metrics.

As a natural environment for Lepage manifolds, Finsler geometry also appears in Chap. 13, where the author illustrates how the geometric properties of Lepage manifolds mirror geometrical, topological, and dynamical properties of differential equations, connecting differential equations with geometry and physics.

Palermo, Italy   Giovanni Falcone

References

1. R. Gilmore, *Lie Groups, Physics, and Geometry: An Introduction for Physicists, Engineers and Chemists* (Cambridge University Press, Cambridge, 2008)
2. A.I. Shirshov, On free Lie rings. Mat. Sb. **45**, 113–122 (1958, in Russian)
3. A.I. Shirshov, On bases of a free Lie algebra. Algebra Logika **1**, 14–19 (1962, in Russian)
4. E. Chibrikov, On free Sabinin algebras. Commun. Algebra **39**, 4014–4035 (2011)
5. L.V. Sabinin, P.O. Mikheev, Infinitesimal theory of local analytic loops. Sov. Math. Dokl. **36**(3), 545–548 (1988)
6. L.V. Sabinin, *Smooth Quasigroups and Loops* (Kluwer Academic Publishers, Dordrecht, 1999)
7. P.J. Olver, *Applications of Lie Groups to Differential Equations*. Graduate Texts in Mathematics, vol. 107 (Springer, New York, 1986)

Contents

List of Contributors

M. Abashidze Batumi Shota Rustaveli State University, Batumi, Georgia

Winfried Auzinger Technische Universität Wien, Vienna, Austria

Rory Biggs Department of Mathematics, Rhodes University, Grahamstown, South Africa

Xinyue Cheng School of Mathematics and Statistics, Chongqing University of Technology, Chongqing, People's Republic of China

D. Devadze Batumi Shota Rustaveli State University, Batumi, Georgia

Giovanni Falcone Dipartimento di Matematica e Informatica, Palermo, Italy

Peter Fiebig Department Mathematik, FAU Erlangen-Nürnberg, Erlangen, Germany

Á. Figula Institute of Mathematics, University of Debrecen, Debrecen, Hungary

Romeo Galdava Sokhumi State University, Tbilisi, Georgia

David Gulua Georgian Technical University, Tbilisi, Georgia

Wolfgang Herfort Technische Universität Wien, Vienna, Austria

Tornike Kadeishvili A. Razmadze Mathematical Institute of Tbilisi State University, Tbilsi, Georgia

Othmar Koch Fakultät für Mathematik, Universität Wien, Vienna, Austria

Shuhua Liu School of Mathematics and Statistics, Chongqing University of Technology, Chongqing, People's Republic of China

Xiaoyu Ma School of Mathematics and Statistics, Chongqing University of Technology, Chongqing, People's Republic of China

M.Z. Menteshashvili Muskhelishvili Institute of Computational Mathematics of the Georgian Technical University, Tbilisi, Georgia

Sokhumi State University, Tbilisi, Georgia

Maria de Lourdes Merlini-Giuliani Universidade Federal do ABC, Santo André, Brazil

Zoltán Muzsnay Institute of Mathematics, University of Debrecen, Debrecen, Hungary

Péter T. Nagy Institute of Applied Mathematics, Óbuda University, Budapest, Hungary

Peter Plaumann Universidad Autónoma "Benito Juárez" de Oaxaca, Oaxaca de Juárez, Mexico

Claudiu C. Remsing Department of Mathematics, Rhodes University, Grahamstown, South Africa

Jemal Rogava I. Vekua Institute of Applied Mathematics, Iv. Javakhishvili Tbilisi State University, Tbilisi, Georgia

Olga Rossi Department of Mathematics, Faculty of Science, The University of Ostrava, Ostrava, Czech Republic

Liudmila Sabinina Universidad Autónoma del Estado de Morelos, Cuernavaca, Mexico

Ts. Sarajishvili Batumi Shota Rustaveli State University, Batumi, Georgia

Yuling Shen School of Mathematics and Statistics, Chongqing University of Technology, Chongqing, People's Republic of China

Karl Strambach Department Mathematik, FAU Erlangen-Nürnberg, Erlangen, Germany

Mechthild Thalhammer Institut für Mathematik, Universität Innsbruck, Innsbruck, Austria

Camillo Trapani Dipartimento di Matematica e Informatica, Palermo, Italy

Ming Xu College of Mathematics, Tianjin Normal University, Tianjin, People's Republic of China

Chapter 1
A Short Survey on Lie Theory and Finsler Geometry

Giovanni Falcone, Karl Strambach, and Ming Xu

The aim of this chapter is to provide readers with a common thread to the many topics dealt with in this book, which is intermediate and which aims at postgraduate students and young researchers, wishing to intrigue those who are not experts in some of the topics. We also hope that this book will contribute in encouraging new collaborations among disciplines which have stemmed from related problems. In this context, we take the opportunity to recommend the book by Hawkins [18]. A basic bibliography is outlined between the lines.

1.1 Generalities

A *Lie algebra* over a field k is a k-vector space $\mathfrak{g}$, endowed with a non-associative binary operation $[\cdot, \cdot] : \mathfrak{g} \times \mathfrak{g} \rightarrow \mathfrak{g}$, called *Lie bracket*, which is bilinear on each factor (that is, $[a\mathbf{x}+b\mathbf{y}, \mathbf{z}] = a[\mathbf{x}, \mathbf{z}]+b[\mathbf{y}, \mathbf{z}]$, and $[\mathbf{z}, a\mathbf{x}+b\mathbf{y}] = a[\mathbf{z}, \mathbf{x}]+b[\mathbf{z}, \mathbf{y}]$), alternating (that is, $[\mathbf{x}, \mathbf{x}] = 0$) and fulfilling the *Jacobi identity*: $[\mathbf{x}, [\mathbf{y}, \mathbf{z}]] + [\mathbf{z}, [\mathbf{x}, \mathbf{y}]] + [\mathbf{y}, [\mathbf{z}, \mathbf{x}]] = 0$. If the characteristic of k is not 2, then $[\mathbf{x}, \mathbf{x}] = 0 \iff [\mathbf{x}, \mathbf{y}] = -[\mathbf{y}, \mathbf{x}]$. The cases where $k = \mathbb{R}, \mathbb{C}, \mathbb{Q}_p$ are of particular interest. A very classic introductory book is the one by Jacobson [21].

G. Falcone (✉)
Dipartimento di Matematica e Informatica, via Archirafi 34, 90123 Palermo, Italy
e-mail: giovanni.falcone@unipa.it

K. Strambach
Department Mathematik, FAU Erlangen-Nürnberg, Cauerstr. 11, 91058 Erlangen, Germany
e-mail: strambach@mi.uni-erlangen.de

M. Xu
College of Mathematics, Tianjin Normal University, 300387 Tianjin, People's Republic of China
e-mail: mgmgmgxu@163.com

© Springer International Publishing AG 2017
G. Falcone (ed.), *Lie Groups, Differential Equations, and Geometry*,
UNIPA Springer Series, DOI 10.1007/978-3-319-62181-4_1

Any associative algebra over a field k, with the Lie bracket defined by

$$[\mathbf{x}, \mathbf{y}] = \mathbf{x}\mathbf{y} - \mathbf{y}\mathbf{x},$$

is a Lie algebra. In particular, the algebra of endomorphisms of a k-vector space V, isomorphic to the algebra of $n \times n$ matrices with entries in k, is a prototypical example, denoted by $\mathfrak{gl}(V)$, or $\mathfrak{gl}(n, \mathsf{k})$. Another basic example is its Lie subalgebra of traceless endomorphisms, denoted by $\mathfrak{sl}(V)$, or $\mathfrak{sl}(n, \mathsf{k})$ (notice that it is closed under Lie bracket, but not closed under addition, nor multiplication). Finally, we want to mention the Heisenberg algebra $\mathfrak{h}_3$ of strictly upper triangular 3×3 matrices, which is generated by three elements

$$\mathbf{x} = \begin{pmatrix} 0 & 1 & 0 \\ 0 & 0 & 0 \\ 0 & 0 & 0 \end{pmatrix}, \quad \mathbf{y} = \begin{pmatrix} 0 & 0 & 0 \\ 0 & 0 & 1 \\ 0 & 0 & 0 \end{pmatrix}, \quad \mathbf{z} = \begin{pmatrix} 0 & 0 & 1 \\ 0 & 0 & 0 \\ 0 & 0 & 0 \end{pmatrix} \tag{1.1}$$

fulfilling the Lie brackets

$$[\mathbf{x}, \mathbf{y}] = \mathbf{z},$$

and which can be generalized, more abstractly, to the Lie algebra $\mathfrak{h}_{2n+1}$, generated by $2n + 1$ elements $\{\mathbf{x}_1, \ldots, \mathbf{x}_n, \mathbf{y}_1, \ldots, \mathbf{y}_n, \mathbf{z}\}$ and defined by the Lie brackets

$$[\mathbf{x}_h, \mathbf{y}_k] = \delta_{hk}\mathbf{z}, \tag{1.2}$$

where δ_{hk} is the Kronecker symbol. We notice, in fact, that, in order to compute the Lie brackets between vectors of a Lie algebra, it is sufficient to know the nonzero Lie brackets between the vectors of a basis.

Similarly to the case of associative algebras, one defines and distinguishes the concept of a Lie *subalgebra* and the one of a Lie *ideal*: a Lie subalgebra I is an *ideal* of $\mathfrak{g}$, if

$$[\mathbf{x}, \mathbf{y}] \in I, \text{ for all } \mathbf{x} \in I \text{ and } \mathbf{y} \in \mathfrak{g}.$$

For instance, $\mathfrak{sl}(n, \mathsf{k})$ is an ideal of $\mathfrak{gl}(n, \mathsf{k})$.

Two basic ideals are the *commutator* ideal $[\mathfrak{g}, \mathfrak{g}]$, generated by all elements $[\mathbf{x}, \mathbf{y}]$, and the *center* $\mathfrak{z}$, consisting of all elements $\mathbf{x}$ such that $[\mathbf{x}, \mathfrak{g}] = 0$. One says that $\mathfrak{g}$ is *Abelian*, if $[\mathfrak{g}, \mathfrak{g}] = 0$ (or, equivalently, if $\mathfrak{z} = \mathfrak{g}$).

As for associative rings, ideals are precisely the kernels of *homomorphisms*. Given a Lie algebra $\mathfrak{h}$ and an ideal I in it, one constructs the factor Lie algebra $\mathfrak{h}/I$: its elements are the cosets $x + I$.

This is probably the right place to mention the following theorem, which was proved in characteristic zero in 1935 by Ado and in arbitrary characteristic in 1948/1949 by Iwasawa and Harish-Chandra:

Every finite-dimensional Lie algebra $\mathfrak{h}$ over a field k is isomorphic to a Lie subalgebra of $\mathfrak{gl}(n, \mathsf{k})$ for some integer n, that is, it has a faithful representation, where a *representation* of a Lie algebra $\mathfrak{h}$ on k^n is a Lie algebra homomorphism

$$\pi : \mathfrak{h} \to \mathfrak{gl}(n, \mathsf{k}),$$

and a representation is said to be *faithful* if its kernel is trivial, i.e., it is injective.

In a nutshell, the structure of finite-dimensional Lie algebras can be summarized as follows. The commutator ideal $[\mathfrak{g}, \mathfrak{g}]$ brings on two series of ideals: the *derived* series

$$\mathfrak{g}^{(k)} = [\mathfrak{g}^{(k-1)}, \mathfrak{g}^{(k-1)}],$$

and the *descending central* series

$$\mathfrak{g}^{[k]} = [\mathfrak{g}, \mathfrak{g}^{[k-1]}],$$

where we put $\mathfrak{g}^{(0)} = \mathfrak{g}^{[0]} = \mathfrak{g}$. A Lie algebra is called *solvable* if the derived series eventually ends to zero, and *nilpotent* if the descending central series eventually ends to zero. Inductively, one sees that $\mathfrak{g}^{(k)} \leqslant \mathfrak{g}^{[k]}$, hence nilpotent Lie algebras are solvable. Moreover, it turns out that $\mathfrak{g}$ is solvable if and only if $[\mathfrak{g}, \mathfrak{g}]$ is nilpotent. In 1876, Sophus Lie proved that a finite-dimensional complex solvable Lie algebra has a common eigenvector, and, by a recursive argument, this yields a matrix representation in upper triangular form. Every Lie algebra has a unique maximal solvable ideal, called the *radical*.

If the only ideals of a non-Abelian Lie algebra $\mathfrak{g}$ are zero and $\mathfrak{g}$ itself, then $\mathfrak{g}$ is called *simple*. The classification of simple complex Lie algebras was essentially carried out by Wilhelm Killing in the late 1880s and sensibly improved by Élie Cartan who extended it to the real case in 1914. If the radical is zero, the Lie algebra is called *semisimple*. Every semisimple Lie algebra turns to decompose as the direct sum of simple Lie algebras. Finally, in 1905 Eugenio Elia Levi proved that every real Lie algebra decomposes as the semidirect sum of its radical with a semisimple subalgebra, called thereafter a *Levi* subalgebra of $\mathfrak{g}$ (which, in general, is not an ideal of $\mathfrak{g}$).

1.1.1 Representations, Derivations, Killing Form

The reader can easily imagine that the same Lie algebra can appear in very different ways. A representative case is the one of the three-dimensional, simple, Lie algebra $\mathfrak{sl}(2, \mathsf{k})$ of traceless 2×2 matrices, which is generated by the matrices

$$X_0 = \begin{pmatrix} 1 & 0 \\ 0 & -1 \end{pmatrix}; \quad X_1 = \begin{pmatrix} 0 & 1 \\ 0 & 0 \end{pmatrix}; \quad X_2 = \begin{pmatrix} 0 & 0 \\ 1 & 0 \end{pmatrix}$$

fulfilling

$$[X_0, X_1] = 2X_1, \quad [X_0, X_2] = -2X_2, \quad [X_1, X_2] = X_0,$$

and which has, for every integer $n \geq 2$, a unique (up to conjugation) faithful representation

$$\psi_n : \mathfrak{sl}(2, \mathsf{k}) \longrightarrow \mathfrak{sl}(n, \mathsf{k}) \leqslant \mathfrak{gl}(n, \mathsf{k}) \tag{1.3}$$

defined, for $0 \leqslant k \leqslant n$, by the following action on the vectors $\{e_0, \ldots, e_{n-1}\}$ of a basis:

$$\psi_n(X_0)e_k = (n - 2k - 1)e_k; \quad \psi_n(X_1)e_k = ke_{k-1}; \quad \psi_n(X_2)e_k = (n - k - 1)e_{k+1}.$$

Notice that each ψ_n is *irreducible*, that is, it has no nontrivial invariant subspace.

We mention here a fundamental theorem of Hermann Weyl in the theory of representations of Lie algebras: any finite-dimensional representation of a semisimple Lie algebra $\mathfrak{g}$ is *completely reducible*, that is, every $\mathfrak{g}$-invariant subspace has a $\mathfrak{g}$-invariant complementary subspace. On the contrary, the upper triangular representation of a solvable Lie algebra $\mathfrak{g}$ determines a flag of $\mathfrak{g}$-invariant subspaces having no $\mathfrak{g}$-invariant complementary subspace.

Physical observables can be represented by operators, acting on the Hilbert space of the states of the system, whose commutating relations are defined through Lie brackets. Two examples: first, if Q represents the position of a particle and P its momentum, then the Heisenberg principle states that $[P, Q] = -i\mathbf{1}$, where $\mathbf{1}$ is the identity operator; second, if the differential operators L_x, L_y, L_z represent the components of the angular momentum, then they satisfy the following commutating relations:

$$[L_x, L_y] = i\hbar L_z, \quad [L_y, L_z] = i\hbar L_x, \quad [L_z, L_x] = i\hbar L_y. \tag{1.4}$$

The three-dimensional real Lie algebra generated by L_x, L_y, L_z is isomorphic to the Lie algebra denoted by $\mathfrak{su}(2)$, and is a subalgebra of the complex Lie algebra $\mathfrak{sl}(2, \mathbb{C})$ described in (1.3), but in Chap. 2 the author looks at these commutation relations within an infinite-dimensional representation in a Hilbert space of unbounded operators.

A fundamental, direct representation is obtained when one considers, for a given $\mathbf{x} \in \mathfrak{h}$, the linear map $\mathrm{ad}_\mathbf{x} : \mathfrak{h} \longrightarrow \mathfrak{h}$ defined by

$$\mathrm{ad}_\mathbf{x}(\mathbf{y}) = [\mathbf{x}, \mathbf{y}],$$

which is called the *adjoint* map of $\mathbf{x}$. Notice that the Jacobi identity itself can be re-written in terms of the adjoint map as

$$\mathrm{ad}_\mathbf{x}([\mathbf{y}, \mathbf{z}]) = [\mathrm{ad}_\mathbf{x}(\mathbf{y}), \mathbf{z}] + [\mathbf{y}, \mathrm{ad}_\mathbf{x}(\mathbf{z})].$$

More generally, a map $\delta : \mathfrak{h} \longrightarrow \mathfrak{h}$ such that $\delta[\mathbf{y}, \mathbf{z}] = [\delta\mathbf{y}, \mathbf{z}] + [\mathbf{y}, \delta\mathbf{z}]$ (and, recursively, such that Leibniz rule

$$\delta^n[\mathbf{y}, \mathbf{z}] = \sum_0^n \binom{n}{k} [\delta^k\mathbf{y}, \delta^{n-k}\mathbf{z}]$$

is fulfilled) is called a *derivation*, and the derivations of the form $\mathrm{ad}_\mathbf{x}$ are called *inner* derivations.

A corroboration of the self-consistency of Lie theory comes from the fact that, whereas the product of two derivations δ_1 and δ_2 is not necessarily a derivation, it is immediate to see that the Lie bracket

$$[\delta_1, \delta_2]$$

is still a derivation, that is, $\mathrm{Der}(\mathfrak{g})$ is a Lie subalgebra of the Lie algebra $\mathfrak{gl}(\mathfrak{g})$ of endomorphisms of the vector space $\mathfrak{h}$. Moreover, if δ is a derivation of $\mathfrak{g}$, then the vector space generated by $\mathfrak{g}$ and δ becomes a Lie algebra putting $[\delta, \mathbf{y}] = \delta(\mathbf{y})$.

The map $\mathrm{ad} : \mathfrak{g} \longrightarrow \mathrm{Der}(\mathfrak{g})$, $\mathbf{x} \mapsto \mathrm{ad}_\mathbf{x}$, is called the *adjoint representation* of $\mathfrak{g}$. Notice that the kernel of ad is precisely the center $\mathfrak{z}$ of $\mathfrak{g}$, therefore, if $\mathfrak{z} = 0$, the adjoint representation is faithful and represents concretely the Lie algebra $\mathfrak{g}$ as a subalgebra of $\mathfrak{gl}(\mathfrak{g})$. This can be seen, but only for Lie algebras with trivial center, as an alternative proof of Ado's theorem.

The adjoint representation provides a fundamental tool in Representation theory: the *Killing form*

$$\kappa(\mathbf{x}, \mathbf{y}) = \mathrm{Tr}(\mathrm{ad}_\mathbf{x} \circ \mathrm{ad}_\mathbf{y}),$$

which has the following properties: it is a symmetric bilinear form, which is zero if and only if $\mathfrak{g}$ is nilpotent, and which is non-degenerate if and only if $\mathfrak{g}$ is semisimple (Cartan criterion). In the latter case, if $\mathsf{k} = \mathbb{R}$, then the Killing form will discriminate the case where the semisimple Lie algebra $\mathfrak{g}$ is *compact* (see p. 9). The κ-orthogonal of any ideal I of $\mathfrak{g}$ is also an ideal, and two ideals with trivial intersection are κ-orthogonal.

For a minimal bibliographical support, we address the reader to one of the most popular introductory books in Representation theory, that is, the one by Humphreys [20].

1.1.2   The Weyl Algebra of Differential Operators

As mentioned in the opening, the composition of differential operators is a non-associative product, thus they often produce a Lie algebra of operators. A good example to be mentioned here is the *Weyl algebra* of differential operators with polynomial coefficients in n variables. Its generators x_k (acting as $f(x) \longmapsto x_k f(x)$) and $\partial/\partial x_k$ satisfy the Lie bracket

$$[\partial/\partial x_h, x_k] = \delta_{hk},$$

because

$$\left(\frac{\partial}{\partial x_h} \circ x_k - x_k \circ \frac{\partial}{\partial x_h}\right) f(x) = \frac{\partial}{\partial x_h}\big(x_k f(x)\big) - x_k \frac{\partial f}{\partial x_h}(x) =$$

$$\delta_{hk} f(x) + x_k \frac{\partial f}{\partial x_h}(x) - x_k \frac{\partial f}{\partial x_h}(x) = \delta_{hk} f(x).$$

It was introduced by Hermann Weyl in his mathematical formulation of quantum mechanics, and it is very closed related to the (generalized) Heisenberg algebra $\mathfrak{h}_{2n+1}$ seen in (1.2). More precisely, the Weyl algebra is a quotient of the *universal enveloping algebra* of $\mathfrak{h}_{2n+1}$ (that is, the free algebra on $\mathfrak{h}_{2n+1}$ modulo the ideal generated by the relations $\mathbf{xy} - \mathbf{yx} - [\mathbf{x}, \mathbf{y}]$), obtained by factoring the ideal generated by $\mathbf{z} - 1$.

Notice, in passing, that also each irreducible representation ψ_n of $\mathfrak{sl}(2, \mathsf{k})$ in (1.3) can be seen as the action of the differential operators

$$\Delta_0 = x_2 \frac{\partial}{\partial x_2} - x_1 \frac{\partial}{\partial x_1}, \quad \Delta_1 = x_2 \frac{\partial}{\partial x_1}, \quad \Delta_2 = x_1 \frac{\partial}{\partial x_2}$$

on the algebra of homogeneous polynomial of degree $n - 1$ in two variables x_1 and x_2, because they act on the vectors of the basis of the form $e_k = x_1^k x_2^{n-k-1}$ mapping

$$\Delta_0(e_k) = (n - 2k - 1)e_k, \quad \Delta_1(e_k) = k e_{k-1}, \quad \Delta_2(e_k) = (n - k - 1)e_{k+1}.$$

1.1.3   Lie Groups, Exponential Map

A Lie group is a group defined over a finite-dimensional smooth manifold, such that the group multiplication and the inversion are smooth maps. The general linear group $GL(n, \mathbb{R})$ of $n \times n$ invertible matrices over $\mathbb{R}$ is a n^2-dimensional smooth manifold where the group operation is differentiable, hence it is a Lie group. This holds also for its topologically closed subgroups (Cartan's theorem), also called *linear* Lie groups.

Similarly to what one does by taking the linear part of the Maclaurin series of a function, one can associate to each connected Lie group G, a Lie algebra $\mathfrak{g}$ whose underlying vector space is the tangent space to the group in the neutral element [27]. This Lie algebra $\mathfrak{g}$ completely captures the local structure of the group G. There can be several Lie groups with the same Lie algebra, so this is not one-to-one correspondence, and one says that two Lie groups are *locally isomorphic*, if their

Lie algebras are isomorphic. All of them, however, are a quotient of the *largest* one, the so-called (simply connected) *universal covering group* associated with $\mathfrak{g}$. In this correspondence, subalgebras (resp. ideals) correspond to closed subgroups (resp. closed normal subgroups).

Two smallest, prototypical examples: first, consider the group $SO_2(\mathbb{R})$ of rotations

$$X = \begin{pmatrix} \cos\theta & \sin\theta \\ -\sin\theta & \cos\theta \end{pmatrix} = \begin{pmatrix} 1 & 0 \\ 0 & 1 \end{pmatrix} + \begin{pmatrix} 0 & \theta \\ -\theta & 0 \end{pmatrix} + \begin{pmatrix} -\frac{1}{2}\theta^2 & 0 \\ 0 & -\frac{1}{2}\theta^2 \end{pmatrix} + \begin{pmatrix} 0 & -\frac{1}{6}\theta^3 \\ \frac{1}{6}\theta^3 & 0 \end{pmatrix} + \cdots,$$

with $\theta \in \mathbb{R}$, which, by the way, are written in complex form as

$$1 + i\theta - \frac{1}{2}\theta^2 - \frac{1}{6}\theta^3 + \cdots = \exp(i\theta).$$

In a neighborhood of the identity, that is, considering the first order approximation, we see that

$$\begin{pmatrix} 1 & \theta_1 \\ -\theta_1 & 1 \end{pmatrix} \begin{pmatrix} 1 & \theta_2 \\ -\theta_2 & 1 \end{pmatrix} = \begin{pmatrix} 1 - \theta_1\theta_2 & \theta_1 + \theta_2 \\ -\theta_1 - \theta_2 & 1 - \theta_1\theta_2 \end{pmatrix},$$

and, disregarding infinitesimal elements of the second order, this says that the multiplication of infinitesimal rotations behaves like the addition of arcs in the tangent space to the circle at the identity, which is

$$I + \theta \cdot \frac{\partial}{\partial\theta}X|_{\theta=0} = I + \begin{pmatrix} 0 & -\theta \\ \theta & 0 \end{pmatrix} \quad \theta \in \mathbb{R}.$$

This explains also why Sophus Lie called *infinitesimal groups* what we now call Lie algebras. Moreover, Lie theory gives a deeper meaning to the concept of exponential map: the usual one, $x \mapsto \exp(x)$, transforms the additive group of the reals onto the multiplicative group of the positive reals. Similarly, the above example of $SO_2(\mathbb{R})$ shows that the exponential map $x \mapsto \exp(ix)$ transforms the additive group of the reals onto the group of rotations. Hence, the Lie groups $\mathbb{R}^*$ (non-connected), $\mathbb{R}_+$ (simply connected), and $SO_2(\mathbb{R})$ (not simply connected) are locally isomorphic, with $SO_2(\mathbb{R})$ being in fact a homomorphic image of the simply connected group $\mathbb{R}_+$ modulo the lattice of periods generated by 2π, and the connected component of $\mathbb{R}^*$ being isomorphic to $\mathbb{R}_+$ via the exponential map.

The second example is the Heisenberg group

$$H_3 = \left\{ X = \begin{pmatrix} 1 & x_1 & x_3 \\ 0 & 1 & x_2 \\ 0 & 0 & 1 \end{pmatrix} : x_1, x_2, x_3 \in \mathbb{R} \right\}.$$

Since for $k = 1, 2, 3$ the three derivatives $\frac{\partial}{\partial x_k} X|_{X=I}$ are, respectively, the three elements $\mathbf{x}$, $\mathbf{y}$, and $\mathbf{z}$ given in (1.1), the tangent space at the identity element I is $\mathfrak{h}_3$.

On the other hand, as the cube of any element $\mathbf{x} \in \mathfrak{h}_3$ is zero, we see that $\exp \mathbf{x} = I + \mathbf{x} + \frac{1}{2}\mathbf{x}^2$, that is,

$$\mathbf{x} = \begin{pmatrix} 0 & a & b \\ 0 & 0 & c \\ 0 & 0 & 0 \end{pmatrix} \longmapsto \exp \mathbf{x} = I + \mathbf{x} + \frac{1}{2}\mathbf{x}^2 = \begin{pmatrix} 1 & a & b + \frac{1}{2}ac \\ 0 & 1 & c \\ 0 & 0 & 1 \end{pmatrix},$$

and

$$\exp \mathbf{x} \cdot \exp \mathbf{y} = \exp\left(\mathbf{x} + \mathbf{y} + \frac{1}{2}[\mathbf{x}, \mathbf{y}]\right). \tag{1.5}$$

Just in the same way as the exponential map $x \mapsto \exp(ix)$ transforms the additive group of the reals onto the group of rotations, the above Eq. (1.5) transforms the additive group of Lie algebra $\mathfrak{h}_3$ onto the non-commutative Lie group H_3.

Finally, for a given Lie algebra $\mathfrak{g}$ and for its Lie algebra of derivations Der $\mathfrak{g}$, one can consider the Lie group $\exp(\text{Der } \mathfrak{g})$ and find out that this is a group of automorphisms of $\mathfrak{g}$, such that

$$\exp(\delta(\mathbf{x})) = \exp(\delta)(\mathbf{x}),$$

confirming, once again, the self-consistency of Lie theory.

The question which arises after these examples is the following: does the exponential map always transform the addition of a suitable, but linear, Lie algebra structure on the tangent space to a Lie group G into the multiplicative structure of the group G itself? Surprisingly, within a series of circumstances which give an account of the richness of the theory, the answer is mostly yes. In order to be applied to the most general case where the group is not commutative, Eq. (1.5) must be completed to the *Baker–Campbell–Hausdorff formula*, whose first terms are

$$\exp(\mathbf{x}) \cdot \exp(\mathbf{y}) = \exp\left(\mathbf{x} + \mathbf{y} + \frac{1}{2}[\mathbf{x}, \mathbf{y}] + \frac{1}{12}\left([\mathbf{x}, [\mathbf{x}, \mathbf{y}]] + [\mathbf{y}, [\mathbf{y}, \mathbf{x}]]\right) + \ldots\right). \tag{1.6}$$

Three-dimensional Lie groups, $(2n + 1)$-dimensional Heisenberg groups, and six-dimensional orthogonal groups are extensively investigated in Chap. 7 as the manifolds on which a *left-invariant affine control system* is defined. A *smooth control system* Σ may be viewed as a family of dynamical systems (or vector fields) Ξ_u on a manifold, (smoothly) parametrized by *control* vectors $u \in \mathbb{R}^l$, and fulfilling

$$\frac{\partial}{\partial t} g = \Xi_u(g).$$

An integral curve of such a vector field is called a *trajectory* of the system, and the main question is whether any two points can be connected by a trajectory, possibly minimizing some cost function.

A significant subclass of control systems, first considered in 1972 by R. W. Brockett and by V. Jurdjevic and H.J. Sussmann, is that of *left-invariant, affine* control systems, where the manifolds are real, finite-dimensional, connected, linear Lie groups G, and the *dynamics* Ξ are invariant under left translations, i.e., $\Xi_u(g) = g\Xi_u(1)$, and affinely parametrized by control vectors $u \in \mathbb{R}^l$, i.e., $\Xi_u(1) = (A + u_1 B_1 + \cdots + u_l B_l)$, where $A, B_1, \ldots, B_l$ are elements of the Lie algebra $\mathfrak{g}$ of G, with $B_1, \ldots, B_l$ linearly independent. Such a system fulfills therefore

$$\frac{\partial}{\partial t} g = g(A + u_1 B_1 + \cdots + u_l B_l),$$

and for such a system the left translation of any trajectory is a trajectory. Additionally, the subclass of *drift-free* systems, i.e., systems with $A = 0$, is reinterpreted in Chap. 7 as invariant sub-Riemannian structures, and some examples are revisited.

1.1.4 Compact Lie Algebras

If the Lie group G is compact, one says that its Lie algebra $\mathfrak{g}$ is compact. But a useful intrinsic criterion for the compactness of $\mathfrak{g}$ comes from the Killing form κ: the Lie algebra $\mathfrak{g}$ is compact if and only if κ is definite negative, as we illustrate with the following example, which is the three-dimensional compact real group SU(2) of special unitary matrices, represented by the compact, simply connected group of quaternions having the norm equal to one:

$$X = \begin{pmatrix} x_1 & x_2 & x_3 & x_4 \\ -x_2 & x_1 & -x_4 & x_3 \\ -x_3 & x_4 & x_1 & -x_2 \\ -x_4 & -x_3 & x_2 & x_1 \end{pmatrix}, \quad x_1^2 + x_2^2 + x_3^2 + x_4^2 = 1.$$

As such, it is the *real form* of the group of complex matrices

$$\begin{pmatrix} z & w \\ -\overline{w} & \overline{z} \end{pmatrix}, \quad z = x_1 + ix_2, \quad w = x_3 + ix_4, \quad |z|^2 + |w|^2 = 1,$$

but, the conjugation being not complex analytic, this is not a complex Lie group (also, its real dimension is odd!).

Putting alternatively $x_3 = x_4 = 0$, $x_2 = x_4 = 0$, and $x_2 = x_3 = 0$, we obtain three one-dimensional subgroups of matrices of the form

$$\begin{pmatrix} \cos\theta & \sin\theta & 0 & 0 \\ -\sin\theta & \cos\theta & 0 & 0 \\ 0 & 0 & \cos\theta & -\sin\theta \\ 0 & 0 & \sin\theta & \cos\theta \end{pmatrix}, \quad \begin{pmatrix} \cos\theta & 0 & \sin\theta & 0 \\ 0 & \cos\theta & 0 & \sin\theta \\ -\sin\theta & 0 & \cos\theta & 0 \\ 0 & -\sin\theta & 0 & \cos\theta \end{pmatrix},$$

$$\begin{pmatrix} \cos\theta & 0 & 0 & \sin\theta \\ 0 & \cos\theta & -\sin\theta & 0 \\ 0 & \sin\theta & \cos\theta & 0 \\ -\sin\theta & 0 & 0 & \cos\theta \end{pmatrix},$$

whence one immediately sees that the tangent space at the identity element I is generated by

$$\begin{pmatrix} 0 & 1 & 0 & 0 \\ -1 & 0 & 0 & 0 \\ 0 & 0 & 0 & -1 \\ 0 & 0 & 1 & 0 \end{pmatrix}, \quad \begin{pmatrix} 0 & 0 & 1 & 0 \\ 0 & 0 & 0 & 1 \\ -1 & 0 & 0 & 0 \\ 0 & -1 & 0 & 0 \end{pmatrix}, \quad \begin{pmatrix} 0 & 0 & 0 & 1 \\ 0 & 0 & -1 & 0 \\ 0 & 1 & 0 & 0 \\ -1 & 0 & 0 & 0 \end{pmatrix}.$$

These matrices generate the real Lie algebra $\mathfrak{su}(2)$ given at p. 4, and are the real forms of the matrices

$$\begin{pmatrix} -i & 0 \\ 0 & i \end{pmatrix}, \quad \begin{pmatrix} 0 & 1 \\ -1 & 0 \end{pmatrix}, \quad \begin{pmatrix} 0 & -i \\ -i & 0 \end{pmatrix},$$

which in turn generate the complex Lie algebra $\mathfrak{sl}_2(\mathbb{C})$ of traceless matrices. Notice that the same complex Lie algebra is obtained as the complexification $\mathfrak{sl}_2(\mathbb{C}) = \mathfrak{sl}_2(\mathbb{R}) \otimes_{\mathbb{R}} \mathbb{C}$ of the real Lie algebra $\mathfrak{sl}_2(\mathbb{R})$. This shows that the Lie algebras $\mathfrak{g}$ of non-isomorphic real Lie groups may have isomorphic complexifications $\mathfrak{g}_{\mathbb{C}} = \mathfrak{g} \otimes_{\mathbb{R}} \mathbb{C}$.

More generally, starting from a real Lie algebra $\mathfrak{g}$, one can consider its complexification $\mathfrak{g}_{\mathbb{C}} = \mathfrak{g} \otimes_{\mathbb{R}} \mathbb{C}$, thus non-isomorphic real Lie groups may have isomorphic complexifications. Notice that the Killing form

$$\kappa(\mathbf{x}, \mathbf{y}) = \text{Tr}(\text{ad}_{\mathbf{x}} \circ \text{ad}_{\mathbf{y}})$$

of a semisimple real Lie algebra $\mathfrak{g}$ determines, by Sylvester's inertia, an *index*, which is the number of positive entries of the diagonal form of κ. It turns out that the index is zero, that is, κ is definite negative, if and only if $\mathfrak{g}$ is the Lie algebra of a compact real Lie group, and the real Lie algebra $\mathfrak{g}$ is said a *compact real form* of $\mathfrak{g}_{\mathbb{C}}$. In the opposite case, where κ is definite positive, the real Lie algebra $\mathfrak{g}$ is said a *split real form* of $\mathfrak{g}_{\mathbb{C}}$. The intermediate cases correspond to the *Cartan decomposition* $\mathfrak{g} = \mathfrak{k} \oplus \mathfrak{p}$ in two κ-orthogonal subspaces, the former being a maximal compact subalgebra, the latter being only a complementary subspace. For instance, in the (not infrequent) case that $\mathfrak{g}$ has a representation as a real Lie algebra of matrices that is closed under transposition, the Cartan decomposition is just the decomposition into the subalgebra of skew-symmetric part and the complementary subspace of symmetric matrices, thus $\mathfrak{g}_\mathbb{C} = (\mathfrak{k} \oplus i\mathfrak{p}) \oplus (i\mathfrak{k} \oplus \mathfrak{p})$. Now, $\mathfrak{k} \oplus i\mathfrak{p}$ is a subalgebra which consists of Hermitian matrices, is a real Lie subalgebra of $\mathfrak{g}_\mathbb{C}$, whose Killing form is negative definite, hence it is compact. Finally, its complexification is clearly $\mathfrak{g}_\mathbb{C}$.

1.2   Non-associative Structures

1.2.1   Loops and Geometry

As one already could see, the non-associative structures are recharged from many sources. One of them which today is almost forgotten is the theory of projective planes. If an affine plane E is non-desarguesian, then E can be coordinatized only by ternary fields, and the addition and multiplication of them are loops. In particular, translation planes are coordinatized by quasifields, whose addition defines an Abelian group but whose multiplication defines a loop such that these two operations are connected by a distributive law. The whole theory of non-associative algebras grew out of this basement. Today a *loop* is defined as a set L with a binary operation $(x, y) \mapsto x \cdot y$, such that there exists an element $e \in L$ with $e \cdot x = x \cdot e = x$ for all $x \in L$, and the equations $a \cdot y = b$ and $x \cdot a = b$ have precisely one solution. The calculations with loops based on this definition are tedious. For this reason one tried to formulate the theory of loops completely within the theory of groups. The possibility for this is delivered, for a loop L, by the left translations $\lambda_a : y \mapsto a \cdot y : L \to L$, which are bijections of L for any $a \in L$. Following Cayley, one can see the loop L as a set of permutations of L acting sharply transitively on L. If G is the group generated by the left translations of L, and H is the stabilizer of the identity 1 of L, then the left translations $\lambda_a : y \mapsto a \cdot y$ form a section $\sigma : G/H \to G$. If $\pi : G \to G/H$ is the natural mapping $x \mapsto xH$, then $\pi \circ \sigma : G/H \to G/H$ is the identity. The image $\sigma(G/H)$ forms a system of representatives for the left cosets of H in G. Conversely, let G be a group and H a subgroup of G containing no nontrivial normal subgroup of G. If $\sigma : G/H \to G$ is a section with $\sigma(H/H) = 1 \in G$ such that $\sigma(G/H)$ generates G and operates *sharply transitively* on G/H (which means that to any xH and yH there exists precisely one $z \in \sigma(G/H)$ with $zxH = yH$), then the multiplication on the factor space G/H given by $xH \cdot yH = \sigma(xH)yH$, respectively the multiplication on the set $\sigma(G/H)$ given by $x \cdot y = \sigma(xy)$, yields a loop having G as the group generated by the left translations $xH \mapsto \sigma(xH)$, respectively $x \mapsto \sigma(xH)$. The advantage of this procedure lies in the enormous power of group theory. Choosing in particular for G a finite group, a Lie group, an algebraic group, or a p-adic Lie group such that H is a closed subgroup and σ is a map in the corresponding category, then many remarkable classification results can be obtained, since the simple groups in these categories are known.

Special classes of loops may be defined by algebraic laws which can be seen as weakenings of the associativity. For loops which are near to groups the algebraic expressions for weak associativity and the validity of the configurations in the corresponding geometries show the same complexity. As an example for this, we can quote the classes of Moufang loops and Bol loops.

But, in general, nicely algebraically defined loops are rare from a geometrical point of view. For instance, the multiplicative loops of quasifields, in which only one distributive law holds, and which coordinatize translation planes, cannot be described in a simple algebraic way.

The real analytic loops possess tangential objects, Malcev algebras, Bol algebras, Sabinin algebras, which describe them near to the identity. Using Malcev algebras it was possible to create, for Moufang loops, a theory having the same level as the theory of Lie groups and to get a complete classification of analytic Moufang loops. Any connected analytic Moufang loop is an almost direct product of a Lie group and of the Moufang loops arising from the multiplicative loop of octonions. The real octonion algebra satisfies Moufang identities and is the only real composition non-associative division algebra arising by the Cayley–Dickson process, which, by the way, is introduced in Chap. 6 to define a sequence of algebras, among which one finds in particular the octonion and the sedenion loops as the multiplicative loop.

Extensive majority of non-commutative Lie groups can be seen as groups of left translations of topological loops. For instance, among the three-dimensional Lie groups, only the group of homotheties of the Euclidean plane $\mathbb{R}^2$ cannot be the group topologically generated by the left translations of a three-dimensional topological loop. The situation changes drastically if one considers Lie groups G which are generated by left translations together by right translations of loops homeomorphic to a Lie group. Already any loop L defined on a connected one-dimensional manifold such that its group generated by the left and right translations is locally compact must be a group isomorphic either to $\mathbb{R}$ or to $SO_2(\mathbb{R})$. In the case of the two-dimensional Lie loop corresponding to the hyperbolic plane and having the three-dimensional group $PSL_2(\mathbb{R})$ as its group generated by the left translations, it is the group generated by left *and right* translations of an infinite-dimensional Lie group. For loops corresponding to higher dimensional hyperbolic spaces one meets the same situation. This indicates that the simplicity of a Lie group is an obstruction for being a group generated by the left and right translations of a topological loop.

The compactness of a Lie group G is a serious obstruction even for the case that G is a group generated by the left translations of a topological loop homeomorphic to a Lie group. For instance, any 1-dimensional connected topological loop having a compact Lie group as the group topologically generated by its left translations is the orthogonal group $SO_2(\mathbb{R})$. Another example showing the restricting power of compact Lie groups is given by topological loops L homeomorphic to the 7-sphere or to the 7-dimensional real projective space. Any such loop L having a compact connected Lie group G as the group topologically generated by the left translations of L is one of the two 7-dimensional compact Moufang loops, G is locally isomorphic to $PSO_8(\mathbb{R})$ and the stabilizer H of the identity $e \in L$ is isomorphic to $SO_7(\mathbb{R})$. Moreover, if the compact Lie group G is topologically generated by the left translations of a proper topological loop, then G is homeomorphic to a connected semisimple Lie group. In a sense, conversely there does not exist any proper topological loop L which is homeomorphic to a connected quasi-simple Lie group and has a compact Lie group as the group topologically generated by its left translations. More generally, if L is homeomorphic to a product of quasi-simple simply connected compact Lie groups and has a compact Lie group G as the group topologically generated by its left translations, then G is at least 14-dimensional. In the case where $\dim G = 14$, the group G is locally isomorphic to $\mathrm{Spin}_3(\mathbb{R}) \times \mathrm{SU}_3(\mathbb{C}) \times \mathrm{Spin}_3(\mathbb{R})$ and L is homeomorphic to a group which is locally isomorphic to $\mathrm{Spin}_3(\mathbb{R}) \times \mathrm{SU}_3(\mathbb{C})$.

1.2.2 Gerstenhaber Algebras

As mentioned above, among non-associative algebras, that is, k-vector spaces with a non-associative, k-bilinear product, we also find Jordan algebras and Gerstenhaber algebras.

In particular, here we limit ourselves to give the definition of a Gerstenhaber algebra, because they are a main object of Chap. 5, and to address the reader to the excellent introduction on this topic given in [4].

A k-vector space A with a k-bilinear product $\cdot : A \times A \longrightarrow A$ and *associator* $(x, y, z) = (x \cdot y) \cdot z - x \cdot (y \cdot z)$ is a Gerstenhaber algebra (or *right-symmetric* algebra), if

$$(x, y, z) = (x, z, y).$$

Whilst the exterior algebra of a Lie algebra is a Gerstenhaber algebra, the commutator

$$[x, y] = x \cdot y - y \cdot x$$

endows a given Gerstenhaber algebra with the structure of a Lie algebra, because the Jacobi identity can be formally written as

$$[[x, y], z] + [[z, x], y] + [[y, z], x] =$$

$$(x, y, z) + (y, z, x) + (z, x, y) - (y, x, z) - (x, z, y) - (z, y, x).$$

The latter identity, holding in any two-sided distributive, not necessarily associative, algebra, shows that Gerstenhaber algebras and left-symmetric algebras deserve the alternate name of pre-Lie algebras.

Gerstenhaber algebras, and their opposite left-symmetric algebras, appeared, not surprisingly, in 1896 in a paper by Arthur Cayley. They were revived in the early 1960s by Vinberg [29], Koszul [24], and Gerstenhaber [17]. More recently, homotopy Gerstenhaber algebras found a place in the investigations of Kontsevich [22], and of Connes and Kreimer [6], among many others.

As pointed out in [23], it was Koszul [25] who realized that an *odd Laplacian*, that is, a second-order differential operator Δ of degree -1 on an associative commutative graded algebra $\mathfrak{A} = \bigcup_k \mathfrak{A}_k$ such that $\Delta^2 = 0$ and $\Delta(1) = 0$, provides $\mathfrak{A}$ with the structure of a Gerstenhaber algebra, by putting:

$$[x, y] = (-1)^k \Big(\Delta(a \cdot b) - \Delta(a) \cdot b - (-1)^k a \cdot \Delta(b) \Big),$$

where $x \in \mathfrak{A}_k$, $y \in \mathfrak{A}$. A slight generalization of the operator Δ defines Batalin–Vilkovisky algebras, a useful tool in the description of certain Lagrangian gauge theories [30].

1.3 Finsler Geometry

1.3.1 Finsler Metric and Minkowski Norm

Finsler geometry is more general than Riemannian geometry in the sense that it does not require the metric to be quadratic for each tangent space. Its definition was proposed by Riemann in his talk in 1854, and the systematical study on Finsler geometry was started by the thesis of Finsler in 1918 [16]. This research field has been quiet for many decades, until its arising in the 1990s, by the full advocation of S.S. Chern. He, together with D. Bao, Z. Shen, and many other geometers, rebuilt the foundations for modern Finsler geometry [2]. Nowadays, Finsler geometry has many applications in biological statistics, electrical science, and general relativity. It also provides viewpoints for us to understand this geometric world beyond intuition and imagination.

We start with the concept of *Finsler metric*, which defines the length of each tangent vector. A Finsler metric on a smooth manifold M is a continuous function $F : TM \to [0, +\infty)$, satisfying the following conditions:

1. Regularity: F is a positive smooth function on the slit tangent bundle $TM \backslash 0$.
2. Positive homogeneity: $F(\lambda y) = \lambda F(y)$ for any tangent vector y and $\lambda \geqslant 0$.
3. Strong convexity: For any *standard local coordinates* on TM, i.e., $x = (x^i) \in M$ and $y = y^i \partial_{x^i} \in T_x M$, the Hessian matrix

$$(g_{ij}(x, y)) = \left(\frac{1}{2} [F^2(x, y)]_{y^i y^j} \right)$$

is positive definite.

We call (M, F) a *Finsler manifold* or *Finsler space*.

In each tangent space, the Finsler metric defines a *Minkowski norm*. Minkowski norm can be more abstractly defined on any real vector space, satisfying similar conditions as above (1)–(3).

Well-known examples of Finsler metrics include the following. A Finsler metric is Riemannian when the Hessian matrix is irrelevant to the y-coordinates for any standard local coordinates. In this case, it defines a smooth global section $g_{ij}(x)dx^idx^j$ of $\mathrm{Sym}^2(T^*M)$ which is usually referred to as the *Riemannian metric*. The most simple and important class of non-Riemannian metrics are *Randers metrics*, which are of the form $F = \alpha + \beta$ where α is a Riemannian metric, and β is a 1-form. The (α, β)-*metrics* are generalizations of Randers metrics, which can be presented as $F = \alpha\phi(\beta/\alpha)$, where α and β are similar as for Randers metrics, and ϕ is a smooth function. Notice for a Randers metric (or an (α, β)-metric) F, α and β (or α, β, and ϕ resp.) must satisfy certain conditions to satisfy the convexity condition (3) [5, 26].

1.3.2 Geodesic and Curvature

The arc lengths of piecewisely smooth curves can be similarly defined in Finsler geometry. Their infimum for all curves from one fixed point to another provides a (possibly irreversible) distance $d_F(\cdot, \cdot)$. Using variational method, geodesics can be defined which satisfies the locally minimizing principle for $d_F(\cdot, \cdot)$.

There is another way to describe the geodesics on a Finsler space. The *geodesic spray* **G** is a global smooth tangent vector field on $TM\backslash 0$, which can be presented as $\mathbf{G} = y^i\partial_{x^i} - 2\mathbf{G}^i\partial_{y^i}$, with $\mathbf{G}^i = \frac{1}{4}g^{il}([F^2]_{x^ky^l}y^k - [F^2]_{x^l})$, for any standard local coordinates on TM. Then a smooth curve $c(t)$ with a nonzero constant speed (i.e., $F(\dot{c}(t)) \equiv \mathrm{const} > 0$) is a geodesic iff its lifting $(c(t), \dot{c}(t))$ in TM is an integration curve of **G**.

The geodesic spray **G** and its coefficients $\mathbf{G}^i$ are very useful for presenting other geometric quantities (curvatures) in Finsler geometry. Curvature is the core concept of Finsler geometry. Some curvatures are generalized from Riemannian geometry, which are called *Riemannian curvatures*. They can help us to detect how a Finsler space curves. Some others vanish in Riemannian geometry, i.e., they only appear for non-Riemannian metrics and can be used to measure how these metrics differ with the Riemannian ones. We call them *non-Riemannian curvatures* [28].

Let us see some curvatures in Finsler geometry. First we look at the Riemannian ones.

In the Jacobi field equation for a smooth family of constant speed geodesics, we can find a linear operator $R_y^F : T_xM \to T_xM$ for any nonzero $y \in T_xM$. We call it the *Riemann curvature*. For any standard local coordinates on TM, it can be presented as $R_y^F = R_k^i(y)\partial_{x^i} \otimes dx^k$, where

$$R_k^i(y) = 2\partial_{x^k}\mathbf{G}^i - y^j\partial^2_{x^jy^k}\mathbf{G}^i + 2\mathbf{G}^j\partial^2_{y^jy^k}\mathbf{G}^i - \partial_{y^j}\mathbf{G}^i\partial_{y^k}\mathbf{G}^j.$$

Using Riemann curvature, we can easily define *flag curvature*, *Ricci curvature*, *scalar curvature*, etc. For example, flag curvature is a natural generalization of sectional curvature in Riemannian geometry. For any $x \in M$, nonzero $y \in T_x M$ (the *pole*), and tangent plane **P** containing y (the *flag*), linearly spanned by y and v, the flag curvature

$$K^F(x, y, y \wedge v) = K^F(x, y, \mathbf{P}) = \frac{\langle R_y v, v \rangle_y^F}{\langle y, y \rangle_y^F \langle v, v \rangle_y^F - (\langle y, v \rangle_y^F)^2},$$

where $\langle \cdot, \cdot \rangle_y^F$ is the inner product on $T_x M$ defined by the Hessian matrix $(g_{ij}(x, y))$. The flag curvature does not depend on the choice of v but only on y and **P**.

Next we look at the non-Riemannian curvatures.

The first degree derivatives of the Hessian matrix $(g_{ij}(x, y))$, i.e., the third degree derivatives of $F^2(x, y)$, define the *Cartan tensor*, the most fundamental non-Riemannian curvature in Finsler geometry. In a global fashion, it can be presented as

$$C_y^F(u, v, w) = \frac{1}{4} \frac{\partial^3}{\partial r \partial s \partial t} F^2(x, y + ru + sv + tw)|_{r=s=t=0},$$

where $y, u, v, w \in T_x M$ and $y \neq 0$.

Notice you may further differentiate the Cartan tensor with respect to the y-coordinates as many times as you like. Then you get infinitely many non-Riemannian curvatures from this thought. They provide obstacles for a (non-compact) Finsler space to be isometrically imbedded in Minkowski spaces. But they are not very useful most of the time.

S-curvature invented by Z. Shen is another important non-Riemannian curvature, with a totally different style. It can be described as the derivative of the distortion function

$$\tau(x, y) = \ln \frac{\sqrt{\det(g_{ij}(x, y))}}{\sigma(x)}$$

in the direction of the geodesic spray **G**. Here $\sigma(x)$ is the coefficient function in some canonical volume form. An important case is to take the Busemann-Hausdorff volume form $dV_{\mathrm{BH}} = \sigma(x) dx^1 \cdots dx^n$, where

$$\sigma(x) = \frac{\omega_n}{\mathrm{Vol}\{(y^i) \in \mathbb{R}^n | F(x, y^i \partial_{x^i}) < 1\}}.$$

Though the definition of $\tau(x, y)$ involves standard local coordinates, it is in fact a globally defined smooth function on $TM \backslash 0$. It is crucial that S-curvature has many important properties which relate it to other curvatures in Finsler geometry.

Why do we need the Lie symmetry for studying Finsler geometry?

1.3.3   Local Homogeneity and Global Homogeneity

There exists a main obstacle for the study of general Finsler geometry. The calculations and formulas are much more complicated than those in Riemannian geometry. Thus we need to call Lie theory for help. The Lie group symmetries could be applied to Finsler geometry in both the local and the global senses.

First look at the Lie symmetries in the local sense. Sometimes we restrict our interest to some types of Finsler metrics with special importance. For example, Riemannian metrics are the most well-known Finsler metrics. In the last century, Riemannian geometry has become one of the most important mathematical branches. Generally speaking, we concern non-Riemannian metrics and spaces in Finsler geometry. Randers and (α, β)-metrics are the most simple and important non-Riemannian Finsler spaces. The majority of research works on Finsler geometry are related to Randers metrics and (α, β)-metrics. The reason behind this phenomenon is a good "local homogeneity," which means for each tangent space T_xM, the connected linear isometry group $L_0(T_xM, F)$ (preserving the Minkowski norm $F(x, \cdot)$ on T_xM) is relatively big. For example, an n-dimensional Riemannian manifold (M, F), $L_0(T_xM, F)$ is isomorphic to $SO(n)$, which is the maximum among all possible ones. For a non-Riemannian Randers space or an (α, β)-space, $L_0(T_xM, F)$ is isomorphic to $SO(n - 1)$, which is a maximum among all non-Riemannian ones. These local homogeneities can significantly reduce the complexity of calculations in Finsler geometry, i.e., make the research subject much easier.

Here are two remarks on the local homogeneity.

First, there is another class of Finsler metrics, generalizing Randers metrics and (α, β)-metrics [15]. Let α be a Riemannian metric on M, with respect to which TM can be orthogonally decomposed as the sum of two sub-bundles $TM = \mathcal{V}_1 \oplus \mathcal{V}_2$. The restrictions $\alpha_i = \alpha|_{\mathcal{V}_i}$ can be naturally regarded as functions on TM. Then a Finsler metric F of the form $F = f(\alpha_1, \alpha_2)$ for some positive smooth function f (which must be positively homogeneous of degree 1) is call an (α_1, α_2)-*metric*.

For an (α_1, α_2)-space (M, F), its connected linear isometry group $L_0(T_xM, F)$ is isomorphic to $SO(n_1) \times SO(n_2)$, where n_i is the real dimension of the fibers of $\mathcal{V}_i$. It reaches another maximum among all non-Riemannian metrics [15]. We guess, the most that works on (α, β)-metrics can be generalized to (α_1, α_2)-metrics.

Second, we may improve the efficiency of the local homogeneities by considering the whole linear isometry group instead of its identity component. For example, reversible metrics have slightly better local homogeneities than irreversible metrics, because of the extra $\mathbb{Z}_2$-isometries for the antipodal map.

Next we look at the Lie symmetries in the global sense. That means a Lie group G acts nontrivially and isometrically on the Finsler space (M, F). We can assume G is a closed connected subgroup of the isometry group $I(M, F)$. Notice the isometry group $I(M, F)$ itself is a Lie transformation group [9]. One of the most typical cases is that G acts transitively on the Finsler space (M, F). Then we call (M, F) a homogeneous Finsler space [8]. Mostly we assume M is connected. Denote

$M = G/H$ a homogeneous Finsler space where G is a closed connected subgroup of $I_0(M, F)$ and H is the compact isotropy subgroup at $o = eH$. Then we have an $\mathrm{Ad}(H)$-invariant decomposition $\mathfrak{g} = \mathfrak{h} + \mathfrak{m}$, where $\mathfrak{g} = \mathrm{Lie}(G)$ and $\mathfrak{h} = \mathrm{Lie}(H)$. The subspace $\mathfrak{m}$ can be canonically identified with $T_o(G/H)$. A G-homogeneous Finsler metric F on G/H is one-to-one determined by an $\mathrm{Ad}(H)$-invariant Minkowski norm on $\mathfrak{m}$. The study on homogeneous Finsler spaces is called homogeneous Finsler geometry.

In homogeneous Finsler geometry, curvatures can be reduced to some tensors on $\mathfrak{m}$, and differential equations are reduced to linear equations. So we may avoid some extremely complicated calculations, and find the intrinsic nature of curvatures from the Lie algebra structures. The most successful examples for this approach include the following. We can use Killing frames (i.e., local frames provided by Killing vector fields) to present the S-curvature, and use the Finslerian submersion technique to deduce the flag curvature formula, for a homogeneous Finsler space [33, 38]. Notice in Finsler geometry, curvatures are mostly defined with standard local coordinates $\{x = (x^i) \in M, y = y^i \partial_{x^i} \in T_x M\}$. But in the homogeneous context, generally speaking, the local coordinates are not compatible with the homogeneous structure, and invariant frames or Killing frames seem more suitable. Applying invariant frames, L. Huang provides explicit formulas for all curvatures in homogeneous Finsler geometry [19].

1.3.4   Some Recent Progress

With the local and global view points, we see an expecting research field in Finsler geometry for which we have both a good local homogeneity and a good global homogeneity. For a homogeneous Finsler space $(G/H, F)$, the isotropy subgroup H is contained in the linear isometry group of $T_o(G/H)$. So $(G/H, F)$ has a good local homogeneity when H is relatively big. Here are some recent progress in homogeneous Finsler geometry which uses the local homogeneity directly or indirectly.

1. The study on the curvatures of homogeneous Randers spaces and homogeneous (α, β)-spaces [1, 7, 12, 13].
2. The study on the affine Berwald symmetric Finsler spaces [11]. Notice the Berwald condition implies many properties of these spaces coincide with those in Riemannian geometry. The Berwald condition can be replaced by assuming the homogeneous space is symmetric [10], i.e., in the $\mathrm{Ad}(H)$-invariant decomposition $\mathfrak{g} = \mathfrak{h} + \mathfrak{m}$, we have $[\mathfrak{m}, \mathfrak{m}] \subset \mathfrak{h}$. On one hand the symmetric condition implies the Lie algebra $\mathfrak{h}$ and thus the isotropy group H be big enough. On the other hand, using Killing frames, its geodesic spray can be explicitly presented, which coincides with that for a Riemannian symmetric metric, and proves the symmetric homogeneous Finsler space G/H is Berwald.

3. The classification of Clifford-Wolf homogeneous Randers spaces [32], which generalizes the corresponding work of Berestovskii and Nikonorov in the Riemannian context [3], and some partial results for left-invariant Clifford-Wolf homogeneous (α, β)-metrics and (α_1, α_2)-metrics on compact Lie groups [14, 15].

4. The classification of positively curved homogeneous Finsler spaces [34, 35, 37, 38]. The main tool for this project is the homogeneous flag curvature formula [19]. To efficiently reduce the classification problem to a solvable algebraic problem, we need to use the local homogeneity encoded in the rank inequality for a positively curved homogeneous Finsler space G/H [38], i.e., when assuming G is compact, $\mathrm{rk}G \leqslant \mathrm{rk}H + 1$. There exists a sufficiently big torus in H, which is important for this classification work. In particular, when $\dim G/H$ is odd, we need a slightly better local homogeneity which requires the positively curved homogeneous metric F on G/H to be reversible.

5. Define the flag-wise positively curved condition and find interesting examples. In Finsler geometry, we can define a special version of non-negatively curved condition. First, the metric is non-negatively curved, i.e., the flag curvature is non-negative everywhere. Second, the metric is flag-wise positively curved, i.e., for any tangent plane $\mathbf{P} \subset T_x M$, we can find a suitable pole $y \in \mathbf{P}$, such that $K^F(x, y, \mathbf{P}) > 0$. Using the Killing navigation process, we can find many compact coset spaces which admit non-negatively and flag-wise positively curved normal homogeneous Randers metrics but no positively curved homogeneous metrics [36]. The flag-wise positively curved condition alone seems to be a much weaker condition. Perturbing a non-negatively curved homogeneous Riemannian metric generically, we have a big chance to finding flag-wise positively curved metrics [31].

Finally, I would like to summarize with our purpose of studying homogeneous Finsler geometry. The curvatures are still in our main consideration. But now they can be described by algebraic data and structures for the Lie algebras. So geometric conditions are translated to algebraic conditions as well. In many cases, those algebraic conditions can be corresponded with explicit classification lists of homogeneous Finsler spaces, which provide precise examples for us to understand the intrinsic nature of Finsler geometry in general.

Acknowledgements G. Falcone and K. Strambach were supported by the European Union's Seventh Framework Programme (FP7/2007–2013) under grant agreement no. 317721.

References

1. H. An, S. Deng, Invariant (α,β)-metrics on homogeneous manifolds. Monatsh. Math. **154**, 89–102 (2008)
2. D. Bao, S.S. Chern, Z. Shen, *An Introduction to Riemannian-Finsler Geometry* (Springer, New York, 2000)

3. V.N. Berestovskii, Y.G. Nikonorov, Killing vector fields of constant length on locally symmetric Riemannian manifolds. Transform. Groups **13**, 25–45 (2008)
4. D. Burde, Left-symmetric algebras, or pre-Lie algebras in geometry and physics. Cen. Eur. J. Math. **4**, 323–357 (2006)
5. S.S. Chern, Z. Shen, *Riemann-Finsler Geometry* (World Scientific Publishers, River Edge, NJ, 2004)
6. A. Connes, D. Kreimer, Hopf algebras, renormalization and noncommutative geometry. Commun. Math. Phys. **199**, 203–242 (1998)
7. S. Deng, The S-curvature of homogeneous Randers spaces. Differ. Geom. Appl. **27**, 75–84 (2009)
8. S. Deng, *Homogeneous Finsler Spaces* (Springer, New York, 2012)
9. S. Deng, Z. Hou, The group of isometries of a Finsler space. Pac. J. Math. **207**, 149–155 (2002)
10. S. Deng, Z. Hou, Minkowski symmetric Lie algebras and symmetric Berwald spaces. Geom. Dedicata **113**, 95–105 (2005)
11. S. Deng, Z. Hou, On symmetric Finsler spaces. Isr. J. Math. **166**, 197–219 (2007)
12. S. Deng, Z. Hou, Homogeneous Einstein-Randers spaces of negative Ricci curvature. C.R. Math. Acad. Sci. Paris **347**, 1169–1172 (2009)
13. S. Deng, Z. Hu, Curvature of homogeneous Randers spaces. Adv. Math. **240**, 194–226 (2013)
14. S. Deng, M. Xu, Left invariant Clifford-Wolf homogeneous (α,β)-metrics on compact semisimple Lie groups. Transform. Groups **20**(2), 395–416 (2015)
15. S. Deng, M. Xu, (a1;a2)-metrics and Clifford-Wolf homogeneity. J. Geom. Anal. **26**, 2282–2321 (2016)
16. P. Finsler, *Über Kurven und Flächen in allgemeinen Räumen* (Dissertation, Göttingen, 1918) (Birkhäuser Verlag, Basel, 1951)
17. M. Gerstenhaber, The cohomology structure of an associative ring. Ann. Math. **78**, 267–288 (1963)
18. Th. Hawkins, *Emergence of the Theory of Lie Groups: An Essay in the History of Mathematics* (Springer, New York, 2000), pp. 1869–1926
19. L. Huang, On the fundamental equations of homogeneous Finsler spaces. Differ. Geom. Appl. **40**, 187–208 (2015)
20. J.E. Humphreys, *Introduction to Lie Algebras and Representation Theory*. Graduate Texts in Mathematics, vol. 9 (Springer, New York, 1978)
21. N. Jacobson, *Lie Algebras* (Dover Publications, New York, 1979)
22. M. Kontsevich, Deformation quantization of Poisson manifolds. Lett. Math. Phys. **66**, 157–216 (2003)
23. Y. Kosmann-Schwarzbach, Exact Gerstenhaber algebras and Lie bialgebroids. Acta Appl. Math. **41**, 153–165 (1995)
24. J.-L. Koszul, Domaines bornés homogènes et orbites de groupes de transformationes affines. Bull. Soc. Math. France **89**, 515–533 (1961)
25. J.-L. Koszul, *Crochet de Schouten-Nijenhuis et cohomologie in Elie Cartan et les mathématiques d'aujourd'hui* (Astérisque, hors série, 1985), pp. 257–271
26. M. Matsumoto, *Foundations of Finsler Geometry and Special Finsler Spaces* (Kaiseisha Press, Otsushi, 1986)
27. J.-P. Serre, *Lie Algebras and Lie Groups* (Springer, Berlin, 2006)
28. Z. Shen, *Lectures on Finsler Geometry* (World Scientific, Singapore, 2001)
29. E.B. Vinberg, Convex homogeneous cones. Transl. Moskow Math. Soc. **12**, 340–403 (1963)
30. S. Weinberg, *The Quantum Theory of Fields*, vol. II (Cambridge University Press, Cambridge, 2005)
31. M. Xu, Examples of flag-wise positively curved spaces. Diff. Geom. Appl. **52**, 42–50 (2017)
32. M. Xu, S. Deng, Clifford-Wolf homogeneous Randers spaces. J. Lie Theory **23**, 837–845 (2013)
33. M. Xu, S. Deng, Killing frames and S-curvature of homogeneous Finsler spaces. Glasg. Math. J. **57**(2), 457–464 (2015)

34. M. Xu, S. Deng, Normal homogeneous Finsler spaces. Transformation Groups (2017). arXiv:1411.3053. https://doi.org/10.1007/s00031-017-9428-7
35. M. Xu, S. Deng, Towards the classification of odd-dimensional homogeneous reversible Finsler spaces with positive flag curvature. Ann. Mat. Pura Appl. **196**(4), 1459–1488 (1923)
36. M. Xu, S. Deng, Homogeneous Finsler spaces and the flag-wise positively curved condition. arXiv:1604.07695 (2016)
37. M. Xu, W. Ziller, Reversible homogeneous Finsler metrics with positive flag curvature. Forum Math. arXiv:1606.02474. https://doi.org/10.1515/forum-2016-0173
38. M. Xu, S. Deng, L. Huang, Z. Hu, Even-dimensional homogeneous Finsler spaces with positive flag curvature, Indiana Univ. Math. J. **66**(3), 949–972 (2017)

Chapter 2
Remarks on Infinite-Dimensional Representations of the Heisenberg Algebra

Camillo Trapani

Abstract Infinite-dimensional representations of Lie algebras necessarily invoke the theory of unbounded operator algebras. Starting with the familiar example of the Heisenberg Lie algebra, we sketch the essential features of this interaction, distinguishing in particular the cases of *integrable* and *nonintegrable* representations. While integrable representations are well understood, nonintegrable representations are quite mysterious objects. We present here a short and didactical-minded overview of the subject.

2.1 Introduction

The theory of representations of Lie algebras by linear operators in infinite-dimensional Hilbert spaces is one of the most fascinating branches of modern mathematics; it involves algebra, topology, functional analysis, operator theory, and differential manifolds and this fact produces at once its charm and its difficulty. Motivations for these studies come from several mathematical subjects and also from physical applications. In analysis, for instance, the theory of representations of Lie algebras is a fundamental tool for the theory of partial differential equations (in certain approaches). On the other hand, it provides a precise mathematical framework for several problems in quantum theory and particle physics. In fact, physical observables (i.e., measurable quantities) are represented, in general, by operators acting on the Hilbert space of the states of the system and observables have a natural structure as Lie algebra whose brackets are realized through commutation relations of the corresponding operators. In fact, the operators A, B representing two observables either commute (i.e., $[A, B] = 0$) or satisfy prescribed *commutation relations*.

For instance, if Q represents the position of a particle and P its momentum, the Heisenberg principle implies that $[P, Q] = -i\mathbf{1}$, where $\mathbf{1}$ stands for the identity operator.

C. Trapani (✉)
Dipartimento di Matematica e Informatica, via Archirafi 34, 90123 Palermo, Italy
e-mail: camillo.trapani@unipa.it

© Springer International Publishing AG 2017   23
G. Falcone (ed.), *Lie Groups, Differential Equations, and Geometry*,
UNIPA Springer Series, DOI 10.1007/978-3-319-62181-4_2

Similarly, the differential operators L_x, L_y, L_z representing the components of the angular momentum satisfy the following commutation relations:

$$[L_x, L_y] = iL_z, \ [L_y, L_z] = iL_x, \ [L_z, L_x] = iL_y.$$

Of course, one can look at these commutation relation as to infinite-dimensional representations of some Lie algebra.

A representation of a Lie algebra $\mathfrak{L}$ is a linear map π from $\mathfrak{L}$ into an operator algebra $\mathfrak{A}$ acting on some vector space V.

Thus if V is taken to be a Hilbert space $\mathscr{H}$, the most natural choice would be taking as $\mathfrak{A}$ the C*-algebra $\mathscr{B}(\mathscr{H})$ of all bounded operators in $\mathscr{H}$.

We recall that a linear operator A in Hilbert space $\mathscr{H}$ is bounded (equivalently, continuous) if there exists $C > 0$ such that

$$\|A\xi\| \leqslant C\|\xi\|, \quad \forall \xi \in \mathscr{H}.$$

If the Hilbert space is finite-dimensional, then every linear operator is bounded. If $\dim(\mathscr{H}) = \infty$, unbounded (and hence everywhere discontinuous) operators are called on the stage.

As we shall see below, the operators involved in representations of Lie algebras are often unbounded (and for some algebra they *need* to be unbounded!); the drawback of involving unbounded operators is that the theory becomes more complicated since one has to deal with sometimes difficult problems of domains, extension of operators, self-adjointness problems, and so on.

But this is not the end of the story. In fact, if we want to represent Lie algebras by unbounded operators we need a precise definition of *commutator* for two unbounded operators and this is quite a difficult task!

In this paper, which has mostly didactical purposes, we quickly review the main features of unbounded representations of Lie algebras, focusing our attentions to the different (and, unfortunately, inequivalent) notions of commutators for unbounded operators. In particular we will consider representations of the Heisenberg Lie algebra $\mathfrak{h}$, which describes the commutation rules of the position Q and of the momentum P in Quantum Mechanics. By the Wiener–Wielandt–von Neumann theorem (Theorem 2.2.1), the representations of this Lie algebra are necessarily unbounded and the celebrated Stone–von Neumann theorem (Theorem 2.2.2) characterizes the so-called *integrable* representations of $\mathfrak{h}$, i.e., representations obtained by differentiating a unitary representation of a Lie group: they are all unitarily equivalent to the Schrödinger representation that can be described, as we shall see below, by the elegant formalism of annihilation and creation operators (see, e.g., [16]). But, *nonintegrable* representations of $\mathfrak{h}$ can also be easily constructed and often occur in applications to Physics. Thus, it is worth discussing them and try to understand what is preserved of the known theory under these very different circumstances.

2.2 Representations of the Heisenberg Algebra

As we mentioned in Sect. 2.1, some Lie algebras do not allow bounded representations in infinite-dimensional Hilbert space. A typical example comes from the
so-called *canonical commutation relations* (CCR, for short) that are nothing but
a special representation π of the Heisenberg Lie Algebra $\mathfrak{h}$ generated by three
elements $a, b, c \in \mathfrak{h}$ whose Lie brackets are defined by

$$[a, b] = c \quad [a, c] = [b, c] = 0$$

The carrier space of π is a Hilbert space $\mathcal{H}$.

The algebra $\mathcal{B}(\mathcal{H})$ of bounded operators in $\mathcal{H}$ can easily be made into a Lie
algebra by defining $[A, B] := AB - BA, A, B \in \mathcal{B}(\mathcal{H})$.

Thus, as a first step, it is natural to suppose that π takes its values in $\mathcal{B}(\mathcal{H})$. So
we should have

$$\pi : \mathfrak{h} \to \mathcal{B}(\mathcal{H}),$$

$$\pi([x, y]) = \pi(x)\pi(y) - \pi(y)\pi(x), \quad x, y \in \{a, b, c\}.$$

The structure itself of the Heisenberg algebra $\mathfrak{h}$ suggests that $\pi([a, b])$ must be a
central element of $\mathcal{B}(\mathcal{H})$, which consists of multiples of the identity $\mathbf{1}$ only. Hence
we require that

$$\pi([a, b]) = \pi(a)\pi(b) - \pi(b)\pi(a) = \mathbf{1}.$$

But, unfortunately,

Theorem 2.2.1 (Wiener, Wielandt, von Neumann) *There exists no bounded
representation π of the Heisenberg algebra satisfying $\pi([a, b]) = \mathbf{1}$.*
A very quick proof of this statement can be made by looking at the *spectra* of the
pretended bounded operators. We recall that, if $A \in \mathcal{B}(\mathcal{H})$, the spectrum $\sigma(A)$ of
A is defined as

$$\sigma(A) = \{\lambda \in \mathbb{C} : A - \lambda\mathbf{1} \notin GL(\mathcal{H})\},$$

where $GL(\mathcal{H})$ denotes the group of invertible elements of $\mathcal{B}(\mathcal{H})$. For every $A \in
\mathcal{B}(\mathcal{H})$, $\sigma(A)$ is a nonempty compact subset of the complex plane (see, e.g., [14]).
The familiar rules of the spectra give

$$\sigma(\pi(a)\pi(b)) \cup \{0\} = \sigma(\pi(a)\pi(b)) \cup \{0\}$$

$$\sigma(\pi(a)\pi(b)) = 1 + \sigma(\pi(b)\pi(a)),$$

contradicting the compactness of the spectra.

This is not a case, since as shown by Doebner and Melsheimer [10], representations of *every* noncompact Lie algebra are *unbounded*.

2.2.1 An Unbounded Representation of the Heisenberg Algebra

We now define the so-called Schrödinger representation of the Heisenberg algebra. The lesson of Theorem 2.2.1 is that we need to introduce *unbounded* representations and, consequently, *unbounded operator algebras*.

For reader's convenience we recall some features of unbounded operators. The first one is that they are not, in general, defined on the whole Hilbert space $\mathscr{H}$: a linear operator A is in fact a linear map defined on a subspace $D(A)$ of $\mathscr{H}$, called the *domain* of A, which one supposes to be dense in $\mathscr{H}$. An unbounded operator is necessarily everywhere discontinuous in its domain. A regularity requirement for an unbounded operator is its *closedness*: this means that the graph $\{(\xi, A\xi) \in D(A) \times \mathscr{H}\}$ is a closed subspace of $\mathscr{H} \times \mathscr{H}$. An operator is called *closable* if it has a closed extension.

Now we turn to unbounded operator algebras.

Let $\mathscr{D}$ be a dense domain in Hilbert space. Then, the family of closable operators

$$\mathscr{L}^{\dagger}(\mathscr{D}) := \{A : \mathscr{D} \to \mathscr{D} : A^* \mathscr{D} \subset \mathscr{D}\}$$

is a *-algebra of (in general) unbounded operators, with the usual operations of addition, multiplication by scalars, multiplication (i.e., composition) of operators and involution defined by $A^{\dagger} := A^* \upharpoonright D$, where A^* denotes the hilbertian *adjoint* of A defined as

$$\begin{cases} D(A^*) = \{\eta \in \mathscr{H} : \exists \eta^* \in \mathscr{H} \text{ such that } \langle A\xi \mid \eta \rangle = \langle \xi \mid \eta^* \rangle, \ \forall \xi \in \mathscr{D}\} \\ A^*\eta = \eta^*, \ \eta \in D(A^*) \end{cases}$$

For reader's convenience, we recall that an operator A, with domain $D(A)$ is *symmetric* if $\langle A\xi \mid \eta \rangle = \langle \xi \mid A\eta \rangle$ for every $\xi, \eta \in D(A)$. This means that the adjoint A^* of A extends A. The operator A is called *self-adjoint* if $A = A^*$ and *essentially self-adjoint* if its minimal closed extension $\overline{A}$ is self-adjoint. For more details we refer to [14, 16].

For the Schrödinger representation of $\mathfrak{h}$, we take as domain the Schwartz space $\mathscr{S}(\mathbb{R})$ ($\subset L^2(\mathbb{R})$, clearly) of C^∞-functions f that are rapidly decreasing; i.e.,

$$\sup_{x \in \mathbb{R}} |x^k (D_x^m f)(x)| < \infty, \quad \forall k, m \in \mathbb{N},$$

where D_x denotes the derivation operator with respect to x. Of course, $\mathscr{S}(\mathbb{R}) \subset L^2(\mathbb{R})$.

We then define the *annihilation* operator A and the *creation* operator $A^\dagger$ as follows:

$$Af = \frac{1}{\sqrt{2}}(xf + D_x f) \quad \text{and } A^\dagger f = \frac{1}{\sqrt{2}}(xf - D_x f), \quad f \in \mathscr{S}(\mathbb{R}).$$

Then, both A and $A^\dagger$ are elements of $\mathscr{L}^\dagger(\mathscr{S}(\mathbb{R}))$ and

$$AA^\dagger f - A^\dagger A f = f, \quad \forall f \in \mathscr{S}(\mathbb{R}).$$

For $a, b, c \in \mathfrak{h}$, as above, we put

$$\pi(a) = A \qquad \pi(b) = A^\dagger$$

Then $[\pi(a), \pi(b)] = \mathbf{1}$. The operator $A^\dagger A$, the so-called *number operator*, is essentially self-adjoint on $\mathscr{S}(\mathbb{R})$ and has discrete simple spectrum; in fact, $\sigma(A^\dagger A) = \mathbb{N}$. Eigenvectors of $A^\dagger A$ are obtained by successive application of $A^\dagger$ to the *vacuum* vector $\psi_0(x) = e^{-\frac{x^2}{2}}$, which is annihilated by A; in other words, an eigenvector associated to the eigenvalue $k \in \mathbb{N}$ is given by $\psi_k = (\sqrt{2}A^\dagger)^k \psi_0$.

The Schrödinger representation can be, equivalently, defined through the following operators:

$$Qf = xf \qquad Pf = -if' \qquad [P, Q]f = -if, \quad f \in \mathscr{S}(\mathbb{R}).$$

Then, one has

$$Af = \frac{1}{\sqrt{2}}(Qf + iPf), \qquad A^\dagger f = \frac{1}{\sqrt{2}}(Qf - iPf)$$

and $\Delta = \frac{1}{2}(Q^2 + P^2 + \mathbf{1})$ is essentially self-adjoint on $\mathscr{S}(\mathbb{R})$.

The Schrödinger representation of $\mathfrak{h}$ is of fundamental importance, since it is essentially the unique *integrable* and irreducible representation of the Heisenberg algebra.

Definition 2.2.1 A representation π of a Lie algebra $\mathfrak{L}$ is called *integrable* if there exists a connected and simply connected Lie group G and a *unitary* representation U of G such that π is the differential of U; i.e. $\pi = dU$.

The integrability of the Schrödinger representation depends clearly from $\mathfrak{h}$ being the Lie algebra of a Lie group G. Let us consider, in fact, as G the set of 3×3-matrices of the form

$$g(\alpha, \beta, \gamma) = \begin{pmatrix} 1 & \alpha & \gamma \\ 0 & 1 & \beta \\ 0 & 0 & 1 \end{pmatrix}, \quad \alpha, \beta \gamma \in \mathbb{R}$$

Checking that G is a Lie group is a straightforward exercise. Then, if we call a, b, c the generators of the one-dimensional subgroups, the Lie bracket of a, b, c

are nothing but $[a, b] = c$; $[a, c] = [b, c] = 0$; that is, the Lie algebra associated to G is exactly $\mathfrak{h}$.

For every $\lambda \in \mathbb{R} \setminus \{0\}$ a unitary representation of G in $L^2(\mathbb{R})$ is obtained by putting, for every $f \in L^2(\mathbb{R})$,

$$(U_\lambda(g(\alpha, \beta, \gamma))f)(x) := \exp(i\lambda(\beta x + \gamma))f(x + \alpha), \quad x \in \mathbb{R}.$$

By differentiating one finds

$$\pi(a) = \frac{d}{dx}, \quad \pi(b) = i\lambda x, \quad \pi(c) = i\lambda.$$

It is not difficult to check that this representation is equivalent to that defined above. The operators $\pi(a)$ and $\pi(b)$ leave $\mathscr{S}(\mathbb{R})$ invariant and the operator $\Delta = -\left(\frac{d}{dt}\right)^2 + \lambda^2 t^2 + \lambda^2$ is essentially self-adjoint on $\mathscr{S}(\mathbb{R})$.

A fundamental result about representations of $\mathfrak{h}$ is the following theorem due to M.H. Stone and J. von Neumann.

Theorem 2.2.2 (Stone, von Neumann) *Every integrable irreducible representation π of the Heisenberg Lie algebra is unitarily equivalent to the Schrödinger representation.*

As a consequence of this theorem, an integrable representation of $\mathfrak{h}$ decomposes into the direct sum of (possibly, infinitely many) representations each one of them being unitarily equivalent to the Schrödinger representation, since every representation can always be decomposed into irreducible ones.

The example discussed above can be cast into the more general situation described by the *Gårding theory*. This approach to the infinite-dimensional representation theory of Lie algebras starts essentially with a work of L. Gårding of 1947 [12]. The cornerstone of the theory is the following theorem [9, Chap. 11].

Theorem 2.2.3 *Let U be a unitary representation of a d-dimensional Lie group G in Hilbert space $\mathscr{H}$ and $\pi = dU$ the corresponding representation of the associated Lie algebra $\mathfrak{L}_G$. Then,*

(i) *there exists a common invariant dense domain D_G for the operators $\pi(X_i)$, where $X_1, \ldots, X_d \in \mathfrak{L}_G$ are the infinitesimal generators of U;*
(ii) *the Nelson operator $X_1^2 + X_2^2 + \cdots + X_d^2$ is essentially self-adjoint on D_G.*

It is clear that π is integrable by definition. The proof of Theorem 2.2.3 is constructive: thus once a unitary representation of the Lie Group G is given, we can explicitly construct the so-called *Gårding domain*. The converse of the previous theorem also holds, provided that D_G consists of *analytic* vectors f for π. A vector $f \in \mathscr{H}$ is analytic for π if, for some $s > 0$,

$$\sum_{n=0}^\infty \frac{1}{n!} \sum_{1 \leqslant i_1, \ldots, i_n \leqslant d} \|\pi(X_{i_1}) \cdots \pi(X_{i_d})f\|s^n < \infty.$$

The notion of analytic vector (that appears first in Harish–Chandra) was introduced, in a slight different context by Nelson [15], but it is with Gårding theory that it develops all its potentiality [13].

2.3   Nonintegrable Representations of the Heisenberg Algebra

Integrable representations of Lie algebras are, as we have seen, quite well understood. But this kind of *well-behavior* is not so frequent in practice and it is not difficult to construct examples where integrability fails (sometimes unexpectedly).

Let us consider the following representation of $\mathfrak{h}$ in the Hilbert space $\mathscr{H} :=$ $L^2([0, 1])$. Similarly, to what we have done before, we define

$$(Qf)(x) = xf(x), \quad f \in \mathscr{H}.$$

$$D(P) = \{f \in C^1(0, 1) : f(0) = f(1) = 0, f' \in L^2([0, 1])\}$$

$$(Pf)(x) = -if'(x)$$

Then Q is bounded and self-adjoint, P is neither bounded nor self-self-adjoint. Performing the same construction as before we put

$$A = \frac{1}{\sqrt{2}}(Q - iP) \quad \text{and } B = \frac{1}{\sqrt{2}}(Q + iP)$$

Of course $[A, B] = 1$; $B = A^\dagger$ but $B \neq A^*$, the hilbertian adjoint of A. By Gårding theorem the integrability of the representation depends on the existence of a common invariant dense domain $\mathscr{D}$ where the Nelson operator $\Delta = \frac{1}{2}(P^2 + Q^2)$ is essentially self-adjoint. The largest domain invariant for P and Q is clearly $C_0^\infty(0, 1)$; but Δ is not essentially self-adjoint on $C_0^\infty(0, 1)$, since P is not. Thus this representation is not integrable. This is, in a sense surprising, since the representation defined in this way can be taken for describing position and momentum of a particle confined to move in an interval of finite length.

On the other hand, it should be mentioned that for several physical models it is not possible to find a common invariant dense domain $\mathscr{D}$ with the prescribed properties and this fact provides a good motivation for considering nonintegrable representations.

2.3.1   Commutation Relations of Operators: Usual Approach

At this point, a discussion about the mathematical meaning to give to the commutation relations of two unbounded operators (or even to the fact that they commute) is in order. In fact, studying operators A, B such that $B \neq A^\dagger$ and $[A, B] = \epsilon \mathbf{1}$ (in a sense to be precise) is an unavoidable step (but just the first one) for the analysis of nonintegrable representations. Of course, the significant cases are $\epsilon = 0$ or $\epsilon = 1$. The latter has been analyzed in [7, 8] moving from the idea of *pseudo-bosons* discussed at length by Bagarello in several papers [4–6].

Let A, B be two closed operators with dense domains, $D(A)$ and $D(B)$, respectively, in Hilbert space $\mathcal{H}$. In order to give a meaning to the formal commutation relation $[A, B] = \epsilon \mathbf{1}$, we require that the identity $AB - BA = \epsilon \mathbf{1}$ holds, at least, on a dense domain $\mathcal{D}$ of Hilbert space $\mathcal{H}$. In other words, we assume that there exists a dense subspace $\mathcal{D}$ of $\mathcal{H}$ such that

(D.1) $\mathcal{D} \subset D(AB) \cap D(BA)$;
(D.2) $AB\xi - BA\xi = \epsilon \xi, \quad \forall \xi \in \mathcal{D}$, where, as usual, $D(AB) = \{\xi \in D(B) : B\xi \in D(A)\}$.

As in [6], we will suppose that

(D.3) $\mathcal{D} \subset D(A^*) \cap D(B^*)$.

2.3.1.1   The Case $\epsilon = 0$

Let A, B be linear operators in Hilbert space $\mathcal{H}$. What we mean when we say that A and B commute? It is clear that, if A, B are both bounded operators, the equation $AB - BA = 0$ has a precise meaning; but the same algebraic relation can be very poor in information if both A and B are unbounded. So, many different mathematical approaches have been developed for handling this case.

Before going forth, we recall some basic facts about self-adjoint operators.

A *spectral family* on the real line $\mathbb{R}$ is a family $\{E(\lambda)\}$ of projection operators, depending on the real parameter λ, satisfying the following conditions:

(a) $E(\lambda) \leqslant E(\mu)$ or, equivalently, $E(\lambda)E(\mu) = E(\lambda)$, if $\lambda \leqslant \mu$;
(b) $\lim_{\epsilon \to 0^+} E(\lambda + \epsilon)\xi = E(\lambda)x$, for every $\xi \in \mathcal{H}$;
(c) $\lim_{\lambda \to -\infty} E(\lambda)\xi = 0$ e $\lim_{\lambda \to +\infty} E(\lambda)\xi = \xi$, for every $\xi \in \mathcal{H}$.

A crucial result of operator theory is the *spectral theorem* for self-adjoint operators:

Theorem 2.3.1 *To every self-adjoint operator A there corresponds a unique spectral family such that*

$$A\xi = \int_{\mathbb{R}} \lambda \, dE(\lambda)\xi,$$

for every $\xi \in D(A)$.

Moreover one has

$$D(A) = \left\{ \xi \in \mathscr{H} : \int_{\mathbb{R}} \lambda^2 d \langle E(\lambda)\xi \mid \xi \rangle < \infty \right\}.$$

A relevant consequence of this theorem is the *functional calculus* for self-adjoint operators which allows to define *functions* of an operator. For instance, if A is self-adjoint and $\{E(\lambda)\}$ is its spectral family, then

$$e^{itA}\xi = \int_{\mathbb{R}} e^{it\lambda} dE(\lambda)\xi, \quad \xi \in \mathscr{H}$$

defines a one-parameter group $U_A(t) := e^{itA}$ of unitary operators in $\mathscr{H}$. The operator A plays the role of *infinitesimal generator* of $U_A(t)$; that is

$$iAf = \lim_{\alpha \to 0} \frac{U_A(\alpha) - \mathbf{1}}{\alpha} f, \quad f \in D(A).$$

Let now A and B be two self-adjoint operators in $\mathscr{H}$. The notion of *strong commutation* of A and B can be expressed by anyone of the following equivalent statements:

1. the associated spectral families $\{E_A(\lambda)\}$, $\{E_B(\mu)\}$, $\lambda, \mu \in \mathbb{R}$ commute;
2. the associated unitary groups $U_A(t) := e^{iAt}$, $U_B(s) := e^{iBs}$ commute, for every $s, t \in \mathbb{R}$;
3. the resolvent functions $(A - \lambda I)^{-1}$, $(B - \mu I)^{-1}$, commute, for every $\lambda \in \rho(A)$, $\mu \in \rho(B)$,

$\rho(X)$ denotes the resolvent set of the operator X; i.e. the complement of the spectrum. We refer again to [14] or [16] for the basic definitions and for the above mentioned results.

Also in the case where the operators $A, B \in \mathscr{L}^\dagger(\mathscr{D})$ the equation $AB = BA$ is well defined, since, as we have seen, $\mathscr{L}^\dagger(\mathscr{D})$ is a *-algebra. But Nelson's example [16] shows that there exist two operators A, B and a dense domain $\mathscr{D}$ such that

1. $A, B \in \mathscr{L}^\dagger(\mathscr{D})$:
2. A, B essentially self-adjoint in $\mathscr{H}$;
3. $AB\xi = BA\xi$, for every $\xi \in \mathscr{D}$

but the spectral families *do not commute*; in contrast with what happens for bounded self-adjoint operators.

On the other hand, strong commutation of self-adjoint operators has a precise meaning when they are supposed to represent physical observables: the two observables are, in this case, compatible (i.e., they can be measured simultaneously) and a joint probability distribution for these observables does exist.

2.3.1.2   The Case $\epsilon = 1$

Parallel to strong commutation is the formulation given by Weyl for the formal commutation relation $[A, B] = \mathbf{1}$ through the groups of unitaries they generate (when they do!).

Thus, if A and B are self-adjoint operators, the equation $[A, B] = \mathbf{1}$ can be translated by the celebrated Weyl commutation relations:

$$U_A(\alpha)U_B(\beta) = e^{\alpha\beta}U_B(\beta)U_A(\alpha), \quad \forall \alpha, \beta \in \mathbb{R}.$$

But for nonself-adjoint operators (excluding trivial cases) this approach is mostly impossible.

2.3.2   Commutation Relations of Operators: A Different Approach

Let A, B be two operators satisfying (D.1)–(D.3). Then, the operators $S := A \upharpoonright \mathscr{D}$ and $T := B \upharpoonright D$ are elements of the partial *-algebra $\mathscr{L}^\dagger(\mathscr{D}, \mathscr{H})$ and satisfy the equality

$$\langle T\xi \mid S^\dagger\eta \rangle - \langle S\xi \mid T^\dagger\eta \rangle = \langle \xi \mid \eta \rangle, \quad \forall \xi, \eta \in \mathscr{D}. \tag{2.1}$$

We recall that $\mathscr{L}^\dagger(\mathscr{D}, \mathscr{H})$ denotes the set of all (closable) linear operators X such that $D(X) = \mathscr{D}$, $D(X^*) \supseteq \mathscr{D}$. The set $\mathscr{L}^\dagger(\mathscr{D}, \mathscr{H})$ is a partial *-algebra with respect to the usual sum $X_1 + X_2$, the scalar multiplication λX, the involution $X \mapsto X^\dagger := X^* \upharpoonright \mathscr{D}$ and the *(weak)* partial multiplication $X_1 \square X_2 = X_1^{\dagger *}X_2$, defined whenever X_2 is a weak right multiplier of X_1 (we shall write $X_2 \in R^w(X_1)$ or $X_1 \in L^w(X_2)$), that is, whenever $X_2\mathscr{D} \subset \mathscr{D}(X_1^{\dagger *})$ and $X_1^*\mathscr{D} \subset \mathscr{D}(X_2^*)$, [2].

The commutation relation (2.1) is the main matter of what follows.

Let us first consider the case where S and/or T are generators of some weakly continuous semigroup $V(t)$ of bounded operators.

More precisely,

Definition 2.3.1 Let $V(t)$, $t \geq 0$, be a semigroup of bounded operators. We say that $X_0 \in \mathscr{L}^\dagger(\mathscr{D}, \mathscr{H})$ is the $\mathscr{D}$-generator of $V(t)$ if

$$\lim_{t \to 0} \left\langle \frac{V(t) - \mathbf{1}}{t}\xi \mid \eta \right\rangle = \langle X_0\xi \mid \eta \rangle, \quad \forall \xi, \eta \in \mathscr{D}.$$

In [7] the following definitions were given.

Definition 2.3.2 Let $S, T \in \mathscr{L}^{\dagger}(\mathscr{D}, \mathscr{H})$. We say that

(CR.1) the commutation relation $[S, T] = \mathbf{1}$ is satisfied (in $\mathscr{L}^{\dagger}(\mathscr{D}, \mathscr{H})$) if, whenever $S \square T$ is well defined, $T \square S$ is well-defined too and $S \square T - T \square S = \mathbf{1}$.

(CR.2) the commutation relation $[S, T] = \mathbf{1}$ is satisfied in *weak sense* if

$$\langle T\xi \mid S^{\dagger}\eta \rangle - \langle S\xi \mid T^{\dagger}\eta \rangle = \langle \xi \mid \eta \rangle, \quad \forall \xi, \eta \in \mathscr{D}.$$

(CR.3) the commutation relation $[S, T] = \mathbf{1}$ is satisfied in *quasi-strong sense* if S is the $\mathscr{D}$-generator of a weakly continuous semigroups of bounded operators $V_S(\alpha)$ and

$$\langle V_S(\alpha)T\xi \mid \eta \rangle - \langle V_S(\alpha)\xi \mid T^{\dagger}\eta \rangle = \alpha \langle V_S(\alpha)\xi \mid \eta \rangle, \quad \forall \xi, \eta \in \mathscr{D}; \alpha \geqslant 0.$$

(CR.4) the commutation relation $[S, T] = \mathbf{1}$ is satisfied in *strong sense* if S and T are $\mathscr{D}$-generators of weakly continuous semigroups of bounded operators $V_S(\alpha), V_T(\beta)$, respectively, satisfying the generalized Weyl commutation relation

$$V_S(\alpha)V_T(\beta) = e^{\alpha\beta}V_T(\beta)V_S(\alpha), \quad \forall \alpha, \beta \geqslant 0.$$

The following implications hold [7]:

$$(\text{CR.4}) \Rightarrow (\text{CR.3}) \Rightarrow (\text{CR.2}) \Rightarrow (\text{CR.1}).$$

Example 1 The implications in the opposite direction do not hold. For instance, as shown by Fulgede and Schmüdgen [11, 17, 18], there exist two essentially self-adjoint operators P, Q with common invariant dense domain $\mathscr{D}$ such that $PQ\xi - QP\xi = -i\xi$, for $\xi \in \mathscr{D}$, but the unitary groups $U_P(t), U_Q(s)$ generated by $\overline{P}, \overline{Q}$ do not satisfy the Weyl commutation relation $U_P(t)U_Q(s) = e^{its}U_Q(s)U_P(t), s, t \in \mathbb{R}$.

2.4 Commutation Relations and Spectral Behavior

It is well known that the traditional commutation relation $[A, A^{\dagger}] = \mathbf{1}$ gives rise (under some additional conditions) to a rich series of spectral results that have plenty of applications [Sect. 2.1]. Here we will shortly discuss analogous results (mostly presented in [7, 8]) for two operators S, T satisfying (in one sense or the other described in Definition 2.3.2) the commutation relation $[S, T] = \mathbf{1}$.

2.4.1   Existence of Eigenvectors

We begin with assuming that $S, T \in \mathcal{L}^\dagger(\mathcal{D}, \mathcal{H})$ satisfy the commutation relation $[S, T] = \mathbf{1}$ in weak sense. Assume that there exists a vector $0 \neq \xi_0 \in \mathcal{D}$ such that $S\xi_0 = 0$. Then

$$\langle T\xi_0 \mid S^\dagger \eta \rangle = \langle \xi_0 \mid \eta \rangle, \quad \forall \eta \in \mathcal{D}. \tag{2.2}$$

This implies that $T\xi_0 \neq 0$ (otherwise, $\xi_0 = 0$) and

$$T\xi_0 \in D(S^{\dagger*}), \quad S^{\dagger*}T\xi_0 = \xi_0 \in \mathcal{D}.$$

Thus T can be applied once more and we get

$$(TS^{\dagger*})T\xi_0 = T(S^{\dagger*}T\xi_0) = T\xi_0. \tag{2.3}$$

Hence $T\xi_0$ is an eigenvector of $TS^{\dagger*}$ with eigenvalue 1.

Let us now assume that $T\xi_0 \in \mathcal{D}$. Then it is easily seen that

$$\langle T^2\xi_0 \mid S^\dagger \eta \rangle = 2 \langle T\xi_0 \mid \eta \rangle, \quad \forall \eta \in \mathcal{D}. \tag{2.4}$$

This, in turn, implies that $T^2\xi_0 \in D(S^{\dagger*})$ and $S^{\dagger*}T^2\xi_0 = 2T\xi_0$. and

$$(TS^{\dagger*})T^2\xi_0 = 2T^2\xi_0.$$

Iterating this procedure we conclude that

Proposition 2.4.1 *Let $S, T \in \mathcal{L}^\dagger(\mathcal{D}, \mathcal{H})$ satisfy the commutation relation $[S, T] = \mathbf{1}$ in weak sense. Assume that there exists a nonzero vector $\xi_0 \in \mathcal{D}$ such that $S\xi_0 = 0$ and the vectors $T\xi_0, T^2\xi_0, \ldots T^{n-1}\xi_0$ all belong to $\mathcal{D}$. Then*

  (i) $T^n\xi_0$ is an eigenvector of $TS^{\dagger*}$ with eigenvalue n;
 (ii) $T^{n-1}\xi_0$ is eigenvector of $S^{\dagger*}T$ with eigenvalue n.

Proposition 2.4.2 *Let $S, T \in \mathcal{L}^\dagger(\mathcal{D}, \mathcal{H})$ satisfy the commutation relation $[S, T] = \mathbf{1}$ in weak sense. Let $\xi \in \mathcal{D}$ and assume that $T^k\xi \in \mathcal{D}$ for $k \leq n$, $n \in \mathbb{N} \cup \{\infty\}$. Then $S\xi \in D((T^{\dagger*})^k)$, $k \leq n$ and*

$$ST^k\xi - (T^{\dagger*})^k S\xi = kT^{k-1}\xi, \quad k \leq n.$$

Remark 1 In particular if $S\xi = 0$ then $ST^k\xi = kT^{k-1}\xi$.

If the assumptions of Proposition 2.4.1 are satisfied one may have that the largest n for which $T^n\xi_0 \in \mathcal{D}$ is finite or infinite. As we have seen the point spectrum $\sigma_p(TS^{\dagger*})$ contains all natural numbers up to n. Let us denote by $\mathcal{N}_0$ the subspace of $\mathcal{D}$ spanned by $\{\xi_o, T\xi_0, \ldots T^n\xi_0\}$ and by $\mathcal{N} := \overline{\mathcal{N}_0}$ its closure in $\mathcal{H}$. Clearly $TS^{\dagger*}$

leaves $\mathcal{N}_0$ invariant. The restriction of $TS^{\dagger*}$ to $\mathcal{N}_0$, denoted by $(TS^{\dagger*})_0$, behaves in quite regular way. We have indeed

Proposition 2.4.3 *In the assumptions of Proposition 2.4.1, the point spectrum $\sigma_p(TS_0^{\dagger*})$ the operator $(TS^{\dagger*})_0$, restriction of $TS^{\dagger*}$ to $\mathcal{N}_0$, consists exactly of the set $\{0, 1, \dots, n\}$, where $n \in \mathbb{N} \cup \{\infty\}$ is the largest natural number such that the vectors $T\xi_0, T^2\xi_0, \dots, T^{n-1}\xi_0$ all belong to $\mathscr{D}$. Each eigenvalue is simple (in $\mathcal{N}_0$).*

Example 2 Let us consider the Hilbert space $L^2(\mathbb{R}, wdx)$ where the weight w is a positive continuously differentiable function with the properties

1. $\lim\limits_{|x| \to +\infty} w(x) = 0$;
2. $\int_{\mathbb{R}} w(x)dx < \infty$.

Let

$$D(\mathsf{p}) = \left\{ f \in L^2(\mathbb{R}, wdx) : \exists g \in L^2(\mathbb{R}, wdx), f(x) = \int_{-\infty}^{x} g(t)dt \right\}.$$

For shortness, we adopt the notation $f'(x) = g(x)$, for $f \in D(\mathsf{p})$.

$$D(\mathsf{q}) = \{ f \in L^2(\mathbb{R}, wdx) : xf(x) \in L^2(\mathbb{R}, wdx) \}.$$

Put $\mathscr{D} = D(\mathsf{q}) \cap D(\mathsf{p})$. Then both the operators S, T defined by

$$(Sf)(x) = f'(x), \qquad (Tf)(x) = xf(x), \quad f \in \mathscr{D}$$

map $\mathscr{D}$ into $L^2(\mathbb{R}, wdx)$.

The operator T is symmetric in $\mathscr{D}$. As for S, we have $g \in D(S^*)$ if, and only if $g \in D(\mathsf{p})$ and $g\frac{w'}{w} \in L^2(\mathbb{R}, wdx)$. In this case

$$(S^*g)(x) = -g'(x) - g(x)\frac{w'(x)}{w(x)}.$$

Hence $S \in \mathscr{L}^{\dagger}(\mathscr{D}, \mathscr{H})$, with $\mathscr{H} = L^2(\mathbb{R}, wdx)$ if, and only if,

$$\int_{\mathbb{R}} |g(x)|^2 \frac{|w'(x)|^2}{w(x)} dx < \infty.$$

This is certainly satisfied if, for instance, w'/w is a bounded function on $\mathbb{R}$. An easy computation shows that the commutation relation $[S, T] = \mathbb{1}$ is satisfied in weak sense.

The function $u_0(x) = 1$, for every $x \in \mathbb{R}$, is clearly in the kernel of S for every function w satisfying the assumptions made so far.

Now we make some particular choice of w.

1. Let us consider $w(x) = w_\alpha(x) = (1 + x^4)^{-\alpha}$, $\alpha > \frac{3}{4}$. It is easily seen that $w_\alpha(x)$ satisfies all the conditions we have required (for instance, w'_α/w_α is bounded). The function $u_0(x) = 1$, which belongs to $L^2(\mathbb{R}, w_\alpha dx)$ for any $\alpha > \frac{3}{4}$, satisfies $Su = 0$ and the largest n for which $T^n u_0$ belongs to $\mathscr{D}$ satisfies $n < 2\alpha - \frac{3}{2}$. Hence, the dimension of the corresponding subspace $\mathscr{N}_0$ is $\left[2\alpha - \frac{3}{2}\right] + 1$.

2. Let us now take $w(x) = e^{-x^2/2}$ and $\mathscr{D}$ the subspace consisting of all polynomials in x. In this case the functions $u_k(x) = x^k$, $k = 1, 2, \ldots$, belong to $\mathscr{D}$ and they satisfy $TS^{\dagger*}u_k = ku_k$ for every $k \in \mathbb{N}$. The subspace $\mathscr{N}_0$ coincides in this case with $\mathscr{D}$. One can readily check that every complex number λ with $\Re\lambda > -\frac{1}{2}$ is an eigenvalue of $TS^{\dagger*}$; but the corresponding eigenvector is in $\mathscr{D}$ if and only if $\Re\lambda$ is a natural number.

As we have seen with the previous examples the subspace $\mathscr{N}_0$ spanned by $\{T^k\xi_0, k \in \mathbb{N}\}$ can be finite dimensional. Thus $N := (TS^{\dagger*})_0$ is a bounded symmetric operator on $\mathscr{N}_0 = \mathscr{N} \cong \mathbb{C}^n$, having the numbers $0, 1, \ldots, n$ as eigenvalues. Hence N is positive and thus there exists an operator $C \in \mathscr{B}(\mathscr{N})$ such that $N = C^\dagger C$. None of the possible solutions of this operator equation can, however, satisfy the commutation relation $[C, C^\dagger] = \mathbf{1}$, due to the Wiener–Wielandt–von Neumann theorem. If $\mathscr{N}_0$ is infinite dimensional, then N may fail to be symmetric, as the last case in Example 2 shows.

2.4.2 Existence of Intertwining Operators

Assume now that the assumptions of Proposition 2.4.1 hold not only for the pair S, T but also for the pair $T^\dagger, S^\dagger$. This means that we assume that there exists also a nonzero vector $\eta_0 \in \mathscr{D}$ such that $T^\dagger\eta_0 = 0$ and the vectors $S^\dagger\eta_0, (S^\dagger)^2\eta_0, \ldots$ $(S^\dagger)^{m-1}\eta_0$ all belong to $\mathscr{D}$, then Proposition 2.4.1 gives that

(i) $(S^\dagger)^m\eta_0$ is an eigenvector of $S^\dagger T^*$ with eigenvalue m;
(ii) $(S^\dagger)^{m-1}\eta_0$ is eigenvector of $T^*S^\dagger$ with eigenvalue m.

Let $m \in \mathbb{N} \cup \{\infty\}$ be the largest natural number such that the vectors $S^\dagger\eta_0$, $(S^\dagger)^2\eta_0, \ldots, (S^\dagger)^{m-1}\eta_0$ all belong to $\mathscr{D}$ and $\mathscr{M}_0$ the linear span of these vectors. Then the point spectrum $\sigma_p((S^\dagger T^*)_0)$, of the operator $(S^\dagger T^*)_0 := S^\dagger T^* \upharpoonright \mathscr{M}_0$, consists, as before, of the numbers $\{0, 1, \ldots, m\}$.

One may wonder if any relation between the two numbers n and m can be established. The answer is negative, in general. Indeed, the operators S, T considered in the second case of Example 2 provide an instance where $n = \infty$ and $m = 0$. Thus it is apparently impossible to find a relationship between $\mathscr{N}_0$ and $\mathscr{M}_0$ without additional assumptions.

Let us now call $\xi_k := \frac{1}{\sqrt{k!}}T^k\xi_0, k = 1, \ldots, n$ and $\eta_r := \frac{1}{\sqrt{r!}}(S^\dagger)^r\eta_0, r = 1, \ldots, m$. It is always possible to choose the normalization of ξ_0 and η_0 in such a way that

$\langle \xi_0 \mid \eta_0 \rangle = 1$. We put $\mathscr{F}_\xi := \{\xi_k; k = 1, \ldots, n\}$ $\mathscr{F}_\eta := \{\eta_r; r = 1, \ldots, m\}$. Then, the sets $\mathscr{F}_\xi$ and $\mathscr{F}_\eta$ are biorthogonal; i.e.,

$$\langle \xi_i \mid \eta_j \rangle = \delta_{i,j},$$

for all $i = 1, \ldots, n$ and $j = 1, \ldots, m$.

Assume now that $n = m = \infty$. Thus the subspaces $\mathscr{N}_0$ and $\mathscr{M}_0$ are both infinite dimensional. Then we can define two operators which obey interesting intertwining relations. More in detail, let us define K_ξ via its action on the basis $\mathscr{F}_\eta$: $K_\xi(\eta_j) = \xi_j$, $j \in \mathbb{N}$. We can also introduce a second operator K_η via its action on the second basis constructed above, $\mathscr{F}_\xi$: $K_\eta(\xi_j) = \eta_j$, $j \in \mathbb{N}$. Both the operators K_ξ and K_η are then extended by linearity to $\mathscr{M}_0$ and $\mathscr{N}_0$, respectively. It is clear that one is the inverse of the other: $K_\eta = K_\xi^{-1}$, but in general neither K_ξ nor K_η are bounded. A direct computation shows that they obey the following intertwining relations:

$$K_\eta \left(TS^{\dagger *}\right) \phi = \left(S^\dagger T^*\right) K_\eta \phi, \quad \forall \phi \in \mathscr{M}_0;$$

$$K_\xi \left(S^\dagger T^*\right) \psi = \left(TS^{\dagger *}\right) K_\xi \psi, \quad \forall \psi \in \mathscr{N}_0.$$

In particular, if $\mathscr{N}_0 = \mathscr{M}_0 = \mathscr{H}$ and K_ξ and K_η are bounded, it is possible to show that $\mathscr{F}_\xi$ and $\mathscr{F}_\eta$ are Riesz bases of $\mathscr{H}$ and an orthonormal basis $e = \{e_j\}$ can be defined by, for instance, $e_j = K_\eta^{1/2}\xi_j$ (see [4]).

It is worth mentioning here that the properties discussed in Sects. 2.4.1 and 2.4.2 depend in an essential way on the nonzero vectors ξ_0 and η_0 satisfying, respectively, $S\xi_0 = 0$ and $T^\dagger \eta_0 = 0$: choosing different elements in the kernels of S and $T^\dagger$ may give drastically different results.

2.4.3 Time Operators and Weyl Extensions

In [18] Schmüdgen studied pairs of operators (T, H) with T symmetric and H self-adjoint such that.

1. $e^{-itH}D(T) \subseteq D(T)$;
2. $Te^{-itH}\xi = e^{-itH}(T + t)\xi$

The operators T, H are regarded as members of $\mathscr{L}^\dagger(\mathscr{D}, \mathscr{H})$ with $\mathscr{D} = D(T) \cap D(H)$. and the two conditions above are equivalent to say that the operators $T, S := iH$ satisfy $[S, T] = \mathbb{1}$ in quasi-strong sense.

Definition 2.4.1 T is time operator for H if (T, H) satisfies the two conditions above.

In many examples, mostly taken from physics, H is a semibounded operator (this often happens, for instance, when H is the Hamiltonian of some physical system).

Theorem 2.4.4 (Arai) *Let H be self-adjoint and semibounded. Then no time operator T of H can be essentially self-adjoint.*

Indeed, as shown in [3], the spectrum $\sigma(T)$ is one of the following sets

1. $\mathbb{C}$, if H is bounded
2. $\mathbb{C}$ or $\overline{\Pi_+} = \{z \in \mathbb{C} : \Im z \geqslant 0\}$ if H is bounded below
3. $\mathbb{C}$ or $\overline{\Pi_-} = \{z \in \mathbb{C} : \Im z \leqslant 0\}$ if H is bounded above.

The assumption $e^{-itH}D(T) \subseteq D(T)$ is quite strong and weaker conditions should be analyzed.

Definition 2.4.2 Let S, T be symmetric operators of $\mathscr{L}^\dagger(\mathscr{D}, \mathscr{H})$. We say that $\{S, T\}$ satisfy the *weak Weyl commutation relation* if there exists a self-adjoint extension H of S such that

(ww_1) $D(\overline{T}) \subset D(H)$;
(ww_2) $\left(e^{-itH}\xi \mid T\eta\right) = \left((T + t)\xi \mid e^{itH}\eta\right), \quad \forall \xi, \eta \in \mathscr{D}; t \in \mathbb{R}.$

Then H is called the weak Weyl extension of S (with respect to T).

Proposition 2.4.5 *Let $\{S, T\}$ satisfy the weak Weyl commutation relation and let H be the weak Weyl extension of S. The following statements hold.*

(i) *Suppose that T is essentially self-adjoint. Then $\{H, \overline{T}\}$ satisfy the Weyl commutation relation, that is,*

$$e^{itH}e^{-is\overline{T}} = e^{-its}e^{-is\overline{T}}e^{itH}, \quad \forall s, t \in \mathbb{R} \qquad (2.5)$$

(ii) *If H is semibounded, then T is not essentially self-adjoint.*

Example 3 Let I be an interval of the real line. Denote by q the multiplication operator on $L^2(I)$ by the variable $x \in I$

For every choice of I, q is a self-adjoint operator.
Let p be the operator on $L^2(I)$ defined as follows:
$D(p) := C_0^\infty(I)$
$(pg)(x) := -ig'(x), \quad g \in D(p)$

Case 1 $I = [0, \infty)$. In this situation, q is positive (but unbounded); $-p$ is a time operator of q and $\sigma(-p) = \overline{\Pi_+}$.
Case 2 $I = (-\ell/2, \ell/2)$, $\ell > 0$. q is a bounded self-adjoint operator and $-p$ is a time operator of q. But, in this case, $\sigma(-p) = \mathbb{C}$.

2.5   Conclusions

The main lesson one can draw from the previous discussion is that infinite-dimensional representations of Lie algebras can be reasonably handled only by invoking the whole apparatus of unbounded operator algebras theory [19]. As it was clear from the very first steps of the theory of unbounded algebras, in the early 1970s, the theory of representations of Lie algebras (or enveloping Lie algebras) was the natural arena for its applications also for the large number of applications in Quantum Physics, where the relevance of invariance under Lie groups of transformations of the dynamical equations was clear all along. This case (integrable representations) was completely under control by the Gårding theory.

But, even if someone says that "Nature knows groups rather than algebras," nonintegrable representations of Lie algebras arise in some recent developments of quantum theories and this is a good reason for them to deserve a more detailed study. The *weak* form (2.1) of the commutation relations involves, as we have discussed so far, more general structures for operators families: partial *-algebras provide, in our opinion, the natural framework where studying a generalized form of the representation theory. The example of nonintegrable representations of the Heisenberg Lie algebra $\mathfrak{h}$ that we have discussed in some detail shows that very few features are preserved in comparison with integrable representations (from the spectral point of view, for instance). Thus, there is a lot of work to be done. We hope to have ignited the interest of some young mathematician to this promising research field.

References

1. J.-P. Antoine, C. Trapani, *Partial Inner Product Spaces: Theory and Applications*. Springer Lecture Notes in Mathematics, vol. 1986 (Berlin, Heidelberg, 2009)
2. J.-P. Antoine, A. Inoue, C. Trapani, *Partial *-Algebras and Their Operator Realizations* (Kluwer, Dordrecht, 2002)
3. A. Arai, Generalized weak Weyl relation and decay of quantum dynamics. Rep. Math. Phys. **17**, 1071–1109 (2005)
4. F. Bagarello, Pseudo-bosons, Riesz bases and coherent states. J. Math. Phys. **50**, 023531 (2010) (10 pp.). doi:10.1063/1.3300804
5. F. Bagarello, Examples of Pseudo-bosons in quantum mechanics. Phys. Lett. A **374**, 3823–3827 (2010)
6. F. Bagarello, Pseudo-bosons, so far. Rep. Math. Phys. **68**, 175–210 (2011)
7. F. Bagarello, A. Inoue, C. Trapani, Weak commutation relations of unbounded operators and applications. J. Math. Phys. **52**, 113508 (2011)
8. F. Bagarello, A. Inoue, C. Trapani, Weak commutation relations of unbounded operators: nonlinear extensions. J. Math. Phys. **53**, 123510 (2012)
9. A.O. Barut, R. Racza, *Theory of Group Representations and Applications* (PWN, Warszawa, 1980)
10. H.D. Doebner, O. Melsheimer, Limitable dynamical groups in quantum mechanics. J. Math. Phys. **9**, 1638–1656 (1968)
11. B. Fuglede, On the relation $PQ - PQ = -iI$. Math. Scand. **20**, 79–88 (1967)
12. L. Gårding, Note on continuous representations of Lie groups. Proc. Natl. Acad. Sci. U.S.A. **33**, 331–332 (1947)
13. L. Gårding, Vecteurs analytiques dans les répresentations des groupes de Lie. Bull. Soc. Math. France **88**, 73–93 (1970)
14. R.V. Kadison, J.R. Ringrose, *Fundamentals of the Theory of Operator Algebras*, vol. I (American Mathematical Society, Providence, 1997)

15. E. Nelson, Analytic vectors. Ann. Math. **70**, 572–615 (1959)
16. M. Reed, B. Simon, *Methods of Modern Mathematical Physics*. Functional Analysis, vol. I (Academic, San Diego, 1980)
17. K. Schmüdgen, On the Heisenberg commutation relations I. J. Funct. Anal. **50**, 8–49 (1983)
18. K. Schmüdgen, On the Heisenberg commutation relations II. Publ. RIMS, Kyoto Univ. **19**, 601–671 (1983)
19. K. Schmüdgen, *Unbounded Operator Algebras and Representation Theory* (Birkhäuser-Verlag, Basel, 1990)

Chapter 3
Character, Multiplicity, and Decomposition Problems in the Representation Theory of Complex Lie Algebras

Peter Fiebig

Abstract This essay is meant as an introduction to a very successful approach towards understanding the structure of certain highest weight categories appearing in algebraic Lie theory. For simplicity, the focus lies on the case of the category $\mathcal{O}$ of representations of a simple complex Lie algebra. We show how the approach yields a proof of the classical Kazhdan–Lusztig conjectures that avoids the theory of D-modules on flag varieties.

3.1 Introduction

This expository article is meant to give an overview on a certain approach towards the representation theory of Lie algebraic structures. This approach is due to W. Soergel and pieces together a couple of ideas that, in one form or the other, make sense in various situations. Its main goal is to understand representation theory from a categorical point of view, i.e. to understand the category of representations rather than the structure of an individual object. Yet sometimes it is strong enough to solve questions that can be formulated categorically. An example is the problem to determine irreducible characters.

For the sake of simplicity only the most classical example of *complex* semisimple Lie algebras and their highest weight modules will be considered. The leitmotif is the famous and most influential Kazhdan–Lusztig conjecture from 1979 on the irreducible highest weight characters. The category that surrounds this conjecture is the not less famous category $\mathcal{O}$ that was introduced in the early 1970s by J. Bernstein, I.M. Gelfand, and S.I. Gelfand. We show how one can translate the character problem into the contexts of Soergel bimodules, of parity sheaves or of moment graph sheaves, where strong additional methods and results to work with are present.

P. Fiebig (✉)

Department Mathematik, FAU Erlangen–Nürnberg, Cauerstraße 11, 91058 Erlangen, Germany
e-mail: fiebig@math.fau.de

© Springer International Publishing AG 2017   41
G. Falcone (ed.), *Lie Groups, Differential Equations, and Geometry*,
UNIPA Springer Series, DOI 10.1007/978-3-319-62181-4_3

Here is a short summary of this article. After describing the *Character Problem* and the *Kazhdan–Lusztig Conjecture* the first problem is translated into the *Multiplicity Problem I* on the Jordan–Hölder multiplicities of Verma modules. Section 3.3 introduces the category $\mathcal{O}$. The BGG-reciprocity relates Jordan–Hölder multiplicities of Verma modules to Verma module multiplicities in Verma flags of projective objects in $\mathcal{O}$. Section 3.3 ends with introducing the *translation functors* which, together with the BGG-reciprocity, allow to translate the Multiplicity Problem I into the *Decomposition Problem I* for Bott-Samelson modules. Section 3.4 describes in more detail the *deformed category $\mathcal{O}$* and provides proofs of the block decomposition and the BGG-reciprocity. Section 3.5 is to encounter W. Soergel's theory that allows to translate the Decomposition Problem I into the *Decomposition Problem II* for a certain iterated tensor product of a commutative algebra $\mathscr{Z}$. In Sect. 3.6 sketches are given of translations of the Decomposition Problem II into the language of Soergel bimodules, moment graph sheaves, or parity sheaves. The article ends with mentioning how one can proceed further: the theory of *intersection cohomology complexes* or, alternatively, the work of B. Elias and G. Williamson on the *Hodge theory of Soergel bimodules* can be used to solve the remaining problem and hence to prove the Kazhdan–Lusztig conjecture. In the short epilogue a few words are said about the modular analogue of Soergel's theory. As prerequisites we assume basic knowledge on Lie algebras and, in particular, the structure theory of semisimple complex Lie algebras. The combinatorial data of the root system, the Weyl group, etc. will be used throughout the article. Due to limitation of space full details cannot be given. However, hopefully the reader will find enough citations to original or review papers and books to be able to delve more deeply into the theory. Surely, as we move along, the arguments become sketchier and sketchier.

This overview has some overlap with Humphreys' book [13] on the category $\mathcal{O}$, but the treatment of the categorical structure deviates at some places from the theory developed there. In particular, no reference is made to Harish–Chandra's work on the center of the universal enveloping algebra. Instead a deformation theory that goes back at least to Jantzen's book [14] is used. The advantage is that almost everything that is developed here can be more or less immediately generalized to the case of symmetrizable Kac–Moody algebras. There are analogous approaches for modular Lie algebras or quantum groups as well (cf. [1]). As mentioned before, the main ideas first appeared in W. Soergel's fundamental paper [19]. However, again this paper deviates a little from this classical paper, and the expert will at certain places see the moment graph theory appear in disguise. There is also some overlap with the paper [11], yet the present article does not focus on the moment graph theory or the theory of parity sheaves. One might view it as a significantly expanded version of Sect. 3 of [11].

3.2   The Character Problem and the Kazhdan–Lusztig Conjecture

Our main motivation is to understand the representation theory of a Lie algebra $\mathfrak{g}$, and a decisive first step is to understand the structure of its *irreducible* modules. For this, we would first like to have a list of all irreducible modules up to isomorphism. In general, this already is a quite hopeless task, which forces us to focus on special cases. For example, we might want to specify the class of Lie algebras that we consider, or we could restrict our attention to modules that are not only irreducible, but also satisfy some additional assumptions. We will do both: assume that $\mathfrak{g}$ is a simple Lie algebra over the field $\mathbb{C}$ of complex numbers, and that the irreducible modules have a "highest weight."

3.2.1   Highest Weights

The notion of "highest weight" depends on some choices. Fix a Cartan subalgebra $\mathfrak{h} \subset \mathfrak{g}$ and a Borel subalgebra $\mathfrak{b} \subset \mathfrak{g}$ that contains $\mathfrak{h}$. A Cartan subalgebra is a maximal abelian subalgebra of $\mathfrak{g}$ such that the adjoint action $\mathrm{ad} \colon \mathfrak{h} \to \mathrm{End}_{\mathbb{C}}(\mathfrak{g})$, $H \mapsto (X \mapsto [H, X])$ is semisimple. Then the Lie algebra $\mathfrak{g}$ decomposes into *weight spaces* under this action, i.e.

$$\mathfrak{g} = \bigoplus_{\lambda \in \mathfrak{h}^\star} \mathfrak{g}_\lambda,$$

where we define for each $\lambda \in \mathfrak{h}^\star = \mathrm{Hom}_{\mathbb{C}}(\mathfrak{h}, \mathbb{C})$ the weight space $\mathfrak{g}_\lambda := \{X \in \mathfrak{g} \mid [H, X] = \lambda(H)X \quad \forall H \in \mathfrak{h}\}$. The maximality of $\mathfrak{h}$ implies $\mathfrak{g}_0 = \mathfrak{h}$, and one calls the set $R := \{\lambda \in \mathfrak{h}^\star \mid \breve{a}\lambda \neq 0, \mathfrak{g}_\lambda \neq \{0\}\}$ the *set of roots* of $\mathfrak{g}$ with respect to $\mathfrak{h}$. Hence

$$\mathfrak{g} = \mathfrak{h} \oplus \bigoplus_{\alpha \in R} \mathfrak{g}_\alpha.$$

A Borel subalgebra is a maximal solvable subalgebra of $\mathfrak{g}$. There always exist Borel subalgebra containing a given Cartan subalgebra $\mathfrak{h}$. Then $\mathfrak{h}$ acts on $\mathfrak{b}$ via the adjoint action as well, and the corresponding weights form a subset R^+ of R. Then $R \subset \mathfrak{h}^\star$ satisfies the axioms of a *root system*, and $R^+ \subset R$ is a choice of a *system of positive roots*. Moreover, our choices yield a partial order on the set $\mathfrak{h}^\star$. It is defined by $\mu \leq \lambda$ if $\lambda - \mu$ can be written as a sum of elements in R^+.

3.2.2   Modules with Highest Weight

We call an $\mathfrak{h}$-module M a *weight module* if it is semisimple. As the simple $\mathfrak{h}$-modules are one-dimensional (and given by a character $\mu \in \mathfrak{h}^\star$), semisimplicity means that M decomposes into a direct sum of its *weight spaces*, i.e.

$$M = \bigoplus_{\mu \in \mathfrak{h}^\star} M_\mu,$$

where for each $\mu \in \mathfrak{h}^\star$ we define $M_\mu = \{m \in M \mid H.m = \mu(H)m \quad \forall H \in \mathfrak{h}\}$. Then μ is called a *weight* of M if $M_\mu \ne \{0\}$ and we call each element of M_μ a *weight vector* of weight μ. Let $\lambda \in \mathfrak{h}^\star$.

Definition 3.2.1 A $\mathfrak{g}$-module M is called a *module of highest weight* λ if the following holds:

1. M is a weight module,
2. all weights of M are smaller or equal to λ, i.e. $M_\mu \ne \{0\}$ implies $\mu \le \lambda$,
3. M is generated by a non-zero vector of weight λ.

(One can show that (3) implies (1).) Here is the classification theorem for the irreducible highest weight modules:

Theorem 3.2.2 (Sect. 1.3 [13]) *Let* $\lambda \in \mathfrak{h}^\star$. *There exists an up to isomorphism unique irreducible module* $L(\lambda)$ *of* $\mathfrak{g}$ *of highest weight* λ. *Moreover,* $L(\lambda) \cong L(\mu)$ *implies* $\lambda = \mu$.
    Set $\rho := \frac{1}{2} \sum_{\alpha \in R^+} \alpha \in \mathfrak{h}^\star$.

Definition 3.2.3 An element λ of $\mathfrak{h}^\star$ is called

1. *integral*, if $\langle \lambda, \alpha^\vee \rangle \in \mathbb{Z}$ for all $\alpha \in R$,
2. *dominant*, if $\langle \lambda + \rho, \alpha^\vee \rangle \not< 0$ for all $\alpha \in R^+$,
3. *antidominant*, if $\langle \lambda + \rho, \alpha^\vee \rangle \not> 0$ for all $\alpha \in R^+$,
4. *regular*, if $\langle \lambda + \rho, \alpha^\vee \rangle \ne 0$ for all $\alpha \in R^+$.

3.2.3   The Character

How can we describe the "structure" of $L(\lambda)$? A precise description could be, at least in the case that $L(\lambda)$ is finite dimensional, to explicitly write down the representing matrices for generators of $\mathfrak{g}$ in terms of a basis of $L(\lambda)$. This might not be very instructional, though. A first interesting piece of information is the dimension of $L(\lambda)$. Here one has the following result:

Theorem 3.2.4 ([13, Sect. 1.6]) *The* $\mathfrak{g}$-*module* $L(\lambda)$ *is finite dimensional if and only if* λ *is integral and dominant.*

A better invariant than the dimension is the *character* of $L(\lambda)$. We will define the character for the class of modules M of $\mathfrak{g}$ that satisfy the following assumptions:

- M is a weight module.
- There exists $\mu_1, \ldots, \mu_n \in \mathfrak{h}^\star$ such that for any weight μ of M there is an $i \in \{1, \ldots, n\}$ with $\mu \leqslant \mu_i$.

Obviously each module with highest weight satisfies these assumptions, and so does each finite direct sum of highest weight modules.

Denote by $\mathbb{Z}[\mathfrak{h}^\star]$ the group algebra of the abelian group $\mathfrak{h}^\star$. Note that it is the free abelian group with basis $\{e^\lambda\}_{\lambda \in \mathfrak{h}^\star}$ and multiplication determined by $e^\lambda \cdot e^\mu = e^{\lambda+\mu}$. Then $\mathbb{Z}[\mathfrak{h}^\star]$ can be viewed as the space of maps $\mathfrak{h}^\star \to \mathbb{Z}$ with finite support, in such a way that the e^λ are the δ-functions at the point λ. Now we can define the subspace $\widehat{\mathbb{Z}[\mathfrak{h}^\star]}$ of $\mathrm{Map}(\mathfrak{h}^\star, \mathbb{Z})$ of maps with "locally bounded support." It contains all $f \in \mathrm{Map}(\mathfrak{h}^\star, \mathbb{Z})$ with the property that there exist $\mu_1, \ldots, \mu_n \in \mathfrak{h}^\star$, depending on f, such that $f(\mu) \neq 0$ implies that there is $i \in \{1, \ldots, n\}$ with $\mu \leqslant \mu_i$. It is not difficult to see that the multiplication on $\mathbb{Z}[\mathfrak{h}^\star]$ has a smooth extension to a multiplication on $\widehat{\mathbb{Z}[\mathfrak{h}^\star]}$.

Definition 3.2.5 Suppose that M satisfies the two assumptions above. Denote by ch M the element in $\widehat{\mathbb{Z}[\mathfrak{h}^\star]}$ that is given by $\mu \mapsto \dim_\mathbb{C} M_\mu$.

Here is the main problem that we are concerned with:

Character Problem *Calculate the character of $L(\lambda)$.*

This is a difficult problem.

3.2.4 The Character Formula of Weyl

In the integral dominant case (i.e., for finite dimensional $L(\lambda)$), the characters have been known for some time now. In order to state the theorem, we introduce the Weyl group of $\mathfrak{g}$. For each $\alpha \in R$ there is a unique element $\alpha^\vee \in [\mathfrak{g}_\alpha, \mathfrak{g}_{-\alpha}]$, called the *coroot* of α, with the property $\alpha(\alpha^\vee) = 2$. The homomorphism $s_\alpha: \mathfrak{h}^\star \to \mathfrak{h}^\star$, $\mu \mapsto \mu - \langle \mu, \alpha^\vee \rangle \alpha$ is called the *reflection* associated with α. Note that it is indeed a reflection, i.e. $s_\alpha^2 = \mathrm{id}_{\mathfrak{h}^\star}$ and the fixed points of s_α form a complex hyperplane. The group $\mathcal{W}$ of linear automorphisms of $\mathfrak{h}^\star$ that is generated by the s_α with $\alpha \in R$ is called the *Weyl group* of $\mathfrak{g}$. Let $\mathcal{S} \subset \mathcal{W}$ be the set of *simple reflections*, i.e. the set of s_α, where $\alpha \in R^+$ is a simple root (i.e., α cannot be written as the sum of at least two positive roots). It turns out that $\mathcal{W}$ is already generated by the simple reflections, and $(\mathcal{W}, \mathcal{S})$ is a *Coxeter system* (cf. [12]). Hence it comes with a *length function* $l: \mathcal{W} \to \mathbb{N}$ and a Bruhat order $\leqslant$ on $\mathcal{W}$.

Theorem 3.2.6 ([13, Sect. 2.4]) *Let λ be an integral and dominant element in $\mathfrak{h}^\star$.*

- *The element* $\sum_{w \in \mathcal{W}} (-1)^{l(w)} e^{w(\lambda+\rho)}$ *is divisible by* $\sum_{w \in \mathcal{W}} (-1)^{l(w)} e^{w(\rho)}$ *in* $\mathbb{Z}[\mathfrak{h}^\star]$. *As* $\mathbb{Z}[\mathfrak{h}^\star]$ *contains no zero-divisors, the element*

$$\chi(\lambda) = \frac{\sum_{w \in \mathcal{W}} (-1)^{l(w)} e^{w(\lambda+\rho)}}{\sum_{w \in \mathcal{W}} (-1)^{l(w)} e^{w(\rho)}}$$

  is hence well defined.
- *It holds that* ch $L(\lambda) = \chi(\lambda)$.

But what about more general λ?

3.2.5   The Hecke Algebra and Kazhdan–Lusztig Polynomials

Before Kazhdan–Lusztig's generalization of Weyl's character formula can be stated, we have to introduce another player, the *Hecke algebra* associated with $\mathfrak{g}$. Let $\mathcal{L} = \mathbb{Z}[v, v^{-1}]$ be the ring of integral Laurent polynomials in the variable v.

Definition 3.2.7 The Hecke algebra associated with $(\mathcal{W}, \mathcal{S})$ is the $\mathcal{L}$-algebra with generators $\{T_w\}_{w \in \mathcal{W}}$ and relations

$$T_x \cdot T_y = T_{xy} \text{ for } x, y \in \mathcal{W} \text{ with } l(xy) = l(x) + l(y),$$

$$T_s^2 = (v^2 - 1)T_s + v^2 T_e \text{ for } s \in \mathcal{S}.$$

It turns out that $\{T_w\}_{w \in \mathcal{W}}$ is an $\mathcal{L}$-basis of $\mathcal{H}$ and T_e is the multiplicative unit in $\mathcal{H}$. Moreover, each T_x is invertible in $\mathcal{H}$. D. Kazhdan and G. Lusztig defined in [16] a $\mathbb{Z}$-linear involution $^\cdot$ on $\mathcal{H}$ as follows: $\overline{v} = v^{-1}$ and $\overline{T_x} = T_{x^{-1}}^{-1}$ and they proved the following result:

Theorem 3.2.8 *For each* $w \in \mathcal{W}$ *there exists a unique element* $C'_w \in \mathcal{H}$ *with the following properties:*

1. C'_w *is self-dual, i.e.* $\overline{C'_w} = C'_w$.
2. $C'_w \in v^{-l(w)} \sum_{x \leqslant w} P_{x,w} T_x$, *where* $P_{x,w}$ *is a polynomial in* v *of degree* $\leqslant l(w) - l(x) - 1$.

The polynomials $P_{x,w}$ are the famous *Kazhdan–Lusztig polynomials* that are ubiquitous in the representation theory and geometry of Lie algebras and Lie groups.

3.2.6   A Generalization of Weyl's Character Formula: The Conjecture of Kazhdan and Lusztig

For all $w \in \mathcal{W}$ set $L_w := L(-w(\rho) - \rho)$ and $\Delta_w := \Delta(-w(\rho) - \rho)$. The *dot-action* of $\mathcal{W}$ on $\mathfrak{h}^\star$ is defined by setting $w.\lambda = w(\lambda + \rho) - \rho$. Then $-w(\rho) - \rho = w.(-2\rho)$.

As $w_0.0 = -2\rho$, where $w_0 \in \mathscr{W}$ is the longest element, it holds that $L_w = L(ww_0.0)$ and $\Delta_w = \Delta(ww_0.0)$.

In [16] one finds the following conjecture:

Conjecture 3.2.9 For $w \in \mathscr{W}$ the character of L_w is given by

$$\mathrm{ch}\, L_w = \sum_{y \leqslant w} (-1)^{l(y)+l(w)} P_{y,w}(1)\, \mathrm{ch}\, \Delta_y.$$

This astounding conjecture is most influential and is the starting point of what sometimes is called *Kazhdan–Lusztig theory*. From the theory of translation functors (cf. Sect. 3.3.4) one can deduce that the above formula is equivalent to the analogous formula where the weight 0 is replaced by an arbitrary integral, dominant, and regular λ. There are similar formulas for dominant and regular, but not necessarily integral weights λ. For those on should replace $\mathscr{W}$ by the so-called *integral Weyl group* $\mathscr{W}_{[\lambda]}$ of λ (cf. Definition 3.3.3).

Conjecture 3.2.9 has been proved by Beilinson and Bernstein in [2] and, independently and in the same year, by Brylinski and Kashiwara in [4] using the theory of D-modules. In this paper we sketch a different proof of the above conjecture that is originally due to Soergel. The first part of this proof is to translate the conjecture into a conjecture involving *Soergel bimodules* (or, equivalently, *moment graph sheaves*). Already this translation is quite a journey! We start with rephrasing the character conjecture of Kazhdan and Lusztig as a conjecture on Jordan–Hölder multiplicities of Verma modules.

3.2.7 The Multiplicity Version of the Character Problem

For each $\lambda \in \mathfrak{h}^\star$ there is a "universal module with highest weight." It is defined as follows. Let $U(\mathfrak{g})$ and $U(\mathfrak{b})$ be the universal enveloping algebras of $\mathfrak{g}$ and $\mathfrak{b}$, respectively. Let $\mathbb{C}_\lambda$ be the one-dimensional module of $\mathfrak{b}$ with weight λ, and obtain by induction (cf. [13, Sect. 1.3]) the Verma module

$$\Delta(\lambda) := U(\mathfrak{g}) \otimes_{U(\mathfrak{b})} \mathbb{C}_\lambda.$$

The subalgebra $\mathfrak{n}^- := \bigoplus_{\alpha \in R^+} \mathfrak{g}_{-\alpha}$ of $\mathfrak{g}$ is nilpotent, and the Poincaré–Birkhoff–Witt Theorem implies that $U(\mathfrak{g}) = U(\mathfrak{n}^-) \otimes_{\mathbb{C}} U(\mathfrak{b})$ (as a left $U(\mathfrak{n}^-)$—and right $U(\mathfrak{b})$-module). Hence $\Delta(\lambda)$ is free over $U(\mathfrak{n}^-)$ of rank 1. Another application of the Poincaré–Birkhoff–Witt Theorem yields the following character formula:

Proposition 3.2.10 ([13, Sect. 1.16]) *The character of a Verma module with highest weight $\lambda \in \mathfrak{h}^\star$ is*

$$\mathrm{ch}\, \Delta(\lambda) = e^\lambda \cdot \prod_{\alpha \in R^+} (1 + e^{-\alpha} + e^{-2\alpha} + \cdots).$$

An easy argument shows that the Character Problem of Sect. 3.2.3 is equivalent to the following:

Multiplicity Problem I *Calculate the Jordan–Hölder multiplicities* $[\Delta(\lambda) : L(\mu)]$ *for all* $\lambda, \mu \in \mathfrak{h}^\star$.

The inversion formula in [16] implies that Conjecture 3.2.9 above is equivalent to the following statement (cf. [13, Sect. 8.4]):

Conjecture 3.2.11 For $\lambda \in \widehat{\mathfrak{h}}^\star$ all subquotients of Δ_w are of the form L_x for some $x \in \mathcal{W}$, and the multiplicities are

$$[\Delta_w : L_x] = P_{w_0w,w_0x}(1).$$

3.3 Category $\mathcal{O}$

As a next step we introduce an abelian category that contains all the objects that were relevant for us up to this point.

Definition 3.3.1 Denote by $\mathcal{O}$ the full subcategory of the category of all $\mathfrak{g}$-modules that contains all objects that are semisimple for the action of $\mathfrak{h}$ and locally finite for the action of $\mathfrak{b}$.

Recall that semisimplicity for the action of $\mathfrak{h}$ is the same as being a weight module. A $\mathfrak{g}$-module M is *locally finite* for the action of $\mathfrak{b}$ if every element of M is contained in a sub-$\mathfrak{b}$-module of M of finite dimension. Sometimes one also requires an object of $\mathcal{O}$ to be finitely generated as a $\mathfrak{g}$-module. For us, this is not necessary and it is not a good assumption in the Kac–Moody case, because the universal enveloping algebra of a Kac–Moody algebra is not Noetherian. All highest weight modules (in particular, the simple highest weight modules $L(\lambda)$ and the Verma modules $\Delta(\lambda)$) belong to category $\mathcal{O}$.

3.3.1 The Block Decomposition of $\mathcal{O}$

We now determine the decomposition of category $\mathcal{O}$ into indecomposable block and the corresponding linkage principle which determines the set of pairs (λ, μ) with $[\Delta(\lambda) : L(\mu)] \neq 0$. We start with an equivalence relation on the set $\mathfrak{h}^\star$.

Definition 3.3.2 Let $\sim$ be the equivalence relation on $\mathfrak{h}^\star$ that is generated by $\lambda \sim \mu$ if there is $\alpha \in R^+$ and $n \in \mathbb{Z}$ such that $\langle \lambda + \rho, \alpha^\vee \rangle = n$ and $\lambda - \mu = n\alpha$.

We can describe the equivalence classes much more explicitly using the action of the Weyl group on $\mathfrak{h}^\star$. Let $\lambda \in \mathfrak{h}^\star$.

Definition 3.3.3 Define

$$R_{[\lambda]} := \{\alpha \in R \mid \langle \lambda, \alpha^\vee \rangle \in \mathbb{Z}\},$$

$$\mathcal{T}_{[\lambda]} := \{s_\alpha \mid \alpha \in R_{[\lambda]}\}$$

$$\mathcal{W}_{[\lambda]} := \langle t \mid t \in \mathcal{T}_{[\lambda]} \rangle.$$

The subgroup $\mathcal{W}_{[\lambda]}$ of $\mathcal{W}$ is called the *integral Weyl group* with respect to λ.
It is now not difficult to prove the following.

Lemma 3.3.4 *Let $\lambda, \mu \in \mathfrak{h}^\star$.*

1. If $\lambda \sim \mu$, then $R_{[\lambda]} = R_{[\mu]}$ and hence $\mathcal{T}_{[\lambda]} = \mathcal{T}_{[\mu]}$ and $\mathcal{W}_{[\lambda]} = \mathcal{W}_{[\mu]}$.
2. We have $\lambda \sim \mu$ if and only if $\mu \in \mathcal{W}_{[\lambda]}.\lambda$.

Hence the equivalence classes are orbits in $\mathfrak{h}^\star$ under certain subgroups of the Weyl group. We now discuss the block decomposition of $\mathcal{O}$. For any equivalence class Λ with respect to $\sim$ let $\mathcal{O}_\Lambda$ be the full subcategory of $\mathcal{O}$ that contains all objects M such that $[M : L(\lambda)] \neq 0$ implies $\lambda \in \Lambda$. Note that even though not all objects in M have a finite Jordan–Hölder series, the statement $[M : L(\lambda)] \neq 0$ makes sense and means that there is a submodule N of M that has $L(\lambda)$ as a quotient.

Theorem 3.3.5 *The functor*

$$\prod_{\Lambda \in \mathfrak{h}^\star/\sim} \mathcal{O}_\Lambda \to \mathcal{O},$$

$$\{M_\Lambda\} \mapsto \bigoplus_{\Lambda \in \mathfrak{h}^\star/\sim} M_\Lambda,$$

is an equivalence of categories.

Theorem 3.3.5 is a special case of the more general Theorem 3.4.21 that we will prove using deformation theory.

3.3.2 Objects Admitting a Verma Flag

Here is an important concept for category $\mathcal{O}$.

Definition 3.3.6 Let M be an object in $\mathcal{O}$. We say that M *admits a Verma flag* if there exists a finite filtration

$$0 = M_0 \subset M_1 \subset \cdots \subset M_n = M$$

such that M_i/M_{i-1} is isomorphic to a Verma module for $i = 1, \ldots n$.

It is not difficult to show that if M admits a Verma flag, the corresponding multiplicities

$$(M : \Delta(\lambda)) = \#\{i \in \{1, \ldots, n\} \mid M_i/M_{i-1} \cong \Delta(\lambda)\}$$

are independent of the chosen filtration.

3.3.3   Projectives in $\mathscr{O}$

The advantage of the categorical framework that $\mathscr{O}$ provides is that we can rephrase the multiplicity problem of Sect. 3.2.7 in terms of *projective* objects in $\mathscr{O}$. In the following we give an account of the theory of projectives in $\mathscr{O}$ that is different from the one presented in [13], but has the advantage that it can be generalized more easily to other situations such as the Kac–Moody case.

Definition 3.3.7 A subset $\mathscr{J}$ of $\mathfrak{h}^\star$ is called *open* if $\lambda \in \mathscr{J}$ and $\mu \in \mathfrak{h}^\star$ with $\mu \leqslant \lambda$ implies $\mu \in \mathscr{J}$. An open subset $\mathscr{J}$ of $\mathfrak{h}^\star$ is called *locally bounded* if for all $\lambda \in \mathscr{J}$ the set $\{\mu \in \mathscr{J} \mid \lambda \leqslant \mu\}$ is finite.

It is easy to show that this defines indeed a topology on the set $\mathfrak{h}^\star$. It has the additional property that arbitrary intersections of open subsets are open again. For any open subset $\mathscr{J} \subset \mathfrak{h}^\star$ denote by $\mathscr{O}^{\mathscr{J}}$ the full subcategory of $\mathscr{O}$ that contains all objects M with the property that $M_\nu \neq 0$ implies $\nu \in \mathscr{J}$. Note that $\Delta(\lambda)$ and $L(\lambda)$ are contained in $\mathscr{O}^{\mathscr{J}}$ if and only if $\lambda \in \mathscr{J}$.

Theorem 3.3.8 *Suppose that $\mathscr{J} \subset \mathfrak{h}^\star$ is open and locally bounded and let $\lambda \in \mathscr{J}$. Then there exists a projective cover $P^{\mathscr{J}}(\lambda) \to L(\lambda)$ of $L(\lambda)$ in $\mathscr{O}^{\mathscr{J}}$. It has the following properties:*

1. *$P^{\mathscr{J}}(\lambda)$ admits a Verma flag.*
2. *It holds that $(P^{\mathscr{J}}(\lambda) : \Delta(\mu)) = [\Delta(\mu) : L(\lambda)]$ (BGG-reciprocity) for all $\mu \in \mathscr{J}$.*

Recall that we call an object P a *projective cover* of a simple object L in an abelian category $\mathscr{A}$ if P is projective and if there is an epimorphism $f: P \to L$ in $\mathscr{A}$ that has the property that for any morphism $g: M \to P$ in $\mathscr{A}$ with the property that $f \circ g: M \to L$ is an epimorphism, g is an epimorphism.

We will prove the above theorem in Sect. 3.4 in a more general situation, i.e. in the context of the *deformed category $\mathscr{O}$*. Using the BGG-reciprocity result we can rephrase the Multiplicity Problem I as follows:

Multiplicity Problem II *Calculate $(P^{\mathscr{J}}(\lambda) : \Delta(\mu))$ for all locally bounded open subsets $\mathscr{J}$ and all $\lambda, \mu \in \mathscr{J}$.*

3.3.4   Translation Functors

This section introduces Jantzen's *translation functors*. To simplify matters, assume that λ is regular and integral, i.e. $\mathscr{W}_{[\lambda]} = \mathscr{W}$ and $\mathrm{Stab}_{\mathscr{W}} = \{e\}$. Suppose also that λ is anti-dominant. Set $\Lambda = [\lambda]$.

Theorem 3.3.9 ([14])   *Let $s \in \mathscr{S}$. There exists a functor $T_s \colon \mathscr{O}_\Lambda \to \mathscr{O}_\Lambda$ with the following properties:*

1. *T_s is exact and self-adjoint.*
2. *For $w \in \mathscr{W}$ it holds that $T_s \Delta(w.\lambda) \cong T_s \Delta(ws.\lambda)$, and if $ws > w$, then there is a short exact sequence*

$$0 \to \Delta(ws.\lambda) \to T_s \Delta(w.\lambda) \to \Delta(w.\lambda) \to 0.$$

Lemma 3.3.10

1. *If M admits a Verma flag, then so does $T_s M$ and for the multiplicities holds*

$$(T_s M : \Delta(w.\lambda)) = (M : \Delta(w.\lambda)) + (M : \Delta(ws.\lambda)).$$

2. *If P is projective in $\mathscr{O}_\Lambda$, then $T_s P$ is also projective in $\mathscr{O}_\Lambda$.*

Proof Theorem 3.3.9 implies that T_s maps each Verma module into a module that admits a Verma flag. As T_s is also exact, it follows that T_s preserves the subcategory in $\mathscr{O}_\Lambda$ of modules that admit a Verma flag. The formula in (1) is another consequence of the exactness of T_s and the short exact sequence in Theorem 3.3.9, (2).

Note that there is a natural isomorphism

$$\mathrm{Hom}_{\mathscr{O}}(T_s P, \cdot) \cong \mathrm{Hom}_{\mathscr{O}}(P, T_s(\cdot)).$$

for any object P in $\mathscr{O}_\Lambda$. If P is projective, then $\mathrm{Hom}_{\mathscr{O}}(P, \cdot)$ is an exact functor, hence $\mathrm{Hom}(T_s P, \cdot)$ is isomorphic to the composition of the exact functors $\mathrm{Hom}_{\mathscr{O}}(P, \cdot)$ and T_s, i.e. $T_s P$ is projective as well.

3.3.5   The Decomposition Problem I

Let $\mathbf{s} = (s_1, \ldots, s_n)$ be a finite sequence of simple reflections. Let $\lambda_0 \in \Lambda$ be dominant, integral, and regular. Then $\Delta(\lambda_0)$ is projective in $\mathscr{O}_\Lambda$. As the translation functors preserve projectivity, the object $BS(\mathbf{s}) := T_{s_1} \circ \cdots \circ T_{s_n} \Delta(\lambda_0)$ is projective in $\mathscr{O}_\Lambda$. Hence there are numbers $a_{\mathbf{s}}^w \in \mathbb{N}$ such that

$$BS(\mathbf{s}) \cong \bigoplus_{w \in \mathscr{W}} P(w.\lambda_0)^{\oplus a_{\mathbf{s}}^w}.$$

The theory of translation functors can be used to prove that the numbers $a_{\mathbf{s}}^w$ do not depend on the choice dominant integral regular element λ.

Decomposition Problem I *Determine all numbers $a_{\mathbf{s}}^w$ for all finite sequences $\mathbf{s}$ and all $w \in \mathscr{W}$.*

Our next problem is to translate Conjecture 3.2.11 into a conjecture about the numbers $a_{\mathbf{s}}(w)$. Recall from Theorem 3.2.8 the self-dual basis $\{C_w'\}_{w \in \mathscr{W}}$ of the Hecke algebra $\mathscr{H}$. For any sequence $\mathbf{s} = (s_1, \dots, s_n)$ we consider the element $C_{s_n}' \cdots C_{s_1}'$. There are unique elements $b_{\mathbf{s}}^w \in \mathbb{Z}[v, v^{-1}]$ such that

$$C_{s_n}' \cdots C_{s_1}' = \sum_{w \in \mathscr{W}} b_{\mathbf{s}}^w C_w'.$$

It is now not difficult to show that Conjecture 3.2.11 is equivalent to the following *Decomposition conjecture*:

Conjecture 3.3.11 For each finite sequence $\mathbf{s}$ in $\mathscr{S}$ and all $w \in \mathscr{W}$ we have $a_{\mathbf{s}}^w = b_{\mathbf{s}}^w(1)$.

3.4 Deforming the Category $\mathscr{O}$

In this section the proofs of Theorems 3.3.5 and 3.3.8 are given. In contrast to the proofs in [13], a deformation theory is used. Let $S = S(\mathfrak{h}) = U(\mathfrak{h})$ be the symmetric algebra of the vector space $\mathfrak{h}$.

Definition 3.4.1 A *deformation algebra* is a commutative, local, Noetherian, and unital S-algebra.

Remark 3.4.2

1. If A is a deformation algebra with (unique) maximal ideal $\mathfrak{m}$, then the residue field $A/\mathfrak{m}$ (with the induced S-algebra structure) is a deformation algebra as well.
2. Let $\widetilde{S}$ be the localization of S at the maximal ideal generated by $\mathfrak{h}$. In geometric terms, S is the $\mathbb{C}$-algebra of regular functions on the complex affine space $\mathfrak{h}^\star$, and $\widetilde{S}$ is the localization at the closed point $0 \in \mathfrak{h}^\star$. Then $\widetilde{S}$ is a deformation algebra. Its residue field is $\mathbb{C}$ with the S-algebra structure that is given by evaluating a local function at the closed point 0.
3. If $\mathfrak{p} \subset \widetilde{S}$ is any prime ideal, then the localization $\widetilde{S}_{\mathfrak{p}}$ of $\widetilde{S}$ at $\mathfrak{p}$ is a deformation algebra as well. In this way we obtain, for example, the quotient field $\operatorname{Quot} S = \widetilde{S}_{(0)}$ of $\widetilde{S}$.

For any deformation algebra A we will now introduce an A-linear version of category $\mathscr{O}$. A more detailed account of the deformed category $\mathscr{O}$ can be found in [7]. Let $\mathfrak{l}$ be a complex Lie algebra. Let M be a $\mathfrak{l}$-A-bimodule. Note that this is an A-module that is endowed with a homomorphism $\rho \colon \mathfrak{l} \to \operatorname{End}_A(M)$ of $\mathbb{C}$-Lie algebras.

Denote by $\tau: \mathfrak{h} \to A$ the *structural map*, i.e. the composition $\mathfrak{h} \subset S \to A$, where the last homomorphism sends f to $f \cdot 1_A$.

Definition 3.4.3

- Let M be an $\mathfrak{h}$-A-bimodule and $\lambda \in \mathfrak{h}^\star$. Define

$$M_\lambda := \{m \in M \mid H.m = (\lambda + \tau)(H)m \quad \forall H \in \mathfrak{h}\}.$$

  Here we consider $(\lambda + \tau)(H) = \lambda(H) \cdot 1_A + \tau(H)$ as an element in A.
- An $\mathfrak{h}$-A-bimodule M is called a *weight module* if $M = \bigoplus_{\lambda \in \mathfrak{h}^\star} M_\lambda$.
- A $\mathfrak{b}$-A-bimodule M is called *locally finite* if for any $m \in M$ the set $U(\mathfrak{b})m \subset M$ generates a finitely generated A-module.
- The category $\mathcal{O}_A$ is the full subcategory of the category of $\mathfrak{g}$-A-bimodules that contains all M that are weight modules for the action of $\mathfrak{h}$ and that are locally finite for the action of $\mathfrak{b}$.

Remark 3.4.4

1. As A is supposed to be Noetherian one can show that $\mathcal{O}_A$ is an abelian category.
2. Note that for $A = \mathbb{C}$ with the S-algebra structure that comes from evaluating at $0 \in \mathfrak{h}^\star$, we have $\tau = 0$ and $\mathcal{O}_\mathbb{C}$ coincides with the classical category $\mathcal{O}$ of Definition 3.3.1.
3. More generally, if $A = \mathbb{K}$ is a field, the deformed category $\mathcal{O}_\mathbb{K}$ is a full subcategory of the usual category $\mathcal{O}$ over the Lie algebra $\mathfrak{g}_\mathbb{K} = \mathfrak{g} \otimes_\mathbb{C} \mathbb{K}$. It contains all objects M with weights contained in the subset $\tau + \mathfrak{h}^\star$ of $\mathfrak{h}_\mathbb{K}^\star = \mathrm{Hom}_\mathbb{K}(\mathfrak{h}_\mathbb{K}, \mathbb{K})$.
4. Let $f: A \to B$ be a homomorphism of deformation algebras. Then the change of scalars functor $\cdot \otimes_A A'$ induces a functor $\mathcal{O}_A \to \mathcal{O}_{A'}$.

3.4.1 Deformed Verma Modules and Verma Flags

Let $\lambda \in \mathfrak{h}^\star$. Let A_λ be the $\mathfrak{b}$-A-bimodule that is free of rank 1 as an A-module and on which $\mathfrak{b}$ acts via the character $\mathfrak{b} \to \mathfrak{h} \xrightarrow{\lambda + \tau} A$.

Definition 3.4.5 The *deformed Verma module* with highest weight λ is defined as

$$\Delta_A(\lambda) := U(\mathfrak{g}) \otimes_{U(\mathfrak{b})} A_\lambda.$$

It is not difficult to check that $\Delta_A(\lambda)$ is contained in $\mathcal{O}_A$. For a homomorphism $f: A \to A'$ of deformation algebras there is an isomorphism $\Delta_A(\lambda) \otimes_A A' \cong \Delta_{A'}(\lambda)$.

The definition of objects that admit a Verma flag in $\mathcal{O}_A$ is the same as Definition 3.3.6 in the case of $\mathcal{O}$. Again, the multiplicity $\#\{i \in \{1, \ldots, n\} \mid \mu_i = \mu\}$ of a subquotient isomorphic to $\Delta_A(\mu)$ is independent of the particular filtration, and is denoted by $(M : \Delta_A(\mu))$. Define

$$\text{supp}_\Delta M := \{\mu \in \mathfrak{h}^\star \mid (M : \Delta_A(\mu)) \neq 0\},$$

the *support* of the Verma flag object M.

For $\lambda \in \mathfrak{h}^\star$ let us denote by $v_\lambda = 1 \otimes 1 \in \Delta_A(\lambda)$ a generator of the highest weight space. The following statement is an immediate consequence of the universal property of induction and the definition of the $\mathfrak{b}$-A-bimodule A_λ:

Lemma 3.4.6 *For all $\lambda \in \mathfrak{h}^\star$ and all objects $M \in \mathcal{O}_A$ the A-linear homomorphism*

$$\text{Hom}_{\mathcal{O}_A^{\mathscr{J}}}(\Delta_A(\lambda), M) \to \{m \in M_\lambda \mid \mathfrak{n}^+ m = 0\},$$

$$f \mapsto f(v_\lambda)$$

is an isomorphism.

Here, by $\mathfrak{n}^+$ we denote the nilpotent subalgebra $\bigoplus_{\alpha \in R^+} \mathfrak{g}_\alpha$.

Remark 3.4.7 One can use the above universal property to reorder a Verma flag $0 = M_0 \subset M_1 \subset \cdots \subset M_n = M$ with $M_{i+1}/M_i \cong \Delta_A(\mu)$ in such a way that $\mu_i \geq \mu_j$ implies $i \leq j$.

3.4.2 Simple Objects in $\mathcal{O}_A$

We can also classify the simple objects in $\mathcal{O}_A$:

Proposition 3.4.8 ([9, Proposition 2.1]) *Let A be a local deformation algebra with residue field $\mathbb{K}$. Then the functor $\cdot \otimes_A \mathbb{K}\colon \mathcal{O}_A \to \mathcal{O}_\mathbb{K}$ induces a bijection on simple isomorphism classes.*

As observed above, $\mathcal{O}_\mathbb{K}$ is just a direct summand of the usual category $\mathcal{O}$ over the Lie algebra $\mathfrak{g}_\mathbb{K}$ that contains all objects with weights of the form $\lambda + \tau$. Hence the simple objects in $\mathcal{O}_\mathbb{K}$ are the irreducible highest weight modules corresponding to these weights. Denote the one corresponding to the highest weight $\lambda + \tau$ by $L_\mathbb{K}(\lambda)$. Then there is an up to isomorphism unique simple object $L_A(\lambda)$ in $\mathcal{O}_A$ with $L_A(\lambda) \otimes_A \mathbb{K} \cong L_\mathbb{K}(\lambda)$.

Our next objective is to study the projective objects in $\mathcal{O}_A$.

3.4.3 Projectives in $\mathcal{O}_A$

Let A be a deformation algebra. Let $\mathscr{J}$ be an open and locally bounded subset of $\mathfrak{h}^\star$ and $\lambda \in \mathscr{J}$. Set $\mathscr{J}' = \mathscr{J} - \lambda = \{\mu - \lambda \mid \mu \in \mathscr{J}\}$. Then $\mathscr{J}'$ is an open and locally bounded subset of $\mathfrak{h}^\star$ containing 0. Denote by $B = U(\mathfrak{b})$ the universal enveloping algebra. Note that it contains $S = U(\mathfrak{h})$ and B is a weight module, i.e. $B = \bigoplus_{\lambda \in \mathfrak{h}^*} B_\lambda$ for the adjoint action of $\mathfrak{h}$. Moreover, $B_\lambda \neq 0$ implies (and is equivalent to) $\lambda \geq 0$ and it holds that $B_0 = S$. So $B' := \bigoplus_{\lambda \in \mathfrak{h}^* \setminus \mathscr{J}'} B_\lambda$ is a two-sided ideal in B (as $\mathscr{J}'$ is open). Define $B^{\mathscr{J}'} = B/B'$ and consider this as a B-module from the left, and as an S-module from the right. Then $B^{\mathscr{J}'} \otimes_S A_\lambda$ is a B-module (B acting on the first factor) and an A-module (here A acts on the right factor). Both actions clearly commute, i.e. we have constructed a B-A-bimodule. Now induce:

$$Q_A^{\mathscr{J}}(\lambda) := U(\mathfrak{g}) \otimes_B (B^{\mathscr{J}'} \otimes_S A_\lambda).$$

This is a $U(\mathfrak{g})$-A-bimodule.

Proposition 3.4.9 ([9, Lemma 2.3])

1. $Q_A^{\mathscr{J}}(\lambda)$ *is an object in* $\mathcal{O}_A^{\mathscr{J}}$ *and it admits a Verma flag with multiplicities*

$$(Q_A^{\mathscr{J}}(\lambda) : \Delta_A(\mu)) = \dim_{\mathbb{C}} B_{\mu-\lambda}/B_0$$

for all $\mu \in \mathscr{J}$.

2. *For any $M \in \mathcal{O}_A^{\mathscr{J}}$ there is an isomorphism*

$$\mathrm{Hom}_{\mathcal{O}_A}(Q_A(\lambda), M) \xrightarrow{\sim} M_\lambda,$$

functorial in M. In particular, $Q_A^{\mathscr{J}}(\lambda)$ is a projective object in $\mathcal{O}_A^{\mathscr{J}}$.

By the above proposition, $(Q_A^{\mathscr{J}}(\lambda) : \Delta_A(\lambda)) = \dim_{\mathbb{C}} B_0/B_0 = 1$ and $(Q_A^{\mathscr{J}}(\lambda) : \Delta_A(\mu)) \neq 0$ implies $\mu \geq \lambda$. From Remark 3.4.7 we deduce that $\Delta_A(\lambda)$, and hence $L_A(\lambda)$, is as a quotient of $Q_A^{\mathscr{J}}(\lambda)$.

Theorem 3.4.10 *Let $\mathscr{J}$ be an open and locally bounded subset of $\mathfrak{h}^*$. For any $\lambda \in \mathscr{J}$ there exists a projective cover $P_A^{\mathscr{J}}(\lambda) \to L_A(\lambda)$ in $\mathcal{O}_A^{\mathscr{J}}$. Moreover, $L_A(\lambda)$ is the only simple quotient of $P_A^{\mathscr{J}}(\lambda)$.*

Proof Let us introduce a convenient notation: write $\mathscr{J}_{\geq \lambda}$ for $\{\mu \in \mathscr{J} \mid \mu \geq \lambda\}$. We prove the theorem by induction on the number of elements in the finite set $\mathscr{J}_{\geq \lambda}$. If this set contains only the element λ, then λ is maximal in $\mathscr{J}$ and the universal property in Lemma 3.4.6 implies that $P_A^{\mathscr{J}}(\lambda) := \Delta_A(\lambda)$ is projective in $\mathcal{O}_A^{\mathscr{J}}$. Clearly it admits a surjection $f: \Delta_A(\lambda) \to L_A(\lambda)$. If $g: M \to P_A^{\mathscr{J}}(\lambda)$ is such that $f \circ g: M \to L_A(\lambda)$ is surjective, then $\mathrm{im}\, g$ contains the λ-weight space of $\Delta_A(\lambda)$ (by Nakayama's lemma). As this weight space generates $\Delta_A(\lambda)$ we obtain $\mathrm{im}\, g = \Delta_A(\lambda)$, so g is surjective. Hence $\Delta_A(\lambda) \to L_A(\lambda)$ is a projective cover. As $\Delta_A(\lambda)$ is a highest weight module of highest weight λ, $L_A(\lambda)$ is the only simple quotient.

Now suppose that the statement for all $\mu \in \mathscr{J}$ with $|\mathscr{J}_{\geq \mu}| < |\mathscr{J}_\lambda|$ is shown. Recall from the preceding paragraph that $\Delta_A(\lambda)$ and hence $L_A(\lambda)$ are quotients of

$Q_A^{\mathscr{I}}(\lambda)$. As the highest weight space of $\Delta_A(\lambda)$ is free of rank 1 over the local ring A, there must be an indecomposable direct summand of $Q_A^{\mathscr{I}}(\lambda)$ that has $\Delta_A(\lambda)$ as a quotient. Let us denote one of those by P. We claim that any surjective homomorphism $f: P \to L_A(\lambda)$ is in fact a projective cover. First we show that $L_A(\lambda)$ is the unique simple quotient of P. So suppose that there is a surjection $f: P \to L_A(\nu)$ and $\nu \neq \lambda$. As P admits a Verma flag (as a direct summand of an object admitting a Verma flag) with subquotients with highest weights in $\mathscr{I}_{\geq \lambda}$, we deduce $\nu > \lambda$. In particular, by induction there already is a projective cover $f': P_A^{\mathscr{I}}(\nu) \to L_A(\nu)$ in $\mathscr{O}_A^{\mathscr{I}}$. The projectivity of P implies that there is a homomorphism $h: P \to P_A^{\mathscr{I}}(\nu)$ such that the diagram

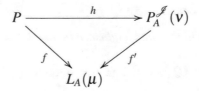

commutes. The projective cover property of $P_A^{\mathscr{I}}(\nu)$ then implies that h is surjective, and hence h must split. As P is indecomposable, we obtain $P \cong P^{\mathscr{I}}(\nu)$ which is impossible, as it is already shown that $P_A^{\mathscr{I}}(\nu)$ has no quotient isomorphic to $L_A(\lambda)$. So $L_A(\lambda)$ is the only simple quotient of P.

It remains to show that $f: P \to L_A(\lambda)$ is a projective cover. So let $g: M \to P$ be any homomorphism in $\mathscr{O}_A^{\mathscr{I}}$ such that $f \circ g$ is surjective. If g is not surjective, then $P/\mathrm{im}\, g$ must have a simple quotient (since it has a finite filtration by highest weight modules). But this simple quotient cannot be isomorphic to $L_A(\lambda)$ (as $(P : \Delta_A(\lambda)) = 1$ and $(P : \Delta_A(\mu)) \neq 0$ implies $\mu \geq \lambda$). But this is a contradiction to what is shown in the preceding paragraph. Hence g must be surjective, so $f: P \to L_A(\lambda)$ is a projective cover.

Remarks 1

1. The proof of the Theorem above shows that any indecomposable projective object in $\mathscr{O}_A^{\mathscr{I}}$ is a direct summand of $Q_A^{\mathscr{I}}(\lambda)$ for some λ.

2. In particular, if P is an indecomposable projective object in $\mathscr{O}_A^{\mathscr{I}}$, then $\mathrm{Hom}_{\mathscr{O}_A}(P, M)$ is a direct summand of $\mathrm{Hom}_{\mathscr{O}_A}(Q_A^{\mathscr{I}}(\lambda), M)$ for some $\lambda \in \mathscr{I}$. Now the latter space can be identified with M_λ. So if M_λ is free as an A-module of finite rank, then $\mathrm{Hom}_{\mathscr{O}_A}(P, M)$ is free of finite rank as an A-module as well.

Proposition 3.4.11 *Let A be a local deformation algebra with residue field $\mathbb{K}$. Let $\mathscr{I} \subset \mathfrak{h}^\star$ be open and locally bounded and let λ be an element in $\mathscr{I}$.*

1. There is an isomorphism $P_A^{\mathscr{I}}(\lambda) \otimes_A \mathbb{K} \cong P_{\mathbb{K}}^{\mathscr{I}}(\lambda)$ in $\mathscr{O}_{\mathbb{K}}^{\mathscr{I}}$.

2. *If P is projective in $\mathscr{O}_A^{\mathscr{J}}$ and $A \to A'$ is a homomorphism of deformation algebras, then for any $M \in \mathscr{O}_A^{\mathscr{J}}$ the natural homomorphism*

$$\mathrm{Hom}_{\mathscr{O}_A^{\mathscr{J}}}(P, M) \otimes_A A' \cong \mathrm{Hom}_{\mathscr{O}_{\mathscr{K}}^{\mathscr{J}}}(P \otimes_A A', M \otimes_A A')$$

is an isomorphism.

Proof We prove part (2) first. Note that for $P = Q_A^{\mathscr{J}}(\lambda)$ the statement follows from the fact that $Q_A^{\mathscr{J}}(\lambda)$ represents the functor $M \mapsto M_\lambda$, and that there is a canonical identification $(M \otimes_A A')_\lambda = (M_\lambda) \otimes_A A'$. As any indecomposable projective object in $\mathscr{O}_A^{\mathscr{J}}$ is a direct summand of some $Q_A^{\mathscr{J}}(\lambda)$, this also implies that the functor $\cdot \otimes_A A'$ preserves projectivity. In order to prove part (1) we check that $L_{\mathbb{K}}(\lambda)$ is the only irreducible quotient of $P_A^{\mathscr{J}}(\lambda) \otimes_A \mathbb{K}$. Indeed, as there is a surjective homomorphism $P_A^{\mathscr{J}}(\lambda) \to P_A^{\mathscr{J}}(\lambda) \otimes_A \mathbb{K}$ (which can be considered in the category $\mathscr{O}_A^{\mathscr{J}}$!), every irreducible quotient of $P_A^{\mathscr{J}}(\lambda) \otimes_A A'$ is an irreducible quotient of $P_A^{\mathscr{J}}(\lambda)$ as well.

3.4.4 The BGG-Reciprocity

The following result is a cornerstone of our approach towards the character problem. Let A be a local deformation algebra with residue field $\mathbb{K}$.

Theorem 3.4.12 (BGG-Reciprocity) *Let $\mathscr{J}$ be an open and locally bounded subset of $\mathfrak{h}^\star$. Then for any $\lambda \in \mathscr{J}$ the object $P_A^{\mathscr{J}}(\lambda)$ admits a Verma flag and for the multiplicities holds*

$$(P_A^{\mathscr{J}}(\lambda) : \Delta_A(\mu)) = \begin{cases} 0, & \text{if } \mu \notin \mathscr{J}, \\ [\Delta_{\mathbb{K}}(\mu) : L_{\mathbb{K}}(\lambda)], & \text{if } \mu \in \mathscr{J}. \end{cases}$$

In order to prove the above result, we need some preparation. First, we introduce the *dual Verma modules* $\nabla_A(\lambda)$. There is a duality functor $(\cdot)^\vee \colon \mathscr{O}_A \to \mathscr{O}_A$ that is defined as follows. For any A-module M consider the dual A-module $M^* = \mathrm{Hom}_A(M, A)$. If M is a $\mathfrak{g}$-A-bimodule, then so is M^*, where the $\mathfrak{g}$-action is given by $(x.f)(m) = -f(x.m)$ for $m \in M, f \in M^*$ and $x \in \mathfrak{g}$. Note that the minus sign is needed to obtain a left action of $\mathfrak{g}$ on M^*. It turns out that $M^\circledast := \bigoplus_{\lambda \in \mathfrak{h}^\star}(M_\lambda)^*$ is a $\mathfrak{g}$-A-submodule of M^*, and it is a weight module with weight spaces

$$(M^\circledast)_\lambda = (M_{-\lambda})^*.$$

Now this is not an object in $\mathscr{O}_A$ in general, as $\mathfrak{b}$ will not act locally finitely in all cases. But if we finally twist the $\mathfrak{g}$-action using the Chevalley involution $\omega \colon \mathfrak{g} \to \mathfrak{g}$

that interchanges the root spaces $\mathfrak{g}_\alpha$ and $\mathfrak{g}_{-\alpha}$ and acts as multiplication by -1 on $\mathfrak{h}$, we obtain the dual object $M^\vee$ in $\mathscr{O}_A$.

Remark 3.4.13 Note that the duality $(\cdot)^\vee$ is involutive only on the subcategory of $\mathscr{O}_A$ that contains the objects M which have the property that each weight space M_λ is *reflexive*, the latter considered as an A-module (i.e., the natural homomorphism $M \to ((M_\lambda)^*)^*$ is an isomorphism).

Set $\nabla_A(\lambda) := \Delta_A(\lambda)^\vee$. As every weight space of $\Delta_A(\lambda)$ is a free A-module of finite rank, it follows that $\Delta_A(\lambda) \cong \nabla_A(\lambda)^\vee$.

Lemma 3.4.14 *Suppose that M is an object in $\mathscr{O}_A^{\mathscr{J}}$ that admits a Verma flag. Then* $\mathrm{Ext}^1_{\mathscr{O}_A^{\mathscr{J}}}(M, \nabla_A(\mu)) = 0$ *for all $\mu \in \mathscr{J}$.*

Proof We prove the statement by induction on the length of a Verma flag of M. Suppose that $M \cong \Delta_A(\lambda)$ and let

$$0 \to \nabla_A(\mu) \to N \overset{f}{\to} \Delta_A(\lambda) \to 0 \qquad (*)$$

be a short exact sequence. If $\mu \not> \lambda$, then λ is a maximal weight of N and the universal property of $\Delta_A(\lambda)$ yields a non-zero homomorphism $g\colon \Delta_A(\lambda) \to N$. Then $f \circ g$ is an endomorphism of $\Delta_A(\lambda)$ that is an automorphism on the highest weight space. So $f \circ g$ is an automorphism. This means that the short exact sequence $(*)$ splits. On the other hand, if $\lambda \not> \mu$, then we can consider the dual sequence

$$0 \to \Delta_A(\lambda)^\vee \to N^\vee \to \nabla_A(\mu)^\vee \to 0. \qquad (*)^\vee$$

This is an exact sequence as well, as all occurring objects are free as A-modules. As $\nabla_A(\mu)^\vee \cong \Delta_A(\mu)$ we can argue as before and see that $(*)^\vee$, hence also $(*)$ splits. Hence $\mathrm{Ext}^1_{\mathscr{O}_A^{\mathscr{J}}}(\Delta_A(\lambda), \nabla_A(\mu)) = 0$ in any case.

If the length of a Verma flag of M is > 1, then we can choose $M' \subset M$ such that both M' and M/M' are non-zero and admit a Verma flag (necessarily of smaller length). The short exact sequence

$$0 \to M' \to M \to M/M' \to 0$$

yields a long exact sequence of $\mathrm{Ext}^i_{\mathscr{O}_A^{\mathscr{J}}}(\cdot, \nabla_A(\mu))$-groups. The $i = 1$ part of this sequence reads

$$\mathrm{Ext}^1_{\mathscr{O}_A^{\mathscr{J}}}(M/M', \nabla_A(\mu)) \to \mathrm{Ext}^1_{\mathscr{O}_A^{\mathscr{J}}}(M, \nabla_A(\mu)) \to \mathrm{Ext}^1_{\mathscr{O}_A^{\mathscr{J}}}(M', \nabla_A(\mu)).$$

As the groups on the left and on the right are trivial by induction assumption, we deduce $\mathrm{Ext}^1_{\mathscr{O}_A^{\mathscr{J}}}(M, \nabla_A(\mu)) = 0$.

Lemma 3.4.15 *Suppose that M is an object in $\mathcal{O}_A^{\mathcal{J}}$ that admits a Verma flag. Then* $\mathrm{Hom}_{\mathcal{O}_A^{\mathcal{J}}}(M, \nabla_A(\mu))$ *is a free A-module for all $\mu \in \mathcal{J}$ and*

$$(M : \Delta_A(\mu)) = \mathrm{rk}_A \mathrm{Hom}_{\mathcal{O}_A^{\mathcal{J}}}(M, \nabla_A(\mu)).$$

Proof Again we prove this by induction on the length of a Verma flag of M. If $M \cong \Delta_A(\lambda)$, then $\mathrm{Hom}_{\mathcal{O}_A^{\mathcal{J}}}(M, \nabla_A(\mu)) = 0$ unless $\lambda = \mu$ (by the universal property of $\Delta_A(\lambda)$ and its dual statement). For $\lambda = \mu$ we have that $\mathrm{Hom}_{\mathcal{O}_A^{\mathcal{J}}}(\Delta_A(\lambda), \nabla_A(\lambda))$ is isomorphic to $\nabla_A(\lambda)_\lambda$, i.e. it is free of rank 1. Hence the claim for $M = \Delta_A(\lambda)$ is settled. Now, if the length of M is > 1, then consider the short exact sequence $0 \to M' \to M \to M/M' \to 0$ as in the proof of Lemma 3.4.14. As $\mathrm{Ext}^1_{\mathcal{O}_A^{\mathcal{J}}}(M/M', \nabla_A(\mu)) = 0$ we obtain a short exact sequence

$$0 \to \mathrm{Hom}_{\mathcal{O}_A^{\mathcal{J}}}(M/M', \nabla_A(\mu)) \to \mathrm{Hom}_{\mathcal{O}_A^{\mathcal{J}}}(M, \nabla_A(\mu))$$
$$\to \mathrm{Hom}_{\mathcal{O}_A^{\mathcal{J}}}(M', \nabla_A(\mu)) \to 0.$$

By the induction hypothesis the end terms are free A-modules, hence so is the middle term. Observing the additivity of the rank with respect to short exact sequences and the fact that

$$(M : \Delta_A(\mu)) = (M' : \Delta_A(\mu)) + (M/M' : \Delta_A(\mu))$$

and using the induction hypothesis on M' and M/M' yield the claim.

Now we can prove the BGG-reciprocity result.

Proof As $P_A^{\mathcal{J}}(\lambda)$ is isomorphic to a direct summand of $Q_A^{\mathcal{J}}(\lambda)$ and as $Q_A^{\mathcal{J}}(\lambda)$ admits a Verma flag, so does $P_A^{\mathcal{J}}(\lambda)$. Then

$$(P_A^{\mathcal{J}}(\lambda) : \Delta_A(\mu)) = \mathrm{rk}_A \mathrm{Hom}_{\mathcal{O}_A^{\mathcal{J}}}(P_A^{\mathcal{J}}(\lambda), \nabla_A(\mu))$$
$$= [\nabla_{\mathbb{K}}(\mu) : L_{\mathbb{K}}(\lambda)]$$
$$= [\Delta_{\mathbb{K}}(\mu) : L_{\mathbb{K}}(\lambda)].$$

Here the first equality results from Lemma 3.4.15, the second one from using the projective cover property of $P_A^{\mathcal{J}}(\lambda)$, and the last identity is a consequence of the fact that the characters of $\Delta_{\mathbb{K}}(\mu)$ and $\nabla_{\mathbb{K}}(\mu)$ coincide.

3.4.5 Another Result of Bernstein–Gelfand–Gelfand

In this section suppose that $A = \mathbb{K}$ is a field.

Definition 3.4.16 Denote by $\uparrow_\mathbb{K}$ the partial order on the set $\mathfrak{h}^\star$ that is generated by $\mu \uparrow_\mathbb{K} \lambda$ if there is $\alpha \in R^+$ such that $\langle \lambda + \tau + \rho, \alpha^\vee \rangle \in \mathbb{Z}$, $\mu = s_\alpha.\lambda$ and $\mu < \lambda$.

Then the following holds (cf. [13, Sect. 5.1]):

Theorem 3.4.17 *For $\lambda, \mu \in \mathfrak{h}^\star$ we have $[\Delta_\mathbb{K}(\lambda) : L_\mathbb{K}(\mu)] \neq 0$ if and only if $\mu \uparrow_\mathbb{K} \lambda$.*

The "if" part was already proven by Verma.

Definition 3.4.18 Let $\sim_A$ be the equivalence on $\mathfrak{h}^\star$ that is generated by $\lambda \sim_A \mu$ if $[\Delta_\mathbb{K}(\lambda) : L_\mathbb{K}(\mu)] \neq 0$.

By definition, $\sim_A$ only depends on the residue field $\mathbb{K}$ (more precisely, on the residue field $\mathbb{K}$ as an S-algebra). From Theorem 3.4.17 it follows that $\sim_A$ is the equivalence relation generated by the partial order $\uparrow_\mathbb{K}$. Using this, we can describe the equivalence classes in more detail.

Definition 3.4.19 For $\lambda \in \mathfrak{h}^\star$ define

$$R_{A,[\lambda]} := \{\alpha \in R \mid \langle \lambda + \tau + \rho, \alpha^\vee \rangle \in \mathbb{Z}\},$$

$$\mathscr{T}_{A,[\lambda]} := \{s_\alpha \mid \alpha \in R_{A,[\lambda]}\}$$

$$\mathscr{W}_{A,[\lambda]} := \langle t \mid t \in \mathscr{T}_{A,[\lambda]} \rangle.$$

Then the following holds:

Lemma 3.4.20

1. *If $\lambda \sim_A \mu$, then $R_{A,[\lambda]} = R_{A,[\mu]}$ and hence $\mathscr{T}_{A,[\lambda]} = \mathscr{T}_{A,[\mu]}$ and $\mathscr{W}_{A,[\lambda]} = \mathscr{W}_{A,[\mu]}$.*
2. *For an equivalence class $\Lambda \in \mathfrak{h}^\star / \sim_A$ we have $\Lambda = \mathscr{W}_{A,[\lambda]}.\lambda$ for all $\lambda \in \Lambda$.*

In particular, $[\Delta_A(\lambda) : L_A(\mu)] \neq 0$ implies $\mu \in \mathscr{W}_{[\lambda]}$ and $\mu \leqslant \lambda$.

3.4.6   The Block Decomposition of $\mathcal{O}_A$

Let A be a local deformation algebra with residue field $\mathbb{K}$. Using the BGG-reciprocity we could replace the condition in the definition of $\sim_A$ by $(P_A^{\mathscr{J}}(\mu) : \Delta_A(\lambda)) \neq 0$ for some open and locally bounded subset $\mathscr{J}$ that contains λ and μ. In particular, it also follows that $\lambda \sim_A \mu$ whenever $[P_A^{\mathscr{J}}(\lambda) : L_A(\mu)] \neq 0$.

Let $\Lambda \in \mathfrak{h}^\star / \sim_A$ be an equivalence class and M an object in $\mathcal{O}_A$. Denote by M_Λ the submodule of M that is generated by the images $\mathrm{im} f$ for all homomorphisms $f : P_A^{\mathscr{J}}(\lambda) \to M$ with $\lambda \in \Lambda$ and all open and locally bounded subsets $\mathscr{J}$ of $\mathfrak{h}^\star$ with $\lambda \in \mathscr{J}$. We also denote by $\mathcal{O}_{A,\Lambda}$ the full subcategory of $\mathcal{O}_A$ that contains all objects N with $[N : L_A(\lambda)] \neq 0$ only if $\lambda \in \Lambda$.

Theorem 3.4.21

1. *The object M_Λ is contained in $\mathcal{O}_\Lambda$.*
2. *We have $M = \bigoplus_\Lambda M_\Lambda$.*

Remark 3.4.22 We can also formulate the above result in the following way: The functor

$$\bigoplus_{\Lambda \in \mathfrak{h}^\star / \sim_A} \mathcal{O}_{A,\Lambda} \to \mathcal{O}_A,$$

$$\{M_\Lambda\}_\Lambda \mapsto \bigoplus_\Lambda M_\Lambda$$

is an equivalence of categories.

Proof Statement (1) follows from the fact that for $\lambda \in \Lambda$ we have that $[P_A^{\mathscr{J}}(\lambda) : L_A(\mu)] \neq 0$ for some $\mathscr{J}$ implies $\mu \in \Lambda$. Clearly M is the sum of its submodules M_Λ. But this must be a direct sum, as there are no non-zero homomorphisms $P_A^{\mathscr{J}}(\lambda) \to N$ for $\lambda \in \Lambda$ and $N \in \mathcal{O}_{\Lambda'}$ if $\Lambda \neq \Lambda'$.

3.4.7 On the Structure of Some Projectives

The following result is one of the reasons why we deform the category $\mathcal{O}$. Recall the localization $\widetilde{S}$ of the symmetric algebra S (cf. Remark 3.4.2).

Proposition 3.4.23 *Let $\mathfrak{p} \subset \widetilde{S}$ be a prime ideal and $A = \widetilde{S}_\mathfrak{p}$.*

1. *If $\tau(\alpha^\vee) \notin \mathfrak{p}$ for all $\alpha \in R$, then each equivalence class Λ with respect to $\sim_A$ is a singleton, i.e. $\Lambda = \{\lambda\}$ for some $\lambda \in \mathfrak{h}^\star$.*
2. *If $\mathfrak{p} = \tau(\alpha^\vee)\widetilde{S}$ for some $\alpha \in R$, then an equivalence class Λ with respect to $\sim_A$ is either a singleton, or contains exactly two elements, i.e. $\Lambda = \{\lambda, \mu\}$ for some $\lambda, \mu \in \mathfrak{h}^\star$. In the latter case, $\mu = s_\alpha.\lambda$.*

Proof For $A = \widetilde{S}_\mathfrak{p}$ with residue field $\mathbb{K}$ one checks that the condition $\langle \lambda + \tau + \rho, \alpha^\vee \rangle \in \mathbb{Z} \subset \mathbb{K}$ is equivalent to $\tau(\alpha^\vee) = 0$, i.e. $\tau(\alpha^\vee) \in \mathfrak{p}$ and $\langle \lambda + \rho, \alpha^\vee \rangle \in \mathbb{Z} \subset \mathbb{C}$.

In a next step we would like to understand the equivalence relation $\sim_A$. We start with a definition.

Definition 3.4.24 Let $\Lambda \in \mathfrak{h}^\star / \sim_A$ be an equivalence class.

1. Λ is called *generic*, if Λ is a singleton, i.e. $\Lambda = \{\lambda\}$ for some $\lambda \in \mathfrak{h}^\star$.
2. Λ is called *subgeneric*, if Λ contains exactly two elements.

Remark 3.4.25 Suppose that $\Lambda = \{\lambda, \mu\}$ is subgeneric. One can then show (for example, using the Jantzen filtration, cf. Sect. 5.3 in [13]) that there is a unique $\alpha \in R$ such that $\langle \lambda + \tau + \rho, \alpha^\vee \rangle \in \mathbb{Z} \subset A$ and $\mu = s_\alpha.\lambda$.

It is now easy to describe the structure of blocks corresponding to a generic equivalence class.

Proposition 3.4.26 *Suppose that* $\Lambda = \{\lambda\}$ *is generic. Then* $P_A(\lambda) \cong \Delta_A(\lambda)$ *and* $\Delta_{\mathbb{K}}(\lambda) = L_{\mathbb{K}}(\lambda)$. *In particular,* $\text{End}_{\mathscr{O}_A}(P_A(\lambda)) = A \cdot \text{id}_{P_A(\lambda)}$.

Proof As $[\Delta_{\mathbb{K}}(\lambda) : L_{\mathbb{K}}(\lambda)] = 1$, the second statement follows readily from the definition of a generic equivalence class. The first statement then is a consequence of BGG-reciprocity.

Next we consider the structure of subgeneric blocks.

Proposition 3.4.27 *Suppose that* $\Lambda = \{\lambda, \mu\}$ *is subgeneric, and suppose that* $\mu = s_\alpha.\lambda > \lambda$. *Then* $P_A(\mu) \cong \Delta_A(\mu)$ *and* $\Delta_{\mathbb{K}}(\lambda) = L_{\mathbb{K}}(\lambda)$. *There are short exact sequences*

$$0 \to \Delta_A(\mu) \to P_A(\lambda) \to \Delta_A(\lambda) \to 0$$

and

$$0 \to L_{\mathbb{K}}(\lambda) \to \Delta_{\mathbb{K}}(\mu) \to L_{\mathbb{K}}(\mu) \to 0.$$

Proof Note that $[\Delta_{\mathbb{K}}(\lambda) : L_{\mathbb{K}}(\mu)] = 0$ as $\mu > \lambda$. We can argue as in the generic case and deduce $\Delta_{\mathbb{K}}(\lambda) = L_{\mathbb{K}}(\lambda)$. So $\Delta_{\mathbb{K}}(\lambda)$ is simple. Clearly $[\Delta_{\mathbb{K}}(\mu) : L_{\mathbb{K}}(\lambda)] \geq 1$. But the maximal submodule of $\Delta_{\mathbb{K}}(\mu)$ must be generated by primitive vectors of weight λ. As $\dim_{\mathbb{K}} \text{Hom}(\Delta_{\mathbb{K}}(\lambda), \Delta_{\mathbb{K}}(\mu)) \leq 1$, this yields $[\Delta_{\mathbb{K}}(\mu) : L_{\mathbb{K}}(\lambda)] \leq 1$, hence $[\Delta_{\mathbb{K}}(\mu) : L_{\mathbb{K}}(\lambda)] = 1$. Now the rest of the statements follow from BGG-reciprocity.

Remark 3.4.28 In the generic and the subgeneric situations we deduce that the projectives $P_A^{\mathscr{J}}(\nu)$ stabilize with respect to $\mathscr{J}$, i.e. $P_A^{\mathscr{J}}(\nu)$ is isomorphic to $P_A^{\mathscr{J}'}(\nu)$ as long as $\mathscr{J}$ and $\mathscr{J}'$ contain the generic or subgeneric equivalence class of ν. Hence $L_A(\nu)$ admits a projective cover already in $\mathscr{O}_A$. Denote it by $P_A(\nu)$.

3.4.8   Endomorphism Rings in the Subgeneric Situations

We now study the categorical structure of subgeneric blocks in detail. Let $\Lambda = \{\lambda, \mu\}$ and suppose that $\lambda < \mu$. First assume that $A = \mathbb{K}$ is a field. Then $[\Delta_{\mathbb{K}}(\mu) : L_{\mathbb{K}}(\mu)] = [\Delta_{\mathbb{K}}(\mu) : L_{\mathbb{K}}(\lambda)] = 1$ and $\Delta_{\mathbb{K}}(\lambda) = L_{\mathbb{K}}(\lambda)$. Moreover, $P_{\mathbb{K}}(\mu) \cong \Delta_{\mathbb{K}}(\mu)$ and there is a short exact sequence

$$0 \to \Delta_{\mathbb{K}}(\mu) \to P_{\mathbb{K}}(\lambda) \to \Delta_{\mathbb{K}}(\lambda) \to 0.$$

For any projective cover P of a simple object L in $\mathscr{O}_\mathbb{K}$ it is easy to show that

$$\dim_\mathbb{K} \operatorname{Hom}_{\mathscr{O}_\mathbb{K}}(P, M) = [M : L]$$

for all $M \in \mathscr{O}_\mathbb{K}$. Then

$$\dim_\mathbb{K} \operatorname{Hom}_{\mathscr{O}_\mathbb{K}}(P_\mathbb{K}(\lambda), P_\mathbb{K}(\mu)) = \dim_\mathbb{K} \operatorname{Hom}_{\mathscr{O}_\mathbb{K}}(P_\mathbb{K}(\mu), P_\mathbb{K}(\lambda))$$
$$= \dim_\mathbb{K} \operatorname{End}_{\mathscr{O}_\mathbb{K}}(P_\mathbb{K}(\mu)) = 1$$

and

$$\dim_\mathbb{K} \operatorname{End}_{\mathscr{O}_\mathbb{K}}(P_\mathbb{K}(\lambda)) = 2.$$

Now choose generators

$$j: P_\mathbb{K}(\mu) \to P_\mathbb{K}(\lambda),$$
$$i: P_\mathbb{K}(\lambda) \to P_\mathbb{K}(\mu)$$

of the respective Hom-spaces. Then j is injective, as $P_\mathbb{K}(\mu)$ is a Verma module and $P_\mathbb{K}(\lambda)$ is free over $U(\mathfrak{n}^-)$. Moreover, $i \circ j = 0$, as otherwise it would be an automorphism of $P_\mathbb{K}(\mu)$, hence i would split. Then $j \circ i: P_\mathbb{K}(\lambda) \to P_\mathbb{K}(\lambda)$ is non-zero, but nilpotent. Hence $\{\operatorname{id}_{P_\mathbb{K}(\lambda)}, j \circ i\}$ is a basis of $\operatorname{End}_{\mathscr{O}_\mathbb{K}}(P_\mathbb{K}(\lambda))$. It is now not difficult to check that $\operatorname{End}_{\mathscr{O}_\mathbb{K}}(P_\mathbb{K}(\lambda) \oplus P_\mathbb{K}(\mu))$ is isomorphic to the path algebra of the quiver

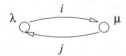

with relation $i \circ j = 0$.

Now let A be an arbitrary deformation algebra with residue field $\mathbb{K}$.

Proposition 3.4.29 *Let $\Lambda = \{\lambda, \mu\}$ as above. Then $\operatorname{End}_{\mathscr{O}_A}(P_A(\mu) \oplus P_A(\lambda))$ is isomorphic to the quiver depicted above with relation $i \circ j = \tau(\alpha^\vee)$.*

Proof Recall that there is a positive root α such that $\lambda - \mu \in \mathbb{Z}\alpha$. From Remark 1 we know that the Hom-spaces between projectives in $\mathscr{O}_A$ are free A-modules of finite rank, and the base change result in Proposition 3.4.11 shows that we can lift the homomorphisms i and j from above to homomorphisms $\tilde{j}: P_A(\mu) \to P_A(\lambda)$ and $\tilde{i}: P_A(\lambda) \to P_A(\mu)$ in such a way that they are bases in the respective Hom-spaces. Using the Jantzen sum formula one can show that $\tilde{j} \circ \tilde{i}$ equals $\tau(\alpha^\vee) \cdot \operatorname{id}_{P_A(\mu)}$ up to an invertible scalar (cf. [6, Proposition 3.4]). Hence $\operatorname{End}_{\mathscr{O}_A}(P_A(\lambda) \oplus P_A(\mu))$ is isomorphic to the path algebra of the quiver depicted above, but now with the relation $i \circ j = \tau(\alpha^\vee)$.

3.5   Soergel's Theory

In his fundamental paper [19], Soergel developed a theory that eventually produced
an alternative proof of the Kazhdan–Lusztig Conjecture 3.2.9. His proof is quite
different from both original proofs of this conjecture, the one given by Beilinson and
Bernstein in [2] and the other one given by Brylinski and Kashiwara in [4]. Indeed,
Soergel's proofs amounts to a detailed study of the combinatorics of translation
functors. A short review of the main ideas will be included in the sequel.

3.5.1   Endomorphisms of Multiplicity Free Projectives

We start with a result on the endomorphism rings of certain projective objects. Let
A be deformation algebra with residue field $\mathbb{K}$ and let Λ be a $\sim_A$-equivalence class
in $\mathfrak{h}^\star$. Let K be a $\mathscr{W}_\Lambda$-set, i.e. a set with a homomorphism $\mathscr{W}_\Lambda \to \mathrm{Aut}(K)$. Define
the following commutative algebra

$$
\mathscr{Z}_A(K) := \left\{ (z_x) \in \bigoplus_{x \in K} A \mid z_x - z_{tx} \in \tau(\alpha_t^\vee) A \quad \forall x \in K, t \in \mathscr{T}_\Lambda \right\}.
$$

Let M be an object in $\mathscr{O}_{A,\Lambda}$ that admits a Verma flag. It is called *multiplicity
free*, if $(M : \Delta_A(\mu)) \leq 1$ for all $\mu \in \mathfrak{h}^\star$. Now choose an enumeration $\Lambda = \{\mu_1, \ldots, \mu_n\}$
of the elements in Λ in such a way that $\mu_i > \mu_j$ implies $i < j$. Remark 3.4.7 implies
that there is a Verma flag $0 = M_0 \subset M_1 \subset \cdots \subset M_n = M$ in such a way that
M_i/M_{i-1} is isomorphic to $\Delta_A(\mu_i)$. Define $M_{[\mu_i]} := M_i/M_{i-1}$.

For the following result assume that $A = \widetilde{S}$. Recall that an $\widetilde{S}$-module M is
called *reflexive* if M is torsion free as an $\widetilde{S}$-module and if the natural inclusion
$M \subset \bigcap_{\mathfrak{p} \in \mathfrak{P}} M_\mathfrak{p}$, where $\mathfrak{P}$ is the set of prime ideals of height one in $\widetilde{S}$ and the
inclusion is taken inside $M_{(0)} = M \otimes_{\widetilde{S}} \mathrm{Quot}(\widetilde{S})$, is a bijection. Now suppose that M
and N are objects in $\mathscr{O}_{\widetilde{S}}$ that are reflexive as S-modules. Then

$$
\mathrm{Hom}_{\mathscr{O}_{\widetilde{S}}}(M, N) = \bigcap_{\mathfrak{p} \in \mathfrak{P}} \mathrm{Hom}_{\mathscr{O}_{\widetilde{S}_\mathfrak{p}}}(M_\mathfrak{p}, N_\mathfrak{p}),
$$

where the intersection is taken inside $\mathrm{Hom}_{\mathscr{O}_{\widetilde{S}_{(0)}}}(M_{(0)}, N_{(0)})$. In particular,
$\mathrm{Hom}_{\mathscr{O}_{\widetilde{S}}}(M, N)$ is a reflexive $\widetilde{S}$-module. If P is projective in $\mathscr{O}_{\widetilde{S},\Lambda}$, then P admits
a Verma flag. So it is free as an $\widetilde{S}$-module, in particular reflexive. The following
result generalizes the Endomorphismensatz of [19]:

Proposition 3.5.1 *Suppose that* $A = \widetilde{S}$ *and that* P *is projective in* $\mathscr{O}_{\widetilde{S},\Lambda}$ *and multiplicity free. Then the natural homomorphism*

$$\mathrm{End}_{\mathscr{O}_{\widetilde{S}}}(P) \to \bigoplus_{\mu \in \mathrm{supp}_\Delta(P)} \mathrm{End}_{\mathscr{O}_{\widetilde{S}}}(P_{[\mu]}) = \bigoplus_{\mu \in \mathrm{supp}_\Delta(P)} \widetilde{S}$$

is injective with image $\mathscr{Z}_{\widetilde{S}}(\mathrm{supp}_\Delta P)$. *In particular,* $\mathrm{End}_{\mathscr{O}_{\widetilde{S}}}(P)$ *is a commutative algebra.*

Proof As P is free over $\widetilde{S}$ (in particular reflexive) it follows that $P = \bigcup_{\mathfrak{p} \in \mathfrak{P}} P_\mathfrak{p} \subset P_{(0)}$ and $\mathrm{End}_{\mathscr{O}_{\widetilde{S}}}(P) = \bigcup_{\mathfrak{p} \in \mathfrak{P}} \mathrm{End}_{\mathscr{O}_{\widetilde{S}_\mathfrak{p}}}(P_\mathfrak{p}) \subset \mathrm{End}_{\mathscr{O}_{\widetilde{S}_{(0)}}}(P_{(0)})$. By Proposition 3.4.23, $P_\mathfrak{p}$ splits into a direct sum of projectives in generic and subgeneric situations, and so does $\mathrm{End}_{\mathscr{O}_{\widetilde{S}_\mathfrak{p}}}(P_\mathfrak{p})$. But in these cases we determined the endomorphism rings already explicitly in Propositions 3.4.26 and 3.4.29. The result follows. $\quad\square$

3.5.2 Big Projectives

A main ingredient for Soergel's approach is the study of the *big projectives* in the category $\mathscr{O}$. They correspond to weights λ that are anti-dominant. They are called *big* because their support is maximal. The following is an important result and one of the cornerstones of the theory.

Lemma 3.5.2 *Suppose that* $\lambda \in \Lambda$ *is anti-dominant. Then* $(P_A(\lambda) : \Delta_A(\mu)) = 1$ *for all* $\mu \in \Lambda$. *In particular, the object* $P_A(\lambda)$ *is multiplicity free.*

Proof Let $s \in \mathscr{S}$. Then it is well known that $[\Delta_{\mathbb{K}}(s.\lambda) : L_{\mathbb{K}}(\lambda)] = 1$. Using BGG-reciprocity and the combinatorics of translation functors one shows that $T_s P_A(\lambda)$ must contain at least two direct summands isomorphic to $P_A(\lambda)$. It follows that $T_s P_A(\lambda) = P_A(\lambda) \oplus P_A(\lambda)$, which can only be the case if $(P_A(\lambda) : \Delta_A(\mu)) = 1$ for all $\mu \in \Lambda$. $\quad\square$

3.5.3 The **Strukturfunktor** and the **Struktursatz**

Here is Soergel's definition:

Definition 3.5.3 Let $\lambda \in \Lambda$ be antidominant. The functor

$$\mathbb{V} = \mathbb{V}_{A,\Lambda} := \mathrm{Hom}_{\mathscr{O}_{A,\Lambda}}(P_A(\lambda), \cdot) \colon \mathscr{O}_{A,\Lambda} \to \mathrm{mod\text{-}End}_{\mathscr{O}_{A,\Lambda}}(P_A(\lambda))$$

is called the *Strukturfunktor* associated to A and Λ.

As seen in Lemma 3.5.2, the object $P_A(\lambda)$ is multiplicity free and $\mathrm{supp}_\Delta P_A(\lambda) = \Lambda$. In particular, $\mathrm{End}_{\mathscr{O}_{A,\Lambda}}(P_A(\lambda)) = \mathscr{Z}_A(\Lambda)$ is commutative. So there is no distinction between left and right $\mathscr{Z}_A(\varLambda)$-modules, and the functor $\mathbb{V}$ is in fact a functor from $\mathscr{O}_{A,\varLambda}$ to $\mathscr{Z}_A(\varLambda)$-mod.

Assume now that $A = \widetilde{S}$ is the localization of S at the maximal ideal generated by $\mathfrak{h}$. We show that $\mathbb{V}_{\widetilde{S},\varLambda}$ is fully faithful on projective objects. Note that $\mathbb{V}(P) = \mathrm{Hom}_{\mathscr{O}_{\widetilde{S}}}(P_{\widetilde{S}}(\lambda), P)$ is reflexive as a $\widetilde{S}$-module.

Theorem 3.5.4 (Soergel's Struktursatz) *The functor* $\mathbb{V}$ *is fully faithful on projectives, i.e. if* $P, P' \in \mathscr{O}_{\widetilde{S},\varLambda}$ *are projective, then* $\mathrm{Hom}_{\mathscr{O}_{\widetilde{S},\varLambda}}(P, P') = \mathrm{Hom}_{\mathscr{Z}_{\widetilde{S}}}(\mathbb{V}(P), \mathbb{V}(P'))$.

Proof By what we observed above,

$$\mathrm{Hom}_{\mathscr{O}_{\widetilde{S},\varLambda}}(P, P') = \bigcap_{\mathfrak{p} \in \mathfrak{P}} \mathrm{Hom}_{\mathscr{O}_{\widetilde{S}\mathfrak{p},\varLambda}}(P_{\mathfrak{p}}, P'_{\mathfrak{p}}).$$

Likewise one shows that

$$\mathrm{Hom}_{\mathscr{Z}_{\widetilde{S}}}(\mathbb{V}P, \mathbb{V}P') = \bigcap_{\mathfrak{p} \in \mathfrak{P}} \mathrm{Hom}_{\mathscr{Z}_{\widetilde{S}\mathfrak{p}}}(\mathbb{V}P_{\mathfrak{p}}, \mathbb{V}P'_{\mathfrak{p}}).$$

Everything is compatible with localizations, hence we only have to check that the natural homomorphism $\mathrm{Hom}_{\mathscr{O}_{\widetilde{S}\mathfrak{p},\varLambda}}(P_{\mathfrak{p}}, P'_{\mathfrak{p}}) \to \mathrm{Hom}_{\mathscr{Z}_{\widetilde{S}\mathfrak{p}}}(\mathbb{V}P_{\mathfrak{p}}, \mathbb{V}P'_{\mathfrak{p}})$ is a bijection for all prime ideals of height one. But these cases split into a direct sum of subgeneric cases, and in the subgeneric case it is an exercise to check the statement using the categorical structure in Sect. 3.4.8.

3.5.4 The Combinatorics of Translation Functors

As $\mathbb{V}$ is a fully faithful functor, the problem of decomposing objects like $T_{s_1} \cdots T_{s_n} \varDelta_{\widetilde{S}}(\lambda_0)$ in $\mathscr{O}_{\widetilde{S},\varLambda}$ is equivalent to decomposing the object $\mathbb{V}(T_{s_1} \cdots T_{s_n} \varDelta_{\widetilde{S}}(\lambda_0))$ in $\mathscr{Z}$-mod. Fortunately, we can describe this $\mathscr{Z}$-module quite explicitly.

Proposition 3.5.5 ([9, Theorem 9]) *There is a natural isomorphism* $\mathbb{V} \circ T_s \cong \mathscr{Z} \otimes_{\mathscr{Z}^s} \cdot \circ \mathbb{V}$ *of functors from* $\mathscr{O}_{\widetilde{S},\varLambda}$ *to* $\mathscr{Z}$-mod.

Before the last problem can be stated we have to understand $\mathbb{V}\varDelta_{\widetilde{S}}(\lambda_0)$ as a $\mathscr{Z}$-module. It is free of rank 1 as an $\widetilde{S}$-module, and from the identification $\mathscr{Z} \cong \mathrm{End}_{\mathscr{O}_{\widetilde{S},\varLambda}}(P_{\widetilde{S}}(\lambda))$ it follows that an element (z_μ) acts on $\mathbb{V}\varDelta_{\widetilde{S}}(\lambda_0)$ as multiplication with z_{λ_0}. Denote this $\mathscr{Z}$-module by $\widetilde{S}_{\lambda_0}$.

As the functor $\mathbb{V}$ is fully faithful on projectives, Proposition 3.5.5 implies that the Decomposition Problem I is equivalent to the following:

Decomposition Problem II *Decompose* $\mathscr{Z} \otimes_{\mathscr{Z}^s} \mathscr{Z} \otimes_{\mathscr{Z}^t} \cdots \otimes_{\mathscr{Z}^u} \widetilde{S}_{\lambda_0}$ *in the category of* $\mathscr{Z}$-modules for all sequences $s, t, \ldots, u$ of simple reflections.

3.6 Soergel Bimodules, Parity Sheaves, and Moment Graph Sheaves

Having started from the Kazhdan–Lusztig Conjecture 3.2.9, through a successive series of translations we moved first to the equivalent Decomposition Problem I and from there, using Soergel's theory, to another equivalent problem, Decomposition Problem II. The latter problem does not refer to modules over a Lie algebra any more, it is rather stated in the representation theory of a commutative, associative algebra $\mathscr{Z}$. The following gives an overview on three approaches to deal with it.

3.6.1 Moment Graph Sheaves

In [8] we studied *sheaves on moment graphs* that were introduced in [3]. A moment graph $\mathscr{G}$ is a graph with a labeling of its edges by element in a lattice X. For any field k one can define a commutative algebra $\mathscr{Z}_k(\mathscr{G})$ over the symmetric algebra S_k of the k-vector space $X \otimes_{\mathbb{Z}} k$. One then has the notion of a k-sheaf on this graph. The space of global sections of every moment graph sheaf gives rise to a module over $\mathscr{Z}_k(\mathscr{G})$.

To any equivalence class Λ in $\mathfrak{h}^*$ with respect to $\sim$ one can associate a moment graph $\mathscr{G}_\Lambda$ in such a way that $\mathscr{Z}_{\mathbb{C}}(\mathscr{G}_\Lambda)$ is the algebra $\mathscr{Z}_S(\Lambda)$ that we considered above. Hence every sheaf on $\mathscr{G}_\Lambda$ gives rise to a $\mathscr{Z}_S(\Lambda)$-module. It is a very nice feature of the theory that one can describe certain sheaves $\mathscr{B}(\mu)$ on $\mathscr{G}_\Lambda$ for any $\mu \in \Lambda$ algorithmically that have the property that the global sections yields the object $\mathbb{V}(P_{\widetilde{S}}(\mu))$ (up to extending scalars from S to $\widetilde{S}$). Hence one can construct the *indecomposable* direct summands that occur in the Decomposition Problem II directly! This gives a way to study the structure of the projectives locally (i.e., on the vertices of the graph $\mathscr{G}_\Lambda$).

3.6.2 Soergel Bimodules

For an integral and regular equivalence class Λ one can define a homomorphism $S \otimes S \to \mathscr{Z}(\Lambda)$ (cf. [9]). Restriction along this homomorphism gives a functor from $\mathscr{Z}_S(\Lambda)$-modules to S-bimodules. Using this functor one can show that the Decomposition Problem II is equivalent to the following:

Decomposition Problem III *Decompose* $S \otimes_{S^s} S \otimes_{S^t} \cdots \otimes_{S^u} S$ *in the category of* S-bimodules for all sequences $s, t, \ldots, u$ of simple reflections.

The indecomposable direct summands that occur are called *Soergel bimodules*. They were introduced in [20]. Soergel bimodules recently found applications in faraway subjects such as knot theory!

3.6.3 Parity Sheaves

The last and probably most important connection of the above problems is to geometry. Again suppose that Λ is integral and regular. Consider the flag variety G/B associated to the Lie algebra[1] $\mathfrak{g}$, and its T-equivariant derived category $D^b_{T,c}(X, \mathbb{C})$ of sheaves of $\mathbb{C}$-vector spaces that are constructible along Schubert cells. In [15], Juteau et al. described the remarkable additive subcategory of *parity sheaves* that are characterized by certain cohomological vanishing properties. For any $w \in \mathscr{W}$ there is an up to shift unique parity sheaf $\mathscr{E}(w)$ whose support is the Schubert variety associated with w inside G/B.

Now there is a hypercohomology functor from $D^b_{T,c}(G/B, \mathbb{C})$ into the category of modules over the equivariant cohomology $H^*_T(G/B)$ of G/B. Again it turns out that $H^*_T(G/B)$ coincides with $\mathscr{Z}_S(\Lambda)$ (canonically if we replace $\mathfrak{g}$ by $\mathfrak{g}^\vee$)! Hence the hypercohomology of parity sheaves yields $\mathscr{Z}_S(\Lambda)$-modules, and again one can show that

$$H^\bullet_T(\mathscr{E}(w)) \otimes_S \widetilde{S} \cong \mathbb{V}(P_{\widetilde{S}}(w.\lambda))$$

for all $w \in \mathscr{W}$. This yields a geometric way to understand the projective objects in $\mathscr{O}_{\widetilde{S},\Lambda}$.

3.6.4 Another Proof of the Kazhdan–Lusztig Conjecture

Apart from the original proofs of Beilinson–Bernstein and Brylinski–Kashiwara, Soergel's approach that we reviewed above also yields a proof of Conjecture 3.2.9. It uses the geometric description of the indecomposable special $\mathscr{Z}$-modules above, and it amounts to showing (and this is extremely difficult) that the parity sheaves (with coefficients in $\mathbb{C}$) are in fact the *intersection cohomology complexes* on the Schubert varieties in the flag variety. They have some remarkable properties, in particular, the ranks of their stalk are given by Kazhdan–Lusztig polynomials (cf. [17]), and this fact implies Conjecture 3.2.9 and hence completes our journey.

There is a new proof due to Ben Elias and Geordie Williamson [5] that directly solves the Decomposition Problem III. It is geometrically inspired, but does not use any geometric result at all. It is quite difficult as well, but it has the advantage that it also works in the more general situation of Coxeter systems, where the Decomposition Problem III makes perfect sense, although a reformulation in terms of geometry and parity sheaves is not available in general.

---

[1]To be precise, one should replace $\mathfrak{g}$ here by its Langlands dual $\mathfrak{g}^\vee$.

3.7 An Epilogue: The Modular Case

One can approach the problem of determining the irreducible rational characters of a reductive algebraic group over a field of positive characteristic in a similar way. However, the analogous results are much more difficult to obtain. For example, the analogue of Soergel's theory, i.e. the translation of the representation theoretic problem into a combinatorial framework, is the main content of the book [1]. There the authors define a combinatorial category over an arbitrary field k, that, in the characteristic zero case, yields Jordan–Hölder multiplicities for quantum groups at a root of unity, and in positive characteristic, Jordan–Hölder multiplicities for modular Lie algebras. In both cases, the multiplicities are encoded in a decomposition problem that very much resembles the decomposition problems that we encountered in this overview article. One can approach this problem again by geometric or moment graph methods (cf. [10]).

A toy example of the modular case is W. Soergel's *modular category* $\mathcal{O}$ introduced in [18]. The combinatorial counterpart of this is the category of Soergel bimodules defined over a field of positive characteristic. It is in this category that G. Williamson recently found counterexamples to Lusztig's modular character formular for characteristics above the Coxeter number (cf. [21]). This is a very exciting, intriguing and challenging result. One might hope that a better understanding of the interplay between representation theoretic, combinatorial, and geometric ideas will lead us to a better understanding of what happens in the case of "small" characteristics.

Acknowledgements The author was partially supported by the DFG grant SP1388.

References

1. H.H. Andersen, J.C. Jantzen, W. Soergel, Representations of quantum groups at a p-th root of unity and of semisimple groups in characteristic p: independence of p. Astérisque **220**, 321 (1994)
2. A. Beilinson, J. Bernstein, Localisation de g-modules. C. R. Acad. Sci. Paris Sér. I **292**, 15–18 (1981)
3. T. Braden, R. MacPherson, From moment graphs to intersection cohomology. Math. Ann. **321**(3), 533–551 (2001)
4. J.-L. Brylinski, M. Kashiwara, Kazhdan-Lusztig conjecture and holonomic systems. Invent. Math. **64**, 387–410 (1981)
5. B. Elias, G. Williamson, The Hodge theory of Soergel bimodules. Ann. Math. (2) **180**(3), 1089–1136 (2014)
6. P. Fiebig, Centers and translation functors for category $\mathcal{O}$ over Kac-Moody algebras. Math. Z. **243**(4), 689–717 (2003)
7. P. Fiebig, The combinatorics of category $\mathcal{O}$ over symmetrizable Kac-Moody algebras. Transform. Groups **11**(1), 29–49 (2006)
8. P. Fiebig, Sheaves on moment graphs and a localization of Verma flags. Adv. Math. **217**, 683–712 (2008)

9. P. Fiebig, The combinatorics of Coxeter categories. Trans. Am. Math. Soc. **360**, 4211–4233 (2008)
10. P. Fiebig, Sheaves on affine Schubert varieties, modular representations and Lusztig's conjecture. J. Am. Math. Soc. **24**, 133–181 (2011)
11. P. Fiebig, Moment graphs in representation theory and geometry (2013, preprint). arXiv:1308.2873
12. J.E. Humphreys, *Reflection Groups and Coxeter Groups*. Cambridge Studies in Advanced Mathematics, vol. 29 (Cambridge University Press, Cambridge, 1990)
13. J.E. Humphreys, *Representations of Semisimple Lie Algebras in the BGG Category 𝒪*. Graduate Studies in Mathematics, vol. 94 (American Mathematical Society, Providence, RI, 2008)
14. J.C. Jantzen, *Moduln mit einem höchsten Gewicht*. Lecture Notes in Mathematics, vol. 750 (Springer, New York, 1979)
15. D. Juteau, C. Mautner, G. Williamson, Parity sheaves. J. Am. Math. Soc. **27**(4), 1169–1212 (2014)
16. D. Kazhdan, G. Lusztig, Representations of Coxeter groups and Hecke algebras. Invent. Math. **53**(2), 165–184 (1979)
17. D. Kazhdan, G. Lusztig, Schubert varieties and Poincaré duality, in *Geometry of the Laplace operator (Proceedings of Symposia in Pure Mathematics)* (University of Hawaii, Honolulu, 1979), pp. 185–203
18. W. Soergel, On the relation between intersection cohomology and representation theory in positive characteristic. J. Pure Appl. Algebra **152**(1–3), 311–335 (2000)
19. W. Soergel, Kategorie 𝒪, perverse Garben und Moduln über den Koinvarianten zur Weylgruppe. J. Am. Math. Soc. **3**(2), 421–445 (1990)
20. W. Soergel, Kazhdan-Lusztig-Polynome und unzerlegbare Bimoduln über Polynomringen. J. Inst. Math. Jussieu **6**(3), 501–525 (2007)
21. G. Williamson; with a joint appendix with Alex Kontorovich and Peter J. McNamara. Schubert calculus and torsion explosion. J. Am. Math. Soc. **30**, 1023–1046 (2017)

Chapter 4
The BCH-Formula and Order Conditions for Splitting Methods

Winfried Auzinger, Wolfgang Herfort, Othmar Koch,
and Mechthild Thalhammer

Abstract As an application of the BCH-formula, order conditions for splitting schemes are derived. The same conditions can be obtained by using noncommutative power series techniques and inspecting the coefficients of Lyndon–Shirshov words.

4.1 Introduction

The main purpose of this note is to present a not so well-known application of the Baker–Campbell–Hausdorff formula (BCH-formula): Computing *order conditions* for *exponential splitting schemes*. There is vast literature, for an overview we particularly refer to [14] and [12, Chap. III] and see [2–4, 5–7, 9, 13, 15–18].

The topic of splitting is a comparatively young field and it is our intention to present only facets—with Lie-theoretic background. We shall first recall a few facts from Lie theory and power series as far as needed. An abstract definition of splitting is given and the computation of order condition is demonstrated with examples. The last section is devoted to an alternative approach for finding the order conditions by inspecting the coefficients of leading Lyndon–Shirshov words in an exponential function of a sum of Lie elements, as currently used by the authors for computationally generating order conditions for exponential splitting schemes.

W. Auzinger • W. Herfort (⊠)
Technische Universität Wien, Wiedner Hauptstraße 8-10/E101, Vienna, Austria
e-mail: Winfried.Auzinger@tuwien.ac.at; Wolfgang.Herfort@tuwien.ac.at

O. Koch
Fakultät für Mathematik, Universität Wien, Oskar Morgensternplatz 1, Vienna, Austria
e-mail: othmar@othmar-koch.org

M. Thalhammer
Institut für Mathematik, Universität Innsbruck, Technikerstrasse 13, Innsbruck, Austria
e-mail: Mechthild.Thalhammer@uibk.ac.at

© Springer International Publishing AG 2017   71
G. Falcone (ed.), *Lie Groups, Differential Equations, and Geometry*,
UNIPA Springer Series, DOI 10.1007/978-3-319-62181-4_4

4.2 Formal Power Series

Let k be a field of characteristic zero. Then $k\langle\langle S\rangle\rangle$ denotes the algebra of formal power series with coefficients in k and S a set of non-commuting variables. The natural grading of $k\langle\langle S\rangle\rangle$ is given as follows: elements in k have degree zero, those in $S^n := \{s_1 \ldots s_n \mid s_j \in S\}$ have degree n, where $n \geq 1$. A *homogeneous* element in $k\langle\langle S\rangle\rangle$ is a k-linear combination of elements of the same degree. Every element f in $k\langle\langle S\rangle\rangle$ allows a unique decomposition into homogeneous components

$$f = \sum_{j=0}^{\infty} f_j$$

where for each j the element f_j is homogeneous of degree j.

In our context the power series ring $R\langle\langle t\rangle\rangle$ in a single variable t and with coefficients in a (not necessarily commutative) ring R will turn out to be useful.

Whenever $f = \sum_j c_j t^j \in k\langle\langle t\rangle\rangle$ and $g \in k\langle\langle S\rangle\rangle$ and g does *not* contain constant terms then one can define the composition

$$f \circ g := \sum_j c_j g^j$$

as, for given degree say n, for $j \geq n + 1$, the power series g^j does not contribute homogeneous elements of degree n.

Example 4.2.1 The univariate formal power series $f := \sum_{j=0}^{\infty} \frac{1}{j!} t^j$ will be denoted by e^t. Hence the composition $f \circ g$ allows to consider $e^g = \sum_{j=0}^{\infty} \frac{1}{j!} g^j$.

The following simple fact will be helpful:

Lemma 4.2.1 *Let* $h = \sum_{j=1}^{\infty} h_j$ *be an element in* $k\langle\langle S\rangle\rangle$, *with each* h_j *homogeneous of degree* j. *Then* $e^h = \sum_{j=0}^{\infty} e_j$ *with homogeneous terms* e_j, *and the following statements are equivalent:*

(i) $h_j = 0$ *for* $j = 1, \ldots, p$;
(ii) $e_j = 0$ *for* $j = 1, \ldots, p$.

Proof Suppose that (i) holds. Then

$$e^h = 1 + \left(h_{p+1} + \cdots\right) + \frac{1}{2}\left(h_{p+1} + \cdots\right)^2 + \cdots$$

shows that there cannot exist homogeneous terms e_j with $1 \leq j \leq p$.

The converse is proved by induction. Suppose for p that $e_1 = \cdots = e_{p-1} = 0$ implies $h_1 = \cdots = h_{p-1} = 0$. Suppose next that also $e_p = 0$. Then

$$e^h = 1 + \left(h_p + \cdots\right) + \frac{1}{2}\left(h_p + \cdots\right)^2 + \cdots = e_0 + e_{p+1} + \cdots$$

From this one concludes that $h_p = 0$ must hold.   □

Corollary 4.2.2 *Let $h = \sum_{j=1}^{\infty} h_j$ and $k = \sum_{j=1}^{\infty} k_j$ be elements in $k\langle\!\langle S \rangle\!\rangle$. Set $e^h = \sum_{j=0}^{\infty} e_j$ and $e^k = \sum_{j=0}^{\infty} f_j$. Then the following statements are equivalent:*

(a) $h_j = k_j$ *for* $j = 1, \ldots, p$;
(b) $e_j = f_j$ *for* $j = 1, \ldots, p$.

Proof Certainly (a) implies (b), as for forming the homogeneous terms in e^h only the terms up to order p contribute.

For proving the converse one again uses induction. Having established that $h_j = k_j$ for $1 \leqslant j \leqslant p - 1$, one observes that $e_p = h_p + \phi(h_1, \ldots, h_{p-1}) = k_p + \phi(k_1, \ldots, k_{p-1}) = f_p$. Here ϕ is a certain multivariate polynomial whose form to know is not needed. Then, as $e_p = f_p$ conclude that $h_p = k_p$.   □

4.3   Reformulation Using Formal Differentiation

Given the algebra $R := k\langle\!\langle S \rangle\!\rangle$, one may use it as the set of coefficients and form the new algebra $k\langle\!\langle S \rangle\!\rangle\langle\!\langle t \rangle\!\rangle$. There is a canonical function $\phi : k\langle\!\langle S \rangle\!\rangle \to k\langle\!\langle S \rangle\!\rangle\langle\!\langle t \rangle\!\rangle$ that sends $f := \sum_{j=0}^{\infty} f_j$ to the element $\phi(f) := \sum_{j=0}^{\infty} f_j t^j$. In $k\langle\!\langle S \rangle\!\rangle\langle\!\langle t \rangle\!\rangle$ we define *formal differentiation* by means of

$$\left(\sum_{j=0}^{\infty} r_j t^j\right)^{\cdot} := \sum_{j=1}^{\infty} j r_{j-1} t^{j-1}.$$

Formal derivatives $f^{(k)}$ of higher order, for elements $f \in k\langle\!\langle S \rangle\!\rangle\langle\!\langle t \rangle\!\rangle$, are defined inductively.

Notation 4.3.1 If an element $f = \sum_{j=0}^{\infty} r_j t^j \in k\langle\!\langle S \rangle\!\rangle\langle\!\langle t \rangle\!\rangle$ has $r_0 = \cdots r_p = 0$, we shall denote this by $f = O(t^{p+1})$ or even by $f = O(p+1)$.
    One proves without difficulty:

Lemma 4.3.2 *For $f = \sum_{j=0}^{\infty} f_j$ the following statements are equivalent:*

(i) $f_0 = \cdots = f_p = 0$;
(ii) $\phi(f) = O(t^{p+1})$.

Next we prove a key lemma:

Lemma 4.3.3 *For $X \in k\langle\langle S\rangle\rangle\langle\langle t\rangle\rangle$ and an element C in $k\langle\langle S\rangle\rangle$ of degree 1, the following statements are equivalent:*

(i) $X - e^{Ct} = O(t^{p+1})$;
(ii) $X(0) = 1$ *and* $D := \dot{X} - CX$ *enjoys* $D = O(t^p)$.

Proof Certainly (i) implies (ii), as can be seen by differentiation. Conversely, if (ii) holds, one only needs to check that $X - e^{Ct}$ vanishes when setting $t = 0$. But this is a consequence of the assumption that $X(0) = 1$.   □

Here is the main observation about the different method to be used in Sect. 4.5 on *splitting schemes*. Namely it will imply that for deriving order conditions it is equivalent to either consider them as the coefficients of a power series or to pass to the logarithm and use thereby the BCH-formula and look at the coefficients of the basic commutators.

Proposition 4.3.4 *Suppose that $h \in k\langle\langle S\rangle\rangle$ has the form $h = e^Z$ with $Z_0 = 0$. Then the following statements about $h = \sum_{j=0}^{\infty} h_j$ and e^C for an element C homogeneous of degree 1 are equivalent:*

(A) $h_0 - 1 = h_1 - C = \cdots = h_p - \frac{C^p}{p!} = 0$;
(B) $\phi(h)^{\cdot} - C\phi(h) = O(t^p)$;
(C) $Z - C = O(t^{p+1})$.

Proof As $\phi(h) = \sum_{j=0}^{\infty} h_j t^j$, the equivalence of (A) and (B) follows from Lemma 4.3.3. The equivalence of (B) and (C) is an immediate consequence of Corollary 4.2.2.   □

4.4 The Baker–Campbell–Hausdorff Formula

The *Baker–Campbell–Hausdorff* formula (BCH-formula, see, for instance, [12]) allows, for given X and Y in $k\langle S\rangle$ without constant terms, to find $Z \in k\langle\langle S\rangle\rangle$ with $e^X e^Y = e^Z$. In fact, Z turns out to be a formal infinite sum of X, Y and homogeneous elements from the Lie algebra $\mathfrak{L}(S)$, generated by the set S and the bracket operation $[l_1, l_2] = l_1 l_2 - l_2 l_1$ for $l_i \in \mathfrak{L}(S)$.

Example 4.4.1 The first terms of Z are

$$Z = X + Y + \frac{1}{2}[X, Y] + \frac{1}{12}\big([X, [X, Y]] + [Y, [Y, X]]\big) + \cdots$$

As noted, Z with the exception of the terms of first order is an infinite sum of Lie elements, i.e., homogeneous elements of the Lie algebra generated by X and Y. One denotes Z by $\log e^X e^Y$. Inductively one can derive an analog for $\log(e^{X_1} \ldots e^{X_n})$, for $X_i \in S$.

4.5   Splitting Schemes

The following abstract definition of a *splitting scheme* will serve our purpose:

Definition 4.5.1 Given the elements A and B of S and suppose there are, for $j = 1, \ldots, s$, $A_j \in \text{span}\{A\}$ and $B_j \in \text{span}\{B\}$, i.e., $A_j = a_j A$, $B_j = b_j B$ with scalar coefficients $a_j, b_j, j = 1, \ldots, s$. Then these data determine a *splitting scheme* of order at least p, provided

$$e^{A_1 t} e^{B_1 t} \ldots e^{A_s t} e^{B_s t} - e^{(A+B)t} = O(t^{p+1})$$

Here is an equivalent formulation of this condition. The proof, in the light of the BCH-formula, is immediate from the definition:

Proposition 4.5.2 *The following statements for given A and B in S and elements $A_i \in \text{span}\{A\}$, $B_i \in \text{span}\{B\}$, where $i = 1, \ldots, s$, are equivalent:*

(A) *The data yield a splitting scheme of order at least p;*
(B) $\log(e^{A_1} e^{B_1} \cdots e^{A_s} e^{B_s}) - (A + B)$ *has homogeneous terms equal to zero for $j = 1, \ldots, p$.*

Remark 4.5.3 Splitting techniques can also be successfully applied to nonlinear evolution equations. The order conditions studied here are also valid for this general case. This follows from an ingenious idea by Gröbner [11], namely formally to express the flow of a nonlinear evolution equation as the exponential of the corresponding Lie derivative; see [12, Sect. III.5].

Let us, as a preparation for Sect. 4.6, compute the logarithm in (A) for $s = 2$ and $s = 3$ up to terms of order $p \leqslant 3$.

The BCH-formula easily yields

$$\log(e^{A_j} e^{B_j}) = A_j + B_j + \frac{1}{2}[A_j, B_j] + \frac{1}{12}\left([A_j, [A_j, B_j]] + [B_j, [B_j, A_j]]\right) + O(4), \quad (4.1)$$

where $O(4)$ stands for all terms in $\mathfrak{L}(S)$ of degree at least 4.

Example 4.5.1 Let $X = X_1 + X_2 + X_3$ and $Y = Y_1 + Y_2 + Y_3$ be a decomposition into homogeneous elements with all nonlinear terms in $\mathfrak{L}(S)$. Then $\log(e^X e^Y) = H$ has first homogeneous terms

$$H_1 = X_1 + Y_1$$

$$H_2 = X_2 + Y_2 + \frac{1}{2}[X_1, Y_1]$$

$$H_3 = X_3 + Y_3 + \frac{1}{2}\left([X_1, Y_2] + [X_2, Y_1]\right)$$

$$+ \frac{1}{12}\left([X_1, [X_1, Y_1]] + [Y_1, [Y_1, X_1]]\right)$$

Lemma 4.5.4 *Let* $X := aA + bB + c[A, B] + d[A, [B, B]] + e[B, [B, A]]$ *and* $X' :=$ $a'A + b'B + c'[A, B] + d'[A, [B, B]] + e'[B, [B, A]]$, *then* $H := \log(e^X e^Y)$ *has first terms*

$$H = H_1 A + H_2 B + H_3[A, B] + H_4[A, [A, B]] + H_5[B, [B, A]]$$

where

$$H_1 = a + a'$$
$$H_2 = b + b'$$
$$H_3 = c + c' + \tfrac{1}{2}(ab' - a'b)$$
$$H_4 = d + d' + \tfrac{1}{2}(ac' - a'c) + \tfrac{1}{12}(ab' - a'b)(a - a')$$
$$H_5 = e + e' - \tfrac{1}{2}(bc' - b'c) - \tfrac{1}{12}(ab' - a'b)(b - b')$$

Proof Using the preceding example for $Y := X'$ and elementary computation yields the result. $\qquad\square$

With the aid of Lemma 4.5.4 one finds:

Corollary 4.5.5 *The first homogeneous terms of* $K := \log(e^{a_1 A} e^{b_1 B} e^{a_2 A} e^{b_2 B} e^{a_3 A} e^{b_3 B})$ *are as follows:*

$$K_1 = (a_1 + a_2 + a_3)A + (b_1 + b_2 + b_3)B$$

$$K_2 = \frac{1}{2}\,(a_1 b_1 + a_2 b_2 + a_3 b_3 + a_1 b_2$$
$$-a_2 b_1 + a_1 b_3 - a_3 b_1 + a_2 b_3 - a_3 b_2)\,[A, B]$$

and $K_3 = \xi[A, [A, B]] + \eta[B, [B, A]]$ *with*

$$\xi = \frac{1}{12}(a_1^2 b_1 + a_2^2 b_2 + a_3^2 b_3)$$

$$+ \frac{1}{4}(a_1 a_2 b_2 - a_1 a_2 b_1 + a_1 a_3 b_3 + a_2 a_3 b_3 - a_1 a_3 b_1 - a_2 a_3 b_2 - a_1 a_3 b_2 + a_2 a_3 b_1)$$

$$+ \frac{1}{12}(a_1^2 b_2 - a_1 a_2 b_1 + a_2^2 b_1 - a_1 a_2 b_2 + a_1^2 b_3 - a_1 a_3 b_1 + a_1 a_2 b_3 - a_1 a_3 b_2$$

$$+ a_1 a_2 b_3 - a_2 a_3 b_1 + a_2^2 b_3 - a_2 a_3 b_2$$

$$+ a_3^2 b_1 - a_1 a_3 b_3 + a_3^2 b_2 - a_2 a_3 b_3)$$

$$\eta = \frac{1}{12}(b_1^2 a_1 + b_2^2 a_2 + b_3^2 a_3)$$

$$+ \frac{1}{4}(b_1 b_2 a_2 - b_1 b_2 a_1 + b_1 b_3 a_3 + b_2 b_3 a_3 - b_1 b_3 a_1 - b_2 b_3 a_2 - b_1 b_3 a_2 + b_2 b_3 a_1)$$

$$+ \frac{1}{12}(b_1^2 a_2 - b_1 b_2 a_1 + b_2^2 a_1 - b_1 b_2 a_2 + b_1^2 a_3 - b_1 b_3 a_1 + b_1 b_2 a_3 - b_1 b_3 a_2$$

$$+ b_1b_2a_3 - b_2b_3a_1 + b_2^2a_3 - b_2b_3a_2$$
$$+ b_3^2b_1 - b_1b_3a_3 + b_3^2a_2 - b_2b_3a_3).$$

Proof One first computes $s_j := \log(e^{a_jA}e^{b_jB}) = a_jA + b_jB + \frac{1}{2}a_jb_j[A,B] + \frac{1}{12}(a_j^2b_j[A,[A,B]] + a_jb_j^2[B,[B,A]]) + O(4)$. Then, using Lemma 4.5.4, compute first $H := \log(e^{s_1}e^{s_2})$ and, again using the lemma, find the desired expressions by computing $K := \log(e^H e^{s_3})$. □

4.6 Computing Order Conditions in Examples

4.6.1 Schemes of Order At Least 1

It follows right from the definition that in this case

$$\sum_{j=1}^{s} A_j = A, \quad \sum_{j=1}^{s} B_j = B$$

i.e., for $A_j = a_jA$ and $B_j = b_jB$ one obtains

$$\sum_{j=1}^{s} a_j = 1, \quad \sum_{j=1}^{s} b_j = 1.$$

4.6.2 The Order Conditions for s = 2 and p = 3

Elementary computation leads to the following observation:

Lemma 4.6.1 *The order conditions for s = 2 and p = 3 are as follows:*

$$1 = a_1 + a_2$$
$$1 = b_1 + b_2$$
$$0 = a_1b_1 + a_2b_2 + a_1b_2 - a_2b_1$$
$$0 = 1 - 6a_1a_2b_1$$
$$0 = 1 - 6b_1b_2a_2$$

Proof The order condition for $s = 1$ in the first line follows from the previous subsection. The higher order conditions result from Lemma 4.5.4. □

4.6.3 The Order Conditions for $s = 3$ and $p = 3$

Making use of Corollary 4.5.5 the conditions on the coefficients a_j and b_j for $j = 1, 2, 3$ in order to let $A_j = a_j A$ and $B_j = b_j B$ determine the necessary and sufficient conditions for a splitting scheme of order p at least 3 when $s = 3$.

Lemma 4.6.2 *The order conditions for $s = p = 3$ read as follows:*

$$1 = a_1 + a_2 + a_3$$
$$1 = b_1 + b_2 + b_3$$
$$\frac{1}{2} = a_2 b_1 + a_3 b_1 + a_3 b_2$$
$$2 = 3(a_2 + a_3) - 6a_2 a_3 b_2$$
$$2 = 3(b_1 + b_2) - 6b_1 b_2 a_2$$

Proof In Corollary 4.5.5, one equates the coefficients of A and B to 1, and those of $[A, B]$, $[A, [A, B]]$ and $[B, [B, A]]$ to zero. Then, using the third equation, terms $a_3 b_2$ have been eliminated from the last two equations. □

To conclude this section let us remark that expanding $e^{X_1 + X_2 + \cdots}$ as a Taylor series, one finds from Lemma 4.2.1 and Proposition 4.3.4:

Proposition 4.6.3 *The following statements for an exponential function e^X for X a sum of homogeneous Lie elements with the exception of the linear term are equivalent:*

(i) $X_1 = X_2 = \cdots = X_p = 0$
(ii) *As a power series in $k\langle\!\langle S \rangle\!\rangle$ the first nonvanishing homogeneous term is* $\frac{1}{(p+1)!} X_{p+1}$.

4.7 Alternative Approach via Taylor Expansion and Computation in the Free Lie Algebra Generated by A, B

According to the ideas from [1, 3] systems of polynomial equations representing order conditions for splitting methods are set up in a different way without making explicit use of the BCH-formula. This is straightforward to implement in computer algebra. The resulting systems of equations are not identical but equivalent to those obtained when implementing the BCH-based procedure described above.

This alternative approach described can also easily be adapted and generalized to cases with various symmetries, pairs of schemes, and more general cases like splitting involving three operators A, B, C, or more, see [3].

To find conditions for the coefficients a_j, b_j such that for $A_j = a_j A$, $B_j = b_j B$ a scheme of order p is obtained,

$$L(t) := e^{A_1 t} e^{B_1 t} \ldots e^{A_s t} e^{B_s t} - e^{(A+B)t} = O(t^{p+1}),$$

we consider the Taylor expansion of $L(t)$, the local error of the splitting scheme applied with stepsize t (satisfying $L(0) = 0$ by construction),[1]

$$L(t) = \sum_{q=1}^{p} \frac{t^q}{q!} \frac{d^q}{dt^q} L(0) + O(t^{p+1}).$$

The method is of order p iff $L(t) = O(t^{p+1})$; thus, the conditions for order p are given by

$$\frac{d}{dt} L(0) = \cdots = \frac{d^p}{dt^p} L(0) = 0. \tag{4.2}$$

Via successive differentiation of $L(t)$ we obtain the following homogeneous representation of $\frac{d^q}{dt^q} L(0)$ in terms of power products in the non-commuting variables A and B: With $\boldsymbol{k} = (k_1, \ldots, k_s) \in \mathbb{N}_0^s$, one obtains

$$\frac{d^q}{dt^q} L(0) = \sum_{|\boldsymbol{k}|=q} \binom{q}{\boldsymbol{k}} \prod_{j=1}^{s} \sum_{l=0}^{k_j} \binom{k_j}{l} A_j^l B_j^{k_j - l} \; - \; (A+B)^q, \quad q = 0, 1, 2, \ldots \tag{4.3}$$

In a computer algebra system, these symbolic expressions can be generated in a straightforward way.

If conditions (4.2) are satisfied up to a given order p, then the leading term of the local error is given by $\frac{t^{p+1}}{(p+1)!} \frac{d^{p+1}}{dt^{p+1}} L(0)$. Proposition 4.6.3 shows that this leading local error term is a homogeneous linear combination of Lie elements. With the terminology

$$LC(q) := \text{'the sum (4.3) is a linear combination of Lie elements of degree } q\text{'}$$

this amounts to $LC(p + 1)$ being true for a scheme of order p. Exploiting this statement allows to design a recursive algorithm for generating a set of order conditions:

(i) By construction, $LC(1) = 0$ holds. But, a priori, for $q > 1$ the expression (4.3) for $\frac{d^q}{dt^q} L(0)$ does *not* enjoy $LC(q)$, see Example 4.7.1 below. On the other hand, from Proposition 4.6.3 we know that

$$LC(1) \wedge \ldots \wedge LC(q - 1) \;\Rightarrow\; LC(q).$$

---

[1] Here, $\frac{d^q}{dt^q} L(0) := \frac{d^q}{dt^q} L(t)\big|_{t=0}$.

By induction over q we see that each solution of (4.2) must satisfy LC(q) for $q = 1, \ldots, p$. (Moreover, for each the resulting solution LC($p + 1$) will hold true.)

(ii) Due to (i), for the purpose of solving the system (4.2) we may assume

$$\frac{d^q}{dt^q} L(0) = \sum_k \lambda_{q,k} B_{q,k}, \quad q = 1, \ldots, p$$

where the $B_{q,k}$ of degree q are elements from a basis of the free Lie algebra generated by A and B. Now the problem is to identify, on the basis of the expressions (4.3), coefficients $\lambda_{q,k} = \lambda_{q,k}(a_j, b_j)$ such that the polynomial system

$$\lambda_{q,k} = 0, \quad q = 1, \ldots, p, \quad k \text{ running over all basis elements of degree } q \tag{4.4}$$

will be equivalent to (4.2). To this end we make use of the one-to-one correspondence between basis elements of degree q represented by non-associative, bracketed words (commutators) and associative words of length q over the alphabet $\{A, B\}$. The implementation described in [1, 3] relies on the Lyndon basis, also called Lyndon–Shirshov basis, which can be generated by an algorithm devised in [10]. With this choice,

each basis element $B_{q,k}$ is uniquely represented by a

Lyndon word of length q associated with the leading term, (4.5)

in lexicographical order, of the expanded version of $B_{q,k}$.

Identifying the coefficients of these "Lyndon monomials" results in the desired polynomial system (4.4).

To be more precise we note that, in general, a given Lyndon monomial shows up in different expanded commutators. Therefore, equating coefficients of Lyndon monomials will not directly result in the system (4.4) but, as a consequence of (4.5), in an equivalent system which is obtained from (4.4) by premultiplication with a regular triangular matrix, see [3].

For the underlying theoretical background concerning Lyndon bases in free Lie algebras, we refer to [8]. For a detailed illustration of our approach for order $p = 5$, see Example 2 from [3]. In the following example we reconsider the case $s = p = 3$.

Example 4.7.1 For $s = 3$ we have [see (4.3)]

$$\frac{d}{dt} L(0) = (a_1 + a_2 + a_3 - 1)A + (b_1 + b_2 + b_3 - 1)B,$$

$$\frac{d^2}{dt^2} L(0) = ((a_1 + a_2 + a_3)^2 - 1)A^2$$

$$+ \left(2a_1(b_1 + b_2 + b_3) + 2a_2(b_2 + b_3) + 2a_3b_3 - 1\right)AB$$
$$+ \left(2a_2b_1 + 2a_3(b_1 + b_2) - 1\right)BA$$
$$+ \left((b_1 + b_2 + b_3)^2 - 1\right)B^2,$$

Assume that $a_1 + a_2 + a_3 = 1$ and $b_1 + b_2 + b_3 = 1$ such that $\frac{d}{dt}L(0) = 0$. Then substituting $a_1 = 1 - a_2 - a_3$ and $b_3 = 1 - b_1 - b_2$ into $\frac{d^2}{dt^2}L(0)$ gives the commutator expression

$$\frac{d^2}{dt^2}L(0) = -\left(2a_2b_1 + 2a_3(b_1 + b_2) - 1\right)[A, B].$$

Therefore the system

$$a_1 + a_2 + a_2 = 1$$
$$b_1 + b_2 + b_3 = 1$$
$$a_2b_1 + a_3(b_1 + b_2) = \frac{1}{2}$$

represents a set of conditions for order $p = 2$.

Extending this computation to $p = 3$ by hand is already somewhat laborious. However, from the above considerations we know that assuming the conditions for order $p = 2$ are satisfied, then

$$\frac{d^2}{dt^2}L(0) = \lambda_{AAB}[A, [A, B]] + \lambda_{ABB}[[A, B], B],$$

where

λ_{AAB} = coefficient of the power product A^2B in the expression (4.3) for $\dfrac{d^2}{dt^2}L(0)$,

λ_{ABB} = coefficient of the power product AB^2 in the expression (4.3) for $\dfrac{d^2}{dt^2}L(0)$.

Here the two independent commutators $[A, [A, B]]$ and $[[A, B], B]$ are represented by the associative Lyndon words "AAB" and "ABB." In computer algebra, extraction of the coefficients λ_{AAB} and λ_{ABB} from the symbolic expression $\frac{d^2}{dt^2}L(0)$ is straightforward. In this way we end up with the system

$$a_1 + a_2 + a_3 = 1$$
$$b_1 + b_2 + b_3 = 1$$

$$a_2 b_1 + a_3(b_1 + b_2) = \frac{1}{2}$$

$$a_2 b_1^2 + a_3(b_1 + b_2)^2 = \frac{1}{3}$$

$$(a_2 + a_3)^2 b_1 + a_3^2 b_2 = \frac{1}{3}$$

representing a set of order conditions for order $p = 3$. The system is equivalent to the one found in Lemma 4.6.2. We note that there is a one-dimensional zero solution manifold containing well-known rational solutions, e.g.,

$$a_1 = \frac{7}{24}, \ a_2 = \frac{3}{4}, \ a_3 = -\frac{1}{24},$$

$$b_1 = \frac{2}{3}, \ b_2 = -\frac{2}{3}, \ b_3 = 1.$$

Acknowledgements Wolfgang Herfort is indebted to UNIPA for generous support in June 2016.

He also would like to thank the Department of Mathematics at the Brigham Young University for the great hospitality during the year 2015. Special thanks to JIM LOGAN for his excellent support during this time with hard- and software.

Othmar Koch acknowledges the support by the Vienna Science and Technology Fund (WWTF) under the grant MA14-002.

References

1. W. Auzinger, W. Herfort, Local error structures and order conditions in terms of Lie elements for exponential splitting schemes. Opuscula Math. **34**(2), 243–255 (2014)
2. W. Auzinger, O. Koch, M. Thalhammer, Defect-based local error estimators for splitting methods, with application to Schrödinger equations, Part II. Higher order methods for linear problems. J. Comput. Appl. Math. **255**, 384–403 (2013)
3. W. Auzinger, H. Hofstätter, D. Ketcheson, O. Koch, Practical splitting methods for the adaptive integration of nonlinear evolution equations. Part I: construction of optimized schemes and pairs of schemes. BIT Numer. Math. Published online: 28 July, 2016. http://dx.doi.org/10.1007/s10543-016-0626-9
4. S. Blanes, F. Casas, On the convergence and optimization of the Baker–Campbell–Hausdorff formula. Linear Algebra Appl. **378**, 135–158 (2004)
5. S. Blanes, P.C. Moan, Practical symplectic partitioned Runge-Kutta and Runge-Kutta-Nyström methods. J. Comput. Appl. Math. **142**, 313–330 (2002)
6. S. Blanes, F. Casas, A. Murua, Splitting and composition methods in the numerical integration of differential equations. Bol. Soc. Esp. Mat. Apl. **45**, 89–145 (2008)
7. S. Blanes, F. Casas, P. Chartier, A. Murua, Optimized high-order splitting methods for some classes of parabolic equations. Math. Comput. **82**, 1559–1576 (2013)
8. L. Bokut, L. Sbitneva, I. Shestakov, Lyndon–Shirshov words, Gröbner-Shirshov bases, and free Lie algebras, in *Non-associative Algebra and its Applications*, Chap. 3 (Chapman & Hall/CRC, Boca Raton, FL, 2006)
9. P. Chartier, A. Murua, An algebraic theory of order. M2AN Math. Model. Numer. Anal. **43**, 607–630 (2009)

10. J.-P. Duval, Géneration d'une section des classes de conjugaison et arbre des mots de Lyndon de longueur bornée. Theor. Comput. Sci. **60**, 255–283 (1988)
11. W. Gröbner, *Die Liereihen und ihre Anwendungen*, 2nd edn. (VEB Deutscher Verlag der Wissenschaften, Berlin, 1967)
12. E. Hairer, C. Lubich, G. Wanner, *Geometrical Numerical Integration – Structure-Preserving Algorithms for Ordinary Differential Equations*, 2nd edn. (Springer, Berlin, 2006)
13. E. Hansen, A. Ostermann, Exponential splitting for unbounded operators. Math. Comput. **78**, 1485–1496 (2009)
14. R. McLachlan, R. Quispel, Splitting methods. Acta Numer. **11**, 341–434 (2002)
15. M. Suzuki, General theory of higher-order decomposition of exponential operators and symplectic integrators. Phys. Lett. A **165**, 387–395 (1992)
16. M. Thalhammer, High-order exponential operator splitting methods for time-dependent Schrödinger equations. SIAM J. Numer. Anal. **46**, 2022–2038 (2008)
17. Z. Tsuboi, M. Suzuki, Determining equations for higher-order decompositions of exponential operators. Int. J. Mod. Phys. B **9**, 3241–3268 (1995)
18. H. Yoshida, Construction of higher order symplectic integrators. Phys. Lett. A **150**, 262–268 (1990)

Chapter 5
Cohomology Operations Defining Cohomology Algebra of the Loop Space

Tornike Kadeishvili

Abstract According to the minimality theorem (Kadeishvili, Russ Math Surv 35(3):231–238, 1980), on the cohomology $H^*(X)$ of a topological space X there exists a sequence of operations $m_i : H^*(X)^{\otimes i} \to H^*(X), i = 2, 3, \ldots$ which form a minimal A_∞-algebra $(H^*(X), \{m_i\})$. This structure defines on the bar construction $BH^*(X)$ a correct differential d_m so that $(BH^*(X), d_m)$ gives cohomology modules of the loop space $H^*(\Omega X)$. In this paper we construct algebraic operations $E_{p,q} : H^*(X)^{\otimes p} \otimes H^*(X)^{\otimes q} \to H^*(X), p, q = 0, 1, 2, 3, \ldots$ which turn $(H^*(X), \{m_i\}, \{E_{p,q}\})$ into a B_∞-algebra. This structure defines on $BH^*(X)$ a correct multiplication, thus determines a cohomology algebra $H^*(\Omega X)$.

5.1 Introduction

The Adams's bar construction $BC^*(X)$ of cochain complex $(C^*(X), d, \smile)$ of a topological space X defines cohomology *modules* of the loop space $H^*(\Omega X)$ [1], but not the cohomology *algebra* $H^*(\Omega X)$.

In [2] Baues constructed a sequence of multioperations

$$E_{1,k} : C^*(X) \otimes C^*(X)^{\otimes k} \to C^*(X), \ k = 1, 2, 3, \ldots$$

so that they determine on $BC^*(X)$ a correct multiplication. That is, this structure $(C^*(X), d, \smile, \{E_{1,k}\})$, which now is called homotopy Gerstenhaber algebra [6, 20] (hGa in short) determines a cohomology algebra $H^*(\Omega X)$.

Our aim is to transfer these structures to homology level, i.e. from $C^*(X)$ to $H^*(X)$.

T. Kadeishvili (✉)
A. Razmadze Mathematical Institute of Tbilisi State University, Tbilsi, Georgia
e-mail: kade@rmi.ge

© Springer International Publishing AG 2017
G. Falcone (ed.), *Lie Groups, Differential Equations, and Geometry*,
UNIPA Springer Series, DOI 10.1007/978-3-319-62181-4_5

Namely, according to the minimality theorem [9], the differential $d : C^*(X) \to C^{*+1}(X)$ and the multiplication (cup product) $\smile : C^*(X) \otimes C^*(X) \to C^*(X)$ determine a sequence of operations

$$m_i : H^*(X)^{\otimes i} \to H^*(X), i = 2, 3, \ldots$$

which form a minimal A_∞-algebra $(H^*(X), \{m_i\})$. This structure defines on the bar construction $BH^*(X)$ a correct differential $d_m : BH^*(X) \to BH^*(X)$ so that $(BH^*(X), d_m)$ and $BC^*(X)$ have isomorphic homology modules, thus this object $(H^*(X), \{m_i\})$ which is called cohomology A_∞-algebra of X, determines cohomology *modules* of the loop space $H^*(\Omega X)$. But not the cohomology *algebra* $H^*(\Omega X)$.

In this paper, as the next step, we transport on cohomology level Baues's operations

$$E_{1,k} : C^*(X) \otimes C^*(X)^{\otimes k} \to C^*(X), k = 1, 2, 3, \ldots \ ,$$

so we obtain algebraic operations

$$E_{p,q} : H^*(X)^{\otimes p} \otimes H^*(X)^{\otimes q} \to H^*(X), p, q = 0, 1, 2, 3, \ldots$$

which turn $(H^*(X), \{m_i\}, \{E_{p,q}\})$ into a B_∞-algebra. These operations define on $BH^*(X)$ a correct multiplication, which determines the cohomology algebra $H^*(\Omega X)$.

Now we describe the algebraic framework for the above, where we will work.

For a differential graded algebra (A, d, μ) (dg algebra in short) the differential $d : A_* \to A_{*-1}$ and the multiplication $\mu : A_p \otimes A_q \to A_{p+q}$ define on the bar construction

$$BA = \sum_{n=0}^{\infty} A^{\otimes n}, \ \dim(a_1 \otimes \cdots \otimes a_n) = \sum_k \dim a_k + n$$

a differential $d_B : BA \to BA$ given by (signs are ignored, let us work over Z_2)

$$d_B(a_1 \otimes \cdots \otimes a_n) = \sum_k a_1 \otimes \cdots \otimes da_k \otimes \cdots \otimes a_n$$

$$+ \sum_k a_1 \otimes \cdots \otimes \mu(a_k, a_{k+1}) \otimes \cdots \otimes a_n \quad (5.1)$$

which turns (BA, d_B, ∇_B) into a dg coalgebra with comultiplication

$$\nabla_B : BA \to BA \otimes BA,$$

$$\nabla_B(a_1 \otimes \cdots \otimes a_n) = \sum_{k=0}^{n} (a_1 \otimes \cdots \otimes a_k) \otimes (a_{k+1} \otimes \cdots \otimes a_n). \quad (5.2)$$

To catch the multiplicative structure, there exists the notion of *homotopy Gerstenhaber algebra*, see [6] and [20] (hGa in short), which we describe below, and which allows to construct a correct multiplication on the bar construction. This is an additional structure on a dg algebra (A, d, μ), consisting of a sequence of operations

$$E_{1,k} : A \otimes A^{\otimes k} \to A, \ k = 1, 2, 3, \ldots$$

which determine on BA a multiplication $\mu_E : BA \otimes BA \to BA$ turning $(BA, d_B, \nabla_B, \mu_E)$ into a dg bialgebra. In [2] Baues has constructed such a structure in $A = C^*(X)$, so that $BC^*(X)$ gives the cohomology *algebra* $H^*(\Omega X)$.

We are going to transfer this structure to the homology level, i.e. from A to $H(A)$ aiming BA and $BH(A)$ to carry the same amount of information.

Note that for a dg algebra (A, d, μ) its homology $H(A)$ is also a dga with trivial differential and induced multiplication $\mu^* : H(A) \otimes H(A) \to H(A)$. But generally the bar constructions BA of dg algebra (A, d, μ) and the bar construction $BH(A)$ of dg algebra $(H(A), d = 0, \mu^*)$ have different homologies. This means that stepping from (A, d, μ) to $(H(A), d = 0, \mu^*)$ we lose part of information.

This lose compensates the *minimality theorem* from [9] which states that the homology $H(A)$ of a dg algebra (A, d, μ) (all $H_i(X)$-s are assumed free) can be equipped with a sequence of multioperations

$$m_i : H(A)^{\otimes i} \to H(A), \ i = 1, 2, 3, \ldots ; \ m_1 = 0, \ m_2 = \mu^*$$

turning $(H(A), \{m_i\})$ into a *minimal A_∞-algebra* in the sense of Stasheff [19] which is weak equivalent to dga (A, d, μ) in a certain category. These A_∞ operations $\{m_i\}$ determine on $BH(A)$ a new, perturbed differential $d_m : BH(A) \to BH(A)$ by

$$d_m(a_1 \otimes \cdots \otimes a_n) = \sum_{k,i} a_1 \otimes \cdots \otimes a_k \otimes m_i(a_{k+1} \otimes \cdots \otimes a_{k+i}) \otimes a_{k+i+1} \otimes \cdots \otimes a_n$$

so that BA and $(BH(A), d_m)$ have isomorphic homology modules.

The aim of this paper is to construct for a hGa $(A, d, \mu, \{E_{1k}\})$ on its homology A_∞-algebra $(H(A), \{m_i\})$ a certain additional structure, the so-called B_∞ algebra [8], consisting of multioperations

$$E_{p,q} : H(A)^{\otimes p} \otimes H(A)^{\otimes q} \to H(A), \ p, q = 0, 1, 2, 3, \ldots ,$$

which determines on $BH(A)$ a correct multiplication so that the bar constructions $BH(A)$ and BA will have isomorphic homology *algebras*.

Remark This can be summarized as follows: If A is a dg algebra, hGa, or a commutative dg algebra, then $H(A)$ becomes, respectively, an A_∞ [9], B_∞ (present paper), or C_∞ [12] algebra (a commutative version of A_∞).

The plan of the paper is as follows. In the section *Preliminaries* we present the notions, facts, and constructions needed for the main goal of the paper. Namely, in Sect. 5.2.1 we present the notions of A_∞-algebra and A_∞-morphisms, their interpretation in terms of bar construction. Then in Sect. 5.2.2 we discuss the notions of twisting cochains and their equivalence, again with interpretation in terms of bar construction, and present the Berikashvili's theorem about the functor D—the set of equivalence classes of twisting cochains. In the next Sect. 5.2.3 we generalize all these notions and facts about twisting cochains from dg algebras to A_∞-algebras and prove the lifting theorem for these A_∞-twisting cochains, which is our main technical tool. The next Sect. 5.2.4 is dedicated to the notion of B_∞-algebra, its particular case hGa, to the interpretation of these notions in terms of twisting cochains; moreover, three remarkable and generic examples of homotopy Gerstenhaber algebras are given. In Sect. 5.3, using this theorem about the lifting of A_∞-twisting cochains, we transport the hGa structure existing on a dg algebra A to the B_∞-algebra structure on its homology $H(A)$. In the final section we use the obtained structures in a topological framework, to obtain a B_∞-algebra structure on the cohomology $H^*(X)$ which determines the cohomology *algebra* of the loop space $H^*(\Omega X)$.

5.2   Preliminaries

In this section we give some notions and constructions needed for the goal of the paper.

5.2.1   Category of A_∞-Algebras

The notion of A_∞-algebra was introduced by James Stasheff in [19]. This is a generalization of the notion of differential graded algebra, where one of the defining conditions, namely the associativity of the multiplication, is satisfied not strictly but up to certain higher homotopies.

The notion of A_∞-algebra gave rise to a series of notions of so-called homotopy algebras, where various axioms are fulfilled not strictly but up to some higher homotopies, such as: C_∞-algebra which is the commutative version of A_∞-algebra; homotopy Gerstenhaber algebra (hGa) where the commutativity is satisfied up to homotopy; its generalization B_∞-algebra; L_∞-algebra, where the Jacobi identity is satisfied up to homotopy; H_∞, the homotopy version of dg Hopf algebra where the connection between multiplication and comultiplication is satisfied up to homotopy, etc.

5.2.1.1 A_∞-Algebras

An A_∞-algebra [19] is a graded module M with a given sequence of operations

$$\{m_i : M^{\otimes i} \to M, \quad i = 1, 2, \ldots, \quad \deg m_i = i - 2\}$$

which satisfies the following conditions

$$\sum_{i+j=n+1} \sum_{k=0}^{n-j} \pm m_i(a_1 \otimes \cdots \otimes a_k \otimes m_j(a_{k+1} \otimes \cdots \otimes a_{k+j}) \otimes \cdots \otimes a_n) = 0. \quad (5.3)$$

In particular, for the operation $m_1 : M \to M$ we have $\deg m_1 = -1$ and $m_1 m_1 = 0$, this m_1 can be regarded as a differential on M. The operation $m_2 : M \otimes M \to M$ has the degree 0 and satisfies

$$m_1 m_2(a_1 \otimes a_2) + m_2(m_1 a_1 \otimes a_2) + m_2(a_1 \otimes m_1 a_2) = 0,$$

i.e., m_2 can be regarded as a multiplication on M and m_1 is a derivation with respect to it. Thus (M, m_1, m_2) is a sort of (possibly nonassociative) dg algebra. For the operation m_3 its degree is 1 and the condition (5.3) gives

$$m_2(m_2(a_1 \otimes a_2) \otimes a_3) + m_2(a_1 \otimes m_2(a_2 \otimes a_3))$$
$$= m_1 m_3(a_1 \otimes a_2 \otimes a_3) + m_3(m_1 a_1 \otimes a_2 \otimes a_3)$$
$$+ m_3(a_1 \otimes m_1 a_2 \otimes a_3) + m_3(a_1 \otimes a_2 \otimes m_1 a_3),$$

this means that the product m_2 is *homotopy associative* and the appropriate chain homotopy is m_3.

5.2.1.2 Bar Construction of an A_∞-Algebra

The meaning of higher operations m_i can be clarified using the notion of bar construction of an A_∞-algebra $(M, \{m_i\})$. It is, as above

$$BM = \sum_{k=0}^{\infty} M^{\otimes k}, \; \dim(a_1 \otimes \cdots \otimes a_n) = \sum_k \dim a_k + n$$

but it can also be represented as

$$BM = T^c(sM) = \Lambda + sM + sM \otimes sM + sM \otimes sM \otimes sM + \cdots$$

here Λ is the ground ring and $(sM)_k = M_{k-1}$ is the standard suspension operator. BA is the graded coalgebra with comultiplication ∇_B, see (5.2). In fact it is the *cofree graded coalgebra cogenerated by M*, that is it has the following universal property:

$$
\begin{array}{c}
M \xleftarrow{\;p\;} T^c(sM) \\
{\scriptstyle\phi}\searrow \quad \uparrow{\scriptstyle f_\phi} \\
K
\end{array}
\tag{5.4}
$$

for each graded coalgebra $(K, \nabla_K : K \otimes K)$ and each homomorphism $\phi : K \to M$ of degree -1 there exists *coextension*, the unique graded coalgebra map $f_\phi : K \to T^c(sM)$ such that $pf_\phi = \phi$ where $p : T^c(sM) \to M$ is the clear projection. In fact f_ϕ is given by

$$
f_\phi = \sum_i (\phi \otimes \cdots \otimes \phi)\nabla_K^i
\tag{5.5}
$$

here $\nabla_K^i : K \to K^{\otimes i}$ is the iteration of comultiplication ∇_K:

$$
\nabla_K^1 = id_K, \quad \nabla_K^2 = \nabla_K, \quad \nabla_K^n = (id \otimes \nabla_K)\nabla_K^{n-1}.
$$

The main meaning of Stashef's defining condition (5.3) of an A_∞-algebra $(M, \{m_i\})$ is the following. The sequence of operations $\{m_i\}$ determines on the bar construction BM a coderivation

$$
d_m(a_1 \otimes \cdots \otimes a_n) = \sum_{k,j} a_1 \otimes \cdots \otimes a_k \otimes m_j(a_{k+1} \otimes \cdots \otimes a_{k+j}) \otimes \cdots \otimes a_n,
$$

and the Stasheffs condition (5.3) is exactly $md_m = 0$ where $m = \sum_i m_i : BM \to M$, and this, since of cofreeness is equivalent to $d_m d_m = 0$. Consequently d_m is a differential and (BM, d_m, ∇) is a dg coalgebra, which is called *bar construction* of A_∞-algebra $(M, \{m_i\})$.

5.2.1.3   Morphisms of A_∞-Algebras

A *morphism of A_∞-algebras* $f : (M, \{m_i\}) \to (M', \{m_i'\})$ can be defined as a dg coalgebra map of bar constructions

$$
f : B(M, \{m_i\}) \to B(M', \{m_i'\}).
$$

Because of the cofreeness of the tensor coalgebra $T^c(sM)$, such f is uniquely determined by the projection

$$
p'f : B(M, \{m_i\}) \xrightarrow{f} B(M', \{m_i'\}) \xrightarrow{p} M',
$$

which, in fact, is a collection of homomorphisms

$$\{f_i : M^{\otimes i} \to M', \quad i = 1, 2, \dots, \quad \deg f_i = i - 1\}. \tag{5.6}$$

Such a collection reconstructs a coalgebra map $f : T^c(sM) \to T^c(M')$ by

$$f(a_1 \otimes \cdots \otimes a_n) = \sum_{t=1}^{i} \sum_{k_1 + \cdots + k_t = i} f_{k_1}(a_1 \otimes \cdots \otimes a_{k_1}) \otimes \cdots \otimes f_{k_t}(a_{i-k_t+1} \otimes \cdots \otimes a_i).$$

If we wish f to be also a chain map $f : BM \to BM'$, that is $fd_m = d_{m'}f$, again because of the cofreeness it suffices to check $p'fd_m = p'd_{m'}f$ and this in terms of components is equivalent to the condition

$$\sum_{k=0}^{i-1} \sum_{j=1}^{i-k} f_{i-j+1}(a_1 \otimes \cdots \otimes a_k \otimes m_j(a_{k+1} \otimes \cdots \otimes a_{k+j}) \otimes \cdots \otimes a_i)$$

$$= \sum_{t=1}^{i} \sum_{k_1 + \cdots + k_t = i} m'_t(f_{k_1}(a_1 \otimes \cdots \otimes a_{k_1}) \otimes \cdots \otimes f_{k_t}(a_{i-k_t+1} \otimes \cdots \otimes a_i)),$$

$$\tag{5.7}$$

see, for example, [9, 12].

So a morphism of A_∞-algebras $f : B(M, \{m_i\}) \to B(M', \{m'_i\})$ can be defined as a collection of homomorphisms $\{f_i\}$ satisfying the condition (5.7).

These conditions particularly imply $f_1 m_1 = m_1 f_1$, i.e. $f_1 : (M, m_1) \to (M', m'_1)$ is a chain map. We define a *weak equivalence of A_∞-algebras* as a morphism $\{f_i\}$ where f_1 induces homology isomorphism.

It follows easily from the definitions that an A_∞-algebra $(M, \{m_i\})$ with $m_{>2} = 0$ is just a dg algebra, and an A_∞-algebra morphism

$$\{f_i\} : (M, \{m_1, m_2, 0, 0, \dots\}) \to (M', \{m'_1, m'_2, 0, 0, \dots\})$$

with $f_{>1} = 0$ is just a multiplicative chain map, thus the category of dg algebras is a subcategory of A_∞-algebras.

Assigning to $(M, \{m_i\})$ its bar construction (BM, d_m, ∇) we obtain a functor from the category of A_∞-algebras to the category of dg coalgebras.

5.2.1.4   Minimal A_∞-Algebras

An A_∞-algebra $(M, \{m_i\})$ we call *minimal* if $m_1 = 0$; in this case, (M, m_2) is *strictly associative* graded algebra, see (5.3) for $n = 3$. Suppose

$$f : (M, \{m_i\}) \to (M', \{m'_i\})$$

is a weak equivalence of minimal A_∞-algebras, then $f_1 : (M, m_1 = 0) \to (M', m_1' = 0)$, which by definition should induce isomorphism of homology, is automatically an isomorphism. It is not hard to check that, in this case, f is an isomorphism in the category of A_∞-algebras, thus each *weak equivalence of minimal A_∞-algebras is an isomorphism.* This fact motivates the word *minimal* in this notion: the Sullivan's minimal model has similar property—a weak equivalence of minimal dg algebras is an isomorphism.

Let us repeat the minimality theorem from [9] mentioned in Introduction.

Theorem 1 *For a dg algebra (A, d, μ) its homology $H(A)$ (all $H_i(X)$-s are assumed free) can be equipped with a sequence of multioperations*

$$m_i : H(A)^{\otimes i} \to H(A), \ i = 1, 2, 3, \dots ; \ m_1 = 0, \ m_2 = \mu^*$$

turning $(H(A), \{m_i\})$ into a minimal A_∞-algebra in sense of Stasheff [19] for which $m_2 = \mu^*$ and there exists a weak equivalence of A_∞-algebras

$$\{f_i\} : (H(A), \{m_i\}) \to (A, \{m_1 = d, m_2 = \mu, m_3 = 0, m_4 = 0, \dots\}).$$

This A_∞-algebra structure is unique up to isomorphism in the category of A_∞-algebras.
In particular, BA and $(BH(A), d_m)$ have isomorphic homology modules.

5.2.2 Twisting Cochains

Let $(K, d_K, \nabla_K : K \to K \otimes K)$ be a dg coalgebra and (A, d_A, μ) be a dg algebra. Consider the *cochain complex* $\mathrm{Hom}(K, A)$ which is a dg algebra with differential $d\phi = d_A\phi + \phi d_K$, multiplication $\phi \smile \psi = \mu(\phi \otimes \psi)\nabla_K$, and grading $\mathrm{Hom}^q(K, A) = \{f : K_* \to A_{*+q}\}$.

A *twisting cochain* is defined as a homomorphism $\phi : K \to A$ of degree -1 (that is, $\phi : K_* \to A_{*-1}$) satisfying the Brown's condition [4]

$$d\phi = \phi \smile \phi \tag{5.8}$$

(Maurer–Cartan equation, master equation in other terms). Let $T(K, A)$ be the set of all twisting cochains.

Bellow assume that (A, d_A, μ) is *connected*, that is nonnegative $A_{<0} = 0$ and $A_0 = \Lambda$, $d_A|A_1 = 0$. Then a twisting cochain $\phi : K \to A$ can be presented as

$$\phi = \phi_2 + \phi_3 + \cdots + \phi_n + \cdots, \ \phi_n = \phi|K_n : K_n \to A_{n-1},$$

and the Brown's condition for components ϕ_n means

$$d_A\phi_n = \phi_{n-1}d_K + \sum_{i=2}^{n-2} \phi_i \smile \phi_{n-i}.$$

5.2.2.1 Twisting Cochains and the Bar Construction

Because of the universal property (5.4) any homomorphism $\phi : K \to A$ of degree -1 induces a graded coalgebra map $f_\phi : K \to T^c(sA)$ by

$$f_\phi = \sum_i (\phi \otimes \cdots \otimes \phi)\nabla_K^i.$$

If, in addition, $\phi : K \to A$ is a twisting cochain, that is satisfies the Brown's condition (5.8), then $f_\phi : K \to B(A)$ is a chain map, i.e. is a map of dg coalgebras: the projection of the condition $d_B f_\phi = f_\phi d_K$ on A gives $p d_B f_\phi = p f_\phi d_K$ and this is exactly $d\phi = \phi \smile \phi$.

Conversely, any dg coalgebra map $f : K \to BA$ is f_ϕ for $\phi = p \circ f : K \to BA \to A$. In fact we have a bijection $\mathrm{Mor}_{\mathrm{dgcoalg}}(K, BA) \leftrightarrow T(K, A)$.

5.2.2.2 Equivalence of Twisting Cochains

Two twisting cochains $\phi, \psi : K \to A$ are equivalent [3] if there exists $c : K \to A$, $\deg c = 0$, such that

$$\psi = \phi + cd_K + d_Ac + \psi \smile c + c \smile \phi, \qquad (5.9)$$

notation $\phi \sim_c \psi$.

This notion of equivalence allows to *perturb* twisting cochains. Let

$$\phi = \phi_2 + \phi_3 + \cdots + \phi_n + \cdots : K \to A$$

be a twisting cochain, and let's take an arbitrary cochain $c = c_n : K_n \to A_n$. Then there exists a twisting cochain $F_{c_n}\phi = \psi : K \to A$ such that $\phi \sim_{c_n} \psi$. Actually the components of the perturbed twisting cochain

$$F_{c_n}\phi = \psi = \psi_2 + \psi_3 + \cdots + \psi_n + \cdots$$

can be solved from (5.9) inductively, and the solution in particular gives that the perturbation $F_{c_n}\phi$ does not change the first components, i.e. $\psi_i = \phi_i$ for $i < n$ and $\psi_n = \phi_n + d_A c_n$.

5.2.2.3 Equivalence Twisting Cochains and Homotopy of Induced Maps

In the category of dg coalgebras there is the following notion of homotopy: two dg coalgebra maps $f, g : (K, d_K, \nabla_K) \to (K', d_{K'}, \nabla_{K'})$ are homotopic, if there exists $D : K \to K'$ of degree $+1$ such that $d_{K'}D + Dd_k = f - g$, i.e. the chain maps f and g are chain homotopic, and additionally the homotopy D is a $f - g$-coderivation, that is $\nabla_{K'}D = (f \otimes D + D \otimes g)\nabla_K$.

If $\phi \sim_c \psi$, then f_ϕ and f_ψ are homotopic in this sense: chain homotopy $D(c)$: $K \to BA$ is given by

$$D(c) = \sum_{i,j}(\psi \otimes \cdots (j\text{-times}) \cdots \otimes \psi \otimes c \otimes \phi \otimes \cdots \otimes \phi)\nabla_K^i, \qquad (5.10)$$

it satisfies $f_\phi - f_\psi = d_B D(c) + D(c)d_K$ because the projection of this condition on A gives $pf\phi - pf\psi = pd_B D(c) + pD(c)d_K$ and this is exactly the condition (5.9). Besides, $D(c)$ is a $f_\phi - f_\psi$-coderivation, that is

$$\nabla_B D(c) = (f_\psi \otimes D(c) + D(c) \otimes f_\phi)\nabla_K,$$

this also follows from the universal property of tensor coalgebra and the extension rule (5.10).

5.2.2.4 Functor D

Berikasvili's functor $D(K, A)$ is defined as the factorset $D(K, A) = \frac{M(K,A)}{\sim}$. Assigning to a twisting chain $\phi : K \to A$ the dg coalgebra map $f_\phi : K \to BA$ and having in mind that $\phi \sim_c \psi$ implies $f_\phi \sim_{D(c)} f_\psi$ we obtain a bijection $D(K, A) \leftrightarrow [K, BA]$ where $[K, BA]$ denotes the set of chain homotopy classes in the category of dg coalgebras.

Any dg algebra map $f : A \to A'$ induces the map $T(K, A) \to T(K, A')$: if $\phi : K \to M$ is a twisting cochain so is the composition $f\phi : K \to A \to A'$. Moreover, if $\phi \sim_c \phi'$, then $f\phi \sim_{fc} f\phi'$. Thus we have a map $D(f) : D(K, A) \to D(K, A')$.

Theorem 2 (Berikashvili [3]) *Let (K, d_K, ∇_K) be a dg coalgebra with free K_is and (A, d_A, μ_A) be a connected dg algebra. If $f : A \to A'$ is a weak equivalence of connected dg algebras (i.e., homology isomorphism), then*

$$D(f) : D(K, A) \to D(K, A')$$

is a bijection.

In fact this theorem means that $[K, BA] \to [K, BA']$ is a bijection. The theorem consists of two parts:

Surjectivity Any twisting cochain $\phi : K \to A'$ can be lifted to a twisting cochain $\psi : K \to A$ so that $\phi \sim f\psi$.

Injectivity If $\psi \sim \psi' \in T(K, A)$, then $f\psi \sim f\psi \in T(K, A')$.

5.2.2.5 Lifting of Twisting Cochains

Below we'll need the surjectivity part of this theorem whose proof we sketch here.

Theorem 3 *Let* (K, d_K, ∇_K) *be a dg coalgebra with free* $K_i s$ *and let* $f :$ $(A, d_A, \mu_A) \rightarrow (A', d_{A'}, \mu_{A'})$ *be a weak equivalence of connected dg algebras, then for an arbitrary twisting cochain*

$$\phi = \phi_2 + \phi_3 + \cdots + \phi_n + \cdots : K \rightarrow A'$$

there exists a twisting cochain

$$\psi = \psi_2 + \psi_3 + \cdots + \psi_n + \cdots : K \rightarrow A$$

such that $\phi \sim f\psi$.

Proof Start with a twisting cochain $\phi = \phi_2 + \phi_3 + \cdots + \phi_n + \cdots : K \rightarrow A'$. The condition (5.8) gives $d_A\phi_2 = 0$, and since $f : A \rightarrow A'$ is a homology isomorphism then there exist $\psi_2 : K_2 \rightarrow A_1$ and $c'_2 : K_2 \rightarrow A'_2$ such that $d_A\psi_2 = 0$ and $f\psi_2 = \phi_2 + d_{A'}c'_2$ (we assume all K_n's are free modules). Perturbing ϕ by this c'_2 we obtain new twisting cochain $F_{c'_2}\phi$ for which $(F_{c'_2}\phi)_2 = \phi_2 + d_{A'}c'_2 = f\psi_2$. So we can assume that $\phi_2 = f\psi_2$.

Assume now that we already have $\psi_2, \psi_3, \ldots, \psi_{n-1}$ which satisfy (5.8) in appropriate dimensions and $f\psi_k = \phi_k$, $k = 2, 3, \ldots, n - 1$. We need the next component $\psi_n : K_n \rightarrow A_{n-1}$ such that

$$d_A\psi_n = \psi_{n-1}d_K + \sum_{i=2}^{n-2} \psi_i \smile \psi_{n-i}$$

and $c'_n : K_n \rightarrow A'_n$ such that $f\psi_n = \phi_n + d_{A'}c'_n$. Then perturbing ϕ by c'_n we obtain new ϕ_n for which $f\psi_n = \phi_n$, and this will complete the proof.

Let us denote by

$$U_n = \psi_{n-1}d_K + \sum_{i=2}^{n-2} \psi_i \smile \psi_{n-i} : K_n \rightarrow A_{n-2},$$

$$U'_n = \phi_{n-1}d_K + \sum_{i=2}^{n-2} \phi_i \smile \phi_{n-i} : K_n \rightarrow A'_{n-2}.$$

So we have $U'_n = d_{A'}\phi_n$ and we want $\psi_n : K_n \rightarrow A_{n-1}$, $c'_n : K_n \rightarrow A'_n$ such that $U_n = d_A\psi_n$ and $f\psi_n = \phi_n + d_{A'}c'_n$.

First, it is not hard to check that $d_A U_n = 0$ that is U_n maps K_n to cycles $Z(A_{n-2}) \subset A_{n-2}$.

Then

$$
f U_n = f \left(\psi_{n-1} d_K + \sum_{i=2}^{n-2} \psi_i \smile \psi_{n-i} \right) = f \psi_{n-1} d_K + \sum_{i=2}^{n-2} f \psi_i \smile f \psi_{n-i}
$$

$$
= \phi_{n-1} d_K + \sum_{i=2}^{n-2} \phi_i \smile \phi_{n-i} = U'_n = d_A \phi_n,
$$

thus, $f U_n$ maps K_n to boundaries $B(A_{n-2}) \subset A_{n-2}$. Since $f : A \to A'$ is a homology isomorphism, there exist $\overline{\psi}_n : K_n \to A_{n-1}$ such that $d_A \overline{\psi}_n = U_n$.

Now we must take care on the condition $f \psi_n = \phi_n$. It is clear that $d_{A'} f \overline{\psi}_n = d_A \phi_n$. Let us denote by $z'_n = f \overline{\psi}_n - \phi_n$; this is the homomorphism which maps K_n to cycles $Z(A'_{n-1})$. Again, since $f : A \to A'$ is a homology isomorphism there exist $z_n : K_n \to Z(A_{n-1})$ and $c'_n : K_n \to A'_n$ such that $f z_n = z'_n - d_{A'} c'_n$. Let us define $\psi_n = \overline{\psi}_n - z_n$. Then $d_A \psi_n = d_A \overline{\psi}_n$ and

$$
f \psi_n = f \overline{\psi}_n - f z_n = (\phi_n + z'_n) - (z'_n - d_{A'} c'_n) = \phi_n + d_{A'} c'_n.
$$

Perturbing ϕ by this c'_n we obtain $F_{c'_n} \phi$ for which $(F_{c'_n} \phi)_n = \phi_n + d'_A c'_n = f \psi_n$.

5.2.3 A_∞-Twisting Cochains

Now we want to replace in the definition of a twisting cochain a dg algebra (A, d_A, μ) with an A_∞-algebra $(M, \{m_i\})$, see [10, 13].

An A_∞-twisting cochain we define as a homomorphism $\phi : K \to M$ of degree -1 satisfying the condition

$$
\sum_{k=1}^{\infty} m_k (\phi \otimes \cdots \otimes \phi) \nabla_K^k = \phi d_K. \tag{5.11}
$$

Let $T_\infty(K, M)$ be the set of all A_∞-twisting cochains.

We assume that our A_∞-algebras $(M, \{m_i\})$ are connected, that is $M = \{M_1, M_2, \dots \}$. In this case any A_∞-twisting cochain $\phi : K \to M$ has components

$$
\phi = \phi_2 + \phi_3 + \cdots + \phi_n + \cdots : K \to M,
$$

here $\phi_n : K_n \to M_{n-1}$ is the restriction of ϕ on K_n. The condition (5.11) restricted on K_n gives

$$\sum_{i=1}^{\text{int}(n/2)} \sum_{k_1+\cdots k_i=n} m_i(\phi_{k_1} \otimes \cdots \otimes \phi_{k_i}) \nabla_K^k = \phi_{n-1} d_K. \tag{5.12}$$

5.2.3.1 A_∞-Twisting Cochains and the Bar Construction

Because of the universal property (5.4) any homomorphism $\phi : K \to M$ of degree -1 induces a graded coalgebra map, $f_\phi : K \to T^c(sM)$ (coextension of ϕ) by

$$f_\phi = \sum_i (\phi \otimes \cdots \otimes \phi) \nabla_K^i.$$

If, in addition, $\phi : K \to M$ is an A_∞-twisting cochain, that is satisfies the condition (5.11), then $f_\phi : K \to B(M)$ is a chain map, i.e. is a map of dg coalgebras: the projection of the condition $d_B f_\phi = f_\phi d_K$ on M gives $p d_B f_\phi = p f_\phi d_K$ and this is exactly (5.11).

Conversely, any dg coalgebra map $f : K \to BM$ is f_ϕ for $\phi = p \circ f : K \to BM \to M$. In fact we have a bijection $\text{Mor}_{\text{dgcoalg}}(K, BM) \leftrightarrow T_\infty(K, M)$.

5.2.3.2 Equivalence A_∞-Twisting Cochains

Two A_∞-twisting cochains $\phi, \psi : K \to M$ we call equivalent if there exists $c : K \to M$, $\deg c = 0$, such that

$$\psi - \phi = c d_K + \sum_{k,j} m_k(\psi \otimes \cdots (j\text{-times}) \cdots \otimes \psi \otimes c \otimes \phi \otimes \cdots \otimes \phi) \nabla^k, \tag{5.13}$$

notation $\phi \sim_c \psi$.

This notion of equivalence allows us to perturb A_∞-twisting cochains. Let

$$\phi = \phi_2 + \phi_3 + \cdots + \phi_n + \cdots : K \to M$$

be an A_∞-twisting cochain and let's take an arbitrary cochain $c_n : K_n \to M_n$. Then there exists a twisting cochain $F_{c_n}\phi = \psi : K \to M$ such that $\phi \sim_{c_n} \psi$. Actually the components of perturbed twisting cochain

$$F_{c_n}\phi = \psi = \psi_2 + \psi_3 + \cdots + \psi_n + \cdots$$

can be solved from (5.13) inductively, and the solution in particular gives $\psi_i = \phi_i$ for $i < n$ and $\psi_n = \phi_n + m_1 c_n$.

5.2.3.3 Equivalence A_∞-Twisting Cochains and Homotopy of Induced Maps

If $\phi \sim_c \psi$, then f_ϕ and f_ψ are homotopic in the category of dg coalgebras: chain homotopy $D_\infty(c) : K \to BM$ is given by

$$D_\infty(c) = \sum_{i,j} (\psi \otimes \cdots (j\text{-times}) \cdots \otimes \psi \otimes c \otimes \phi \otimes \cdots \otimes \phi)\nabla_K^i, \qquad (5.14)$$

which automatically is a $f_\phi - f_\psi$-coderivation, that is

$$\nabla_B D_\infty(c) = (f_\psi \otimes D_\infty(c) + D_\infty(c) \otimes f_\phi)\nabla_K,$$

and it satisfies $f_\phi - f_\psi = d_B D_\infty(c) + D_\infty(c)d_K$ since the projection of this condition on M gives $pf\phi - pf_\psi = pd_B D_\infty(c) + pD_\infty(c)d_K$ and this is exactly the condition (5.13).

5.2.3.4 Functor D_∞

We denote by $D_\infty(K, M)$ the factorset $D_\infty(K, M) = \frac{T_\infty(K,M)}{\sim}$. Thus we have a bijection $[K, BM] \leftrightarrow D(K, M)$.

Assigning to an A_∞-twisting cochain $\phi : K \to M$ the dg coalgebra map $f_\phi : K \to BM$ and having in mind that $\phi \sim_c \psi$ implies $f_\phi \sim_{D_\infty(c)} f_\psi$ we obtain a bijection $D_\infty(K, A) \leftrightarrow [K, BM]$.

Suppose $f = \{f_i\} : (M, \{m_i\}) \to (M', \{m_i'\})$ is a morphism of A_∞-algebras and $\phi : K \to M$ is an A_∞-twisting cochain. Then it is possible to show that $f(\phi) : K \to M'$ given by

$$f(\phi) = \sum_i f_i(\phi \otimes \cdots \otimes \phi)\nabla_K^i \qquad (5.15)$$

is an A_∞-twisting cochain too. Actually this follows from bar construction interpretation: the composition of dg coalgebra maps

$$B(f) \circ f_\phi : K \to BM \to BM'$$

is the coextension of the twisting cochain $pB(f) \circ f_\phi$ which is exactly $f(\phi)$.

Moreover, if $\phi \sim_c \psi$, then $f(\phi) \sim_{c'} f(\psi)$ with $c' : K \to M'$ given by

$$c' = \sum_{i,j} f_i(\psi \otimes \cdots (j\text{-times}) \cdots \otimes \psi \otimes c \otimes \phi \otimes \cdots \otimes \phi)\nabla_K^i.$$

Again this follows from bar construction interpretation: If $\phi \sim_c \psi$, then $f_\phi, f_\psi : K \to BM$ are homotopic dg coalgebra maps, and the corresponding homotopy

$D_\infty(c)$ is given by (5.14). This implies that the compositions $B(f) \circ f_\phi$, $B(f) \circ f_\psi$: $K \to BM'$ are homotopic by homotopy $B(f) \circ D_\infty(c) : K \to M'$ and the projection $p \circ B(f) \circ D_\infty(c) : K \to BM \to BM' \to M'$ is exactly c'. This gives

$$f(\phi) = pB(f)f_\phi \sim_{c'} pB(f)f_\psi = f(\psi).$$

Thus we have a map $D_\infty(f) : D_\infty(K, M) \to D_\infty(K, M')$.

The following theorem is an analog of Berikashvilis's Theorem 2 for A_∞-algebras and was proved in [10].

Theorem 4 *Let (K, d_K, ∇_K) be a dg coalgebra with free K_is and $(M, \{m_i\})$ be a connected dg algebra. If $f = \{f_i\} : (M, \{m_i\}) \to (M', \{m'_i\})$ is a weak equivalence of A_∞-algebras, then*

$$D_\infty(f) : D_\infty(K, M) \to D_\infty(K, M')$$

is a bijection.

In fact this theorem means that $[K, BM] \to [K, BM']$ is a bijection. The theorem consists of two parts:

Surjectivity Any A_∞-twisting cochain $\phi : K \to M'$ can be lifted to an A_∞-twisting cochain $\psi : K \to M$ so that $\phi \sim f\psi$.

Injectivity If $\psi \sim \psi' \in T_\infty(K, M)$, then $f\psi \sim f\psi \in T_\infty(K, M')$.

5.2.3.5 Lifting of A_∞-Twisting Cochains

Below we'll use the surjectivity part of this theorem whose proof we sketch here.

Theorem 5 *Let (K, d_K, ∇_K) be a dg coalgebra with free K_is and let $f : (M, \{m_i\}) \to (M', \{m'_i\})$ be a weak equivalence of connected A_∞-algebras (that is, $M_0 = M'_0 = 0$). Then for an arbitrary A_∞-twisting cochain*

$$\phi = \phi_2 + \phi_3 + \cdots + \phi_n + \cdots : K \to M'$$

there exists an A_∞-twisting cochain

$$\psi = \psi_2 + \psi_3 + \cdots + \psi_n + \cdots : K \to M$$

such that $\phi \sim f\psi$.

Proof Start with an A_∞-twisting cochain

$$\phi = \phi_2 + \phi_3 + \cdots + \phi_n + \cdots : K \to M'.$$

The condition (5.11) gives $m_1\phi_2 = 0$, and since $f : (M, m_1) \to (M'm'_1)$ is homology isomorphism then there exist $\psi_2 : K_2 \to M_1$ and $c'_2 : K_2 \to M'_2$ such that $m_1\psi_2 = 0$

and $f\psi_2 = \phi_2 + m_1'c_2'$. Perturbing ϕ by this c_2' we obtain $F_{c_2'}\phi$ for which $(F_{c_2'}\phi)_2 = \phi_2 + m_1'c_2' = f\psi_2$.

Assume now that we already have $\psi_2, \psi_3, \ldots, \psi_{n-1}$ which satisfy (5.11) and (5.15) in appropriate dimensions, that is

$$m_1\psi_k = \psi_{k-1}d_K + \sum_{i=2}^{\mathrm{int}(k/2)} \sum_{k_1+\cdots k_i=k} m_i(\psi_{k_1} \otimes \cdots \otimes \psi_{k_i})\nabla_K^k, \quad k = 2, 3, \ldots, n-1$$

$$\tag{5.16}$$

and

$$\phi_k = \sum_{i=1}^{\mathrm{int}(k/2)} \sum_{k_1+\cdots+k_i=k} f_i(\psi_{k_1} \otimes \cdots \otimes \psi_{k_i})\nabla_K^i, \quad k = 1, 2, \ldots, n-1. \tag{5.17}$$

We need the next component $\psi_n : K_n \to A_{n-1}$ satisfying the condition (5.16) for $k = n$

$$m_1\psi_n = \psi_{n-1}d_K + \sum_{i=2}^{\mathrm{int}(n/2)} \sum_{k_1+\cdots k_i=n} m_i(\psi_{k_1} \otimes \cdots \otimes \psi_{k_i})\nabla_K^k, \tag{5.18}$$

and $c_n' : K_n \to A_n'$ such that

$$m_1c_n' + \phi_n = \sum_{i=1}^{\mathrm{int}(n/2)} \sum_{k_1+\cdots+k_i=n} f_i(\psi_{k_1} \otimes \cdots \otimes \psi_{k_i})\nabla_K^i. \tag{5.19}$$

Then perturbing ϕ by c_n' we obtain new ϕ_n for which the condition (5.17) will be satisfied for $k = n$ and this will complete the proof.

Let us put

$$U_n = \psi_{n-1}d_K + \sum_{i=2}^{\mathrm{int}(n/2)} \sum_{k_1+\cdots k_i=n} m_i(\psi_{k_1} \otimes \cdots \otimes \psi_{k_i})\nabla_K^k,$$

$$U_n' = \phi_{n-1}d_K + \sum_{i=2}^{\mathrm{int}(n/2)} \sum_{k_1+\cdots k_i=n} m_i(\phi_{k_1} \otimes \cdots \otimes \phi_{k_i})\nabla_K^k$$

$$V_n = \sum_{i=2}^{\mathrm{int}(n/2)} \sum_{k_1+\cdots+k_i=n} f_i(\psi_{k_1} \otimes \cdots \otimes \psi_{k_i})\nabla_K^i.$$

Then the needed conditions (5.18) and (5.19) look as

$$U_n = m_1\psi_n, \quad m_1'c_n' + \phi_n = f_1\psi_n + V_n.$$

First, it is possible to check that $m_1U_n = 0$ and $f_1U_n = U_n' + m_1'V_n$. Having in mind that $U_n' = m_1'\phi_n$ the last condition means $f_1U_n = m_1'(\phi_n + V_n)$. So U_n is a map to m_1-cycles

$$U_n : K_n \to Z(M_{n-2}) \subset M_{n-2}$$

and $f_1 U_n$ maps K_n to boundaries

$$f_1 U_n : K_n \to B(M'_{n-2}) \subset M_{n-2}.$$

Then since $f_1 : (M, m_1) \to (M', m'_1)$ is a homology isomorphism, there exist $\overline{\psi}_n :$ $K_n \to M_{n-1}$ such that $m_1 \overline{\psi}_n = U_n$. So our $\overline{\psi}_n$ satisfies (5.18). Now let us perturb this $\overline{\phi}$ in order to catch the condition (5.19) too.

Using the above equality $f_1 U_n = m'_1(\phi_n + V_n)$, we have

$$m'_1 f_1 \overline{\psi}_n = f_1 m_1 \overline{\psi}_n = f_1 U_n = m'_1 (\phi_n + V_n)$$

i.e., $m'_1(f_1 \overline{\psi}_n - (\phi_n + V_n)) = 0$. Thus

$$z'_n = (f_1 \overline{\psi}_n - (\phi_n + V_n)) : K_n \to Z(M'_{n-1}),$$

and again, since $f_1 : (M, m_1) \to (M', m'_1)$ is a homology isomorphism there exist $z_n : K_n \to Z(M_{n-1})$ and $c'_n : K_n \to M'_n$ such that $z'_n = f_1 z_n - m'_1 c'_n$. Let's define $\psi_n = \overline{\psi}_n - z_n$. Then

$$f_1 \psi_n = f_1 \overline{\psi}_n - f_1 z_n = (z'_n + \phi_n + V_n) - (z'_n - m'_1 c'_n) = \phi_n + V_n.$$

5.2.4 B_∞-Algebras

The notion of B_∞-algebra was introduced in [2, 8] as an additional structure on a dg module (A, d) which turns the tensor coalgebra $T^c(sA)$ (i.e., the bar construction BA) into a dg bialgebra. So it requires differential

$$\widetilde{d} : BA \to BA,$$

which should be a coderivation with respect to standard coproduct of BA, and a (maybe nonassociative) multiplication

$$\widetilde{\mu} : (BA, \widetilde{d}) \otimes (BA, \widetilde{d}) \to (BA, \widetilde{d})$$

which should be a map of dg coalgebras, with $1_\Lambda \in \Lambda \subset BA$ as a unit element.

It is mentioned above (see also, for example, [9, 12, 18]) that the specification of such $\widetilde{d}$ is equivalent to the specification on A of a structure of A_∞-algebra, namely of a sequence of operations $\{m_i : \otimes^i A \to A, i = 1, 2, 3, \ldots\}$ subject of Stasheff's conditions (5.3).

As for the multiplication

$$\widetilde{\mu} : B(A, \{m_i\}) \otimes B(A, \{m_i\}) \to B(A, \{m_i\})$$

by definition of a dg bialgebra it must be a map of dg coalgebras. Consequently it is uniquely determined by an A_∞-twisting cochain, say

$$E_{*,*} : B(A, \{m_i\}) \otimes B(A, \{m_i\}) \to (A, \{m_i\}). \tag{5.20}$$

In turn such a twisting cochain is represented by a sequence of operations

$$\{E_{pq} : A^{\otimes p} \otimes A^{\otimes q} \to A, \ p, q = 0, 1, 2, 3, \ldots\}$$

satisfying certain coherency conditions together with A_∞ operations $\{m_i\}$ which follow from the Brown's condition (5.11).

So a B_∞-algebra is a graded module equipped with two sets of algebraic multioperations $(A, \{m_i\}, \{E_{p,q}\})$.

The particular case of a B_∞-algebra of type $m_{\geq 3} = 0$ is called a Hirsch algebra, and the particular case of a Hirsch algebra with $E_{>1,q} = 0$ satisfying certain additional conditions is called homotopy Gerstenhaber algebra, see below. We address the reader for more explanations of these structures to [13, 15]. In fact the present description is enough for this paper.

5.2.4.1 Homotopy G-Algebras

A particular case of B_∞-algebra is the *homotopy G-algebra* (hGa in short), the notion was introduced in [6, 20] by M. Gerstenhaber and A. Voronov. This is a dg object whose homology carries a Gerstenhaber algebra structure, that is a commutative multiplication and a Lie bracket well connected to each other. From another point of view a hGa is a dg algebra with "good" $\smile_1$ product. In fact the mentioned Lie bracket is induced by the commutator of $\smile_1$.

Definition A homotopy G-algebra is defined as a dg algebra $(A, d, \cdot)$ with a given sequence of operations

$$E_{1,k} : A \otimes (A^{\otimes k}) \to A, \quad k = 0, 1, 2, 3, \ldots$$

(the value of the operation $E_{1,k}$ on $a \otimes b_1 \otimes \cdots \otimes b_k \in A \otimes (A \otimes \cdots \otimes A)$ we write as $E_{1,k}(a; b_1, \ldots, b_k)$) which satisfies the conditions

$$E_{1,0} = id, \tag{5.21}$$

$$dE_{1,k}(a; b_1, \ldots, b_k) + E_{1,k}(da; b_1, \ldots, b_k) + \sum_i E_{1,k}(a; b_1, \ldots, db_i, \ldots, b_k)$$

$$= b_1 \cdot E_{1,k-1}(a; b_2, \ldots, b_k) + E_{1,k-1}(a; b_1, \ldots, b_{k-1}) \cdot b_k$$

$$+ \sum_i E_{1,k-1}(a; b_1, \ldots, b_i \cdot b_{i+1}, \ldots, b_k), \tag{5.22}$$

$$E_{1,k}(a_1 \cdot a_2; b_1, \ldots, b_k)$$

$$= a_1 \cdot E_{1,k}(a_2; b_1, \ldots, b_k) + E_{1,k}(a_1; b_1, \ldots, b_k) \cdot a_2$$

$$+ \sum_{p=1}^{k-1} E_{1,p}(a_1; b_1, \ldots, b_p) \cdot E_{1,m-p}(a_2; b_{p+1}, \ldots, b_k), \tag{5.23}$$

$$E_{1,n}(E_{1,m}(a; b_1, \ldots, b_m); c_1, \ldots, c_n)$$

$$= \sum_{0 \leq i_1 \leq j_1 \leq \ldots \leq i_m \leq j_m \leq n}$$

$$E_{1,n-(j_1+\cdots+j_m)+(i_1+\cdots+i_m)+m}(a; c_1, \ldots, c_{i_1}, E_{1,j_1-i_1}(b_1; c_{i_1+1}, \ldots, c_{j_1}),$$

$$c_{j_1+1}, \ldots, c_{i_2}, E_{1,j_2-i_2}(b_2; c_{i_2+1}, \ldots, c_{j_2}), c_{j_2+1}, \ldots, c_{i_m},$$

$$E_{1,j_m-i_m}(b_m; c_{i_m+1}, \ldots, c_{j_m}), c_{j_m+1}, \ldots, c_n). \tag{5.24}$$

Let us present these conditions in low dimensions.
The condition (5.22) for $k = 1$ looks as

$$dE_{1,1}(a; b) + E_{1,1}(da; b) + E_{1,1}(a; db) = a \cdot b + b \cdot a. \tag{5.25}$$

So the operation $E_{1,1}$ is a sort of $\smile_1$ product: it is the chain homotopy which measures the noncommutativity of A. Below we denote $E_{1,1}(a; b) = a \smile_1 b$.
The condition (5.23) for $k = 1$ looks as

$$(a \cdot b) \smile_1 c + a \cdot (b \smile_1 c) + (a \smile_1 c) \cdot b = 0, \tag{5.26}$$

this means that the operation $E_{1,1} = \smile_1$ satisfies the *left Hirsch formula*.
The condition (5.22) for $k = 2$ looks as

$$dE_{1,2}(a; b, c) + E_{1,2}(da; b, c) + E_{1,2}(a; db, c) + E_{1,2}(a; b, dc)$$

$$= a \smile_1 (b \cdot c) + (a \smile_1 b) \cdot c + b \cdot (a \smile_1 c), \tag{5.27}$$

this means that this $\smile_1$ satisfies the *right Hirsch formula* just up to homotopy and the appropriate homotopy is the operation $E_{1,2}$.

The condition (5.24) for $n = m = 2$ looks as

$$(a \smile_1 b) \smile_1 c + a \smile_1 (b \smile_1 c) = E_{1,2}(a; b, c) + E_{1,2}(a; c, b), \qquad (5.28)$$

this means that the same operation $E_{1,2}$ measures also the deviation from the associativity of the operation $E_{1,1} = \smile_1$.

5.2.4.2 hGa as a $B(\infty)$-Algebra

Here we show that a hGa structure on A is a particular $B(\infty)$-algebra structure: it induces on $B(A) = (T^c(sA), d_B)$ an *associative* multiplication but does not change the differential d_B (see [5, 8, 13, 15]).

Let us extend our sequence $\{E_{1,k}, \ k = 0, 1, 2, \ldots\}$ to the sequence

$$\{E_{p,q} : (A^{\otimes p}) \otimes (A^{\otimes q}) \to A, \ p, q = 0, 1, \ldots\}$$

adding

$$E_{0,1} = id, \ E_{0,q>1} = 0, \ E_{1,0} = id, \ E_{p>1,0} = 0, \qquad (5.29)$$

and $E_{p>1,q} = 0$.

This sequence defines a map $E : B(A) \otimes B(A) \to A$ by $E([a_1, \ldots, a_m] \otimes [b_1, \ldots, b_n]) = E_{p,q}(a_1, \ldots, a_m; b_1, \ldots, b_n)$. The conditions (5.22) and (5.23) mean exactly $dE + E(d_B \otimes id + id \otimes d_B) = E \smile E$, i.e. E is a twisting cochain. Thus its coextension is a dg coalgebra map, i.e. a correct multiplication

$$\mu_E : B(A) \otimes B(A) \to B(A).$$

The condition (5.24) can be rewritten as $E(\mu_E \otimes id) = E(id \otimes \mu_E)$ and the equality of these twisting cochains implies the equality of their coextensions, that is $\mu_E(\mu_E \otimes id = \mu_E(id \otimes \mu_E)$ so this condition means that the multiplication μ_E is strictly associative.

And the condition (5.29) implies that $[\ ] = 1 \in \Lambda \subset B(A)$ is the unit for this multiplication.

Finally we obtained that $(B(A), d_B, \Delta, \mu_E)$ is a dg bialgebra, thus a hGa is a $B(\infty)$-algebra with strictly associative multiplication.

5.2.4.3 Three Examples of hGa-s

There are three remarkable examples of homotopy G-algebras, see [16].

The first one is the cochain complex of 1-reduced simplicial set $C^*(X)$. The operations $E_{1,k}$ here are dual to cooperations defined by Baues in [2], and the starting operation $E_{1,1}$ is the classical Steenrod's $\smile_1$ product.

The second example is the Hochschild cochain complex $C^*(U, U)$ of an associative algebra U. The operations $E_{1,k}$ here were defined in [11] with the purpose of describing $A(\infty)$-algebras in terms of Hochschild cochains although the properties of those operations which were used as defining ones for the notion of homotopy G-algebra in [6] did not appear there. These operations were defined also in [7]. Again the starting operation $E_{1,1}$ is the classical Gerstenhaber's circle product which is sort of $\smile_1$-product in the Hochschild complex. These operations, denoted as $E_{1,k}(a; , b_1, \ldots, b_k) = a\{b_1, \ldots, b_k\}$ and called *braces*, were used in [17] in the proof of Deligne's hypothesis.

The third example is the cobar construction ΩC of a dg-bialgebra C. The cobar construction ΩC of a DG-coalgebra $(C, d : C \to C, \Delta : C \to C \otimes C)$ is, by definition, a DG-algebra. Suppose now that C is additionally equipped with a multiplication $\mu : C \otimes C \to C$ turning (C, d, Δ, μ) into a DG-bialgebra. How this multiplication μ reflects on the cobar construction ΩC? There arises natural hGa structure, the operations $E_{1,k}$ are constructed in [14]. And again the starting operation $E_{1,1}$ is classical, it is Adams's $\smile_1$-product defined for ΩC in [1].

5.3 B_∞-Algebra Structure in Homology of a hGa

Here we turn to the problem of determining correct multiplication on the bar construction $B(H(A), \{m_i\})$.

Suppose that $(A, d, \mu, \{E_{1,k}\})$ is a hGa. Note that by Sect. 5.2.4.2 the sequence of operations $\{E_{1,k}\}$ determines a twisting cochain $E : BA \otimes BA \to A$.

By the Minimality Theorem [9] there exists on $H(A)$ a structure of minimal A_∞-algebra $(H(A), \{m_i\})$ and a weak equivalence of A_∞-algebras

$$f = \{f_i\} : (H(A), \{m_i\}) \to (A, \{m_1 = d, m_2 = \mu, m_3 = 0, m_4 = 0, \ldots\}).$$
$$(5.30)$$

This weak equivalence induces a weak equivalence of dg coalgebras

$$B(f) : B(H(A), \{m_i\}) \to BA.$$

Composing the tensor product $B(f) \otimes B(f)$ with the twisting cochain $E : BA \otimes BA \to A$ determined by the given hGa structure $\{E_{1,k}\}$ we obtain a twisting cochain

$$E \circ (B(f) \otimes B(f)) : B(H(A), \{m_i\}) \otimes B(H(A), \{m_i\}) \to BA \otimes BA \to A. \quad (5.31)$$

Now, using the lifting Theorem 5, let us lift this twisting cochain along the weak equivalence of A_∞-algebras (5.30). We obtain an A_∞-twisting cochain

$$E_{*,*} : B(H(A), \{m_i\}) \otimes B(H(A), \{m_i\}) \to (H(A), \{m_i\})$$

which in turn defines multiplication on $B(H(A), \{m_i\})$, a dg coalgebra map

$$\mu_{E_{*,*}} : B(H(A), \{m_i\}) \otimes B(H(A), \{m_i\}) \to B(H(A), \{m_i\})$$

i.e., a B_∞-algebra structure on $(H(A), \{m_i\})$.

These twisting cochains can be observed from the diagram

$$
\begin{array}{ccc}
B(H(A), \{m_i\}) \otimes B(H(A), \{m_i\}) & \xrightarrow{E_{*,*}} & (H(A), \{m_i\})) \\
B(f) \otimes B(f) \downarrow & & \downarrow f \\
BA \otimes BA & \xrightarrow{E} & A.
\end{array}
$$

This diagram does not commute but the twisting cochains $f \circ E_{*,*}$ and $E \circ (B(f) \otimes B(f))$ are equivalent. Consequently the following diagram of induced dg coalgebra maps commutes up to homotopy

$$
\begin{array}{ccc}
B(H(A), \{m_i\}) \otimes B(H(A), \{m_i\}) & \xrightarrow{\mu_{E_{*,*}}} & B(H(A), \{m_i\})) \\
B(f) \otimes B(f) \downarrow & & \downarrow B(f) \\
BA \otimes BA & \xrightarrow{\mu_E} & BA.
\end{array}
$$

Thus the dg coalgebra weak equivalence

$$B(f) : B(H(A), \{m_i\}, \{E_{p,q}\}) \to B(A, d, \mu, \{E_{1,k}\})$$

is multiplicative up to homotopy. Consequently it induces multiplicative isomorphism of their homology algebras

$$H(B(H(A), \{m_i\}, \{E_{p,q}\})) \approx H(B(A, d, \mu, \{E_{1,k}\})).$$

Finally we have the

Theorem 6 *Let $(A, d, \mu, \{E_{1,k}\})$ be a hGa. Then on its homology $H(A)$ there exists a structure of B_∞-algebra $(H(A), \{m_i\}, \{E_{p,q}\})$ such that homology algebras $H(B(H(A), \{m_i\}, \{E_{p,q}\}))$ and $H(B(A, d, \mu, \{E_{1,k}\}))$ are isomorphic.*

Remark For a hGa $(A, d, \mu, \{E_{1,k}\})$ the twisting cochain $E : BA \otimes BA \to A$ satisfies additional conditions (5.24) which guarantee that the induced multiplication on BA is associative. The twisting cochain $E_{*,*}$ we have obtained satisfies only Brown's condition, but not that condition for associativity, so the obtained multiplication

$$\mu_{E_{*,*}} : B(H(A), \{m_i\}) \otimes B(H(A), \{m_i\}) \to B(H(A), \{m_i\})$$

is a chain map, but nonassociative generally. Thus the bar construction $B(H(A), \{m_i\})$ is a nonassociative bialgebra. We expect that this nonassociative multiplication will be a part of certain A_∞ algebra structure on $B(H(A), \{m_i\})$ which will allow to iterate the process.

5.4 Cohomology Algebra of the Loop Space

Here we turn to the main goal of this paper, cohomology *algebra* $H^*(\Omega X)$.

Start with a topological space X and take $(A, d, \mu) = (C^*(X), d, \smile)$, the cochain complex of X. Then, by Adams [1] we have that $H(BA) = H(BC^*(X))$ and $H^*(\Omega X)$ are isomorphic as graded *modules*.

As for the multiplicative structure of $H^*(\Omega X)$, in [2] Baues constructed a sequence of multioperations

$$E_{1,k} : C^*(X) \otimes C^*(X)^{\otimes k} \to C^*(X), \ k = 1, 2, 3, \ldots$$

turning $(C^*(X), d, \smile, \{E_{1,k}\})$ into a hGa, which determines on the bar construction $BA = BC^*(X)$ correct multiplication, so that we have isomorphism of graded algebras

$$H(B(C^*(X), d, \smile, \{E_{1,k}\})) \approx H^*(\Omega X).$$

Now we are going to transfer these structures to cohomology level, i.e. from $C^*(X)$ to $H^*(X)$.

Firstly, according to minimality theorem [9], on the homology of dg algebra $(C^*(X), d, \smile)$ appears a structure of minimal A_∞-algebra $(H^*(X), \{m_i\})$ and a weak equivalence of A_∞-algebras

$$\{f_i\} : (H^*(X), \{m_i\}) \to (C^*(X), \{m_1 = d, \ m_2 = \smile, \ m_3 = 0, \ \ldots \}),$$

which induces weak equivalences of their bar constructions and isomorphism of graded modules

$$H(B(H^*(X), \{m_i\})) \approx H(BC^*(X)) \approx H^*(\Omega X).$$

Thus this object $(H^*(X), \{m_i\})$, which is called cohomology A_∞-algebra of X, determines cohomology *modules* of the loop space $H^*(\Omega X)$. But not the cohomology *algebra* $H^*(\Omega X)$.

Now, to obtain correct multiplication on $B(H^*(X), \{m_i\})$ we use the above theorem, which for $(A, d, \mu) = (C^*(X), d, \smile)$ gives

Theorem 7 *Let* $(C^*(X), d, \smile, \{E_{1,k}\})$ *be a hGa structure on the cochain complex of a topological space X equipped with Baues's operations $\{E_{1,k}\}$. Then on there*

exists on $H^*(X)$ structure of B_∞-algebra $(H^*(X), \{m_i\}, \{E_{p,q}\})$ which determines cohomology algebra of the loop space, that is there exists isomorphism of graded algebras

$$H(B(H^*(X), \{m_i\}, \{E_{p,q}\})) \approx H^*(\Omega X).$$

Acknowledgements The author was supported by the European Union's Seventh Framework Programme (FP7/2007–2013) under grant agreement no. 317721.

References

1. J. Adams, On the non-existence of elements of Hpof invariant one. Ann. Math. **72**, 20–104 (1960)
2. H. Baues, The double bar and cobar construction. Compos. Math. **43**, 331–341 (1981)
3. N. Berikashvili, On the differentials of spectral sequence. Proc. Tbil. Math. Inst. **51**, 1–105 (1976)
4. E. Brown, Twisted tensor products. Ann. Math. **69**, 223–246 (1959)
5. M. Franz, Tensor product of homotopy Gerstenhaber algebras (2011). ArXiv:1009.1116v2 [math.AT]
6. M. Gerstenhaber, A. Voronov, Higher-order operations on the Hochschild complex. Funct. Anal. i Prilojen. **29**(1), 1–6 (1995)
7. E. Getzler, Cartan homotopy formulas and the Gauss-Manin connection in cyclic homology. Isr. Math. Conf. Proc. **7**, 65–78 (1993)
8. E. Getzler, J.D. Jones, Operads, homotopy algebra, and iterated integrals for double loop spaces. Preprint (1994)
9. T. Kadeishvili, On the homology theory of fibrations. Russ. Math. Surv. **35**(3), 231–238 (1980)
10. T. Kadeishvili, Predifferential of a fiber bundle. Russ. Math. Surv. **41**(6), 135–147 (1986)
11. T. Kadeishvili, The $A(\infty)$-algebra structure and cohomology of Hochschild and Harrison. Proc. Tbil. Math. Inst. **91**, 19–27 (1988)
12. T. Kadeishvili, $A(\infty)$ -algebra structure in cohomology and rational homotopy type. Proc. Tbil. Mat. Inst. **107**, 1–94 (1993)
13. T. Kadeishvili, Measuring the noncommutativity of DG-algebras. J. Math. Sci. **119**(4), 494–512 (2004)
14. T. Kadeishvili, On the cobar construction of a bialgebra. Homol. Homotopy Appl. **7**(2), 109–122 (2005). Preprint at math.AT/0406502
15. T. Kadeishvili, Twisting elements in homotopy G-algebras, in *Higher Structures in Geometry and Physics. Honor of Murray Gerstenhaber and Jim Stasheff*. Progress in Mathematics, vol. 287 (Birkhauser, Boston, 2011), pp. 181–200. arXiv:math/0709.3130
16. T. Kadeishvili, Homotopy Gerstenhaber algebras: examples and applications. J. Math. Sci. **195**(4), 455–459 (2013)
17. J. McLure, J. Smith, A solution of Deligne's conjecture. Preprint (1999). math.QA/9910126
18. V. Smirnov, Homology of fiber spaces. Russ. Math. Surv. **35**(3), 294–298 (1980)
19. J.D. Stasheff, Homotopy associativity of H-spaces. I, II. Trans. Am. Math. Soc. **108**, 275–312 (1963)
20. A. Voronov, Homotopy Gerstenhaber algebras (1999). QA/9908040

Chapter 6
Half-Automorphisms of Cayley–Dickson Loops

Maria de Lourdes Merlini-Giuliani, Peter Plaumann, and Liudmila Sabinina

To the memory of Karl Strambach

In the year 1957 W.R. Scott introduced in [28] the concept of a half-homomorphism between semi-groups. He calls a mapping $\varphi : S_1 \to S_2$ between semi-groups S_1 and S_2 a *half-homomorphism* if the alternative

$$\forall x, y \in S_1 : \ \varphi(xy) = \varphi(x)\varphi(y) \text{ or } \varphi(xy) = \varphi(y)\varphi(x) \qquad (6.1)$$

holds and proves the following

Theorem *Every half-isomorphism of a cancellation semi-group S_1 into a cancellation semi-group S_2 is either an isomorphism or an anti-isomorphism.*

We call a half-isomorphism *proper* if it is neither an isomorphism nor an anti-isomorphism. Then one finds in Scott's paper the following result which in this article we call **Scott's Theorem**.

Theorem *There is no proper half-homomorphism from a group G_1 into a group G_2.* In the article of Scott references are given to Hua's Theorem [16] and to Jordan homomorphisms [17] which give a hint where the origins of the notion of a half-automorphism are to be searched (see also [26]).

At the end of his short article Scott gives an example of a loop S of order 8 which possesses a proper half-automorphism. Recently a systematic research on the

M.d.L. Merlini-Giuliani
Universidade Federal do ABC, Rua Santa Adélia 166, Santo André, SP 09210-170, Brazil
e-mail: maria.giuliani@ufabc.edu.br

P. Plaumann
Universidad Autónoma "Benito Juárez" de Oaxaca, Oaxaca de Juárez, Mexico
e-mail: peter.plaumann@mi.uni-erlangen.de

L. Sabinina (✉)
Universidad Autónoma del Estado de Morelos, Av. Universidad 1001, Cuernavaca 62209, Mexico
e-mail: liudmila@uaem.mx

© Springer International Publishing AG 2017   109
G. Falcone (ed.), *Lie Groups, Differential Equations, and Geometry*,
UNIPA Springer Series, DOI 10.1007/978-3-319-62181-4_6

existence of proper half-automorphisms of Moufang loops has begun (see [9, 10, 12, 13]). In a new manuscript [19] Kinyon et al. have shown that Scott's Theorem holds for a class of Moufang loops which contains all automorphic Moufang loops.

Petr Vojtechovsky gave us an example of a loop V in which the inversion map J satisfying $x^J x = 1 = xx^J$ for all $x \in V$ exists (Vojtechovsky, July 2015, private communication). The Cayley table of V is

1	2	3	4	5	6	7	8
2	1	4	7	8	5	3	6
3	5	8	1	6	4	2	7
4	3	1	8	2	7	6	5
5	4	2	6	7	8	1	3
6	7	5	3	4	1	8	2
7	8	6	2	1	3	5	4
8	6	7	5	3	2	4	1

It is easy to check that the inversion mapping J is a proper half-automorphism of the loop V. Note that V is not a diassociative loop since in diassociative loops the inversion always is an anti-automorphism.

In this note we exclusively study half-automorphisms of diassociative 2-loops. A special type of half-automorphisms, which we call elementary maps [see (6.5)], plays a particular role. We will use them as a part of a generating system of the group of all half-automorphisms of the loops considered, if this is possible.

Sections 6.1 and 6.3 of this article are used to describe topics which are background knowledge. In Sect. 6.1 we collect fundamentals of the Theory of Loops which we need. Would this be an article on Group Theory, this part would not be necessary since these facts belong to common knowledge for the variety of groups. We try to write there only things which are necessary for our considerations. In Sect. 6.3 we present the well-known so-called Cayley–Dickson construction of a sequence of algebras A_n of dimension $2^n, n > 0$. We restrict ourselves to the classical case of algebras over the field of real numbers where the multiplication is given by (6.9). The well-known generalizations of this definition we do not need here. The material in Sect. 6.3 is included in our text mostly to fix notations in such a way that for the iterated doubling process one obtains canonical bases $\mathfrak{B}_n$ for the algebras A_n which allow to assume that $A_n < A_{n+1}$ for all n.

In Sect. 6.4 we turn our view to the fact that sets $C_n = \mathfrak{B}_n \cup (-\mathfrak{B}_n)$ are closed under multiplication in A_n. Hence they form an infinite sequence of loops C_n which are called Cayley–Dickson loops. Note that this does not contradict the fact that Cayley–Dickson algebras of dimension $\geqslant 2^4$ have zero divisors. Cayley–Dickson loops in particular the octonion loop C_4 and the sedenion loop C_5 have been considered by several authors. We mention here publications by Kivunge and Smith [21], Kinyon et al. [18, Sect. 7], Cawagas et al. [5] and Kirshtein [20].

Already before, in Sect. 6.2, we had presented the octonion loop C_4 of order 16 which is small enough to be treated by direct computation. Theorem 6.2.2, which includes the fact that C_4 is a Moufang loop, allows us to describe the group of all half-automorphisms of C_4 (Theorem 6.2.5). In the proof of this theorem mappings $C_n \to C_n$ which we call elementary mappings play a particular role.

For all Cayley–Dickson loops we have with Theorem 6.4.3 a result analogous to Theorem 6.2.2. This allowed us to prove the existence of elementary half-automorphisms for every Cayley–Dickson loop in Theorem 6.4.5. We could, however, not determine the full structure of the group of half-automorphisms for all Cayley–Dickson loops.

In Sect. 6.4 we also discuss the existence of proper half-automorphisms of code loops (see [11]). Since the octonion loop is an example of a code loop (see [14, 15]) the elementary half-automorphisms should play a great role in the class of code loops.

6.1 Basic Facts About Loops

The fundamental definitions and facts from the Theory of Loops are found in the monographs of Bruck [4] and of Pflugfelder [23]. We give a short description of facts that we need. We call a set $(S, \circ)$ with a multiplication

$$(a, b) \mapsto a \circ b : S \times S \to S$$

a *magma*. Consider the *left translations*

$$L_a = (x \mapsto a \circ x) : S \to S$$

and the *right translations*

$$R_b = (x \mapsto x \circ b) : S \to S.$$

Definition 6.1.1 A **loop** $(L, \circ)$ is a magma such that

(1) There is an identity element $1 \in L$ such that $1 \circ x = x = x \circ 1$ for all $x \in L$.
(2) All mappings L_a, R_b are bijections.

Given a loop L put $x/y := R_y^{-1}(x)$ and $y \backslash x := L_y^{-1}(x)$. Then one has a universal algebra $(L, \circ, /, \backslash, 1)$ of type $(3, 0, 1)$.

Definition 6.1.2 A loop L is a **Moufang loop** if it satisfies the equivalent identities

$$((xy)x)z = x(y(xz)) \tag{6.2}$$

$$((xy)z)y = x(y(zy)) \tag{6.3}$$

$$(xy)(zx) = (x(yz))x \tag{6.4}$$

A loop L is called *diassociative* if all subloops $\langle x, y \rangle$ generated by elements $x, y \in L$ are groups. Similarly a loop Q is called *power-associative* if all subloops of the form $\langle x \rangle$, $x \in Q$, are groups. In [23] one finds the

Proposition 6.1.3 *All Moufang loops are diassociative.*

Let L be a diassociative loop. Then $\langle x \rangle$ is a group. Hence there is a unique element $x' \in L$ such that $xx' = 1 = x'x$. Considering the group $\langle x, y \rangle$ one sees $L_y^{-1} = L_{y'}$ and $R_y^{-1} = R_{y'}$. It follows that $x/y = xy'$ and $y \backslash x = y'x$. Thus a diassociative loop can be considered as a universal algebra of type $(1, 1, 1)$ as one usually does with groups. In this article we shall do this permanently.

Definition 6.1.4 A loop Q is called a **C-loop** if the identity

$$x(y \cdot yz) = (xy \cdot y)z$$

holds in Q.

Information about C-loops one finds in [24] and [18]. Remembering that a loop is called *flexible* if the identity $x(yx) = (xy)x$ holds one finds there the

Proposition 6.1.5 *A flexible C-loop is diassociative.*

For a loop Q one considers the *multiplication groups*

$$
\begin{aligned}
&\mathsf{LMult}(Q) = \langle L_a \mid a \in Q \rangle && \text{(left multiplication group of } Q\text{)}, \\
&\mathsf{RMult}(Q) = \langle R_b \mid b \in Q \rangle && \text{(right multiplication group of } Q\text{)}, \\
&\mathsf{Mult}(Q) = \langle \mathsf{LMult}(Q) \cup \mathsf{RMult}(Q) \rangle && \text{(multiplication group of } Q\text{)}.
\end{aligned}
$$

These groups act transitively on Q and give us reason to consider the *inner mapping groups*

$$
\begin{aligned}
&\mathsf{LInt}(Q) = \{\Phi \in \mathsf{LMult}(Q) \mid \Phi(1) = 1\} && \text{(left inner mapping group of } Q\text{)}, \\
&\mathsf{RInt}(Q) = \{\Phi \in \mathsf{RMult}(Q) \mid \Phi(1) = 1\} && \text{(right inner mapping group of } Q\text{)}, \\
&\mathsf{Int}(Q) = \{\Phi \in \mathsf{Mult}(Q) \mid \Phi(1) = 1\} && \text{(inner mapping group of } Q\text{)}.
\end{aligned}
$$

Definition 6.1.6 A subloop of a loop Q is **normal** in Q if and only if it is invariant under the group $\mathsf{Int}(Q)$.

We use the notation $N \lhd Q$ to indicate that N is a normal subloop of Q.

Definition 6.1.7 Let Q be a loop and let η be a loop homomorphism defined on Q. Then the set

$$\ker \eta = \{x \in Q \mid \eta(x) = 1\}$$

is called the **kernel** of η.

Proposition 6.1.8 *A subloop of a loop is normal if and only if it is the kernel of a homomorphism.*

Definitions 6.1.6 and 6.1.7 together with Proposition 6.1.8 are the fundament for the homomorphism theory of loops which resembles the homomorphism theory for groups (see [1, 3]).

If the loop Q is a group, then the elements of $\mathsf{Int}(Q)$ are just the inner automorphisms of Q. For Moufang loops in general the inner mappings are *pseudo-automorphisms* (see [23]).

Definition 6.1.9 Let Q be a diassociative loop and let $\varphi : Q \to Q$ be a bijective mapping. If

$$(\text{SA}) \quad \varphi(yxy) = \varphi(y)\varphi(x)\varphi(y)$$

holds for all $x, y \in Q$, then φ is called a **semi-automorphism** of Q.

All inner mappings of a Moufang loop are semi-automorphisms (see [23]). It is easy to see that for a diassociative loop all half-automorphisms are semi-automorphisms as well (see also [9]).

In particular, characteristic subloops, that is subloops which are invariant under all automorphisms, in general not are normal.

We denote the group of automorphisms of a loop L by $\mathsf{Aut}(L)$.

Definition 6.1.10 A loop L is called **left (right) automorphic** if all left (right) inner mappings of L are automorphisms. If $\mathsf{Int}(L) \subseteq \mathsf{Aut}(L)$, then L is called **automorphic**.

In an automorphic loop all characteristic subloops are normal, a fact which makes many considerations easier.

We now define some important characteristic respectively normal characteristic subloops of a given loop L. Detailed information is found in [4] and [23].

1. The *commutator subloop* of a loop L.
   For $x, y \in L$ the *commutator* is $[x, y] = xy \backslash yx$
   $[L, L]$—**normal** subloop generated by all commutators.
   Note that $[L, L]$ is the smallest normal subloop N of L such that L/N is a commutative loop and that $xy/yx \in [L, L]$ for all $x, y \in L$.
2. The *associator subloop* of a loop L.
   For $x, y, z \in L$ the *associator* is $(x, y, z) = x(yz) \backslash (xy)z$
   (L, L, L)—**normal** subloop generated by all associators.
   Note that (L, L, L) is the smallest normal subloop N of L such that L/N is a group and that $(xy)z \backslash x(yz) \in (L, L, L)$ for all $x, y, z \in L$.
3. We call $\Delta(L) = (L, L, L) \cap [L, L]$ the *radical-defect*.
4. The nuclei of a loop.
   The *left nucleus* is $N_l = \{u \in L \mid (u, x, y) = 1 \text{ for all } x, y \in L\}$.
   Analogously one defines the *middle nucleus* N_m and the *right nucleus* N_r .
   The *nucleus* of L is $N(L) = N_l \cap N_m \cap N_r$
5. The *commutant* of a loop L is $C(L) = \{c \in L \mid (\forall x \in L) \, [c, x] = 1\}$
6. The *center* of a loop L
   $_3L$—**normal** closure of the subloop $C(L) \cap N(L)$.

7. The *Frattini subloop* of a loop L.

$\Phi(L)$—**normal** closure of the intersection of all maximal subloops.

Note 1 For a Moufang loop M one can show that the center and $N(M) = N_l(M) = N_m(M) = N_r(M)$ are normal [4]. The same holds for the commutant (see [8]).

At the end of this section we cannot resist the temptation to cite R. Baer who writes in the article [2]: "It is not the object of this address to introduce you to new theories or to tell of great discoveries."

6.2 Some Small Moufang Loops

1. Chein Loops In 1974 O. Chein found a construction to obtain from a given group G a Moufang loop $M(G, 2)$. He used this construction efficiently in the classification of finite Moufang loops of small order. For a finite group G and a group of order two $\langle u \rangle$ such that $u \notin G$ define on the set $M(G, 2)) = G \times \langle u \rangle$ a multiplication by

$$(g, 1)(h, 1) = (gh, 1)$$

$$(g, u)(h, 1) = (gh^{-1}, u)$$

$$(g, 1)(h, u) = (hg, u)$$

$$(g, u)(h, u) = (h^{-1}g, 1)$$

Theorem 6.2.1 (O. Chein [6]) *If G is a finite group, then $M(G, 2)$ is a Moufang loop and $G \times \langle 1 \rangle$ is a normal subloop of $M(G, 2)$. Furthermore, $M(G, 2)$ is a group if and only if the group G is abelian.* **q.e.d.**

In [10] one finds much information about half-automorphisms of Chein loops.

2. Moufang Loops of Order 16 The LOOPS package of GAP[1] contains a library of loops of small order. There one finds 5 Moufang loops of order 16 which are not groups.

MoufangLoop(16,1)	-	the Chein double of the dihedral group D_8
MoufangLoop(16,2)	-	the Chein double of the quaternion group Q_8
MoufangLoop(16,3)	-	the "octonion loop" o
MoufangLoop(16,4)	-	
MoufangLoop(16,5)	-	

For 4 of these Moufang, the loops MoufangLoop(16,1), MoufangLoop(16,2), MoufangLoop(16,4), and MoufangLoop(16,5), we have no satisfactory theory of the group of half-automorphisms. In [10, Theorem 7] it is

---

[1]The GAP–Groups, Algorithms, and Programming, Version 4.4.10.

shown that a Moufang loop of even order which admits a proper half-automorphism contains elements of order 4, but this does not help us. For MoufangLoop(16,2) in [10, Example 8] the existence of a proper half-automorphism is shown.

We can, however, easily determine the full group of half-automorphisms of MoufangLoop(16,3) in detail.

3. The Octonion Loop o. The octonion loop MoufangLoop(16,3) carries its name since it is closely connected with the classical alternative division algebra $\mathbb{O}$ of dimension 8 over the field $\mathbb{R}$ of real numbers. To obtain this finite loop one considers the standard basis $\mathfrak{B} = (1 = b_1, \ldots, b_8)$ of the vector space $\mathbb{O}$ over $\mathbb{R}$. Put $L = \{\pm b_1, \ldots, \pm b_8\}$. Then L is closed under the octonion multiplication; below one sees the multiplication table of L in some ordering of $\mathfrak{B}$.

Studying Table 6.1 it is easy to get the following

Theorem 6.2.2 *In the octonion loop* o *the following statements are true:*

(1) o *is a loop of order* 16
(2) o *is a Moufang loop.*
(3) *In* o *there are one element* $z = -1$ *of order* 2 *and* 14 *elements of order* 4.
(4) $zx = xz$ *for all* $x \in$ o.
(5) *If H is a subloop of* o, *then* $z \in H$.
(6) *Z is a normal subloop of* o *which coincides with the nucleus, the center, the associator subloop, the commutator subloop and the Frattini subloop of* o.
(7) o/Z *is an elementary abelian group of order* 8.   **q.e.d.**

Some details of the following construction hold for arbitrary diassociative loops.

Table 6.1 Cayley table of octonion loop

o	1	2	3	4	5	6	7	8	9	10	11	12	13	14	15	16
1	1	2	3	4	5	6	7	8	9	10	11	12	13	14	15	16
2	2	4	8	6	3	1	5	7	14	9	16	10	11	12	13	15
3	3	5	4	7	6	8	1	2	15	13	9	11	14	16	12	10
4	4	6	7	1	8	2	3	5	12	14	15	9	16	10	11	13
5	5	7	2	8	4	3	6	1	13	11	14	16	12	15	10	9
6	6	1	5	2	7	4	8	3	10	12	13	14	15	9	16	11
7	7	8	1	3	2	5	4	6	11	16	12	15	10	13	9	14
8	8	3	6	5	1	7	2	4	16	15	10	13	9	11	14	12
9	9	10	11	12	16	14	15	13	4	6	7	1	5	2	3	8
10	10	12	16	14	15	9	13	11	2	4	5	6	3	1	8	7
11	11	13	12	15	10	16	9	14	3	8	4	7	6	5	1	2
12	12	14	15	9	13	10	11	16	1	2	3	4	8	6	7	5
13	13	15	10	16	9	11	14	12	8	7	2	5	4	3	6	1
14	14	9	13	10	11	12	16	15	6	1	8	2	7	4	5	3
15	15	16	9	11	14	13	12	10	7	5	1	3	2	8	4	6
16	16	11	14	13	12	15	10	9	5	3	6	8	1	7	2	4

Definition 6.2.3 Let Q be a diassociative loop and let c be an element of Q. Then the mapping $\tau_c : Q \to Q$ defined by

$$\tau_c(x) = \begin{cases} x^{-1} & \text{if } x \in \{c, c^{-1}\} \\ x & \text{if } x \notin \{c, c^{-1}\} \end{cases} \tag{6.5}$$

is called an **elementary mapping**.

Obviously the following proposition holds.

Proposition 6.2.4 *For a diassociative loop Q of exponent 4 put*

$$X = \{x \in Q \mid x^2 \neq 1\}.$$

Then

(1) If $c \in X$, then the mapping τ_c is an involution.
(2) For $a \neq b \in X$ one has $\tau_a = \tau_b$ if and only if $ab = 1$.
(3) For $a, b \in X$ the group $\langle \tau_a, \tau_b \rangle$ is abelian.   **q.e.d.**

In $\mathfrak{o}$ there are 14 elements of order 4 which give us precisely 7 different sets $\{x, x^{-1}\}$ with 2 elements. The set $\mathscr{I}$ of these 2-element sets consists of

$$\{2, 6\}, \{3, 7\}, \{8, 5\}, \{9, 12\}, \{14, 10\}, \{15, 11\}, \{16, 13\}. \tag{6.6}$$

To study one typical example choose the pair $\{8, 5\} \in \mathscr{I}$ and consider the transposition $\tau = (5, 8)$ in the symmetric group S_{14}.

One computes
$$\tau(2 \cdot 7) = \tau(5) = 8 = \tau(7) \cdot \tau(2) \neq \tau(2) \cdot \tau(7) = 5,$$
$$\tau(3 \cdot 9) = \tau(15) = 15 = \tau(3) \cdot \tau(9) \neq \tau(9) \cdot \tau(3) = 11.$$
Thus τ is a proper half-automorphism of $\mathfrak{o}$.

Because of the symmetry of $\mathfrak{o}$ every transposition of a pair from the list (6.6) is a proper half-automorphism. One can verify this fact directly using Table 6.1. This way one obtains seven commuting proper half-automorphisms of $\mathfrak{o}$ of order 2. Denote these seven transpositions by $\rho_1, \ldots, \rho_7$. They are elementary mappings and generate a group $\Gamma = \Gamma(\mathfrak{o})$ which is an elementary abelian 2-group of order 2^7. Thus an inspection of Table 6.1 gives us the

Theorem 6.2.5 *Let c be an element of order 4 of the octonion loop $\mathfrak{o}$. Then the elementary map τ_c is a proper half-automorphism of $\mathfrak{o}$.*   **q.e.d.**

Theorem 6.2.6 *Let $\mathfrak{o}$ be the octonion loop and Γ be the group generated by all elementary maps. Then every half-automorphism of $\mathfrak{o}$ is the product of an automorphism of $\mathfrak{o}$ and an element $\gamma \in \Gamma$.*

Proof Suppose that $\varphi : \mathfrak{o} \to \mathfrak{o}$ is a proper half-automorphism. Since $\mathfrak{o}$ is a Moufang loop by Gagola III and Merlini Giuliani [9] thus there exists a triple (x, y, z) such that φ restricted to $\langle x, y \rangle$ is an automorphism, φ restricted to $\langle x, z \rangle$ is an anti-automorphism, and $[x, y] \neq 1, [x, z] \neq 1$. Since φ is proper, the loop $\langle x, y, z \rangle$ cannot be a group by the theorem of Scott. Hence $\langle x, y, z \rangle = \mathfrak{o}$. In what follows we always will consider these generators x, y, z of the loop $\mathfrak{o}$. In particular $yz \neq zy$. In $\mathfrak{o}$ there are 16 elements:

$$\{\pm 1, \pm x, \pm y, \pm z, \pm xy, \pm xz, \pm yz, \pm x(yz)\}.$$

Note that for any half-automorphism ψ of $\mathfrak{o}$ and for any $t \in \mathfrak{o}$ one has:

$$\psi(1) = 1, \quad \psi(-t) = -\psi(t)$$

since -1 is the unique element of order 2 in $\mathfrak{o}$ and central. Note also that the following relations hold:

$$xy = -yx, \quad xz = -zx, \quad yz = -zy,$$

$$x(yz) = z(xy) = y(zx) = -(yz)x = -(xy)z = -(zx)y.$$

In order to understand the behavior of proper half-automorphism φ on $\mathfrak{o}$ it is enough to consider only the set $S = \{x, y, z, xy, xz, yz, x(yz)\}$ of seven elements and to analyze the behavior of φ on all possible subgroups generated by two elements of S. There are 21 different pairs, some of them generate the same subgroups. Indeed:

(i) $\langle x, y \rangle = \langle x, xy \rangle = \langle xy, y \rangle$

(ii) $\langle x, z \rangle = \langle x, xz \rangle = \langle xz, z \rangle$

(iii) $\langle y, z \rangle = \langle y, yz \rangle = \langle yz, z \rangle$

(iv) $\langle x, yz \rangle = \langle x, x(yz) \rangle = \langle yz, x(yz) \rangle$

(v) $\langle y, xz \rangle = \langle y, x(yz) \rangle = \langle xz, x(yz) \rangle$

(vi) $\langle z, yx \rangle = \langle z, x(yz) \rangle = \langle yx, x(yz) \rangle$

(vii) $\langle xy, yz \rangle = \langle xy, xz \rangle = \langle xz, yz \rangle$

One obtains (vii) using the fact that $u^2 = [u, v] = (u, v, t) = -1$ and the Moufang identity in $\mathfrak{o}$ in the following form. For any elements $u, v, t \in S$ we have:

$$ut \cdot vt = -tu \cdot vt = -(t(uv)t) = t^2 \cdot (uv) = -(uv) \tag{6.7}$$

Thus we get seven different 2-generated subgroup of $\mathfrak{o}$, every one of them is isomorphic to Q_8. For example, $\langle x, y \rangle = \{\pm 1, \pm x, \pm y, \pm xy\}$.

There are two possibilities: φ restricted to $\langle y, z \rangle$ is an automorphism or φ restricted to $\langle y, z \rangle$ is an antiautomorphism.

Case 1 φ restricted to $\langle y, z \rangle$ is an automorphism.

Consider the mapping $\varphi \circ \tau_{xz} = \varphi_1$.

We see that φ_1 is an automorphism restricted to the following three subgroups:
$\langle x, y \rangle \cong \langle x, z \rangle \cong \langle z, y \rangle$

Let us analyze the behavior of φ_1 restricted to $\langle xy, yz \rangle$.
By (6.7) one has

$$\varphi_1(xy \cdot yz) = \varphi_1(yx \cdot zy) = \varphi_1(y(xz)y) = \varphi_1(xz) = \varphi_1(x)\varphi_1(z).$$

Suppose that φ_1 restricted to $\langle xy, yz \rangle$ is an anti-automorphism.
Then we get

$$\varphi_1(xy \cdot yz) = \varphi_1(yz) \cdot \varphi_1(xy) = (\varphi_1(y)\varphi_1(z)) \cdot (\varphi_1(x)\varphi_1(y))$$
$$= \varphi_1(z)\varphi_1(x) \neq \varphi_1(x)\varphi_1(z).$$

This is a contradiction. Thus φ_1 is an automorphism restricted to the group $\langle xy, yz \rangle$.

On the subgroup $\langle x, yz \rangle$ the map φ_1 can behave as an automorphism or as an anti-automorphism. We will show that in both cases the behavior of φ_1 restricted to subgroup $\langle x, yz \rangle$ determines the behavior of φ_1 restricted to subgroups $\langle y, xz \rangle$ and $\langle z, yx \rangle$. In other words there are two possibilities: the first one is the case that φ_1 is an automorphism restricted to all three subgroups

$$\langle x, yz \rangle, \quad \langle y, xz \rangle, \quad \langle z, yx \rangle$$

and the second one is the case that φ_1 is an anti-automorphism restricted to these subgroups.

Suppose that φ_1 restricted to $\langle x, yz \rangle$ is an automorphism. Then

$$\varphi_1(x(yz)) = \varphi_1(x)(\varphi_1(y)\varphi_1(z)).$$

Let us analyze the behavior of φ_1 restricted to $\langle y, xz \rangle$.

$$\varphi_1(y(xz)) = -\varphi_1(x(yz)) = -\varphi_1(x)(\varphi_1(y)\varphi_1(z)).$$

Assume that φ_1 is an anti-automorphism restricted to $\langle y, xz \rangle$. Then we get

$$\varphi_1(y(xz)) = \varphi_1(xz)\varphi_1(y) = (\varphi_1(x)\varphi_1(z))\varphi_1(y) = \varphi_1(x)(\varphi_1(y)\varphi_1(z)).$$

This is a contradiction. Thus φ_1 restricted to $\langle y, xz \rangle$ is an automorphism too.

By the same arguments one can show that if φ_1 is an automorphism restricted to $\langle x, yz \rangle$ then it is automorphism restricted to $\langle z, xy \rangle$ as well.

In the same way one can see that if φ_1 is an anti-automorphism restricted to one of the subgroups of $\{\langle x, yz \rangle, \langle y, xz \rangle, \langle z, xy \rangle\}$ then φ_1 is an anti-automorphism restricted to any of two others, too. Hence if φ_1 restricted to all of subgroups $\{\langle x, yz \rangle, \langle y, xz \rangle, \langle z, xy \rangle\}$ is an automorphism then $\varphi_1 \in \mathsf{Aut}(\mathfrak{o})$. If φ_1 restricted to all of subgroups $\{\langle x, yz \rangle, \langle y, xz \rangle, \langle z, xy \rangle\}$ is an anti-automorphism then $\varphi_2 = \varphi_1 \circ \tau_{x(yz)} \in \mathsf{Aut}(\mathfrak{o})$.

Therefore our Theorem is proved for Case 1.

Case 2 φ restricted to $\langle y, z \rangle$ is an anti-automorphism.

Now we can introduce the map $\bar{\varphi} = \varphi \circ \tau_{xz} \circ \tau_{yz}$ which is an automorphism on $\langle x, y \rangle \cong \langle x, z \rangle \cong \langle z, y \rangle \cong Q_8$ and can repeat all considerations from the Case 1.

For a diassociative loop Q the inversion mapping

$$(J : x \mapsto x^{-1}) : Q \to Q$$

is an anti-automorphism of Q. Above we had defined seven elementary half-automorphisms $\{\tau_1, \ldots \tau_7\}$ which correspond to the set S of the elements of $\mathfrak{o}$. Note that in the octonion loop $\mathfrak{o}$ one has

$$\prod_{i=1}^{7} \tau_i = J.$$

Obviously, if φ is an anti-automorphism, then $\varphi \circ J$ is an automorphism.

Thus we have proved that for any half-automorphism $\varphi : \mathfrak{o} \to \mathfrak{o}$ there exists an element $\gamma \in \Gamma$ such that $\alpha = \varphi \circ \gamma$ is an automorphism.   **q.e.d.**

Note that the elementary maps of the octonion loop $\mathfrak{o}$ considered as permutations of the set of elements of $\mathfrak{o}$ generate the elementary abelian group $\Gamma(\mathfrak{o})$ of order 2^7. But not all elements of Γ are proper half-automorphisms. Here we do not discuss the intersection $\mathsf{HAut}(\mathfrak{o}) \cap \Gamma(\mathfrak{o})$.

For a diassociative loop Q we consider the group

$$\mathsf{SAut}(Q) = \mathsf{Aut}(Q) \times \langle J \rangle$$

of all automorphisms and all anti-automorphisms of Q which is a subgroup of $\mathsf{HAut}(Q)$.

Proposition 6.2.7 *If Q is a diassociative loop of exponent 4 and $\Gamma(Q)$ is a subgroup of $\mathsf{HAut}(Q)$, then $\Gamma(Q)$ is normal in $\mathsf{HAut}(Q)$.*

Proof For an element $c \in Q$ of order 4 and for $\alpha \in \mathsf{Aut}(Q)$ one has
$\alpha \circ (c, c^{-1}) \circ \alpha^{-1} = (\alpha(c), \alpha(c^{-1}))$. Hence $\alpha \circ \tau_c \circ \alpha^{-1} = \tau_{\alpha(c)}$. The proposition follows from Lemma 6.2.6   **q.e.d.**

6.3   The Cayley–Dickson Extension Process for Algebras

An algebra A with an involution over a field F is a non-associative F-algebra with a linear mapping $a \mapsto a^*$ satisfying the identity

$$(ab)^* = b^* a^*. \tag{6.8}$$

In [27, III.4] for a given algebra A of finite dimension n with an involution it is described how one can construct F-algebras $\mathsf{CD}(A)$ of finite dimension $2n$ with an involution which contains an isomorphic image of A. For our purposes it is enough to have this construction for the field $\mathbb{Q}$ of the rational numbers or the field $\mathbb{R}$ of real numbers.

Consider the vector space $\mathsf{CD}(A) = A \oplus A$ with the operations

$$(x_1, y_1)(x_2, y_2) = (x_1 x_2 - y_2^* y_1, y_2 x_1 + y_1 x_2^*), \tag{6.9}$$

$$(x, y)^* = (x^*, -y). \tag{6.10}$$

Then an elementary computation shows that $\mathsf{CD}(A)$ is an algebra with an involution. Using (6.9) one obtains the identities

$$(u, 0)(v, 0) = (uv - v^* 0, 0u + v0) = (uv, 0), \tag{6.11}$$

$$(u, 0)(0, v) = (u0 + v^* 0, vu + 00) = (0, vu), \tag{6.12}$$

$$(0, v)(u, 0) = (0u - v^* 0, 0 + vu^*) = (0, vu^*), \tag{6.13}$$

$$(0, u)(0, v) = (0 - v^* u, v0 + u0) = -(v^* u, 0) \tag{6.14}$$

and from (6.10) one obtains

$$(u, 0)^* = (u^*, 0), \quad (0, v)^* = -(0, v). \tag{6.15}$$

Consider the linear mappings

$$\mu = (a \mapsto (a, 0)) : A \to A \oplus A,$$

$$\nu = (b \mapsto (0, b)) : A \to A \oplus A.$$

Then $\mu : \mathsf{A} \to \mathsf{CD}(A)$ is an injective homomorphism of algebras with an involution. If the algebra A has a neutral element e, then $\mu(e) = (e, 0)$ is the neutral element of $\mathsf{CD}(A)$.

If $(b_1, \ldots, b_m)$ is an F-basis of A, then

$$\big(\mu(b_1) \ldots, \mu(b_m), \nu(b_1), \ldots, \nu(b_m)\big) \tag{6.16}$$

is an F-basis of $\mathsf{CD}(A)$.

Given data $\big(\mathsf{A}, \mathfrak{B}\big)$, where A is an algebra over F with an involution which has a neutral element e and

$$\mathfrak{B} = \big(b_1, \ldots, b_m\big)$$

is an F-basis of A, we have defined by (6.9) and (6.10) on the vector space $A \oplus A$ the structure of a non-associative algebra $CD(A)$ with an involution and injective linear mappings

$$\mu, \nu : A \to A \oplus A$$

such that

(a) μ is a monomorphism of algebras with an involution,
(b) $e^{(\mu, \nu)} = (\mu(e), 0)$ is a neutral element of the algebra $A \oplus A$,
(c) $\mathfrak{B}^{(\mu, \nu)} = \big(\mu(b_1) \ldots, \mu(b_m), \nu(b_1), \ldots, \nu(b_m)\big)$ in an F-basis of $A \oplus A$.

In our situation for a given F-algebra A with an involution and $x \in A$ we put $q_A(x) = \|x\| = xx^*$. Then

$$q_A(x)^* = (xx^*)^* = (x^*)^* x^* = xx^* = q_A(x) \qquad (6.17)$$

We now specialize $F = \mathbb{R}$ and define inductively a sequence $(A_k)_{k=1}^{\infty}$ of algebras with an involution of dimension 2^{k-1} having bases $\mathfrak{B}^{(k)}$ and neutral elements $e^{(k)}$ such that

(i) $A_1 = \mathbb{R}$, $\mathfrak{B}^{(1)} = (1)$,
(ii) $A_{k+1} = CD(A_k)$, $\mathfrak{B}^{(k+1)} = \big(\mu(\mathfrak{B}^{(k)}), \nu(\mathfrak{B}^{(k)})\big)$.

Here μ and ν denote the mappings $A \to CD(A)$ defined above. The algebras $A_k, k \geqslant 1$ are called *Cayley–Dickson algebras*. The injective homomorphisms $\mu : A_k \to A_{k+1}$ allow to assume that

$$A_1 < A_2 < \cdots < A_k < \cdots$$

and to form the infinite-dimensional Cayley–Dickson algebra $A_\infty = \bigcup_k A_k$ with its basis $\mathfrak{B}^\infty = \bigcap_k \mathfrak{B}^{(k)}$. Put $q(a) = aa^*$ for $a \in A_\infty$. Equation (6.17) says that $q(a)^* = q(a)$ for all $a \in A$. One shows by induction the

Lemma 6.3.1 *In* A_∞ *the following statements hold:*

(i) $0 < aa^* \in A_1 = \mathbb{R}$ *for all* $0 \neq a \in A_\infty$.
(ii) $b^*b = 1 = bb^*$ *for all* $b \in \mathfrak{B}^\infty$.   **q.e.d.**

6.4   Half-Automorphisms of Cayley–Dickson Loops and of Code Loops

We now use the Cayley–Dickson algebra A_m over $\mathbb{R}$ of dimension m to construct a finite loop C_m of order 2^m. Consider the set

$$C_m = \{\pm b_1, \ldots, \pm b_m\},$$

where $b_1, \ldots, b_m$ are the elements of the basis $\mathfrak{B}^{(m)}$ constructed for A_m. Here we choose $b_1 = 1$, the neutral element of A_m. By the definition of the multiplication of A_m the set C_m is closed under this multiplication.

Proposition 6.4.1 *The multiplication of C_m is diassociative.*

Proof One can check this easily looking at the multiplication rules (6.11)–(6.14).

q.e.d.

Note 2 Observe that the Cayley–Dickson algebras $A_n, n \geqslant 5$ are flexible but not diassociative.

Now observing that by the conventions made before $C_m \leqslant C_{m+1}$ for all $m \geqslant 1$ we define $C_\infty = \bigcup_m C_m$.

Corollary 6.4.2 *With respect to the multiplication in A_∞ the set C_∞ is an infinite diassociative loop.*

q.e.d.

Remember that $C_m = \mathfrak{B}^{(m)} \cup (-\mathfrak{B}^{(m)})$, where $\mathfrak{B}^{(m)}$ is the standard basis of the vector space A_m. Given a half-automorphism η of the loop C_m the set $\eta(\mathfrak{B}^{(m)})$ is a basis of A_m, too. Hence we have the linear extension $\tilde{\eta} : A_m \to A_m$. Obviously $\tilde{\eta}$ is a linear map of the algebra A_m such that

$$\forall x, y \in C_m : \ \tilde{\eta}(xy) = \tilde{\eta}(x)\tilde{\eta}(y) \text{ or } \tilde{\eta}(xy) = \tilde{\eta}(y)\tilde{\eta}(x) \tag{6.18}$$

is satisfied. However $\tilde{\eta}$ is not a half- automorphism of the algebra A_m.

Let us collect fundamental properties of the loop C_∞, respectively of the loops C_n in analogy to Theorem 6.2.2.

Theorem 6.4.3 *In the Cayley–Dickson loops $C_n, n \geqslant 4$ and C_∞ the following statements are true:*

(1) C_n is a loop of order 2^n.
(2) C_∞ is a diassociative loop.
(3) In C_n there are one element $z = -1$ of order 2 and $2^n - 2$ elements of order 4.
(4) $zx = xz$ for all $x \in C_\infty$.
(5) If H is a subloop of C_∞, then $z \in H$.
(6) $Z = \langle z \rangle$ is a normal subloop of C_∞ which coincides with the nucleus, the center, the associator subloop, the commutator subloop, and the Frattini subloop of C_∞.
(7) C_n/Z is an elementary abelian group of order $2^n - 1$.

Proof The proof is a straightforward induction over n beginning with Theorem 6.2.2 for $n = 4$.

q.e.d.

Furthermore, all the loops C_n are *Hamiltonian loops*, i.e. they have the property that all subloops are normal. For $\mathfrak{o}$ this fact has been discussed in detail by D. Norton [22] in the year 1951.

We now give a proof that Theorem 6.2.5 holds for arbitrary Cayley–Dickson loops. Note that in the quaternion group C_3 for all $c \in C_3$ with $c^2 \neq 1$ the elementary mappings τ_c are anti-automorphisms. A. Grishkov suggested to study the half-automorphisms of the class $\mathfrak{L}$ of all loops L satisfying the following conditions:

(L.1) L is a diassociative loop of exponent 4.
(L.2) In L there is a central subloop $Z = \langle z \rangle$ of order 2.
(L.3) $x^2 = z$ for all $x \notin Z$ and $[x, y] = z$ for all $y \notin \{1, z, x, xz\}$.

Lemma 6.4.4 *Given a loop $L \in \mathfrak{L}$ and an element $c \in L$ of order 4 the elementary map τ_c is a half-automorphism.*

Proof Let us show that for all $x, y \in L$

$$\tau_c(xy) = xy = \tau_c(x)\tau_c(y) \quad \text{or} \quad \tau_c(xy) = xy = \tau_c(y)\tau_c(x).$$

To do this it is sufficient to assume that $xy \neq yx$ and to consider the following three cases.

First case: $x \notin \{c, c^{-1}\}$, $y \notin \{c, c^{-1}\}$, $xy \notin \{c, c^{-1}\}$
    Then by definition of τ_c one has $\tau_c(xy) = xy = \tau_c(x)\tau_c(y)$
Second case: $x \notin \{c, c^{-1}\}$, $y \in \{c, c^{-1}\}$
    Then $xy \notin \{c, c^{-1}\}$ and we get

$$\tau_c(xy) = xy = [x^{-1}, y^{-1}]yx = y^2 \cdot yx = y^{-1}x = \tau_c(y)\tau_c(x).$$

and by the same arguments we get

$$\tau_c(yx) = \tau_c(x)\tau_c(y).$$

Third case: $x \notin \{c, c^{-1}\}$, $y \notin \{c, c^{-1}\}$, $xy \in \{c, c^{-1}\}$
    By definition of τ_c we have:

$$\tau_c(xy) = (xy)^{-1} = (xy)^3 = (xy)^2(xy) = [y^{-1}, x^{-1}](xy) = yx = \tau_c(y)\tau_c(x).$$

q.e.d.

Theorem 6.4.5 *In a not associative Cayley–Dickson loop all elementary mappings are proper half-automorphisms.*

Proof If C a Cayley–Dickson loop, it follows from Theorem 6.4.3 that C belongs to $\mathcal{L}$. Hence all elementary mappings $\tau_c, c \in C$ are half-automorphisms. If C is not associative for $c \in C$ there exist quaternion subgroups $Q_{(+)}$ and $Q_{(-)}$ such that $c \in Q_{(-)}$ and $c \notin Q_{(+)}$. Hence τ_c restricted to $Q_{(-)}$ is an anti-automorphism and τ_c restricted to $Q_{(+)}$ is an automorphism. Thus the Theorem is proved. **q.e.d.**
    One can easily show that for the quaternion group $Q_8 = C_3$ the equality $\mathsf{HAut}(Q_8) = \mathsf{SAut}(Q_8) = \Gamma(Q_8)\mathsf{Aut}(Q_8)$ holds. For octonion loop $\mathfrak{o} = C_4$ by our Theorem 6.2.6 we have the factorization $\mathsf{HAut}(\mathfrak{o}) = \Gamma(\mathfrak{o})\mathsf{Aut}(\mathfrak{o})$. For $r \geqslant 5$ one only knows the inclusion $\mathsf{HAut}(C_r) \geqslant \Gamma(C_r)\mathsf{Aut}(C_r)$. The reason for this lies in the fact that our proof of Theorem 6.2.6 uses the known subloop structure of $\mathfrak{o}$: All

subloops of o are Cayley–Dickson loops. The sedenion loop C_5 of order 32 contains a subloop õ of order 16 which is not a Cayley–Dickson loop (see [5, 21]).

Question Is the statement $\mathsf{HAut(C)} = \Gamma\mathsf{(C)Aut(C)}$ true for every Cayley–Dickson loop C?

Turning to code loops (see [11]) and using their description by Chein and Goodaire [7] and Hsu [14, 15] we note that the unique non-associative code loop belongs to the class $\mathfrak{L}$ is o.

To decide when for an element c of an arbitrary code loop the elementary mapping τ_c is a half-automorphism stays an open question.

In [25] Pires and Grishkov study the variety of loops $\mathfrak{E}$ generated by all finite code loops and determine the free elements of $\mathfrak{E}$. To us it seems a demanding task to discuss half-automorphisms of loops in $\mathfrak{E}$.

On page 114 we discussed the five Moufang loops of order 16 which are not groups, among them the octonion loop. It is known that all these loops are code loops. Two of these loops are Chein loops. In [10] necessary conditions are given that for finite group G of even order the Chein loop $M(G, 2)$ admits a proper half-automorphism and it is shown that on $M(Q_8, 2)$ the only three elementary mappings really are half-automorphisms [10, Example 8].

Question For which finite groups G of order 2^n there exist elementary mappings of the Chein loop $M(G, 2)$ such that they are half-automorphisms?

Acknowledgements The first author thanks to SRE, Mexico. The second author is grateful to SNI, Mexico. The third author expresses her gratitude to the University of Sao Paulo for hospitality during the July–August of 2015 and to FAPESP, proceso 2015/07245-4.

References

1. A.A. Albert, Quasigroups II. Trans. Am. Math. Soc. **55**, 401–419 (1944)
2. R.Baer, The higher commutator subgroups of a group. Bull. Am. Math. Soc. **50**, 143–160 (1944)
3. R. Baer, The Homomorphism theorems for loops. Am. J. Math. **67**(3), 450–460 (1945)
4. R. Bruck, *A Survey of Binary Systems* (Springer, Berlin, 1966)
5. R.E. Cawagas, A.S. Carrascal, L.A. Bautista, J.P. Sta. Maria, J.D. Urrutia, B. Nobles, The subalgebra structure of the Cayley–Dickson algebra of dimension 32 (2009). arXiv: 0907.2047v3
6. O. Chein, Moufang loops of small order. I. Trans. Am. Math. Soc. **188**, 31–51 (1974)
7. O. Chein, E. Goodaire, Moufang loops with a unique nonidentity commutator (associator, square). J. Algebra **130**(2), 369–384 (1990)
8. S. Gagola III, A Moufang loop's commutant. Math. Proc. Camb. Philos. Soc. **152**(2), 193–212 (2012)
9. S. Gagola III, M.L. Merlini Giuliani, Half-isomorphisms of Moufang loops of odd order. J. Algebra Appl. **11**(5), 194–199 (2012)
10. S. Gagola III, M.L. Merlini Giuliani, On half-automorphisms of certain Moufang loops with even order. J. Algebra **386**, 131–141 (2013)
11. R. Griess, Code loops. J. Algebra **100**, 224–234 (1986)

12. A. Grishkov, M.L. Merlini Giuliani, M. Rasskazova, L. Sabinina, Half-isomorphisms of finite automorphic Moufang loops. Commun. Algebra **44**(10), 4252–4261 (2016)
13. A. Grishkov, P. Plaumann, M. Rasskazova, L. Sabinina, Half-automorphisms of free automorphic Moufang loops. Math. Notes **98**(2), 133–135 (2015)
14. T. Hsu, Explicit constructions of code loops as centrally twisted products. Math. Proc. Camb. Philos. Soc. **128**(2), 223–232 (2000)
15. T. Hsu, Class 2 Moufang loops, small Frattini Moufang loops, and code loops. arXiv:math/9611214v1 [math.GR] (1996)
16. L.K. Hua, On the automorphisms of a Sfield. Proc. Natl. Acad. Sci. USA **35**, 386–389 (1949)
17. N. Jacobson, C.E. Rickart, Jordan homomorphisms of rings. Trans. Am. Math. Soc. **69**, 479–502 (1950)
18. M.K. Kinyon, J.D. Phillips, P. Vojtechovsky, C-loops: extensions and constructions. J. Algebra Appl. **6**(1), 1–20 (2007)
19. M. Kinyon, I. Stuhl, P. Vojtechovsky, Half-automorphisms of Moufang loops. J. Algebra **450**, 152–161 (2016)
20. J. Kirshtein, Automorphism groups of Cayley-Dickson loops. J. Gen. Lie Theory Appl. **6** (2012). Article ID G110901
21. B.M. Kivunge, J.D.H. Smith, Subloops of sedenions. Commentat. Math. Univ. Carol. **45**(2), 295–302 (2004)
22. D.A. Norton, Hamiltonian loops. Proc. Am. Math. Soc. **3**, 56–65 (1952)
23. H. Pflugfelder, *Quasigroups and Loops: Introduction*. Sigma Series in Pure Mathematics, vol. 7 (Heldermann Verlag, Berlin, 1990)
24. J.D. Philips, P. Vojtechovsky, C-loops: an introduction. Publ. Math. Debrecen **68**, 115–137 (2006)
25. R. Pires, A. Grishkov, Code loops: automorphisms and representations (in preparation). arXiv: 1412 2185 (2014)
26. S. Saeki, On square-preserving isometries of convolution algebras. Trans. Am. Math. Soc. **346**, 707–718 (1994)
27. R.D. Schafer, *An Introduction to Nonassociative Algebras*. Pure and Applied Mathematics, vol. 22 (Academic, New York, 1966)
28. W.R. Scott, Half-homomorphisms of groups. Proc. Am. Math. Soc. **8**, 1141–1144 (1957)

Chapter 7
Invariant Control Systems on Lie Groups

Rory Biggs and Claudiu C. Remsing

Abstract This is a survey of our research (conducted over the last few years) on invariant control systems, the associated optimal control problems, and the associated Hamilton–Poisson systems. The focus is on equivalence and classification. State space and detached feedback equivalence of control systems are characterized in simple algebraic terms; several classes of systems (in three dimensions, on the Heisenberg groups, and on the six-dimensional orthogonal group) are classified. Equivalence of cost-extended systems is shown to imply equivalence of the associated Hamilton–Poisson systems. Cost-extended systems of a certain kind are reinterpreted as invariant sub-Riemannian structures. A classification of quadratic Hamilton–Poisson systems in three dimensions is presented. As an illustrative example, the stability and integration of a typical system is investigated.

7.1 Introduction

Geometric control theory began in the late 1960s with the study of (nonlinear) control systems by using concepts and methods from differential geometry (cf. [10, 64, 96]). A smooth control system may be viewed as a family of vector fields (or dynamical systems) on a manifold, smoothly parametrized by controls. As soon as a control function is fixed, one has a nonautonomous vector field (or, in coordinates, a nonautonomous ordinary differential equation). An integral curve of such a vector field is called a trajectory of the system. The first basic question one asks of a control system is whether or not any two points can be connected by a trajectory: this is known as the *controllability problem*. Once one has established that two points can be connected by a trajectory, one may wish to find a trajectory that minimizes some (practical) cost function: this is known as the *optimality problem*.

R. Biggs (✉)
Department of Mathematics and Applied Mathematics, University of Pretoria, 0002 Pretoria, South Africa
e-mail: rory.biggs@up.ac.za

C.C. Remsing
Department of Mathematics, Rhodes University, 6140 Grahamstown, South Africa
e-mail: c.c.remsing@ru.ac.za

© Springer International Publishing AG 2017
G. Falcone (ed.), *Lie Groups, Differential Equations, and Geometry*,
UNIPA Springer Series, DOI 10.1007/978-3-319-62181-4_7

A significant subclass of control systems rich in symmetry are those evolving on Lie groups and invariant under left translations; for such a system the left translation of any trajectory is a trajectory. This class of systems was first considered in 1972 by Brockett [43] and by Jurdjevic and Sussmann [68]; it forms a natural geometric framework for various (variational) problems in mathematical physics, mechanics, elasticity, and dynamical systems (cf. [10, 40, 64, 66]). In the last few decades substantial work on applied nonlinear control has drawn attention to invariant control affine systems evolving on matrix Lie groups of low dimension (see, e.g., [67, 89, 92, 93] and the references therein).

This paper serves as a survey of our recent research on left-invariant control affine systems, the associated optimal control problems, and the associated Hamilton–Poisson systems. Ideas and key results from several papers published over the last couple of years are reexamined and restructured; some elements are also reinterpreted.

The first aspect we address is the *equivalence of control systems*. Simple algebraic characterizations of state space equivalence and detached feedback equivalence are presented. Using these characterizations, several classes of systems are classified. Specifically, classifications (under detached feedback equivalence) of the full-rank systems on the three-dimensional Lie groups, the $(2n + 1)$-dimensional Heisenberg groups, and the six-dimensional orthogonal group are treated.

The second aspect we address is the *equivalence of invariant optimal control problems*, or rather, their associated cost-extended systems. If two cost-extended systems are equivalent, then their optimal trajectories (resp. extremal trajectories) are related by a Lie group isomorphism. The classification problem is briefly discussed and a few examples are presented in some detail. One associates to each cost-extended system, via the Pontryagin Maximum Principle, a quadratic Hamilton–Poisson system on the associated Lie–Poisson space. It is shown that equivalence of cost-extended systems implies equivalence of the associated Hamilton–Poisson systems. Additionally, the subclass of drift-free systems with homogeneous cost are reinterpreted as invariant sub-Riemannian structures; some examples are revisited.

The third and last aspect we address is the *equivalence, stability, and integration of quadratic Hamilton–Poisson systems*. A classification of the homogeneous systems in three dimensions is presented; a classification of the inhomogeneous systems on the three-dimensional orthogonal Lie–Poisson space $\mathfrak{so}(3)^*_-$ is also exhibited. As an illustrative example, we investigate the stability and the explicit integration of a typical inhomogeneous system on the Euclidean Lie–Poisson space $\mathfrak{se}(2)^*_-$.

Throughout, we make use of the classification of three-dimensional Lie groups and Lie algebras; relevant details are given in the Appendix.

7.1.1   Invariant Control Affine Systems

An ℓ-input left-invariant control affine system Σ on a (real, finite-dimensional, connected) Lie group G consists of a family of left-invariant vector fields Ξ_u on

G, affinely parametrized by controls $u \in \mathbb{R}^\ell$. Such a system is written, in classical notation, as

$$\dot{g} = \Xi_u(g) = \Xi(g, u) = g(A + u_1 B_1 + \cdots + u_\ell B_\ell), \qquad g \in \mathsf{G}, \ u \in \mathbb{R}^\ell.$$

Here $A, B_1, \ldots, B_\ell$ are elements of the Lie algebra $\mathfrak{g}$ with $B_1, \ldots, B_\ell$ linearly independent. The "product" gA denotes the left translation $T_1 L_g \cdot A$ of $A \in \mathfrak{g}$ by the tangent map of $L_g : \mathsf{G} \rightarrow \mathsf{G}$, $h \mapsto gh$; in fact, $T\mathsf{G}$ has trivialization $T\mathsf{G} \cong \mathsf{G} \times \mathfrak{g}$, $gA \leftrightarrow (g, A)$. (When G is a matrix Lie group, this product is simply matrix multiplication.) Note that the "dynamics" $\Xi : \mathsf{G} \times \mathbb{R}^\ell \rightarrow T\mathsf{G}$ are invariant under left translations, i.e., $\Xi(g, u) = g\,\Xi(\mathbf{1}, u)$. A system Σ is completely determined by the specification of its *state space* G and its *parametrization map* $\Xi(\mathbf{1}, \cdot) : \mathbb{R}^\ell \rightarrow \mathfrak{g}$. When G is fixed, we specify Σ by simply writing

$$\Sigma : A + u_1 B_1 + \cdots + u_\ell B_\ell.$$

The *trace* Γ of a system Σ is the affine subspace $\Gamma = A + \Gamma^0 = A + \langle B_1, \ldots, B_\ell \rangle$ of $\mathfrak{g}$. (Here $\Gamma^0 = \langle B_1, \ldots, B_\ell \rangle$ is the subspace of $\mathfrak{g}$ spanned by $B_1, \ldots, B_\ell$.) A system Σ is called *homogeneous* if $A \in \Gamma^0$, and *inhomogeneous* otherwise; Σ is said to be *drift free* if $A = 0$. Also, Σ is said to have *full rank* if its trace generates the whole Lie algebra, i.e., $\mathsf{Lie}(\Gamma) = \mathfrak{g}$.

Remark 1 We have the following characterization of the full-rank condition for systems on G when $\dim \mathsf{G} = 3$. No homogeneous single-input system has full rank. An inhomogeneous single-input system has full rank if and only if A, B_1, and $[A, B_1]$ are linearly independent. A homogeneous two-input system has full rank if and only if B_1, B_2, and $[B_1, B_2]$ are linearly independent. Any inhomogeneous two-input or (homogeneous) three-input system has full rank.

The admissible controls are piecewise continuous maps $u(\cdot) : [0, T] \rightarrow \mathbb{R}^\ell$. A *trajectory* for an admissible control $u(\cdot)$ is an absolutely continuous curve $g(\cdot) : [0, T] \rightarrow \mathsf{G}$ such that $\dot{g}(t) = g(t)\,\Xi(\mathbf{1}, u(t))$ for almost every $t \in [0, T]$. We say that a system Σ is *controllable* if for any $g_0, g_1 \in \mathsf{G}$, there exists a trajectory $g(\cdot) : [0, T] \rightarrow \mathsf{G}$ such that $g(0) = g_0$ and $g(T) = g_1$. The *attainable set* (from the identity $\mathbf{1} \in \mathsf{G}$) is the set $\mathscr{A}$ of all terminal points $g(T)$ of all trajectories $g(\cdot) : [0, T] \rightarrow \mathsf{G}$ starting at $\mathbf{1}$. (A system Σ is controllable if and only if $\mathscr{A} = \mathsf{G}$.) If Σ is controllable, then it has full rank.

For more details about invariant control systems, see, e.g., [10, 64, 68, 94].

7.1.2   Examples of Invariant Optimal Control Problems

We present three illustrative examples of optimal control problems. Numerous such problems have been investigated over the last few decades; these problems include the dynamic equations of the rigid body, various versions of Euler and Kirchhoff elastic rod problem, Dubins' problem, the motion of a particle in a magnetic or

Yang–Mills field, the control of quantum systems, and the sub-Riemannian geodesic problem (see, e.g., the monographs [10, 40, 64] as well as [89]).

The Plate-Ball Problem [63] The plate-ball problem consists of the following kinematic situation: a ball rolls without slipping between two horizontal plates; the lower plate is fixed and the ball is rolled through the horizontal movement of the upper plate. The problem is to transfer the ball from a given initial position and orientation to a prescribed final position and orientation along a path which minimizes $\int_0^T \|v(t)\| dt$. Here v denotes the velocity of the moving plate and T is the time of transfer.

This problem can be regarded as an invariant optimal control problem on the five-dimensional Lie group

$$\mathbb{R}^2 \times \mathsf{SO}\,(3) = \left\{ \begin{bmatrix} e^{x_1} & 0 & 0 \\ 0 & e^{x_2} & 0 \\ 0 & 0 & R \end{bmatrix} : x_1, x_2 \in \mathbb{R},\ R \in \mathsf{SO}\,(3) \right\}$$

specified by (the dynamical law)

$$\dot{g} = g \left(u_1 \begin{bmatrix} 1 & 0 & 0 & 0 & 0 \\ 0 & 0 & 0 & 0 & 0 \\ 0 & 0 & 0 & 0 & -1 \\ 0 & 0 & 0 & 0 & 0 \\ 0 & 0 & 1 & 0 & 0 \end{bmatrix} + u_2 \begin{bmatrix} 0 & 0 & 0 & 0 & 0 \\ 0 & 1 & 0 & 0 & 0 \\ 0 & 0 & 0 & 0 & 0 \\ 0 & 0 & 0 & 0 & -1 \\ 0 & 0 & 0 & 1 & 0 \end{bmatrix} \right), \qquad g \in \mathbb{R}^2 \times \mathsf{SO}\,(3)$$

the boundary conditions $g(0) = g_0$, $g(T) = g_1$, and the cost functional

$$\int_0^T \left(u_1(t)^2 + u_2(t)^2 \right) dt$$

to be minimized.

The Airplane Problem [100] The airplane landing problem consists of finding a trajectory for an airplane moving—at a constant velocity—from some starting point and orientation to some final point and orientation; the airplane is constrained to arrive at the final point and orientation at (some fixed) time T. A simplified kinematic model of the airplane entails that it always fly forward and the controls may yaw, pitch, and roll the aircraft; the cost assigned to a particular solution will be the sum square of the inputs (this cost encodes the requirement to minimize the amount of maneuvering an aircraft will do).

Such a problem may be formulated as an optimal control problem on the six-dimensional Euclidean group

$$\mathsf{SE}\,(3) = \mathbb{R}^3 \rtimes \mathsf{SO}\,(3) = \left\{ \begin{bmatrix} 1 & 0 \\ x & R \end{bmatrix} : x \in \mathbb{R}^3,\ R \in \mathsf{SO}\,(3) \right\}$$

specified by (the equations of motion)

$$\dot{g} = g\,(A + u_1 B_1 + u_2 B_2 + u_3 B_3)\,, \qquad g \in \mathsf{SE}\,(3)$$

with

$$A = \begin{bmatrix} 0 & 0 & 0 & 0 \\ 1 & 0 & 0 & 0 \\ 0 & 0 & 0 & 0 \\ 0 & 0 & 0 & 0 \end{bmatrix}, \quad B_1 = \begin{bmatrix} 0 & 0 & 0 & 0 \\ 0 & 0 & 0 & 0 \\ 0 & 0 & 0 & -1 \\ 0 & 0 & 1 & 0 \end{bmatrix}, \quad B_2 = \begin{bmatrix} 0 & 0 & 0 & 0 \\ 0 & 0 & 0 & 1 \\ 0 & 0 & 0 & 0 \\ 0 & -1 & 0 & 0 \end{bmatrix}, \quad B_3 = \begin{bmatrix} 0 & 0 & 0 & 0 \\ 0 & 0 & -1 & 0 \\ 0 & 1 & 0 & 0 \\ 0 & 0 & 0 & 0 \end{bmatrix}$$

the boundary conditions $g(0) = g_0$, $g(T) = g_1$ and the cost functional

$$\mathscr{J} = \frac{1}{2} \int_0^T \left(c_1 u_1(t)^2 + c_2 u_2(t)^2 + c_3 u_3(t)^2 \right) dt, \qquad c_1, c_2, c_3 > 0.$$

If the cost weights c_1, c_2, c_3 are equal, then the resulting system of equations is related to Lagrange's top (see, e.g., [64]); the Kovalevskaya's top is obtained for a different ratio of weights (see, e.g., [65]).

Control of Serret–Frenet Systems [64] A differentiable curve γ in the Euclidean plane $\mathbb{E}^2$, parametrized by arc-length, can be lifted to the group of motions of $\mathbb{E}^2$ by means of a positively oriented orthonormal moving frame v_1, v_2 defined by

$$\dot{\gamma} = v_1, \qquad \dot{v}_1 = \kappa\, v_2, \qquad \dot{v}_2 = -\kappa\, v_1 \tag{7.1}$$

where κ is the signed curvature of γ. The moving frame can be expressed by a rotation matrix R whose columns consist of the coordinates of v_1 and v_2 relative to a fixed orthonormal frame $e_1, e_2 \in \mathbb{E}^2$. Omitting any notational distinctions between vectors and their coordinate vectors, then $R e_i = v_i$. The curve γ along with its moving frame can be represented as an curve $g(\cdot)$ on the group of motions of $\mathbb{E}^2$, namely

$$\mathsf{SE}\,(2) = \left\{ \begin{bmatrix} 1 & 0 & 0 \\ \gamma_1 & & \\ \gamma_2 & R & \end{bmatrix} : \gamma_1, \gamma_2 \in \mathbb{R}, R \in \mathsf{SO}\,(2) \right\}.$$

Interpreting the curvature κ as a control function, the Serret–Frenet differential system (7.1) can then be written as an (inhomogeneous) invariant control system

$$\dot{g} = g \left(\begin{bmatrix} 0 & 0 & 0 \\ 1 & 0 & 0 \\ 0 & 0 & 0 \end{bmatrix} + \kappa \begin{bmatrix} 0 & 0 & 0 \\ 0 & 0 & -1 \\ 0 & 1 & 0 \end{bmatrix} \right), \qquad g \in \mathsf{SE}\,(2).$$

In this way, many classic variational problems in geometry become problems in optimal control. For example, the problem of finding a curve γ that will satisfy the given boundary conditions $\gamma(0) = a$, $\dot{\gamma}(0) = \dot{a}$, $\gamma(T) = b$, $\dot{\gamma}(T) = \dot{b}$, and will minimize $\int_0^T \kappa^2(t)\, dt$ goes back to Euler; its solutions are known as the elastica (see, e.g., [92, 93]).

7.2   Equivalence of Control Systems

The most natural equivalence relation for control systems is equivalence up to coordinate changes in the state space. This is called *state space equivalence* (see [61]). State space equivalence is well understood; it establishes a one-to-one correspondence between the trajectories of the equivalent systems. However, this equivalence relation is very strong. For analytic control systems on manifolds, Krener characterized local state space equivalence in terms of the existence of a linear isomorphism preserving iterated Lie brackets of the system's vector fields ([72], see also [10, 95, 96]).

Another fundamental equivalence relation for control systems is that of *feedback equivalence*. Two feedback equivalent control systems have the same set of trajectories (up to a diffeomorphism in the state space) which are parametrized differently by admissible controls. Feedback equivalence has been extensively studied in the last few decades (see [90] and the references therein). There are a few basic methods used in the study of feedback equivalence. These methods are based either on (studying invariant properties of) associated distributions or on Cartan's method of equivalence [52] or inspired by the Hamiltonian formalism [61]; also, another fruitful approach is closely related to Poincaré's technique for linearization of dynamical systems. Feedback transformations play a crucial role in control theory, particularly in the important problem of *feedback linearization* [62]. The study of feedback equivalence of general control systems can be reduced, by a simple trick, to the case of control affine systems [61]. For a thorough study of the equivalence and classification of control affine systems, see [50].

We consider state space equivalence and feedback equivalence in the context of left-invariant control affine systems ([38], see also [25]). Characterizations of state space equivalence and (detached) feedback equivalence are obtained in terms of Lie group isomorphisms. A complete classification of systems on the (simply connected) three-dimensional Lie groups is discussed; the controllability of these systems is also briefly treated. Classifications of the systems on the $(2n + 1)$-dimensional Heisenberg groups and the six-dimensional orthogonal group are also covered.

7.2.1 State Space Equivalence

Two systems Σ and Σ' are called *state space equivalent* if there exists a diffeomorphism $\phi : \mathsf{G} \to \mathsf{G}'$ such that, for each control value $u \in \mathbb{R}^\ell$, the vector fields Ξ_u and Ξ'_u are ϕ-related, i.e., $T_g\phi \cdot \Xi(g, u) = \Xi'(\phi(g), u)$ for $g \in \mathsf{G}$ and $u \in \mathbb{R}^\ell$. We have the following simple algebraic characterization of this equivalence.

Theorem 1 ([38], see also [72]) *Two full-rank systems Σ and Σ' are state space equivalent if and only if there exists a Lie group isomorphism $\phi : \mathsf{G} \to \mathsf{G}'$ such that $T_1\phi \cdot \Xi(1, u) = \Xi'(1, u)$ for all $u \in \mathbb{R}^\ell$.*

Proof (Sketch) Suppose that Σ and Σ' are state space equivalent. By composition with a left translation, we may assume $\phi(1) = 1$. As the elements $\Xi_u(1)$, $u \in \mathbb{R}^\ell$ generate $\mathfrak{g}$ and the push-forward $\phi_* \Xi_u$ of left-invariant vector fields Ξ_u are left invariant, it follows that ϕ is a Lie group isomorphism satisfying the requisite property (cf. [25]). Conversely, suppose that $\phi : \mathsf{G} \to \mathsf{G}'$ is a Lie group isomorphism as prescribed. Then $T_g\phi \cdot \Xi(g, u) = T_1(\phi \circ L_g) \cdot \Xi(1, u) = T_1(L_{\phi(g)} \circ \phi) \cdot \Xi(1, u) = \Xi'(\phi(g), u)$. $\qquad\square$

Remark 2 If ϕ is defined only between some neighbourhoods of identity of G and G', then Σ and Σ' are said to be locally state space equivalence. A characterization similar to that given in Theorem 1, in terms of Lie algebra automorphisms, holds. In the case of simply connected Lie groups, local and global equivalence are the same (as $d\,\mathsf{Aut}(\mathsf{G}) = \mathsf{Aut}(\mathfrak{g})$).

State space equivalence is quite a strong equivalence relation. Hence, there are so many equivalence classes that any general classification appears to be very difficult if not impossible. However, there is a chance for some reasonable classification in low dimensions. We give an example to illustrate this point.

Example 1 ([2]) Any two-input inhomogeneous full-rank control affine system on the Euclidean group $\mathsf{SE}(2)$ is state space equivalent to exactly one of the following systems

$$\Sigma_{1,\alpha\beta\gamma} : \alpha E_3 + u_1(E_1 + \gamma_1 E_2) + u_2(\beta E_2)$$

$$\Sigma_{2,\alpha\beta\gamma} : \beta E_1 + \gamma_1 E_2 + \gamma_2 E_3 + u_1(\alpha E_3) + u_2 E_2$$

$$\Sigma_{3,\alpha\beta\gamma} : \beta E_1 + \gamma_1 E_2 + \gamma_2 E_3 + u_1(E_2 + \gamma_3 E_3) + u_2(\alpha E_3).$$

Here $\alpha > 0, \beta \neq 0$ and $\gamma_1, \gamma_2, \gamma_3 \in \mathbb{R}$, with different values of these parameters yielding distinct (non-equivalent) class representatives.

Proof (Sketch) The group of linearized automorphisms of $\mathsf{SE}(2)$ is given by

$$d\,\mathsf{Aut}(\mathsf{SE}(2)) = \left\{ \begin{bmatrix} x & y & v \\ -\varsigma y & \varsigma x & w \\ 0 & 0 & \varsigma \end{bmatrix} : x, y, v, w \in \mathbb{R}, \, x^2 + y^2 \neq 0, \, \varsigma = \pm 1 \right\}.$$

Indeed, it is straightforward to show that every automorphism of the simply connected Lie group $\widetilde{\mathsf{SE}}(2)$ preserves the kernel of the universal covering q : $\widetilde{\mathsf{SE}}(2) \to \mathsf{SE}(2)$ [33]; hence, $d\,\mathsf{Aut}(\mathsf{SE}(2)) = \mathsf{Aut}(\mathfrak{se}(2))$ (see the Appendix).

Let $\Sigma = (\mathsf{SE}(2), \varXi)$, $\varXi(\mathbf{1}, u) = \sum_{i=1}^{3} a_i E_i + u_1 \sum_{i=1}^{3} b_i E_i + u_2 \sum_{i=1}^{3} c_i E_i$, or in *matrix form*

$$\Sigma : \quad \begin{bmatrix} a_1 & b_1 & c_1 \\ a_2 & b_2 & c_2 \\ a_3 & b_3 & c_3 \end{bmatrix}.$$

It is then straightforward to show that there exists $\psi \in d\,\mathsf{Aut}(\mathsf{SE}(2))$ such that

$$\psi \cdot \begin{bmatrix} a_1 & b_1 & c_1 \\ a_2 & b_2 & c_2 \\ a_3 & b_3 & c_3 \end{bmatrix} = \begin{bmatrix} 0 & 1 & 0 \\ 0 & \gamma_1 & \beta \\ \alpha & 0 & 0 \end{bmatrix} \quad \text{if} \quad b_3 = 0,\ c_3 = 0$$

$$\psi \cdot \begin{bmatrix} a_1 & b_1 & c_1 \\ a_2 & b_2 & c_2 \\ a_3 & b_3 & c_3 \end{bmatrix} = \begin{bmatrix} \beta & 0 & 0 \\ \gamma_1 & 0 & 1 \\ \gamma_2 & \alpha & 0 \end{bmatrix} \quad \text{if} \quad b_3 \neq 0,\ c_3 = 0$$

$$\text{or} \quad \psi \cdot \begin{bmatrix} a_1 & b_1 & c_1 \\ a_2 & b_2 & c_2 \\ a_3 & b_3 & c_3 \end{bmatrix} = \begin{bmatrix} \beta & 0 & 0 \\ \gamma_1 & 1 & 0 \\ \gamma_2 & \gamma_3 & \alpha \end{bmatrix} \quad \text{if} \quad c_3 \neq 0.$$

Thus Σ is state space equivalent to $\Sigma_{1,\alpha\beta\gamma}$, $\Sigma_{2,\alpha\beta\gamma}$, or $\Sigma_{3,\alpha\beta\gamma}$. It is a simple matter to verify that these class representatives are non-equivalent.   □

Note 1 A full classification (under state space equivalence) of systems on $\mathsf{SE}(2)$ appears in [2], whereas a classification of systems on $\mathsf{SE}(1,1)$ appears in [17]. For a classification of systems on $\mathsf{SO}(2,1)_0$, see [39].

7.2.2 Detached Feedback Equivalence

We specialize feedback equivalence in the context of invariant systems by requiring that the feedback transformations are compatible with the Lie group structure. Two systems Σ and Σ' are called *detached feedback equivalent* if there exist diffeomorphisms $\phi : \mathsf{G} \to \mathsf{G}'$ and $\varphi : \mathbb{R}^\ell \to \mathbb{R}^\ell$ such that for each control value $u \in \mathbb{R}^\ell$ the vector fields $\varXi_u$ and $\varXi'_{\varphi(u)}$ are ϕ-related, i.e., $T_g\phi \cdot \varXi(g, u) = \varXi'(\phi(g), \varphi(u))$ for $g \in \mathsf{G}$ and $u \in \mathbb{R}^\ell$. We have the following simple algebraic characterization of this equivalence in terms of the traces $\Gamma = \mathrm{im}\,\varXi(\mathbf{1}, \cdot)$ and $\Gamma' = \mathrm{im}\,\varXi'(\mathbf{1}, \cdot)$ of Σ and Σ'.

Theorem 2 ([38]) *Two full-rank systems Σ and Σ' are detached feedback equivalent if and only if there exists a Lie group isomorphism $\phi : G \to G'$ such that $T_1\phi \cdot \Gamma = \Gamma'$.*

Proof (Sketch) Suppose Σ and Σ' are detached feedback equivalent. By composing ϕ with an appropriate left translation, we may assume $\phi(1) = 1'$. Hence $T_1\phi \cdot \Xi(1, u) = \Xi'(1', \varphi(u))$ and so $T_1\phi \cdot \Gamma = \Gamma'$. Moreover, as the elements $\Xi_u(1)$, $u \in \mathbb{R}^\ell$ generate $\mathfrak{g}$ and the push-forward of the left-invariant vector fields Ξ_u are left invariant, it follows that ϕ is a group isomorphism (cf. [25]). On the other hand, suppose there exists a group isomorphism $\phi : G \to G'$ such that $T_1\phi \cdot \Gamma = \Gamma'$. Then there exists a unique affine isomorphism $\varphi : \mathbb{R}^\ell \to \mathbb{R}^{\ell'}$ such that $T_1\phi \cdot \Xi(1, u) = \Xi'(1', \varphi(u))$. As with state space equivalence, by left-invariance and the fact that ϕ is a Lie group isomorphism, it then follows that $T_g\phi \cdot \Xi(g, u) = \Xi'(\phi(g), \varphi(u))$. $\qquad\square$

Remark 3 If ϕ is defined only between some neighbourhoods of identity of G and G', then Σ and Σ' are said to be locally detached feedback equivalent. A characterization similar to that given in Theorem 2, in terms of Lie algebra automorphisms, holds. As for state space equivalence, in the case of simply connected Lie groups local and global equivalence are the same (as $d\,\mathsf{Aut}(G) = \mathsf{Aut}(\mathfrak{g})$).

Detached feedback equivalence is notably weaker than state space equivalence. To illustrate this point, we give a classification, under detached feedback equivalence, of the same class of systems considered in Example 1.

Example 2 ([29]) Any two-input inhomogeneous full-rank control affine system on $\mathsf{SE}\,(2)$ is detached feedback equivalent to exactly one of the following systems

$$\Sigma_1 : E_1 + u_1 E_2 + u_2 E_3$$

$$\Sigma_{2,\alpha} : \alpha E_3 + u_1 E_1 + u_2 E_2.$$

Here $\alpha > 0$ parametrizes a family of class representatives, each different value corresponding to a distinct non-equivalent representative.

Proof (Sketch) As argued in Example 1, we have $d\,\mathsf{Aut}(\mathsf{SE}\,(2)) = \mathsf{Aut}(\mathfrak{se}\,(2))$. Let $\Sigma = (\mathsf{SE}\,(2), \Xi)$ be an inhomogeneous system with trace

$$\Gamma = \sum_{i=1}^{3} a_i E_i + \left\langle \sum_{i=1}^{3} b_i E_i, \sum_{i=1}^{3} c_i E_i \right\rangle.$$

If $c_3 \neq 0$ or $b_3 \neq 0$, then $\Gamma = a_1' E_1 + a_2' E_2 + \langle b_1' E_1 + b_2' E_2, c_1' E_1 + c_2' E_2 + E_3 \rangle$. Now either $b_1' \neq 0$ or $b_2' \neq 0$, and so

$$\begin{bmatrix} b_2' & -b_1' \\ b_1' & b_2' \end{bmatrix} \begin{bmatrix} v_1 \\ v_2 \end{bmatrix} = \begin{bmatrix} a_2' \\ a_1' \end{bmatrix}$$

has a unique solution (with $v_2 \neq 0$). Therefore

$$\psi = \begin{bmatrix} v_2 b_2' & v_2 b_1' & c_1' \\ -v_2 b_1' & v_2 b_2' & c_2' \\ 0 & 0 & 1 \end{bmatrix}$$

is a (Lie algebra) automorphism such that $\psi \cdot \Gamma_1 = \psi \cdot (E_1 + \langle E_2, E_3 \rangle) = \Gamma$. Thus Σ is detached feedback equivalent to Σ_1. On the other hand, suppose $b_3 = 0$ and $c_3 = 0$. Then $\Gamma = a_3 E_3 + \langle E_1, E_2 \rangle$. Hence $\psi = \operatorname{diag}(1, 1, \operatorname{sgn}(a_3))$ is an automorphism such that $\psi \cdot \Gamma = \alpha E_3 + \langle E_1, E_2 \rangle$ with $\alpha > 0$. Thus Σ is detached feedback equivalent to $\Sigma_{2,\alpha}$. As the subspace $\langle E_1, E_2 \rangle$ is invariant (under automorphisms), Σ_1 and $\Sigma_{2,\alpha}$ cannot be equivalent. It is easy to show that $\Sigma_{2,\alpha}$ a $\Sigma_{2,\alpha'}$ are equivalent only if $\alpha = \alpha'$.   □

Remark 4 Σ and Σ' are feedback equivalent if there exists a *diffeomorphism* ϕ : $\mathsf{G} \to \mathsf{G}'$ such that (the push-forward) $\phi_* \mathscr{F} = \mathscr{F}'$. Here $g \mapsto \mathscr{F}(g) = g\,\Gamma$ is the field of admissible velocities. The specialization to detached feedback equivalence corresponds to the existence of a *Lie group isomorphism* ϕ such that $\phi_* \mathscr{F} = \mathscr{F}'$. Thus feedback equivalence is weaker than detached feedback equivalence. For example, suppose $\Gamma = \mathfrak{g}$, $\Gamma' = \mathfrak{g}'$, and G is diffeomorphic to G'. Then Σ and Σ' are feedback equivalent. However, Σ and Σ' will be detached feedback equivalent only if G and G' are, in addition, isomorphic as Lie groups.

7.2.3 Classification in Three Dimensions

We exhibit a classification, under detached feedback equivalence, of the full-rank systems evolving on three-dimensional *simply connected* Lie groups. A representative is identified for each equivalence class. Systems on the Euclidean group and the orthogonal group are discussed as typical examples. The three-dimensional Lie groups, their Lie algebras, and automorphism groups are listed in the Appendix. For a classification of invariant control affine systems on *all* three-dimensional *matrix* Lie groups, see [33].

7.2.3.1 Solvable Groups

The classification procedure is as follows. First the group of automorphisms is determined (see the Appendix). Equivalence class representatives are then constructed by considering the action of an automorphism on the trace of a typical system. Lastly, one verifies that none of the representatives are equivalent.

Theorem 3 ([28, 29], see also [26, 31])   *Suppose Σ is a full-rank system evolving on a simply connected solvable Lie group* G. *Then* G *is isomorphic to one of the groups listed below and* Σ *is detached feedback equivalent to exactly one of accompanying (full-rank) systems on that group.*

1. *On* $\mathbb{R}^3$ *we have the systems*

$$\Sigma^{(2,1)} : E_1 + u_1 E_2 + u_2 E_3 \qquad \Sigma^{(3,0)} : u_1 E_1 + u_2 E_2 + u_3 E_3.$$

2. *On* H_3 *we have the systems*

$$\Sigma^{(1,1)} : E_2 + u E_3 \qquad\qquad \Sigma^{(2,0)} : u_1 E_2 + u_2 E_3$$

$$\Sigma_1^{(2,1)} : E_1 + u_1 E_2 + u_2 E_3 \qquad \Sigma_2^{(2,1)} : E_3 + u_1 E_1 + u_2 E_2$$

$$\Sigma^{(3,0)} : u_1 E_1 + u_2 E_2 + u_3 E_3.$$

3. *On* $\mathsf{G}_{3.2}$ *we have the systems*

$$\Sigma_1^{(1,1)} : E_2 + u E_3 \qquad\qquad \Sigma_{2,\beta}^{(1,1)} : \beta E_3 + u E_2$$

$$\Sigma^{(2,0)} : u_2 E_2 + u_2 E_3 \qquad\qquad \Sigma_1^{(2,1)} : E_1 + u_1 E_2 + u_2 E_3$$

$$\Sigma_2^{(2,1)} : E_2 + u_1 E_3 + u_2 E_1 \qquad \Sigma_{3,\beta}^{(2,1)} : \beta E_3 + u_1 E_1 + u_2 E_2$$

$$\Sigma^{(3,0)} : u_1 E_1 + u_2 E_2 + u_3 E_3.$$

4. *On* $\mathsf{G}_{3.3}$ *we have the systems*

$$\Sigma_1^{(2,1)} : E_1 + u_1 E_2 + u_2 E_3 \qquad \Sigma_{2,\beta}^{(2,1)} : \beta E_3 + u_1 E_1 + u_2 E_2$$

$$\Sigma^{(3,0)} : u_1 E_1 + u_2 E_2 + u_3 E_3.$$

5. *On* $\mathsf{SE}\,(1,1)$ *we have the systems*

$$\Sigma_1^{(1,1)} : E_2 + u E_3 \qquad\qquad \Sigma_{2,\alpha}^{(1,1)} : \alpha E_3 + u E_2$$

$$\Sigma^{(2,0)} : u_1 E_2 + u_2 E_3 \qquad\qquad \Sigma_1^{(2,1)} : E_1 + u_1 E_2 + u_2 E_3$$

$$\Sigma_2^{(2,1)} : E_1 + u_1(E_1 + E_2) + u_2 E_3 \qquad \Sigma_{3,\alpha}^{(2,1)} : \alpha E_3 + u_1 E_1 + u_2 E_2$$

$$\Sigma^{(3,0)} : u_1 E_1 + u_2 E_2 + u_3 E_3.$$

6. *On* $\mathsf{G}_{3.4}^a$ *(resp. on* $\mathsf{Aff}\,(\mathbb{R})_0 \times \mathbb{R})$ *we have the systems*

$$\Sigma_1^{(1,1)} : E_2 + uE_3 \qquad\qquad \Sigma_{2,\beta}^{(1,1)} : \beta E_3 + uE_2$$

$$\Sigma^{(2,0)} : u_1E_2 + u_2E_3 \qquad\qquad \Sigma_1^{(2,1)} : E_1 + u_1E_2 + u_2E_3$$

$$\Sigma_2^{(2,1)} : E_1 + u_1(E_1 + E_2) + u_2E_3 \qquad \Sigma_3^{(2,1)} : E_1 + u_1(E_1 - E_2) + u_2E_3$$

$$\Sigma_{4,\beta}^{(2,1)} : \beta E_3 + u_1E_1 + u_2E_2 \qquad\qquad \Sigma^{(3,0)} : u_1E_1 + u_2E_2 + u_3E_3.$$

7. *On* $\widetilde{\mathsf{SE}}\,(2)$ *we have the systems*

$$\Sigma_1^{(1,1)} : E_2 + uE_3 \qquad\qquad \Sigma_{2,\alpha}^{(1,1)} : \alpha E_3 + uE_2$$

$$\Sigma^{(2,0)} : u_1E_2 + u_2E_3 \qquad\qquad \Sigma_1^{(2,1)} : E_1 + u_1E_2 + u_2E_3$$

$$\Sigma_{2,\alpha}^{(2,1)} : \alpha E_3 + u_1E_1 + u_2E_2 \qquad\qquad \Sigma^{(3,0)} : u_1E_1 + u_2E_2 + u_3E_3.$$

8. *On* $\mathsf{G}_{3.5}^a$ *we have the systems*

$$\Sigma_1^{(1,1)} : E_2 + uE_3 \qquad\qquad \Sigma_{2,\beta}^{(1,1)} : \beta E_3 + uE_2$$

$$\Sigma^{(2,0)} : u_1E_2 + u_2E_3 \qquad\qquad \Sigma_1^{(2,1)} : E_1 + u_1E_2 + u_2E_3$$

$$\Sigma_{2,\beta}^{(2,1)} : \beta E_3 + u_1E_1 + u_2E_2 \qquad\qquad \Sigma^{(3,0)} : u_1E_1 + u_2E_2 + u_3E_3.$$

Here $\alpha > 0$ *and* $\beta \neq 0$ *parametrize families of distinct (non-equivalent) class representatives.*

Proof We treat, as typical case, only item (7). The group of linearized automorphisms of $\widetilde{\mathsf{SE}}\,(2)$ is given by

$$d\,\mathsf{Aut(SE)}\,(2) = \left\{ \begin{bmatrix} x & y & u \\ -\sigma y & \sigma x & v \\ 0 & 0 & \sigma \end{bmatrix} : x, y, u, v \in \mathbb{R},\, x^2 + y^2 \neq 0,\, \sigma = \pm 1 \right\}.$$

Let Σ be a single-input inhomogeneous system with trace $\Gamma = A + \Gamma^0 \subset \widetilde{\mathfrak{se}}\,(2)$. Suppose $E_3^*(\Gamma^0) \neq \{0\}$. (Here E_3^* is the corresponding element of the dual basis.) Then $\Gamma = a_1E_1 + a_2E_2 + \langle b_1E_1 + b_2E_2 + E_3 \rangle$. Thus

$$\psi = \begin{bmatrix} a_1 & a_1 & b_1 \\ -a_1 & a_2 & b_2 \\ 0 & 0 & 1 \end{bmatrix}$$

is an automorphism such that $\psi \cdot \Gamma_1^{(1,1)} = \Gamma$. So Σ is equivalent to $\Sigma_1^{(1,1)}$. On the other hand, suppose $E_3^*(\Gamma^0) = \{0\}$. Then $\Gamma = a_1E_1 + a_2E_2 + a_3E_3 +$

$\langle b_1 E_1 + b_2 E_2 \rangle$ with $a_3 \neq 0$. Hence

$$\psi = \begin{bmatrix} b_2 \ \mathrm{sgn}(a_3) & b_1 & \frac{a_1}{a_3 \ \mathrm{sgn}(a_3)} \\ -b_1 \ \mathrm{sgn}(a_3) & b_2 & \frac{a_2}{a_3 \ \mathrm{sgn}(a_3)} \\ 0 & 0 & \mathrm{sgn}(a_3) \end{bmatrix}$$

is an automorphism such that $\psi \cdot \Gamma_{2,\alpha}^{(1,1)} = \Gamma$, where $\alpha = a_3 \ \mathrm{sgn}(a_3)$.

Let Σ be a two-input homogeneous system with trace $\Gamma = \langle B_1, B_2 \rangle$. Then $\widehat{\Sigma} : B_1 + \langle B_2 \rangle$ is a (full-rank) single-input inhomogeneous system. Therefore, there exists an automorphism ψ such that $\psi \cdot (B_1 + \langle B_2 \rangle)$ equals either $E_2 + \langle E_3 \rangle$ or $\alpha E_3 + \langle E_2 \rangle$. Hence, in either case, we get $\psi \cdot \langle B_1, B_2 \rangle = \langle E_2, E_3 \rangle$. Thus Σ is equivalent to $\Sigma^{(2,0)}$.

The two-input inhomogeneous systems were covered in Example 2. If Σ is a three-input system, then it is equivalent to $\Sigma^{(3,0)}$.

Again, most pairs of systems cannot be equivalent due to different homogeneities or different number of inputs. As the subspace $\langle E_1, E_2 \rangle$ is invariant (under the action of automorphisms), $\Sigma_1^{(1,1)}$ is not equivalent to any system $\Sigma_{2,\alpha}^{(1,1)}$. For $A \in \widetilde{\mathfrak{se}}\,(2)$ and $\psi \in d\,\mathsf{Aut}(\mathsf{SE})\,(2)$, we have that $E_3^*(\psi \cdot \alpha E_3) = \pm \alpha$. Thus $\Sigma_{2,\alpha}^{(1,1)}$ and $\Sigma_{2,\alpha'}^{(1,1)}$ are equivalent only if $\alpha = \alpha'$. For the two-input inhomogeneous systems, similar arguments hold. $\qquad\square$

7.2.3.2 Semisimple Groups

The procedure for classification is similar to that of the solvable groups. However, here we employ an invariant bilinear product ω (the Lorentzian product and the dot product, respectively); the inhomogeneous systems are (partially) characterized by the level set $\{A \in \mathfrak{g} : \omega(A, A) = \alpha\}$ that their trace is tangent to.

Theorem 4 ([30], see also [26, 31]) *Suppose Σ is a full-rank system evolving on a simply connected semisimple Lie group G. Then G is isomorphic to one of the groups listed below and Σ is detached feedback equivalent to exactly one of accompanying (full-rank) systems on that group.*

1. On $\widetilde{\mathsf{SL}}\,(2, \mathbb{R})$ we have the systems

$$\Sigma_1^{(1,1)} : E_3 + u(E_2 + E_3) \qquad\qquad \Sigma_{2,\alpha}^{(1,1)} : \alpha E_2 + u E_3$$

$$\Sigma_{3,\alpha}^{(1,1)} : \alpha E_1 + u E_2 \qquad\qquad \Sigma_{4,\alpha}^{(1,1)} : \alpha E_3 + u E_2$$

$$\Sigma_1^{(2,0)} : u_1 E_1 + u_2 E_2 \qquad\qquad \Sigma_2^{(2,0)} : u_1 E_2 + u_2 E_3$$

$$\Sigma_1^{(2,1)} : E_3 + u_1 E_1 + u_2(E_2 + E_3) \qquad \Sigma_{2,\alpha}^{(2,1)} : \alpha E_1 + u_1 E_2 + u_2 E_3$$

$$\Sigma_{3,\alpha}^{(2,1)} : \alpha E_3 + u_1 E_1 + u_2 E_2 \qquad\qquad \Sigma^{(3,0)} : u_1 E_1 + u_2 E_2 + u_3 E_3.$$

2. *On* $\mathsf{SU}\,(2)$ *we have the systems*

$$\Sigma_\alpha^{(1,1)} : \alpha E_2 + u E_3 \qquad\qquad \Sigma^{(2,0)} : u_1 E_2 + u_2 E_3$$

$$\Sigma_\alpha^{(2,1)} : \alpha E_1 + u_1 E_2 + u_2 E_3 \qquad\qquad \Sigma^{(3,0)} : u_1 E_1 + u_2 E_2 + u_3 E_3.$$

Here $\alpha > 0$ parametrizes families of distinct (non-equivalent) class representatives.

Proof We consider only item (2), i.e., systems on the unitary group $\mathsf{SU}\,(2)$. (The proof for item (1), although more involved, is similar.) The group of linearized automorphisms of $\mathsf{SU}\,(2)$ is $d\,\mathsf{Aut}(\mathsf{SU}\,(2)) = \mathsf{SO}\,(3) = \{g \in \mathbb{R}^{3 \times 3} : gg^\top = \mathbf{1}, \det g = 1\}$. The dot product $\bullet$ on $\mathfrak{su}\,(2)$ is given by $A \bullet B = a_1 b_1 + a_2 b_2 + a_3 b_3$. (Here $A = \sum_{i=1}^3 a_i E_i$ and $B = \sum_{i=1}^3 b_i E_i$.) The level sets $\mathscr{S}_\alpha = \{A \in \mathfrak{su}\,(2) : A \bullet A = \alpha\}$ are spheres of radius $\sqrt{\alpha}$ (and are preserved by automorphisms). The group of automorphisms acts transitively on each sphere $\mathscr{S}_\alpha$. The *critical point* $\mathfrak{C}^\bullet(\Gamma)$, at which an affine subspace $\Gamma = A + \langle B \rangle$ or $A + \langle B_1, B_2 \rangle$ is tangent to a sphere $\mathscr{S}_\alpha$, is given by

$$\mathfrak{C}^\bullet(\Gamma) = A - \frac{A \bullet B}{B \bullet B} B$$

$$\mathfrak{C}^\bullet(\Gamma) = A - \begin{bmatrix} B_1 & B_2 \end{bmatrix} \begin{bmatrix} B_1 \bullet B_1 & B_1 \bullet B_2 \\ B_1 \bullet B_2 & B_2 \bullet B_2 \end{bmatrix}^{-1} \begin{bmatrix} A \bullet B_1 \\ A \bullet B_2 \end{bmatrix}$$

respectively. Critical points behave well under the action of automorphisms, i.e., $\psi \cdot \mathfrak{C}^\bullet(\Gamma) = \mathfrak{C}^\bullet(\psi \cdot \Gamma)$ for any automorphism ψ.

Let Σ be a single-input inhomogeneous system with trace Γ. There exists an automorphism ψ such that $\psi \cdot \Gamma = \alpha \sin \theta\, E_1 + \alpha \cos \theta\, E_2 + \langle E_3 \rangle$, where $\alpha = \sqrt{\mathfrak{C}^\bullet(\Gamma) \bullet \mathfrak{C}^\bullet(\Gamma)}$. Hence

$$\psi' = \begin{bmatrix} \cos \theta & -\sin \theta & 0 \\ \sin \theta & \cos \theta & 0 \\ 0 & 0 & 1 \end{bmatrix}$$

is an automorphism such that $\psi' \cdot \psi \cdot \Gamma = \Gamma_\alpha^{(1,1)}$.

Let Σ be a two-input homogeneous system with trace $\Gamma = \langle B_1, B_2 \rangle$. Then $\widehat{\Sigma} : B_1 + \langle B_2 \rangle$ is a (full-rank) single-input inhomogeneous system. Therefore, there exists an automorphism ψ such that $\psi \cdot (B_1 + \langle B_2 \rangle) = \alpha E_2 + \langle E_3 \rangle$. Hence, $\psi \cdot \langle B_1, B_2 \rangle = \langle E_2, E_3 \rangle$. Thus Σ is equivalent to $\Sigma^{(2,0)}$.

Let Σ be a two-input inhomogeneous system with trace Γ. We have $\mathfrak{C}^\bullet(\Gamma) \bullet \mathfrak{C}^\bullet(\Gamma) = \alpha^2$ for some $\alpha > 0$. As $\mathfrak{C}^\bullet(\Gamma_{1,\alpha}) \bullet \mathfrak{C}^\bullet(\Gamma_{1,\alpha}) = \alpha^2$, there exists an automorphism ψ such that $\psi \cdot \mathfrak{C}^\bullet(\Gamma) = \mathfrak{C}^\bullet(\Gamma_{1,\alpha})$. Hence $\psi \cdot \Gamma$ and $\Gamma_{1,\alpha}$ are both equal to the tangent plane of $\mathscr{S}_{\alpha^2}$ at $\psi \cdot \mathfrak{C}^\bullet(\Gamma)$, and are therefore identical.

If Σ is a three-input system, then it is equivalent to $\Sigma^{(3,0)}$.

Lastly we note that none of the representatives obtained are equivalent. (Again, we first distinguish representatives in terms of homogeneity and number of inputs.) As $\alpha^2 = \mathfrak{C}^\bullet(\Gamma_\alpha^{(1,1)}) \bullet \mathfrak{C}^\bullet(\Gamma_\alpha^{(1,1)})$ (resp. $\alpha^2 = \mathfrak{C}^\bullet(\Gamma_\alpha^{(2,1)}) \bullet \mathfrak{C}^\bullet(\Gamma_\alpha^{(2,1)})$) is an invariant quantity, the systems $\Sigma_\alpha^{(1,1)}$ and $\Sigma_{\alpha'}^{(1,1)}$ (resp. $\Sigma_\alpha^{(2,1)}$ and $\Sigma_{\alpha'}^{(2,1)}$) are equivalent only if $\alpha = \alpha'$. $\qquad\qquad\qquad\qquad\qquad\qquad\qquad\qquad\qquad\qquad\qquad\qquad\quad\square$

7.2.3.3 Controllability

The controllability property is invariant under detached feedback equivalence. The following results, concerning control systems on Lie groups, are well known. If a system is controllable, then it has full rank. Moreover, if a system is homogeneous or if its state space is compact, then the full-rank condition implies controllability. Each of the following conditions is sufficient for a full-rank system to be controllable [68]:

- the direction space Γ^0 generates $\mathfrak{g}$, i.e., $\mathsf{Lie}\,(\Gamma^0) = \mathfrak{g}$;
- the identity element $\mathbf{1}$ is in the interior of the attainable set

$$\mathscr{A} = \{g(t_1) \;:\; g(\cdot) \text{ is a trajectory such that } g(0) = \mathbf{1},\; t_1 \geqslant 0\};$$

- there exists $C \in \Gamma$ such that $t \mapsto \exp(tC)$ is periodic.

If the state space is simply connected and completely solvable, then the condition $\mathsf{Lie}\,(\Gamma^0) = \mathfrak{g}$ is necessary for controllability [94].

In several cases, these results allow one to immediately determine the controllability nature of a given system; for instance, for a system on a compact Lie group one need only check whether or not $\mathsf{Lie}(\Gamma) = \mathfrak{g}$. In some cases determining the controllability of a system can be much more involved; for instance, we could only show that the system $\Sigma_1^{(1,1)}$ on $\mathsf{G}_{3.5}^a$ is controllable by explicitly constructing a family of trajectories showing that $\mathbf{1} \in \mathrm{int}\,\mathscr{A}$.

Theorem 5 ([33], see also [34, 42, 94]) *For full-rank systems on three-dimensional matrix Lie groups, the following criteria for controllability hold.*

1. *On each of the groups* $\mathsf{Aff}\,(\mathbb{R})_0 \times \mathbb{R}$, H_3, $\mathsf{G}_{3.2}$, $\mathsf{G}_{3.3}$, $\mathsf{SE}\,(1,1)$, *and* $\mathsf{G}_{3.4}^a$, *a system is controllable if and only if* $\mathsf{Lie}\,(\Gamma^0) = \mathfrak{g}$.
2. *On each of the groups* $\mathsf{SE}_n(2)$, $n \in \mathbb{N}$, $\mathsf{SO}\,(3)$, *and* $\mathsf{SU}\,(2)$, *all systems are controllable.*
3. *On each of the groups* $\mathsf{Aff}\,(\mathbb{R}) \times \mathbb{T}$, $\mathsf{SL}\,(2,\mathbb{R})$, *and* $\mathsf{SO}\,(2,1)_0$, *a system is controllable if and only if it is homogeneous or there exists* $A \in \Gamma$ *such that* $t \mapsto \exp(tA)$ *is periodic.*
4. *On each of the groups* $\widetilde{\mathsf{SE}}\,(2)$ *and* $\mathsf{G}_{3.5}^a$, *a system is controllable if and only if* $E_3^*(\Gamma^0) \neq \{0\}$. *(Here* E_3^* *denotes the corresponding element of the dual basis.)*

7.2.4   Classification Beyond Three Dimensions

Detached feedback equivalence has proved useful and feasible not only in three
dimensions, but also in some higher dimensional cases. In this spirit we classify
controllable systems on the Heisenberg groups H_{2n+1} and revisit a classification of
homogeneous systems on the six-dimensional orthogonal group $SO(4)$.

Note 2 Homogeneous systems on four-dimensional Lie groups have been classified
on the oscillator group [a central extension of $SE(2)$] in [36], and on the four-
dimensional central extension of $SE(1, 1)$ in [19]. Recently, subspaces of the
four-dimensional Lie algebras were classified up to automorphism in [37]; from this
classification one can easily derive a classification (under detached feedback equivalence) of the homogeneous systems on the simply connected four-dimensional Lie
groups.

7.2.4.1   The Heisenberg Groups H_{2n+1}

The $(2n + 1)$-dimensional Heisenberg group may be realized as a matrix Lie group

$$
H_{2n+1} = \left\{
\begin{bmatrix}
1 & x_1 & x_2 & \cdots & x_n & z \\
0 & 1 & 0 & & 0 & y_1 \\
0 & 0 & 1 & & 0 & y_2 \\
\vdots & & & \ddots & & \vdots \\
0 & & \cdots & & 1 & y_n \\
0 & & \cdots & & 0 & 1
\end{bmatrix}
: x_i, y_i, z \in \mathbb{R}
\right\}.
$$

H_{2n+1} is a simply connected nilpotent Lie group with one-dimensional center; its
Lie algebra

$$
\mathfrak{h}_{2n+1} = \left\{
\begin{bmatrix}
0 & x_1 & x_2 & \cdots & x_n & z \\
0 & 0 & 0 & & 0 & y_1 \\
0 & 0 & 0 & & 0 & y_2 \\
\vdots & & & \ddots & & \vdots \\
0 & & \cdots & & 0 & y_n \\
0 & & \cdots & & 0 & 0
\end{bmatrix}
= zZ + \sum_{i=1}^{n}(x_i X_i + y_i Y_i) : x_i, y_i, z \in \mathbb{R}
\right\}
$$

has non-zero commutators $[X_i, Y_j] = \delta_{ij} Z$.

The automorphisms of $\mathfrak{h}_{2n+1}$ are exactly those linear isomorphisms that preserve
the centre $\mathfrak{z}$ of $\mathfrak{h}_{2n+1}$ and for which the induced map on $\mathfrak{h}_{2n+1}/\mathfrak{z}$ preserves an
appropriate symplectic structure (cf. [53]). More precisely, let ω be the skew-
symmetric bilinear form on $\mathfrak{h}_{2n+1}$ specified by $[A, B] = \omega(A, B)Z$ for $A, B \in$

$\mathfrak{h}_{2n+1}$. A linear isomorphism $\psi : \mathfrak{h}_n \to \mathfrak{h}_{2n+1}$ is a Lie algebra automorphism if and only if $\psi \cdot Z = cZ$ and $\omega(\psi \cdot A, \psi \cdot B) = c\,\omega(A, B)$ for some $c \neq 0$.

We give a matrix representation for the group of automorphisms. Throughout, we use the ordered basis

$$(Z, X_1, Y_1, X_2, Y_2, \ldots, X_n, Y_n)$$

for $\mathfrak{h}_{2n+1}$. The bilinear form ω takes the form

$$\omega = \begin{bmatrix} 0 & 0 \\ 0 & J \end{bmatrix}, \quad \text{where} \quad J = \begin{bmatrix} 0 & 1 & & & 0 \\ -1 & 0 & & & \\ & & \ddots & & \\ & & & 0 & 1 \\ 0 & & & -1 & 0 \end{bmatrix}.$$

We note that the linear map

$$\varsigma = \begin{bmatrix} -1 & 0 & \cdots & & 0 \\ 0 & 0 & 1 & & 0 \\ & 1 & 0 & & \\ \vdots & & & \ddots & \\ & & & & 0 & 1 \\ 0 & 0 & & & 1 & 0 \end{bmatrix} \tag{7.2}$$

is an automorphism.

Proposition 1 (cf. [24, 91]) *The group of automorphisms* $\mathsf{Aut}(\mathfrak{h}_{2n+1})$ *is given by*

$$\left\{ \begin{bmatrix} r^2 & v \\ 0 & rg \end{bmatrix}, \ \varsigma \begin{bmatrix} r^2 & v \\ 0 & rg \end{bmatrix} : r > 0, \ v \in \mathbb{R}^{2n}, \ g \in \mathsf{Sp}\,(n, \mathbb{R}) \right\}$$

where

$$\mathsf{Sp}\,(n, \mathbb{R}) = \left\{ g \in \mathbb{R}^{2n \times 2n} : g^\top J g = J \right\}$$

is the $n(2n + 1)$-*dimensional symplectic group over* $\mathbb{R}$.

Theorem 6 (cf. [24]) *Every controllable system on* H_{2n+1} *is detached feedback equivalent to one of the following three systems*

$$\Sigma^{(2n,0)} \ : \ u_1 X_1 + \cdots + u_n X_n + u_{n+1} Y_1 + \cdots + u_{2n} Y_n$$

$$\Sigma^{(2n,1)} \ : \ Z + u_1 X_1 + \cdots + u_n X_n + u_{n+1} Y_1 + \cdots + u_{2n} Y_n$$

$$\Sigma^{(2n+1,0)} \ : \ u_1 X_1 + \cdots + u_n X_n + u_{n+1} Y_1 + \cdots + u_{2n} Y_n + u_{2n+1} Z.$$

Proof Let Σ be a system on H_{2n+1} with trace $\Gamma = A + \Gamma^0$. As H_{2n+1} is completely solvable and simply connected, controllability of Σ is equivalent to the condition $\mathsf{Lie}(\Gamma^0) = \mathfrak{h}_{2n+1}$ [94]. We have $\mathsf{Lie}(\Gamma^0) \subseteq \mathrm{span}(\Gamma^0, Z)$. Therefore, if Σ is controllable, then Γ^0 has codimension zero or one. Moreover, if $\mathsf{Lie}(\Gamma^0) = \mathfrak{h}_{2n+1}$ and Γ^0 has codimension one, then

$$\Gamma^0 = \langle X_1 + v_1 Z, \dots, X_n + v_n Z, Y_1 + v_{n+1} Z, \dots, Y_n + v_{2n} Z \rangle$$

for some $v_1, \dots, v_{2n} \in \mathbb{R}$.

Let Σ be a homogeneous $2n$-input system on H_{2n+1} with trace $\Gamma = \Gamma^0$. As Σ is controllable, we have $\mathsf{Lie}(\Gamma) = \mathfrak{h}_{2n+1}$. As H_{2n+1} is simply connected, we have $d\,\mathsf{Aut}(\mathsf{H}_{2n+1}) = \mathsf{Aut}(\mathfrak{h}_{2n+1})$. Hence, it suffices to show that there exists an (inner) automorphism $\psi \in \mathsf{Aut}(\mathfrak{h}_n)$ such that $\psi \cdot \Gamma^0 = \langle X_1, \dots, X_n, Y_1, \dots, Y_n \rangle$. By the above argument

$$\Gamma = \langle X_1 + v_1 Z, \dots, X_n + v_n Z, Y_1 + v_{n+1} Z, \dots, Y_n + v_{2n} Z \rangle .$$

Accordingly,

$$\psi = \begin{bmatrix} 1 & -v \\ 0 & I_{2n} \end{bmatrix}, \quad v = \begin{bmatrix} v_1 & v_{n+1} & v_2 & v_{n+2} \cdots v_n & v_{2n} \end{bmatrix}$$

is an inner automorphism such that $\psi \cdot \Gamma = \langle X_1, \dots, X_n, Y_1, \dots, Y_n \rangle$.

Let Σ be a inhomogeneous $2n$-input system on H_{2n+1} with trace $\Gamma = A + \Gamma^0$. As Σ is controllable, we have $\mathsf{Lie}(\Gamma^0) = \mathfrak{h}_{2n+1}$. By the above argument, there exists an automorphism ψ such that $\psi \cdot \Gamma^0 = \langle X_1, \dots, X_n, Y_1, \dots, Y_n \rangle$. Thus $\psi \cdot \Gamma = A + \langle X_1, \dots, X_n, Y_1, \dots, Y_n \rangle = aZ + \langle X_1, \dots, X_n, Y_1, \dots, Y_n \rangle$ for some $a \neq 0$. Accordingly, $\psi' = \mathrm{diag}(\frac{1}{|a|}, \frac{1}{\sqrt{|a|}}, \dots, \frac{1}{\sqrt{|a|}})$ or $\psi' = \varsigma \, \mathrm{diag}(\frac{1}{|a|}, \frac{1}{\sqrt{|a|}}, \dots, \frac{1}{\sqrt{|a|}})$ is an automorphism such that $\psi \cdot \Gamma = Z + \langle X_1, \dots, X_n, Y_1, \dots, Y_n \rangle$.

Clearly, any $(2n + 1)$-input system is equivalent to $\Sigma^{(2n+1,0)}$.   □

7.2.4.2   The Orthogonal Group $\mathsf{SO}\,(4)$

The orthogonal group

$$\mathsf{SO}\,(4) = \{g \in \mathsf{GL}\,(4, \mathbb{R}) : g^\top g = \mathbf{1}, \ \det g = 1\}$$

is a six-dimensional semisimple compact connected Lie group. Its Lie algebra $\mathfrak{so}\,(4) = \{A \in \mathbb{R}^{4 \times 4} : A^\top + A = \mathbf{0}\}$ is isomorphic to $\mathfrak{so}\,(3) \oplus \mathfrak{so}\,(3)$. Let $(\mathbf{E}_1, \mathbf{E}_2, \mathbf{E}_3)$ denote the standard (ordered) basis for $\mathfrak{so}\,(3)$ (see the Appendix). The map $\varsigma : \mathfrak{so}\,(3) \oplus \mathfrak{so}\,(3) \to \mathfrak{so}\,(4)$, given by

$$\left(\begin{bmatrix} 0 & -x_3 & x_2 \\ x_3 & 0 & -x_1 \\ -x_2 & x_1 & 0 \end{bmatrix}, \begin{bmatrix} 0 & -y_3 & y_2 \\ y_3 & 0 & -y_1 \\ -y_2 & y_1 & 0 \end{bmatrix}\right) \longmapsto \frac{1}{2}\begin{bmatrix} 0 & x_3-y_3 & x_2-y_2 & x_1-y_1 \\ -x_3+y_3 & 0 & x_1+y_1 & -x_2-y_2 \\ -x_2+y_2 & -x_1-y_1 & 0 & x_3+y_3 \\ -x_1+y_1 & x_2+y_2 & -x_3-y_3 & 0 \end{bmatrix}$$

is a Lie algebra isomorphism. We define an ordered basis $(E_1,\dots,E_6)$ for $\mathfrak{so}\,(4)$ by

$$E_i = \varsigma \cdot (\mathbf{E}_i, \mathbf{0}), \quad i = 1,2,3 \qquad E_j = \varsigma \cdot (\mathbf{0}, \mathbf{E}_{j-3}), \quad j = 4,5,6.$$

The group of inner automorphisms of $\mathfrak{so}\,(4)$ is given by

$$\mathsf{Int}\,(\mathfrak{so}\,(4)) = \left\{ \begin{bmatrix} \psi_1 & \mathbf{0} \\ \mathbf{0} & \psi_2 \end{bmatrix} : \psi_1,\, \psi_2 \in \mathsf{SO}\,(3) \right\}.$$

The group of automorphisms $\mathsf{Aut}(\mathfrak{so}\,(4))$ is generated by $\mathsf{Int}\,(\mathfrak{so}\,(4))$ and the swap automorphism $\zeta = \begin{bmatrix} \mathbf{0} & I_3 \\ I_3 & \mathbf{0} \end{bmatrix}$ [3]. Moreover, the group of automorphisms decomposes as a semi-direct product $\mathsf{Aut}(\mathfrak{so}\,(4)) = \mathsf{Int}\,(\mathfrak{so}\,(4)) \rtimes \{\mathbf{1}, \zeta\}$ of the inner automorphisms (normal) and the two-element subgroup generated by the involution ζ.

Proposition 2 $d\,\mathsf{Aut}(\mathsf{SO}\,(4)) = \mathsf{Aut}(\mathfrak{so}(4))$.

Proof As $\mathsf{Aut}(\mathfrak{so}\,(4))$ is generated by $\mathsf{Int}\,(\mathfrak{so}\,(4))$ and ζ, and clearly $\mathsf{Int}\,(\mathfrak{so}\,(4)) \subseteq d\,\mathsf{Aut}(\mathsf{SO}\,(4))$, it suffices to show that $\zeta \in d\,\mathsf{Aut}(\mathsf{SO}\,(4))$.

The mapping $\psi : \mathfrak{su}\,(2) \oplus \mathfrak{su}\,(2) \to \mathfrak{so}\,(4)$ given by

$$\left(\begin{bmatrix} \frac{i}{2}x_1 & \frac{1}{2}(ix_3+x_2) \\ \frac{1}{2}(ix_3-x_2) & -\frac{i}{2}x_1 \end{bmatrix}, \begin{bmatrix} \frac{i}{2}y_1 & \frac{1}{2}(iy_3+y_2) \\ \frac{1}{2}(iy_3-y_2) & -\frac{i}{2}y_1 \end{bmatrix}\right)$$

$$\longmapsto \frac{1}{2}\begin{bmatrix} 0 & x_3-y_3 & x_2-y_2 & x_1-y_1 \\ -x_3+y_3 & 0 & x_1+y_1 & -x_2-y_2 \\ -x_2+y_2 & -x_1-y_1 & 0 & x_3+y_3 \\ -x_1+y_1 & x_2+y_2 & -x_3-y_3 & 0 \end{bmatrix}$$

is a Lie algebra isomorphism. Accordingly, as $\mathsf{SU}\,(2) \times \mathsf{SU}\,(2)$ is simply connected, there exists a unique Lie group homomorphism $q : \mathsf{SU}\,(2) \times \mathsf{SU}\,(2) \to \mathsf{SO}\,(4)$ with $T_1 q = \psi$ (see, e.g., [53]). It is not difficult to see that q is in fact an epimorphism and so q is a universal covering of $\mathsf{SO}\,(4)$. Consequently, $\ker q$ is a subgroup of the centre $Z(\mathsf{SU}(2) \times \mathsf{SU}(2)) = \{(I_2, I_2), (I_2, -I_2), (-I_2, I_2), (-I_2, -I_2)\}$. Let $C = \begin{bmatrix} i\pi & 0 \\ 0 & -i\pi \end{bmatrix} \in \mathfrak{su}(2)$. We have $\exp(C) = -I_2$, $\exp(\psi \cdot (C, 0)) = \exp(\psi \cdot (0, C)) = -I_4$, and $\exp(\psi \cdot (C, C)) = I_4$. Hence, as $q(\exp(A)) = \exp(\psi \cdot A)$ for $A \in \mathfrak{su}(2) \oplus \mathfrak{su}(2)$, it follows that $\ker q = \{(I_2, I_2), (-I_2, -I_2)\}$. Let ϕ be the unique automorphism of $\mathsf{SU}(2) \times \mathsf{SU}(2)$ such that $T_1\phi = \psi^{-1} \circ \zeta \circ \psi$. We have that $\exp(T_1\phi \cdot (C, C)) = \exp((C, C))$. Hence $\phi(\ker q) = \ker q$. It follows that $\zeta \in d\,\mathsf{Aut}(\mathsf{SO}(4))$ (see Proposition 5). $\qquad\square$

Theorem 7 ([5], see also [3]) *Any homogeneous system on* $\mathsf{SO}(4)$ *is detached feedback equivalent to exactly one of the following systems*

$\Sigma_\beta^{(1,0)} : u_1(E_1 + \beta E_4), \quad 0 \leqslant \beta \leqslant 1$

$\Sigma_1^{(2,0)} : u_1 E_1 + u_2 E_4$

$\Sigma_{2,\alpha}^{(2,0)} : u_1(E_1 + \alpha_1 E_4) + u_2(E_2 + \alpha_2 E_5),$

$\quad (0 = \alpha_2 \leqslant \alpha_1) \vee (1 \leqslant \tfrac{1}{\alpha_2} \leqslant \alpha_1) \vee (0 < \alpha_2 \leqslant \alpha_1 < 1)$

$\Sigma_{1,\beta}^{(3,0)} : u_1(E_1 + \beta E_4) + u_2 E_2 + u_3 E_6, \quad 0 \leqslant \beta \leqslant 1$

$\Sigma_{2,\alpha}^{(3,0)} : u_1(E_1 + \alpha_1 E_4) + u_2(E_2 + \alpha_2 E_5) + u_3(E_3 + \alpha_3 E_6),$

$\quad (0 = \alpha_3 \leqslant \alpha_2 \leqslant \alpha_1) \vee (0 < |\alpha_3| \leqslant \alpha_2 < 1 \wedge \alpha_2 \leqslant \alpha_1) \vee (\alpha_2 = 1 \leqslant \tfrac{1}{|\alpha_3|} \leqslant \alpha_1)$

$\Sigma_1^{(4,0)} : u_1 E_2 + u_2 E_3 + u_3 E_5 + u_4 E_6$

$\Sigma_{2,\alpha}^{(4,0)} : u_1(E_4 - \alpha_1 E_1) + u_2(E_5 - \alpha_2 E_2) + u_3 E_3 + u_4 E_6,$

$\quad (0 = \alpha_2 \leqslant \alpha_1) \vee (1 \leqslant \tfrac{1}{\alpha_2} \leqslant \alpha_1) \vee (0 < \alpha_2 \leqslant \alpha_1 < 1)$

$\Sigma_\beta^{(5,0)} : u_1(E_4 - \beta E_1) + u_2 E_2 + u_3 E_3 + u_4 E_5 + u_5 E_6$

$\Sigma^{(6,0)} : u_1 E_1 + u_2 E_2 + u_3 E_3 + u_4 E_4 + u_5 E_5 + u_6 E_6.$

Here $\boldsymbol\alpha$ *and* β *parametrize families of class representatives.*

Remark 5 As $\mathsf{SO}(4)$ is compact, a system on $\mathsf{SO}(4)$ is controllable if and only if it has full rank. For the homogeneous systems on $\mathsf{SO}(4)$ we have the following. No single input system has full rank; every five- or six-input system has full rank. $\Sigma_1^{(2,0)}$ does not have full rank; $\Sigma_{2,\alpha}^{(2,0)}$ does not have full rank exactly when $\alpha_2 = 0$ or $\alpha_1 = \alpha_2 = 1$. $\Sigma_{1,\beta}^{(3,0)}$ has full rank exactly when $\beta > 0$; $\Sigma_{2,\alpha}^{(3,0)}$ does not have full rank exactly when $\alpha_1 = \alpha_2 = \alpha_3 = 1$ or $\alpha_2 = 0$. $\Sigma_1^{(4,0)}$ has full rank; $\Sigma_{2,\alpha}^{(4,0)}$ does not have full rank exactly when $\alpha_1 = \alpha_2 = 0$.

Proof (Sketch) Any linearized automorphism $\psi \in d\,\mathsf{Aut}(\mathsf{SO}\,(4))$ preserves the dot product $A \bullet B = \sum_{i=1}^{6} a_i b_i$. (Here $A = \sum_{i=1}^{6} a_i E_i$ and $B = \sum_{i=1}^{6} b_i E_i$.) Let $\Gamma^{\perp}$ denote the orthogonal complement of a subspace $\Gamma \subset \mathfrak{so}\,(4)$. It is then easy to show that, if Γ, $\widetilde{\Gamma}$ are subspaces of $\mathfrak{so}\,(4)$ and $\psi \in \mathsf{Aut}(\mathfrak{so}\,(4))$, then $\psi \cdot \Gamma = \widetilde{\Gamma}$ if and only if $\psi \cdot \Gamma^{\perp} = \widetilde{\Gamma}^{\perp}$. The classification of the $(6-\ell)$-input systems therefore follows from the classification of the ℓ-input systems. Hence, we need only classify the single-input, two-input, and three-input systems. In order to find normal forms for the four- and five-input systems we need only take the orthogonal complements of the traces of the two- and one-input systems, respectively. (The classification for the six-input systems is trivial.)

We write an ℓ-input homogeneous system Σ : $u_1 \sum_{i=1}^{6} b_1^i E_i + \cdots + u_\ell \sum_{i=1}^{6} b_\ell^i E_i$ (in matrix form) as

$$\Sigma : \begin{bmatrix} M_1 \\ M_2 \end{bmatrix} = \begin{bmatrix} b_1^1 & \cdots & b_\ell^1 \\ \vdots & & \vdots \\ b_1^6 & \cdots & b_\ell^6 \end{bmatrix}.$$

Here M_1, $M_2 \in \mathbb{R}^{3 \times \ell}$. The evaluation $\psi \cdot \varXi\,(1, \boldsymbol{u})$ then becomes a matrix multiplication. Accordingly, two ℓ-input homogeneous systems $\Sigma : \begin{bmatrix} M_1 \\ M_2 \end{bmatrix}$ and $\Sigma' : \begin{bmatrix} M_1' \\ M_2' \end{bmatrix}$ are equivalent if and only if there exist an automorphism $\psi \in d\,\mathsf{Aut}(\mathsf{SO}\,(4))$ and $K \in \mathsf{GL}\,(\ell, \mathbb{R})$ such that $\psi \cdot \begin{bmatrix} M_1 \\ M_2 \end{bmatrix} K = \begin{bmatrix} M_1' \\ M_2' \end{bmatrix}$. More precisely, Σ and Σ' are equivalent if and only if there exist $R_1, R_2 \in \mathsf{SO}\,(3)$ and $K \in \mathsf{GL}\,(\ell, \mathbb{R})$ such that

$$\left(R_1 M_1 K = M_1' \quad \text{and} \quad R_2 M_2 K = M_2' \right)$$

$$\text{or} \quad \left(R_1 M_2 K = M_1' \quad \text{and} \quad R_2 M_1 K = M_2' \right).$$

The singular value decomposition (see, e.g., [80]) plays a key role in our argument. For any matrix $M \in \mathbb{R}^{m \times n}$ of rank r, there exist orthogonal matrices $U \in \mathbb{R}^{m \times m}$, $V \in \mathbb{R}^{n \times n}$ and a diagonal matrix $D \in \mathbb{R}^{r \times r} = \mathrm{diag}(\sigma_1, \ldots, \sigma_r)$ such that $M = U \begin{bmatrix} D & 0 \\ 0 & 0 \end{bmatrix} V^{\mathsf{T}}$ with $\sigma_1 \geqslant \cdots \geqslant \sigma_r > 0$. Using this result, together with the above characterization of equivalence (in matrix form), it is fairly straightforward to compute normal forms for the one-, two-, and three-input homogeneous systems on $\mathsf{SO}\,(4)$. $\qquad\square$

Note 3 The inhomogeneous systems on $\mathsf{SO}\,(4)$ were also classified in [5]; the approach is similar though the computations are more involved.

7.3   Invariant Optimal Control

We consider the class of left-invariant optimal control problems on Lie groups with fixed terminal time, affine dynamics, and affine quadratic cost. Formally, such a problem is written as

$$\dot{g} = g\left(A + u_1 B_1 + \cdots + u_\ell B_\ell\right), \qquad g \in \mathsf{G}, \ u \in \mathbb{R}^\ell \tag{7.3}$$

$$g(0) = g_0, \quad g(T) = g_1 \tag{7.4}$$

$$\mathscr{J} = \int_0^T \left(u(t) - \mu\right)^\top Q\left(u(t) - \mu\right) dt \longrightarrow \min. \tag{7.5}$$

Here G is a (real, finite-dimensional, connected) Lie group with Lie algebra $\mathfrak{g}$, A, $B_1, \ldots, B_\ell \in \mathfrak{g}$ (with $B_1, \ldots, B_\ell$ linearly independent), $u = (u_1, \ldots, u_\ell) \in \mathbb{R}^\ell$, $\mu \in \mathbb{R}^\ell$, and Q is a positive definite $\ell \times \ell$ matrix. To each such problem, we associate a *cost-extended system* (Σ, χ). Here Σ is the control system (7.3) and the cost function $\chi : \mathbb{R}^\ell \to \mathbb{R}$ has the form $\chi(u) = (u - \mu)^\top Q\left(u - \mu\right)$. Each cost-extended system corresponds to a family of invariant optimal control problems; by specification of the boundary data (g_0, g_1, T), the associated problem is uniquely determined.

Optimal control problems of this kind have received considerable attention in the last few decades. Various physical problems have been modelled in this manner (cf. Sect. 7.1.2), such as optimal path planning for airplanes, motion planning for wheeled mobile robots, spacecraft attitude control, and the control of underactuated underwater vehicles [77, 89, 100]; also, the control of quantum systems and the dynamic formation of DNA [48, 54]. Many problems (as well as sub-Riemannian structures) on various low-dimensional matrix Lie groups have been considered by a number of authors (see, e.g., [22, 23, 41, 63, 67, 69, 81, 83, 88, 92, 93]).

We introduce a form of equivalence for problems of the form (7.3)–(7.4)–(7.5), or rather, for the associated cost-extended systems (cf. [27, 35]). *Cost equivalence* establishes a one-to-one correspondence between the associated optimal trajectories, as well as the associated extremal curves. We associate to each cost-extended system, via the Pontryagin Maximum Principle, a quadratic Hamilton–Poisson systems on the associated Lie–Poisson space. We show that cost equivalence of cost-extended systems implies equivalence of the associated Hamiltonian systems. In addition, we reinterpret drift-free cost-extended systems (with homogeneous cost) as invariant sub-Riemannian structures.

A number of illustrative examples concerning classification (some quite extensive) are included.

7.3.1 Pontryagin Maximum Principle

The Pontryagin Maximum Principle provides necessary conditions for optimality which are naturally expressed in the language of the geometry of the cotangent bundle T^*G of G (see [10, 51, 64]). The cotangent bundle T^*G can be trivialized (from the left) such that $T^*G = G \times \mathfrak{g}^*$; here $\mathfrak{g}^*$ is the dual of the Lie algebra $\mathfrak{g}$. To an optimal control problem (7.3)–(7.4)–(7.5) we associate, for each real number λ and each control parameter $u \in \mathbb{R}^\ell$ a Hamiltonian function on $T^*G = G \times \mathfrak{g}^*$:

$$H_u^\lambda(\xi) = \lambda\chi(u) + \xi\,(\Xi_u(g))$$
$$= \lambda\chi(u) + p\,(\Xi_u(\mathbf{1})), \qquad \xi = (g,p) \in T^*G. \tag{7.6}$$

We denote by $\overrightarrow{H}_u^\lambda$ the corresponding Hamiltonian vector field (with respect to the symplectic structure on T^*G).

Lemma 1 ([64], see also [10]) *A curve $\xi(\cdot) = (g(\cdot), p(\cdot))$ is an integral curve of (the time varying Hamiltonian vector field) $\overrightarrow{H}_{u(t)}^\lambda$ if and only if*

$$\dot{g}(t) = \Xi(g(t), u(t)) \qquad\qquad \dot{p}(t) = \mathrm{ad}^*\,\Xi_{u(t)}(\mathbf{1}) \cdot p(t).$$

Here $(\mathrm{ad}^*\,A \cdot p)(B) = p([A, B])$ *for* $A, B \in \mathfrak{g}$ *and* $p \in \mathfrak{g}^*$.

In terms of the above Hamiltonians, the Maximum Principle can be stated as follows.

Theorem 8 (Maximum Principle) *Suppose the controlled trajectory $(\bar{g}(\cdot), \bar{u}(\cdot))$ defined over the interval $[0, T]$ is a solution for the optimal control problem (7.11). Then, there exists a curve $\xi(\cdot) : [0, T] \to T^*G$ with $\xi(t) \in T_{\bar{g}(t)}^*G$, $t \in [0, T]$, and a real number $\lambda \le 0$, such that the following conditions hold for almost every $t \in [0, T]$:*

$$(\lambda, \xi(t)) \not\equiv (0, 0) \tag{7.7}$$

$$\dot{\xi}(t) = \overrightarrow{H}_{\bar{u}(t)}^\lambda(\xi(t)) \tag{7.8}$$

$$H_{\bar{u}(t)}^\lambda\,(\xi(t)) = \max_u H_u^\lambda\,(\xi(t)) = \text{constant}. \tag{7.9}$$

An optimal trajectory, $g(\cdot) : [0, T] \to G$ is the projection of an integral curve $\xi(\cdot)$ of the (time-varying) Hamiltonian vector field $\overrightarrow{H}_{\bar{u}(t)}^\lambda$. A trajectory-control pair $(\xi(\cdot), u(\cdot))$ is said to be an extremal pair if $\xi(\cdot)$ satisfies the conditions (7.7)–(7.9). The projection $\xi(\cdot)$ of an extremal pair is called an extremal. An extremal curve is called normal if $\lambda < 0$ and abnormal if $\lambda = 0$.

For the class of optimal control problems under consideration, the maximum condition (7.9) eliminates the parameter u from the family of Hamiltonians (H_u); as a result, we obtain a smooth G-invariant function H on $T^*G = G \times \mathfrak{g}^*$. This Hamilton–Poisson system on T^*G can be reduced to a Hamilton–Poisson system on the (minus) Lie–Poisson space $\mathfrak{g}_-^*$, with Poisson bracket given by

$$\{F, G\} = -p([dF(p), dG(p)]).$$

Here $F, G \in C^\infty(\mathfrak{g}^*)$ and $dF(p), dG(p)$ are elements of the double dual $\mathfrak{g}^{**}$ which is canonically identified with the Lie algebra $\mathfrak{g}$.

7.3.2 Equivalence of Cost-Extended Systems

Let (Σ, χ) and (Σ', χ') be two cost-extended systems. (Σ, χ) and (Σ', χ') are said to be *cost equivalent* if there exist a Lie group isomorphism $\phi : \mathsf{G} \to \mathsf{G}'$ and an affine isomorphism $\varphi : \mathbb{R}^\ell \to \mathbb{R}^\ell$ such that

$$T_g\phi \cdot \Xi(g, u) = \Xi'(\phi(g), \varphi(u)) \qquad \text{and} \qquad \chi' \circ \varphi = r\chi$$

for $g \in \mathsf{G}$, $u \in \mathbb{R}^\ell$ and some $r > 0$. Equivalently, (Σ, χ) and (Σ', χ') are cost equivalent if and only if there exist a Lie group isomorphism $\phi : \mathsf{G} \to \mathsf{G}'$ and an affine isomorphism $\varphi : \mathbb{R}^\ell \to \mathbb{R}^\ell$ such that $T_\mathbf{1}\phi \cdot \Xi(\mathbf{1}, u) = \Xi'(\mathbf{1}', \varphi(u))$ and $\chi' \circ \varphi = r\chi$ for some $r > 0$. Accordingly:

- If (Σ, χ) and (Σ', χ') are cost equivalent, then Σ and Σ' are detached feedback equivalent.
- If two full-rank systems Σ and Σ' are state space equivalent, then (Σ, χ) and (Σ', χ) are cost equivalent for any cost χ.
- If two full-rank systems Σ and Σ' are detached feedback equivalent with respect to a feedback transformation φ, then $(\Sigma, \chi \circ \varphi)$ and (Σ', χ) are cost equivalent for any cost χ.

Remark 6 The cost-preserving condition $\chi' \circ \varphi = r\chi$ is partially motivated by the following considerations. Each cost χ on $\mathbb{R}^\ell$ induces a strict partial ordering $u < v \iff \chi(u) < \chi(v)$. It turns out that χ and χ' induce the same strict partial ordering on $\mathbb{R}^\ell$ if and only if $\chi = r\chi'$ for some $r > 0$. The dynamics-preserving condition $T_g\phi \cdot \Xi(g, u) = \Xi'(\phi(g), \varphi(u))$ is just that of detached feedback equivalence (on full-rank systems).

Let $(g(\cdot), u(\cdot))$ be a controlled trajectory, defined over an interval $[0, T]$, of a cost-extended system (Σ, χ). We say that $(g(\cdot), u(\cdot))$ is a *virtually optimal controlled trajectory* (shortly VOCT) if it is a solution for the associated optimal control problem with boundary data $(g(0), g(T), T)$. Similarly, we say that $(g(\cdot), u(\cdot))$ is an *extremal controlled trajectory* (shortly ECT) if it satisfies the necessary conditions of the Pontryagin Maximum Principle (with $\lambda \leqslant 0$). Clearly, any VOCT is an ECT. A map $\phi \times \varphi$ defining a cost equivalence between two cost-extended systems establishes a one-to-one correspondence between their respective VOCTs (and ECTs).

Theorem 9 ([27, 35]) *Suppose $\phi \times \varphi$ defines a cost equivalence between (Σ, χ) and (Σ', χ'). Then*

1. *$(g(\cdot), u(\cdot))$ is a VOCT if and only if $(\phi \circ g(\cdot), \varphi \circ u(\cdot))$ is a VOCT;*
2. *$(g(\cdot), u(\cdot))$ is an ECT if and only if $(\phi \circ g(\cdot), \varphi \circ u(\cdot))$ is an ECT.*

Proof

(1) Suppose $(g(\cdot), u(\cdot))$ is a controlled trajectory of (Σ, χ) and $(\phi \circ g(\cdot), \varphi \circ u(\cdot))$ is a VOCT of (Σ', χ'). Further suppose $(g(\cdot), u(\cdot))$ is not a VOCT of (Σ, χ). Then there exists another controlled trajectory $(h(\cdot), v(\cdot))$ such that $h(0) = g(0)$, $h(T) = g(T)$, and

$$\mathscr{J}(v(\cdot)) = \int_0^T \chi(v(t))\, dt < \int_0^T \chi(u(t))\, dt = \mathscr{J}(u(\cdot)).$$

Hence $\big(\phi \circ h(\cdot), \varphi \circ v(\cdot)\big)$ is a controlled trajectory of (Σ', χ') such that (for some $r > 0$)

$$\mathscr{J}'(\varphi \circ v(\cdot)) = \int_0^T (\chi' \circ \varphi)(v(t))\, dt = r \int_0^T \chi(v(t))\, dt$$

$$< r \int_0^T \chi(u(t))\, dt = \int_0^T (\chi' \circ \varphi)(u(t))\, dt = \mathscr{J}'(\varphi \circ u(\cdot)).$$

However, this contradicts the fact that $(\phi \circ g(\cdot), \varphi \circ u(\cdot))$ is a VOCT of (Σ', χ'). Thus if $(\phi \circ g(\cdot), \varphi \circ u(\cdot))$ is a VOCT, then so is $(g(\cdot), u(\cdot))$. As $\phi^{-1} \times \varphi^{-1}$ defines a cost equivalence between (Σ', χ') and (Σ, χ), the same argument can be used to show the converse.

(2) The Hamiltonian functions (7.6), associated to (Σ, χ) and (Σ', χ'), are given by

$$H_u(g, p) = p(\Xi_u(\mathbf{1})) + \lambda\chi(u) \quad \text{and} \quad H'_{u'}(g', p') = p'(\Xi'_{u'}(\mathbf{1})) + \lambda\chi'(u)$$

respectively (for a fixed λ). Assume $(g(\cdot), u(\cdot))$ is a controlled trajectory of Σ such that $(g'(\cdot), u'(\cdot)) = (\phi \circ g(\cdot), \varphi \circ u(\cdot))$ is an ECT of (Σ', χ'). Then there exists $p'(\cdot) : [0, T] \to \mathfrak{g}^*$ such that $(\xi'(\cdot), u'(\cdot))$, $\xi'(t) = (g'(t), p'(t))$ satisfies (7.7)–(7.8)–(7.9). As $\dot{\xi}'(t) = \vec{H}'_{u'(t)}(\xi'(t))$ (7.8) we have (see Lemma 1)

$$\dot{g}'(t) = \Xi'(g'(t), u'(t)) \qquad \dot{p}'(t) = \mathrm{ad}^* \, \Xi'_{u'(t)}(\mathbf{1}) \cdot p'(t).$$

Let $p(\cdot) = \frac{1}{r}(T_1\phi)^* \cdot p'(\cdot)$ and $\xi(t) = (g(t), p(t))$. (Here $r > 0$ is the unique constant associated to $\phi \times \varphi$.)

We show that $(\xi(\cdot), u(\cdot))$ satisfies (7.8)–(7.9). By assumption we have that $\dot{g}(t) = \Xi(g(t), u(t))$. Thus, in order to satisfy (7.8), we are left to show that

$\dot{p}(t) = \text{ad}^* \, \Xi_{u(t)}(\mathbf{1}) \cdot p(t)$. Indeed, for $A \in \mathfrak{g}^*$ we have

$$\dot{p}(t) \cdot A = \tfrac{1}{r} \left((T_1\phi)^* \cdot \dot{p}'(t) \right) \cdot A$$

$$= \tfrac{1}{r} \left(\text{ad}^* \, \Xi'_{u'(t)}(\mathbf{1}) \cdot p'(t) \right) \cdot (T_1\phi \cdot A)$$

$$= \tfrac{1}{r} p'(t) \cdot \left[T_1\phi \cdot \Xi_{u(t)}(\mathbf{1}), T_1\phi \cdot A \right]$$

$$= \left(\text{ad}^* \, \Xi_{u(t)}(\mathbf{1}) \cdot p(t) \right) \cdot A.$$

In order to show that $(\xi(\cdot), u(\cdot))$ satisfies (7.9), we first show that $H_u(\xi(t)) = \tfrac{1}{r} H'_{\varphi(u)}(\xi'(t))$ for $u \in \mathbb{R}^k$. Indeed,

$$H_u(\xi(t)) = p(t) \cdot \Xi_u(\mathbf{1}) + \lambda \chi(u)$$

$$= (\tfrac{1}{r}(T_1\phi)^* \cdot p'(t)) \cdot \Xi_u(\mathbf{1}) + \lambda \chi(u)$$

$$= \tfrac{1}{r} p'(t) \cdot \Xi'_{\varphi(u)}(\mathbf{1}) + \tfrac{1}{r} \lambda \chi'(\varphi(u))$$

$$= \tfrac{1}{r} H'_{\varphi(u)}(\xi'(t)).$$

Hence, it follows that

$$H_{u(t)}(\xi(t)) = \tfrac{1}{r} H'_{u'(t)}(\xi'(t))$$

$$= \max_{u' \in \mathbb{R}^{k'}} \tfrac{1}{r} H'_{u'}\left(\xi'(t)\right) \qquad (= \text{constant})$$

$$= \max_{u \in \mathbb{R}^k} \tfrac{1}{r} H'_{\varphi(u)}\left(\xi'(t)\right)$$

$$= \max_{u \in \mathbb{R}^k} H_u\left(\xi(t)\right) = \text{constant}.$$

It is left to show that (7.7) holds. If $(g'(\cdot), u'(\cdot))$ is normal, then $\lambda < 0$ and so $(\lambda, \xi(\cdot)) \not\equiv 0$. Suppose $(g'(\cdot), u'(\cdot))$ is abnormal. We have that $(T_1\phi)^*$ is injective and as $(\lambda, \xi'(\cdot)) \not\equiv 0$, i.e., $p'(\cdot) \not\equiv 0$, it follows that $p(\cdot) = \tfrac{1}{r}(T_1\phi)^* \cdot p'(t) \not\equiv 0$, i.e., $(\lambda, \xi(\cdot)) \not\equiv 0$. Again, the converse may be found by a similar argument (utilizing $\phi^{-1} \times \varphi^{-1}$). $\qquad \Box$

7.3.2.1 Classification

The classification of a given class of cost-extended systems may be approached as follows. First, we classify the underlying systems under detached feedback equivalence and obtain normal forms Σ_i, $i \in I$. Any cost-extended system must then be equivalent to a cost-extended system (Σ_i, χ) for some $i \in I$ and some cost χ. Note that (Σ_i, χ) and (Σ_j, χ') cannot be equivalent unless $i = j$. Therefore the classification problem reduces to finding normal forms of the cost χ for each

Σ_i. Accordingly, we wish to characterize cost equivalence between cost-extended systems for which the underlying system is identical.

Let (Σ, χ) and (Σ, χ') be two cost-extended systems (with identical underlying systems). Let $\mathscr{T}_\Sigma$ be the group of feedback transformations leaving Σ invariant. More precisely,

$$\mathscr{T}_\Sigma = \{\varphi \in \mathsf{Aff}(\mathbb{R}^\ell) : \exists \psi \in d\,\mathsf{Aut}(G),\ \psi \cdot \Gamma = \Gamma,\ \psi \cdot \Xi(\mathbf{1}, u) = \Xi(\mathbf{1}, \varphi(u))\}.$$

(Here $\mathsf{Aff}(\mathbb{R}^\ell)$ is the group of affine isomorphisms of $\mathbb{R}^\ell$.) The following result is easy to prove.

Proposition 3 (Σ, χ) *and* (Σ, χ') *are cost equivalent if and only if there exists an element* $\varphi \in \mathscr{T}_\Sigma$ *such that* $\chi' = r\chi \circ \varphi$ *for some* $r > 0$.

Four examples of the classification of cost-extended systems (on the Euclidean, semi-Euclidean and Heisenberg groups) are exhibited. In the first two cases the classification procedure (as outlined above) is carried out in some detail.

Example 3 ([27]) On $\mathsf{SE}(2)$, any full-rank two-input drift-free cost-extended system (Σ, χ) with homogeneous cost (i.e., $\chi(0) = 0$) is cost equivalent to

$$(\Sigma^{(2,0)}, \chi^{(2,0)}) : \begin{cases} \Sigma : u_1 E_2 + u_2 E_3 \\ \chi(u) = u_1^2 + u_2^2. \end{cases}$$

Proof Let (Σ, χ) be a cost-extended system as prescribed. By Theorem 3 and the fact that $d\,\mathsf{Aut}(\mathsf{SE}(2)) = \mathsf{Aut}(\mathfrak{se}(2))$ (see Example 1), we have that Σ is detached feedback equivalent to the system $\Sigma^{(2,0)}$. Moreover, it turns out that the feedback transformation φ for this equivalence must be linear. Thus (Σ, χ) is cost equivalent to a cost-extended system $(\Sigma^{(2,0)}, \chi')$ for some $\chi' : u \mapsto u^\top Q' u$.

A straightforward calculation shows that the group $\mathscr{T}_{\Sigma^{(2,0)}}$ of feedback transformations leaving $\Sigma^{(2,0)}$ invariant is given by

$$\mathscr{T}_{\Sigma^{(2,0)}} = \left\{ u \mapsto \begin{bmatrix} \sigma x & w \\ 0 & \sigma \end{bmatrix} u : x \neq 0,\ w \in \mathbb{R},\ \sigma = \pm 1 \right\}.$$

Let $Q' = \begin{bmatrix} a_1 & b \\ b & a_2 \end{bmatrix}$. Now $\varphi_1 = \begin{bmatrix} 1 & -\frac{b}{a_1} \\ 0 & 1 \end{bmatrix} \in \mathscr{T}_{\Sigma_1}$ and $(\chi' \circ \varphi_1)(u) =$

$u^\top \mathrm{diag}(a_1, a_2 - \frac{b^2}{a_1}) u$. Let $a_2' = a_2 - \frac{b^2}{a_1}$ and let $\varphi_2 = \mathrm{diag}(\sqrt{\frac{a_2'}{a_1}}, 1) \in \mathscr{T}_{\Sigma_1}$. Then $(\chi' \circ (\varphi_1 \circ \varphi_2))(u) = a_2' u^\top u = a_2' \chi^{(2,0)}(u)$. Consequently, by Proposition 3, (Σ, χ) is cost equivalent to $(\Sigma^{(2,0)}, \chi^{(2,0)})$. $\square$

Example 4 ([21, 35]) Any controllable two-input inhomogeneous cost-extended system on H_3 is cost equivalent to exactly one of

$$(\Sigma^{(2,1)}, \chi_\alpha^{(2,1)}) \;:\; \begin{cases} \Sigma^{(2,1)} \;:\; E_1 + u_1 E_2 + u_2 E_3 \\[2mm] \chi_\alpha^{(2,1)}(u) = (u_1 - \alpha)^2 + u_2^2. \end{cases}$$

Here $\alpha \geq 0$ parametrizes a family of (non-equivalent) class representatives.

Proof Let (Σ, χ) be a cost-extended system as prescribed. By Theorem 6, Σ is detached feedback equivalent to the system $\Sigma^{(2,1)}$. Thus, (Σ, χ) is cost equivalent to $(\Sigma^{(2,1)}, \chi_0)$ for some cost χ_0.

As H_3 is simply connected, $d\,\mathsf{Aut}(H_3) = \mathsf{Aut}(\mathfrak{h}_3)$. A simple calculation shows that $\mathscr{T}_{\Sigma^{(2,1)}} = \mathsf{SL}(2,\mathbb{R})$. Now $\chi_0 : u \mapsto (u - \mu)^\top Q\,(u - \mu)$ for some positive definite matrix $Q = \begin{bmatrix} a_1 & b \\ b & a_2 \end{bmatrix}$ and $\mu \in \mathbb{R}^2$. We transform χ_0 into multiple of $\chi_\alpha^{(2,1)}$ by successively composing with elements of $\mathscr{T}_{\Sigma^{(2,1)}}$. We have $\varphi_1 = \begin{bmatrix} 1 & -\frac{b}{a_1} \\ 0 & 1 \end{bmatrix} \in \mathscr{T}_{\Sigma^{(2,1)}}$ and

$$\chi_1(u) = (\chi_0 \circ \varphi_1)(u) = (u - \mu')^\top \; \mathrm{diag}(a_1, a_2 - \tfrac{b^2}{a_1})\,(u - \mu')$$

for some $\mu' \in \mathbb{R}^2$. Let $a_2' = a_2 - \tfrac{b^2}{a_1}$. Then $\varphi_2 = \sqrt[4]{a_1 a_2'}\;\mathrm{diag}\left(\tfrac{1}{\sqrt{a_1}}, \tfrac{1}{\sqrt{a_2'}}\right) \in \mathscr{T}_{\Sigma^{(2,1)}}$ and

$$\chi_2(u) = \tfrac{1}{\sqrt{a_1 a_2'}}(\chi_1 \circ \varphi_2)(u) = (u - \mu'')^\top (u - \mu'')$$

for some $\mu'' \in \mathbb{R}^2$. There exist $\alpha \geq 0$ and $\theta \in \mathbb{R}$ such that $\mu_1'' = \alpha \cos \theta$ and $\mu_2'' = \alpha \sin \theta$. Hence $\varphi_3 = \begin{bmatrix} \cos \theta & -\sin \theta \\ \sin \theta & \cos \theta \end{bmatrix} \in \mathscr{T}_{\Sigma^{(2,1)}}$ and

$$\chi_\alpha^{(2,1)}(u) = (\chi_2 \circ \varphi_3)(u) = \left(u - \begin{bmatrix} \alpha \\ 0 \end{bmatrix}\right)^\top \left(u - \begin{bmatrix} \alpha \\ 0 \end{bmatrix}\right).$$

Therefore $\chi_\alpha^{(2,1)} = \tfrac{1}{\sqrt{a_1 a_2'}}\,\chi_0 \circ (\varphi_1 \circ \varphi_2 \circ \varphi_3)$. Consequently, by Proposition 3, (Σ, χ) is cost equivalent to $(\Sigma^{(2,1)}, \chi_\alpha^{(2,1)})$. A simple verification shows that $\chi_\alpha^{(2,1)} \circ \varphi \neq r\chi_{\alpha'}^{(2,1)}$ for any $\alpha \neq \alpha'$, $r > 0$, and $\varphi \in \mathscr{T}_{\Sigma^{(2,1)}}$. $\qquad\square$

Example 5 ([18]) On $\mathsf{SE}(1,1)$, any controllable drift-free cost-extended system (Σ, χ) with homogeneous cost (i.e., $\chi(0) = 0$) is cost equivalent to exactly one of the systems

$$(\Sigma^{(2,0)}, \chi^{(2,0)}) : \begin{cases} \Sigma^{(2,0)} : u_1 E_1 + u_2 E_3 \\ \chi^{(2,0)}(u) = u_1^2 + u_2^2 \end{cases}$$

$$(\Sigma^{(3,0)}, \chi_\lambda^{(3,0)}) : \begin{cases} \Sigma^{(3,0)} : u_1 E_1 + u_2 E_2 + u_3 E_3 \\ \chi_\lambda^{(3,0)}(u) = u_1^2 + \lambda u_2^2 + u_3^2. \end{cases}$$

Here $0 < \lambda \leqslant 1$ parametrizes a family of cost-extended systems, with each distinct value corresponding to a distinct (non-equivalent) class representative.

Example 6 ([24]) On H_{2n+1}, any drift-free controllable cost-extended system with homogeneous cost (i.e., $\chi(0) = 0$) is cost-equivalent to exactly one of the following cost-extended systems:

$$(\Sigma^{(2n,0)}, \chi_\lambda^{(2n,0)}) : \begin{cases} \Sigma^{(2n,0)} : \sum_{i=1}^{n} (u_i X_i + u_{n+i} Y_i) \\ \chi_\lambda^{(2n,0)}(u) = \sum_{i=1}^{n} \lambda_i (u_i^2 + u_{n+i}^2) \end{cases}$$

$$(\Sigma^{(2n+1,0)}, \chi_\lambda^{(2n+1,0)}) : \begin{cases} \Sigma^{(2n+1,0)} : \sum_{i=1}^{n} (u_i X_i + u_{n+i} Y_i) + u_{2n+1} Z \\ \chi_\lambda^{(2n+1,0)}(u) = \sum_{i=1}^{n} \lambda_i (u_i^2 + u_{n+i}^2) + u_{2n+1}^2. \end{cases}$$

Here $1 = \lambda_1 \geqslant \lambda_2 \geqslant \cdots \geqslant \lambda_n > 0$ parametrize families of (non-equivalent) class representatives.

7.3.3 Pontryagin Lift

To any cost-extended system (Σ, χ) on a Lie group G we associate, via the Pontryagin Maximum Principle, a Hamilton–Poisson system on the associated Lie–Poisson space $\mathfrak{g}_-^*$ (cf. [10, 64, 94]). We show that equivalence of cost-extended systems implies equivalence of the associated Hamilton–Poisson systems.

Note 4 The Pontryagin lift may be realized as a contravariant functor between the category of cost-extended control systems and the category of Hamilton–Poisson systems ([35], see also [51]).

A *quadratic Hamilton–Poisson system* $(\mathfrak{g}_-^*, H_{A,\mathscr{Q}})$ is specified by

$$H_{A,\mathscr{Q}} : \mathfrak{g}^* \to \mathbb{R}, \quad p \mapsto p(A) + \mathscr{Q}(p).$$

Here $A \in \mathfrak{g}$ and $\mathscr{Q}$ is a quadratic form on $\mathfrak{g}^*$. If $A = 0$, then the system is called *homogeneous*; otherwise, it is called *inhomogeneous*. (When $\mathfrak{g}_-^*$ is fixed, a system $(\mathfrak{g}_-^*, H_{A,\mathscr{Q}})$ is identified with its Hamiltonian $H_{A,\mathscr{Q}}$.) To each function $H \in$

$C^\infty(\mathfrak{g}^*)$, we associate a *Hamiltonian vector field* $\vec{H}$ on $\mathfrak{g}^*$ specified by $\vec{H}[F] = \{F, H\}$. A function $C \in C^\infty(\mathfrak{g}^*)$ is a *Casimir function* if $\{C, F\} = 0$ for all $F \in C^\infty(\mathfrak{g}^*)$, or, equivalently $\vec{C} = 0$. A linear map $\psi : \mathfrak{g}^* \to \mathfrak{h}^*$ is a *linear Poisson morphism* if $\{F, G\} \circ \psi = \{F \circ \psi, G \circ \psi\}$ for all $F, G \in C^\infty(\mathfrak{h}^*)$. Linear Poisson morphisms are exactly the dual maps of Lie algebra homomorphisms.

Let $(E_1, \ldots, E_n)$ be a ordered basis for the Lie algebra $\mathfrak{g}$ and let $(E_1^*, \ldots, E_n^*)$ denote the corresponding dual basis for $\mathfrak{g}^*$. We write elements $B \in \mathfrak{g}$ as column vectors and elements $p \in \mathfrak{g}^*$ as row vectors. The equations of motion for the integral curve $p(\cdot)$ of the Hamiltonian vector field $\vec{H}$ corresponding to $H \in C^\infty(\mathfrak{g}^*)$ then take the form $\dot{p}_i = -p([E_i, dH(p)])$. Whenever convenient, linear maps will be identified with their matrices. Writing elements $u \in \mathbb{R}^\ell$ as column vectors as well, we can express $\Xi_u(1) = A + u_1 B_1 + \cdots + u_\ell B_\ell$ as $\Xi_u(1) = A + \mathbf{B} u$, where $\mathbf{B} = \begin{bmatrix} B_1 \cdots B_\ell \end{bmatrix}$ is a $n \times \ell$ matrix.

Let (Σ, χ) be a cost-extended system with

$$\Xi_u(1) = A + \mathbf{B} u, \qquad \chi(u) = (u - \mu)^\top Q(u - \mu).$$

By the Pontryagin Maximum Principle we have the following result.

Theorem 10 ([27]) *Any normal ECT $(g(\cdot), u(\cdot))$ of (Σ, χ) is given by*

$$\dot{g}(t) = \Xi(g(t), u(t)), \qquad u(t) = Q^{-1} \mathbf{B}^\top p(t)^\top + \mu$$

where $p(\cdot) : [0, T] \to \mathfrak{g}^*$ is an integral curve for the Hamilton–Poisson system on $\mathfrak{g}_-^*$ specified by

$$H(p) = p(A + \mathbf{B} \mu) + \tfrac{1}{2} p \mathbf{B} Q^{-1} \mathbf{B}^\top p^\top. \tag{7.10}$$

Proof The Hamiltonian (7.6) is given by $H_u(g, p) = pA + p \mathbf{B} u + \lambda (u - \mu)^\top Q (u - \mu)$. Now, $\frac{\partial H_u}{\partial u}(g, p) = p \mathbf{B} + 2\lambda (u - \mu)^\top Q$. Hence the maximum $\max_u H_u(g, p)$ occurs at $u^\top = -\frac{1}{2\lambda} p \mathbf{B} Q^{-1} + \mu^\top$ (for a normal ECT we have $\lambda < 0$). Therefore the maximized Hamiltonian is given by

$$H(g, p) = \max_{u \in \mathbb{R}^k} H_u(g, p)$$

$$= p A + p \mathbf{B} (-\tfrac{1}{2\lambda} Q^{-1} \mathbf{B}^\top p^\top + \mu) + \tfrac{\lambda}{4\lambda^2} p \mathbf{B} Q^{-1} Q Q^{-1} \mathbf{B}^\top p^\top$$

$$= p (A + \mathbf{B} \mu) - \tfrac{1}{4\lambda} p \mathbf{B} Q^{-1} \mathbf{B}^\top p^\top.$$

As the maximized Hamiltonian H is defined and smooth, it follows that if $(\xi(\cdot), u(\cdot))$ satisfies (7.7)–(7.8)–(7.9), then $\dot{\xi}(t) = \vec{H}(\xi(t))$ and $u(\cdot) = -\frac{1}{2\lambda} Q^{-1} \mathbf{B}^\top p(\cdot) + \mu$; moreover, if $\xi(\cdot) = (g(\cdot), p(\cdot))$ is an integral curve of $\vec{H}$, then the pair $(\xi(\cdot), \tilde{u}(\cdot))$, $\tilde{u}(\cdot) = -\frac{1}{2\lambda} Q^{-1} \mathbf{B}^\top p(\cdot) + \mu$ satisfies (7.7)–(7.8)–(7.9) (see [10]). Accordingly the normal ECTs are the projection $(g(\cdot), p(\cdot))$ of the pairs $(\xi(\cdot), u(\cdot))$ where $u(\cdot) = -\frac{1}{2\lambda} Q^{-1} \mathbf{B}^\top p(\cdot) + \mu$ and $\xi(\cdot) = (g(\cdot), p(\cdot))$ is an integral curve of the Hamiltonian vector field $\vec{H}$ on $T^*\mathsf{G} = \mathsf{G} \times \mathfrak{g}^*$. However, as the Hamiltonian H is G-invariant, the integral curves $(g(\cdot), p(\cdot))$ of H are given by (see, e.g., [64])

$$\dot{g}(t) = g(t)\, dH(p(t)) \qquad \text{and} \qquad \dot{p}(t) = \mathrm{ad}^* dH(p(t)) \cdot p(t)$$

where H is viewed as a function on $\mathfrak{g}^*$ and $dH(p) \in \mathfrak{g}^{**}$ is canonically identified with an element in $\mathfrak{g}$. Accordingly we get $\dot{g}(t) = \Xi(g(t), u(t))$ where $u(t) = -\frac{1}{2\lambda}Q^{-1}\,\mathbf{B}^\top\, p(t)^\top + \mu$ and $p(t)$ is an integral curve of the Hamiltonian system $(\mathfrak{g}^*_-, H)$. Finally, as the pair $(\lambda, \xi(\cdot))$ can be multiplied by any positive number [10], we take $\lambda = -\frac{1}{2}$ for convenience. $\square$

We say that two quadratic Hamilton–Poisson systems $(\mathfrak{g}^*_-, G)$ and $(\mathfrak{h}^*_-, H)$ are *linearly equivalent* if there exists a linear isomorphism $\psi : \mathfrak{g}^* \to \mathfrak{h}^*$ such that the Hamiltonian vector fields $\vec{G}$ and $\vec{H}$ are ψ-related, i.e., $T_p\psi \cdot \vec{G}(p) = \vec{H}(\psi(p))$ for $p \in \mathfrak{g}^*$.

Proposition 4 *The following pairs of Hamilton–Poisson systems (on $\mathfrak{g}^*_-$, specified by their Hamiltonians) are linearly equivalent:*

1. $H_{A,Q} \circ \psi$ *and* $H_{A,Q}$, *where* $\psi : \mathfrak{g}^*_- \to \mathfrak{g}^*_-$ *is a linear Lie–Poisson automorphism;*
2. $H_{A,Q}$ *and* $H_{A,rQ}$, *where* $r \neq 0$;
3. $H_{A,Q}$ *and* $H_{A,Q} + C$, *where* C *is a Casimir function.*

Proof

(1) As ψ is a (linear) Poisson isomorphism, it follows that $T\psi \circ \overrightarrow{H_{A,Q} \circ \psi} = \overrightarrow{H_{A,Q}} \circ \psi$ (see, e.g., [79]), i.e., $\psi_* \overrightarrow{H_{A,Q} \circ \psi} = \overrightarrow{H_{A,Q}}$.

(2) Let $p(\cdot)$ be an integral curve of $\overrightarrow{H_{A,Q}}$, i.e., suppose $\dot{p}_i(t) = -p(t) \cdot [E_i, A + p(t)^\top Q]$. Let $\bar{p}(\cdot) = \frac{1}{r}p(\cdot)$. Then $\dot{\bar{p}}_i(t) = -\frac{1}{r}p(t) \cdot [E_i, A + p(t)^\top Q] = -\bar{p}(t) \cdot [E_i, A + \bar{p}(t)^\top rQ]$. Thus $\bar{p}(\cdot)$ is an integral curve of $\overrightarrow{H_{A,rQ}}$. Thus, $\overrightarrow{H_{A,Q}}$ and $\overrightarrow{H_{A,rQ}}$ are compatible with the map $\delta_{1/r} : p \mapsto \frac{1}{r}p$ (as $\delta_{1/r}$ maps the flow of $\overrightarrow{H_{A,Q}}$ to the flow of $\overrightarrow{H_{A,rQ}}$).

(3) We have $\{H_{A,Q}+C, F\} = \{H_{A,Q}, F\}$ for $F \in C^\infty(\mathfrak{g}^*)$. Thus $\overrightarrow{H_{A,Q} + C} = \overrightarrow{H_{A,Q}}$, i.e., these vector fields are compatible with the identity map. $\square$

Theorem 11 *If two cost-extended systems are cost equivalent, then their associated Hamilton–Poisson systems, given by (7.10), are linearly equivalent.*

Proof Let (Σ, χ) and (Σ', χ') be cost-extended systems with $\Xi_u(\mathbf{1}) = A + \mathbf{B}u$ and $\Xi'_u(\mathbf{1}) = A' + \mathbf{B}'u'$, respectively. The associated Hamilton–Poisson systems (on $\mathfrak{g}^*$ and $(\mathfrak{g}')^*$, respectively) are given by

$$H_{(\Sigma,\chi)}(p) = p\,(A + \mathbf{B}\,\mu) + \tfrac{1}{2}p\,\mathbf{B}\,Q^{-1}\,\mathbf{B}^\top\,p^\top$$

$$H_{(\Sigma',\chi')}(p) = p\,(A' + \mathbf{B}'\,\mu') + \tfrac{1}{2}p\,\mathbf{B}'\,Q'^{-1}\,\mathbf{B}'^\top\,p^\top.$$

Suppose $\phi \times \varphi$ defines a cost equivalence between (Σ, χ) and (Σ', χ'), where $\varphi(u) = Ru + \varphi_0$ and $R \in \mathbb{R}^{\ell \times \ell}$. We have $\chi' \circ \varphi = r\chi$ for some $r > 0$. A simple calculation yields

$$T_1\phi \cdot A = A' + \mathbf{B}' \varphi_0, \quad R\mu + \varphi_0 = \mu', \quad T_1\phi \cdot \mathbf{B} = \mathbf{B}'R, \quad R Q^{-1} R^{\mathsf{T}} = r(Q')^{-1}.$$

Thus $(H_{(\Sigma,\chi)} \circ (T_1\phi)^*)(p) = p(A' + \mathbf{B}' \mu') + \frac{r}{2} p \, \mathbf{B}' (Q')^{-1} \mathbf{B}'^{\mathsf{T}} p^{\mathsf{T}}$. Here $(T_1\phi)^*$: $(\mathfrak{g}')^* \to \mathfrak{g}^*$ is the dual of the linear map $T_1\phi$. Hence, the vector fields associated with $H_{(\Sigma',\chi')}$ and $H_{(\Sigma,\chi)} \circ (T_1\phi)^*$, respectively, are related by the dilation $\delta_{1/r}$: $(\mathfrak{g}')^* \to (\mathfrak{g}')^*, \, p \mapsto \frac{1}{r}p$ (Proposition 4). Moreover, the vector fields associated with $H_{(\Sigma,\chi)} \circ (T_1\phi)^*$ and $H_{(\Sigma,\chi)}$, respectively, are related by the linear Poisson isomorphism $(T_1\phi)^*$ (Proposition 4). Consequently $\frac{1}{r}(T_1\phi)^*$ defines a linear equivalence between $((\mathfrak{g}')^*_-, H_{(\Sigma',\chi')})$ and $(\mathfrak{g}^*_-, H_{(\Sigma,\chi)})$.   □

Remark 7 The converse of the above statement is not true in general. In fact, one can construct cost-extended systems with different number of inputs but equivalent Hamiltonians (cf. [35]).

Accordingly, from a classification of cost-extended systems, one can generally get a (partial) classification of the corresponding quadratic Hamilton–Poisson systems. For instance, we have the following result.

Example 7 (cf. Example 6) Any quadratic Hamilton–Poisson systems $((\mathfrak{h}_3)^*_-, H)$, where H is a positive definite quadratic form, is linearly equivalent to the system on $(\mathfrak{h}_3)^*_-$ with Hamiltonian $H'(p) = \frac{1}{2}(p_1^2 + p_2^2 + p_3^2)$.

Proof Let $((\mathfrak{h}_3)^*_-, H)$ be a quadratic Hamilton–Poisson system with Hamiltonian $H(p) = p Q p^{\mathsf{T}}$ being a positive-definite quadratic form. For the cost-extended system (Σ, χ) on H_3 specified by

$$(\Sigma', \chi') : \quad \begin{cases} \Sigma' : u_1 X_1 + u_2 Y_1 + u_3 Z = u_1 E_2 + u_2 E_3 + u_3 E_1 \\ \chi(u) = \frac{1}{2} u^{\mathsf{T}} Q^{-1} u \end{cases}$$

we have that the associated Hamiltonian system (7.10) is $((\mathfrak{h}_3)^*_-, H)$. Furthermore, by Example 6, we have that (Σ, χ) is cost equivalent to the system

$$\left(\Sigma^{(2,0)}, \chi_1^{(2,0)}\right) : \quad \begin{cases} \Sigma^{(2,0)} : u_1 Y_1 + u_2 X_1 + u_3 Z \\ \chi_1^{(2,0)}(u) = u_1^2 + u_2^2 + u_3^2 \end{cases}$$

which has associated Hamiltonian system $((\mathfrak{h}_3)^*_-, H')$. Consequently, by Theorem 11, we have that $((\mathfrak{h}_3)^*_-, H)$ is L-equivalent to $((\mathfrak{h}_3)^*_-, H')$.   □

Remark 8 It is a simple matter to expand the above result to positive definite quadratic Hamilton–Poisson systems on $(\mathfrak{h}_{2n+1})^*_-$.

7.3.4 Sub-Riemannian Structures

Left-invariant sub-Riemannian (and, in particular, Riemannian) structures on Lie groups can naturally be associated to drift-free cost-extended systems with homogeneous cost. We show that if two cost-extended systems are cost equivalent, then the associated sub-Riemannian structures are isometric up to rescaling. At least for some classes (namely Riemannian structures on nilpotent groups and sub-Riemannian Carnot groups), the converse holds as well.

A *left-invariant sub-Riemannian manifold* is a triplet $(\mathsf{G}, \mathscr{D}, \mathbf{g})$, where G is a (real, finite dimensional) connected Lie group, $\mathscr{D}$ is a nonintegrable left-invariant distribution on G, and $\mathbf{g}$ is a left-invariant Riemannian metric on $\mathscr{D}$. More precisely, $\mathscr{D}(\mathbf{1})$ is a linear subspace of the Lie algebra $\mathfrak{g}$ of G and $\mathscr{D}(g) = g\mathscr{D}(\mathbf{1})$; the metric $\mathbf{g}_1$ is a positive definite symmetric bilinear from on $\mathfrak{g}$ and $\mathbf{g}_g(gA, gB) = \mathbf{g}_1(A, B)$ for $A, B \in \mathfrak{g}$, $g \in \mathsf{G}$. When $\mathscr{D} = T\mathsf{G}$ (i.e., $\mathscr{D}(\mathbf{1}) = \mathfrak{g}$) then one has a left-invariant Riemannian structure. An absolutely continuous curve $g(\cdot) : [0, T] \to \mathsf{G}$ is called a *horizontal curve* if $\dot{g}(t) \in \mathscr{D}(g(t))$ for almost all $t \in [0, T]$. We shall assume that $\mathscr{D}$ satisfies the bracket generating condition, i.e., $\mathscr{D}(\mathbf{1})$ has full rank; this condition is necessary and sufficient for any two points in G to be connected by a horizontal curve. An *isometry* between two left-invariant sub-Riemannian manifolds $(\mathsf{G}, \mathscr{D}, \mathbf{g})$ and $(\mathsf{G}', \mathscr{D}', \mathbf{g}')$ is a diffeomorphism $\phi : \mathsf{G} \to \mathsf{G}'$ such that $\phi_* \mathscr{D} = \mathscr{D}'$ and $\mathbf{g} = \phi^* \mathbf{g}'$, i.e., $T_g \phi \cdot \mathscr{D}(g) = \mathscr{D}'(\phi(g))$ and $\mathbf{g}_g(gA, gB) = \mathbf{g}'_{\phi(g)}(T_g \phi \cdot gA, T_g \phi \cdot gB)$ for $g \in \mathsf{G}$ and $A, B \in \mathfrak{g}$. If the isometry ϕ is a Lie group isomorphism, we shall say it is an $\mathfrak{L}$-*isometry*.

A standard argument shows that the length minimization problem

$$\dot{g}(t) \in \mathscr{D}(g(t)), \qquad g(0) = g_0, \quad g(T) = g_1,$$

$$\int_0^T \sqrt{\mathbf{g}(\dot{g}(t), \dot{g}(t))} \longrightarrow \min$$

is equivalent to the energy minimization problem, or invariant optimal control problem:

$$\dot{g} = \Xi_u(g), \ u \in \mathbb{R}^\ell \qquad g(0) = g_0, \ g(T) = g_1$$

$$\int_0^T \chi(u(t)) \, dt \longrightarrow \min. \tag{7.11}$$

Here $\Xi_u(\mathbf{1}) = u_1 B_1 + \cdots + u_\ell B_\ell$ where $B_1, \ldots, B_\ell$ are some linearly independent elements of the Lie algebra $\mathfrak{g}$ such that $\langle B_1, \ldots, B_\ell \rangle = \mathscr{D}(\mathbf{1})$; $\chi(u(t)) = u(t)^\top Q u(t) = \mathbf{g}_1(\Xi_{u(t)}(\mathbf{1}), \Xi_{u(t)}(\mathbf{1}))$ for some $\ell \times \ell$ positive definite (symmetric) matrix Q. More precisely, energy minimizers are exactly those length minimizers which have constant speed. In other words, the VOCTs of the cost-extended system (Σ, χ) associated with (7.11) are exactly the (constant speed) minimizing geodesics of the sub-Riemannian structure $(\mathsf{G}, \mathscr{D}, \mathbf{g})$; the normal (resp. abnormal) ECTs of (Σ, χ) are the normal (resp. abnormal) geodesics of $(\mathsf{G}, \mathscr{D}, \mathbf{g})$.

Accordingly, to a (full-rank) cost-extended system (Σ, χ) on G of the form

$$\Sigma : u_1 B_1 + \cdots + u_\ell B_\ell, \qquad \chi(u) = u^\top Q u$$

we associate a sub-Riemannian structure $(\mathsf{G}, \mathscr{D}, \mathbf{g})$ specified by

$$\mathscr{D}(\mathbf{1}) = \Gamma = \langle B_1, \ldots, B_\ell \rangle, \qquad \mathbf{g_1}(u_1 B_1 + \cdots + u_\ell B_\ell, u_1 B_1 + \cdots + u_\ell B_\ell) = \chi(u).$$

Let $(\mathsf{G}, \mathscr{D}, \mathbf{g})$ and $(\mathsf{G}', \mathscr{D}', \mathbf{g}')$ be two sub-Riemannian structures associated to (Σ, χ) and (Σ', χ'), respectively.

Theorem 12 (Σ, χ) and (Σ', χ') are cost equivalent if and only if $(\mathsf{G}, \mathscr{D}, \mathbf{g})$ and $(\mathsf{G}', \mathscr{D}', \mathbf{g}')$ are $\mathfrak{L}$-isometric up to rescaling.

Proof Suppose $\phi \times \varphi$ defines a cost equivalence between (Σ, χ) and (Σ', χ'), i.e., $\phi_* \Xi_u = \Xi'_{\varphi(u)}$ and $\chi' \circ \varphi = r\chi$ for some $r > 0$. As $T_1\phi \cdot \Xi_u(\mathbf{1}) = \Xi_{\varphi(u)}(\mathbf{1})$, it follows that $T_1\phi \cdot \mathscr{D}(\mathbf{1}) = \mathscr{D}'(\mathbf{1})$. Hence, as ϕ is a Lie group isomorphism, by left invariance we have $\phi_* \mathscr{D} = \mathscr{D}'$. Furthermore

$$r\,\chi(u) = \chi'(\varphi(u))$$
$$\Longleftrightarrow \qquad r\,\mathbf{g_1}(\Xi_u(\mathbf{1}), \Xi_u(\mathbf{1})) = \mathbf{g}'_1(\Xi'_{\varphi(u)}(\mathbf{1}), \Xi'_{\varphi(u)}(\mathbf{1}))$$
$$\Longleftrightarrow \qquad r\,\mathbf{g_1}(\Xi_u(\mathbf{1}), \Xi_u(\mathbf{1})) = \mathbf{g}'_1(T_1\phi \cdot \Xi_u(\mathbf{1}), T_1\phi \cdot \Xi_u(\mathbf{1})). \qquad (7.12)$$

Hence, as ϕ is a Lie group isomorphism, by left invariance we have $r\,\mathbf{g} = \phi^* \mathbf{g}'$.

Conversely, suppose $\phi_* \mathscr{D} = \mathscr{D}'$ and $\mathbf{g} = r\,\phi^* \mathbf{g}'$. We have $T_1\phi \cdot \mathscr{D}(\mathbf{1}) = \mathscr{D}'(\mathbf{1})$ and so $T_1\phi \cdot \Gamma = \Gamma'$. Hence there exists a unique linear map $\varphi : \mathbb{R}^\ell \to \mathbb{R}^\ell$ such that $T_1\phi \cdot \Xi_u(\mathbf{1}) = \Xi'_{\varphi(u)}(\mathbf{1})$. Thus $\phi \times \varphi$ defines a detached feedback equivalence between Σ and Σ'. By (7.12), it follows that $\chi' \circ \varphi = r\chi$. Thus (Σ, χ) and (Σ', χ') are cost equivalent. $\qquad \square$

Clearly every $\mathfrak{L}$-isometry is an isometry. For some classes of sub-Riemannian structures, every isometry is the composition of a left translation and an $\mathfrak{L}$-isometry. Hence, for these classes, if $(\mathsf{G}, \mathscr{D}, \mathbf{g})$ and $(\mathsf{G}', \mathscr{D}', \mathbf{g}')$ are isometric (up to rescaling), then (Σ, χ) and (Σ', χ') are cost equivalent.

Theorem 13 (cf. [102], also [74]) *Let $(\mathsf{G}, \mathbf{g})$ and $(\mathsf{G}', \mathbf{g}')$ be two invariant Riemannian structures on simply connected nilpotent Lie groups G and G', respectively. A diffeomorphism $\phi : \mathsf{G} \to \mathsf{G}'$ is an isometry between $(\mathsf{G}, \mathbf{g})$ and $(\mathsf{G}', \mathbf{g}')$ if and only if ϕ is the composition $\phi = L_{\phi(1)} \circ \phi'$ of a left translation $L_{\phi(1)}$ on G' and a Lie group isomorphism $\phi' = L_{\phi(1)^{-1}} \circ \phi : \mathsf{G} \to \mathsf{G}'$ such that $\mathbf{g_1}(A, B) = \mathbf{g}'_{1'}(T_1\phi' \cdot A, T_1\phi' \cdot B)$.*

Corollary 1 *Two invariant Riemannian structures on simply connected nilpotent Lie groups are isometric if and only if they are $\mathfrak{L}$-isometric.*

A k-step *Carnot* group G is a simply connected nilpotent Lie group whose Lie algebra $\mathfrak{g}$ has stratification $\mathfrak{g} = \mathfrak{g}_1 \oplus \mathfrak{g}_2 \oplus \cdots \oplus \mathfrak{g}_k$ with $[\mathfrak{g}_1, \mathfrak{g}_j] = \mathfrak{g}_{1+j}$, $j = 1, \ldots, k-1$ and $[\mathfrak{g}_j, \mathfrak{g}_k] = \{0\}$, $j = 1, \ldots, k$. Let $\mathscr{D}$ be the left-invariant distribution on G specified by $\mathscr{D}(\mathbf{1}) = \mathfrak{g}_1$. Once we fix a left-invariant metric $\mathbf{g}$ on $\mathscr{D}$ we have a left-invariant structure on G (note that $\mathscr{D}$ is bracket generating). We shall refer to such a structure as a *sub-Riemannian Carnot group*. We likewise have the following characterization of the isometries between such structures.

Theorem 14 (cf. [56, 70], see also [44, 76]) *Let $(\mathsf{G}, \mathscr{D}, \mathbf{g})$ and $(\mathsf{G}', \mathscr{D}', \mathbf{g}')$ be two sub-Riemannian Carnot groups. A diffeomorphism $\phi : \mathsf{G} \to \mathsf{G}'$ is an isometry between $(\mathsf{G}, \mathscr{D}, \mathbf{g})$ and $(\mathsf{G}', \mathscr{D}', \mathbf{g}')$ if and only if ϕ is the composition $\phi = L_{\phi(\mathbf{1})} \circ \phi'$ of a left translation $L_{\phi(\mathbf{1})}$ on G' and a Lie group isomorphism $\phi' = L_{\phi(\mathbf{1})^{-1}} \circ \phi : \mathsf{G} \to \mathsf{G}'$ such that $T_1\phi' \cdot \mathscr{D}(\mathbf{1}) = \mathscr{D}'(\mathbf{1}')$ and $\mathbf{g}_1(A, B) = \mathbf{g}'_{1'}(T_1\phi' \cdot A, T_1\phi' \cdot B)$.*

Corollary 2 *Two sub-Riemannian Carnot groups are isometric if and only if they are $\mathfrak{L}$-isometric.*

Remark 9 Among the three-dimensional simply connected Lie groups, only the Abelian $\mathbb{R}^3$ and Heisenberg H_3 groups are nilpotent. Among the four-dimensional simply connected Lie groups, only the Abelian group $\mathbb{R}^4$, the trivial extension $\mathsf{H}_3 \times \mathbb{R}$ of H_3, and the Engel group are nilpotent (see, e.g., [82, 86, 87]). Except for the Abelian groups, all these groups admit sub-Riemannian Carnot group structures.

Analogous to Example 3, we have the following classification of sub-Riemannian structures on the Euclidean group $\mathsf{SE}\,(2)$.

Example 8 On $\mathsf{SE}\,(2)$, any left-invariant sub-Riemannian structure $(\mathscr{D}, \mathbf{g})$ is isometric (up to rescaling) to the structure $(\bar{\mathscr{D}}, \bar{\mathbf{g}})$ specified by

$$
\begin{cases}
\mathscr{D}(\mathbf{1}) = \langle E_2, E_3 \rangle \\[2mm]
\mathbf{g}_1 = \begin{bmatrix} 1 & 0 \\ 0 & 1 \end{bmatrix}
\end{cases}
$$

i.e., with orthonormal frame (E_2, E_3).

Analogous to Example 6, we have the following classification of structures on the Heisenberg groups.

Example 9 ([24]) Any left-invariant sub-Riemannian structure $(\mathscr{D}, \mathbf{g})$ on H_{2n+1} is isometric to exactly one of the structures $(\mathscr{D}, \mathbf{g}^\lambda)$ specified by

$$
\begin{cases}
\mathscr{D}(\mathbf{1}) = \langle X_1, Y_1, \ldots, X_n, Y_n \rangle \\[2mm]
\mathbf{g}_1^\lambda = \Lambda = \mathrm{diag}(\lambda_1, \lambda_1, \lambda_2, \lambda_2, \ldots, \lambda_n, \lambda_n)
\end{cases}
$$

i.e., with orthonormal frame

$$\left(\tfrac{1}{\sqrt{\lambda_1}}X_1, \tfrac{1}{\sqrt{\lambda_1}}Y_1, \tfrac{1}{\sqrt{\lambda_2}}X_2, \tfrac{1}{\sqrt{\lambda_2}}Y_2, \ldots, \tfrac{1}{\sqrt{\lambda_n}}X_n, \tfrac{1}{\sqrt{\lambda_n}}Y_n\right).$$

Here $1 = \lambda_1 \geq \lambda_2 \geq \cdots \geq \lambda_n > 0$ parametrize a family of (non-equivalent) class representatives.

Example 10 ([24]) Any left-invariant Riemannian structure $\mathbf{g}$ on H_{2n+1} is isometric to exactly one of the structures

$$\mathbf{g}_1^\lambda = \begin{bmatrix} 1 & 0 \\ 0 & \Lambda \end{bmatrix}, \quad \Lambda = \mathrm{diag}(\lambda_1, \lambda_1, \lambda_2, \lambda_2, \ldots, \lambda_n, \lambda_n)$$

i.e., with orthonormal frame

$$\left(Z, \tfrac{1}{\sqrt{\lambda_1}}X_1, \tfrac{1}{\sqrt{\lambda_1}}Y_1, \tfrac{1}{\sqrt{\lambda_2}}X_2, \tfrac{1}{\sqrt{\lambda_2}}Y_2, \ldots, \tfrac{1}{\sqrt{\lambda_n}}X_n, \tfrac{1}{\sqrt{\lambda_n}}Y_n\right).$$

Here $\lambda_1 \geq \lambda_2 \geq \cdots \geq \lambda_n > 0$ parametrize a family of (non-equivalent) class representatives.

7.4 Quadratic Hamilton–Poisson Systems

The dual space of a Lie algebra admits a natural Poisson structure, namely the Lie–Poisson structure; these structures are in a one-to-one correspondence with linear Poisson structures [73]. Lie–Poisson structures arise naturally in a variety of fields of mathematical physics and engineering such as classical dynamical systems, robotics, fluid dynamics, and superconductivity, to name but a few. Many interesting dynamical systems are naturally expressed as quadratic Hamilton–Poisson systems on Lie–Poisson spaces; prevalent examples are Euler's classic equations for the rigid body, its extensions and its generalizations [59, 60, 64, 79, 97].

As described in Sect. 7.3 (Theorem 10), quadratic Hamilton–Poisson systems arise in the study of invariant optimal control problems. In this vein, a number of quadratic Hamilton–Poisson systems on lower-dimensional Lie–Poisson spaces have been considered (see, e.g., [4, 7, 23, 47, 81, 93]). Several classes of quadratic (and especially homogeneous) systems have also recently been studied in their own right (see, e.g., [1, 6, 8, 12–15, 20, 32, 97, 98]). Equivalence of quadratic Hamilton–Poisson systems on Lie–Poisson spaces has been investigated by a few authors [6, 8, 20, 27, 35, 49, 97]; in these papers, two systems are considered equivalent if their Hamiltonian vector fields are compatible with either an affine isomorphism, a linear isomorphism, or a Lie–Poisson isomorphism.

In this section, we shall give a brief overview of our classification of the (positive semidefinite) homogeneous systems in three dimensions; a classification of the inhomogeneous systems on $\mathsf{so}\,(3)^*_-$ is also exhibited. Finally, for a typical inhomogeneous system (on the Euclidean space $\mathsf{se}\,(2)^*_-$), we determine the stability nature of its equilibria and find explicit expressions for some of its integral curves.

7.4.1 Classification in Three Dimensions

7.4.1.1 Homogeneous Systems

We consider those three-dimensional Lie–Poisson spaces that admit global Casimir functions. Up to linear Poisson isomorphisms, this class is comprised of

$$(\mathfrak{g}_{2.1} \oplus \mathfrak{g}_1)^*_-, \quad (\mathfrak{g}_{3.1})^*_-, \quad (\mathfrak{g}_{3.4}^0)^*_-, \quad (\mathfrak{g}_{3.5}^0)^*_-, \quad (\mathfrak{g}_{3.6})^*_-, \quad \text{and} \quad (\mathfrak{g}_{3.7})^*_-.$$

We present a classification of the positive semidefinite homogeneous quadratic Hamilton–Poisson systems on these spaces, i.e., Hamiltonians of the form $H(p) = p^\top Q p$ with matrix Q positive semidefinite. (The quadratic Hamilton–Poisson system associated to any cost-extended system is positive semidefinite, see Theorem 10.)

We omit any mention of the Abelian Lie–Poisson space $(3\mathfrak{g}_1)^*_-$ as any system on $(3\mathfrak{g}_1)^*_-$ has trivial dynamics. We find it convenient here to use a basis for $\mathfrak{g}_{2.1} \oplus \mathfrak{g}_1$ different from the one listed in Table 7.1 in the Appendix. More precisely, we use the basis $E'_1 = \frac{1}{2}(E_1 - E_2)$, $E'_2 = -\frac{1}{2}E_3$, $E'_3 = \frac{1}{2}(E_1 + E_2)$; the only non-zero commutator is then $[E'_1, E'_2] = E'_1$.

The classification is presented in two stages. First, a classification of the Hamilton–Poisson systems on each Lie–Poisson space $\mathfrak{g}^*_-$ is given.

Theorem 15 ([32], see also [6, 8, 20, 27, 35]) *Let* $(\mathfrak{g}^*_-, H_\mathscr{Q})$ *be a homogeneous positive semidefinite quadratic Hamilton–Poisson system.*

1. *If* $\mathfrak{g}^*_- \cong (\mathfrak{g}_{2.1} \oplus \mathfrak{g}_1)^*_-$, *then* $(\mathfrak{g}^*_-, H_\mathscr{Q})$ *is equivalent to exactly one of the following systems on* $(\mathfrak{g}_{2.1} \oplus \mathfrak{g}_1)^*_-$

$$H_1(p) = p_1^2, \qquad H_2(p) = p_2^2, \qquad H_3(p) = p_1^2 + p_2^2,$$

$$H_4(p) = (p_1 + p_3)^2, \qquad H_5(p) = p_2^2 + (p_1 + p_3)^2.$$

2. *If* $\mathfrak{g}^*_- \cong (\mathfrak{g}_{3.1})^*_-$, *then* $(\mathfrak{g}^*_-, H_\mathscr{Q})$ *is equivalent to exactly one of the following systems on* $(\mathfrak{g}_{3.1})^*_-$

$$H_1(p) = p_3^2, \qquad H_2(p) = p_2^2 + p_3^2.$$

3. *If $\mathfrak{g}_-^* \cong (\mathfrak{g}_{3.4}^0)_-^*$, then $(\mathfrak{g}_-^*, H_2)$ is equivalent to exactly one of the following systems on $(\mathfrak{g}_{3.4}^0)_-^*$*

$$H_1(p) = p_1^2, \qquad H_2(p) = p_3^2, \qquad H_3(p) = p_1^2 + p_3^2,$$
$$H_4(p) = (p_1 + p_2)^2, \qquad H_5(p) = (p_1 + p_2)^2 + p_3^2.$$

4. *If $\mathfrak{g}_-^* \cong (\mathfrak{g}_{3.5}^0)_-^*$, then $(\mathfrak{g}_-^*, H_2)$ is equivalent to exactly one of the following systems on $(\mathfrak{g}_{3.5}^0)_-^*$*

$$H_1(p) = p_2^2, \qquad H_2(p) = p_3^2, \qquad H_3(p) = p_2^2 + p_3^2.$$

5. *If $\mathfrak{g}_-^* \cong (\mathfrak{g}_{3.6})_-^*$, then $(\mathfrak{g}_-^*, H_2)$ is equivalent to exactly one of the following systems on $(\mathfrak{g}_{3.6})_-^*$*

$$H_1(p) = p_1^2, \qquad H_2(p) = p_3^2, \qquad H_3(p) = p_1^2 + p_3^2,$$
$$H_4(p) = (p_2 + p_3)^2, \qquad H_5(p) = p_2^2 + (p_1 + p_3)^2.$$

6. *If $\mathfrak{g}_-^* \cong (\mathfrak{g}_{3.7})_-^*$, then $(\mathfrak{g}_-^*, H_2)$ is equivalent to exactly one of the following systems on $(\mathfrak{g}_{3.7})_-^*$*

$$H_1(p) = p_1^2, \qquad H_2(p) = p_1^2 + \tfrac{1}{2} p_2^2.$$

Proof (Sketch) The classification on each Lie–Poisson space $\mathfrak{g}^*$ may be obtained as follows. First one calculates the group of linear Poisson automorphisms; these are exactly the dual maps of the Lie algebra automorphisms. Using Proposition 4, one then reduces the Hamiltonian $H(p) = p^\top Q p$ as much as possible, yielding a collection $\{H_i : i \in I\}$ of class representatives. (In many cases, this produces the given normal forms.) Thereafter additional normalization is performed by determining whether or not there exists a linear isomorphism ψ such that $\psi \cdot \vec{H}_i = \vec{H}_j$ for $i, j \in I$. This amounts to solving a system of equations in nine variables; a computational software program such as Mathematica can be used to facilitate these computations. A classification is obtained once the indexing set I has been refined to the extent that there is no solution for any $i, j \in I$, $i \neq j$. $\square$

Systems on three-dimensional Lie–Poisson spaces can be partitioned into three classes based on the geometry of the integral curves. A system $(\mathfrak{g}_-^*, H)$ is said to be *ruled*, if for each integral curve of $\vec{H}$ there exists a line containing its trace. Likewise, $(\mathfrak{g}_-^*, H)$ is called *planar* if it is not ruled and for each integral curve of $\vec{H}$ there exists a plane containing its trace. Otherwise, $(\mathfrak{g}_-^*, H)$ is called *non-planar*. The ruled, planar, and non-planar properties are each invariant under equivalence, i.e., if two systems are equivalent, then they must belong to the same class.

By determining whether or not any of the normal forms (on distinct Lie–Poisson space) are equivalent to one another, we then get the following classification of system in the context of all three-dimensional Lie–Poisson spaces admitting global Casimirs.

Theorem 16 ([32])   *Let* $(\mathfrak{g}_-^*, H_{\mathscr{Q}})$ *be a homogeneous positive semidefinite quadratic Hamilton–Poisson system.*

1. *If* $(\mathfrak{g}_-^*, H)$ *is linear, then it is equivalent to exactly one of the systems*

$$((\mathfrak{g}_{2.1} \oplus \mathfrak{g}_1)_-^*, p_2^2), \qquad ((\mathfrak{g}_{3.4}^0)_-^*, (p_1 + p_2)^2), \qquad ((\mathfrak{g}_{3.5}^0)_-^*, p_2^2).$$

2. *If* $(\mathfrak{g}_-^*, H)$ *is planar, then it is equivalent to exactly one of the systems*

$$((\mathfrak{g}_{2.1} \oplus \mathfrak{g}_1)_-^*, p_1^2 + p_2^2), \qquad ((\mathfrak{g}_{2.1} \oplus \mathfrak{g}_1)_-^*, p_2^2 + (p_1 + p_3)^2),$$

$$((\mathfrak{g}_{3.4}^0)_-^*, p_3^2), \qquad ((\mathfrak{g}_{3.5}^0)_-^*, p_3^2), \qquad ((\mathfrak{g}_{3.6})_-^*, (p_2 + p_3)^2).$$

3. *If* $(\mathfrak{g}_-^*, H)$ *is non-planar, then it is equivalent to exactly one of the systems*

$$((\mathfrak{g}_{3.4}^0)_-^*, (p_1 + p_2)^2 + p_3^2), \qquad ((\mathfrak{g}_{3.5}^0)_-^*, p_2^2 + p_3^2).$$

Remark 10 Any system on $(\mathfrak{g}_{3.1})_-^*$ or $(\mathfrak{g}_{3.7})_-^*$ is equivalent to one on $(\mathfrak{g}_{3.5}^0)_-^*$. Any system on $(\mathfrak{g}_{2.1} \oplus \mathfrak{g}_1)_-^*$ or $(\mathfrak{g}_{3.1})_-^*$ is a planar (or linear) one. Every system on $(\mathfrak{g}_{3.1})_-^*$, $(\mathfrak{g}_{3.4}^0)_-^*$, $(\mathfrak{g}_{3.5}^0)_-^*$, or $(\mathfrak{g}_{3.7})_-^*$ may be realized on more than one Lie–Poisson space (for $(\mathfrak{g}_{3.6})_-^*$, the only exceptions are those systems equivalent to $((\mathfrak{g}_{3.6})_-^*, (p_2 + p_3)^2)$).

7.4.1.2   Inhomogeneous Systems

In the context of inhomogeneous systems we find it desirable to weaken the equivalence relation under consideration. We say that two quadratic Hamilton–Poisson systems $(\mathfrak{g}_-^*, G)$ and $(\mathfrak{h}_-^*, H)$ are *affinely* equivalent if there exists an affine isomorphism $\psi : \mathfrak{g}^* \to \mathfrak{h}^*$ such that the Hamiltonian vector fields $\vec{G}$ and $\vec{H}$ are ψ-related, i.e., $T_p\psi \cdot \vec{G}(p) = \vec{H}(\psi(p))$ for $p \in \mathfrak{g}^*$. Clearly, if two systems are linearly equivalent, then they are affinely equivalent. It turns out that for homogeneous systems the converse holds as well.

We exhibit a classification of the inhomogeneous systems on the orthogonal Lie–Poisson space $(\mathfrak{g}_{3.7})_-^* = \mathfrak{so}(3)_-^*$. The classification procedure is very similar to that used for homogeneous systems.

Theorem 17 ([8]) *Any inhomogeneous quadratic system* $((\mathfrak{g}_{3.7})_-^*, H)$ *is affinely equivalent to exactly one of the systems:*

$$H_{1,\alpha}^0(p) = \alpha p_1, \quad \alpha > 0$$

$$H_0^1(p) = \tfrac{1}{2}p_1^2$$

$$H_1^1(p) = p_2 + \tfrac{1}{2}p_1^2$$

$$H_{2,\alpha}^1(p) = p_1 + \alpha p_2 + \tfrac{1}{2}p_1^2, \quad \alpha > 0$$

$$H^2_{1,\alpha}(p) = \alpha p_1 + p_1^2 + \tfrac{1}{2}p_2^2, \quad \alpha > 0$$

$$H^2_{2,\alpha}(p) = \alpha p_2 + p_1^2 + \tfrac{1}{2}p_2^2, \quad \alpha > 0$$

$$H^2_{3,\alpha}(p) = \alpha_1 p_1 + \alpha_2 p_2 + p_1^2 + \tfrac{1}{2}p_2^2, \quad \alpha_1, \alpha_2 > 0$$

$$H^2_{4,\alpha}(p) = \alpha_1 p_1 + \alpha_3 p_3 + p_1^2 + \tfrac{1}{2}p_2^2, \quad \alpha_1 \geq \alpha_3 > 0$$

$$H^2_{5,\alpha}(p) = \alpha_1 p_1 + \alpha_2 p_2 + \alpha_3 p_3 + p_1^2 + \tfrac{1}{2}p_2^2,$$

$$\alpha_2 > 0, \; \alpha_1 > |\alpha_3| > 0 \; or \; \alpha_2 > 0, \; \alpha_1 = \alpha_3 > 0.$$

Here $\alpha, \alpha_1, \alpha_2, \alpha_3$ *parametrize families of class representatives, each different value corresponding to a distinct (non-equivalent) representative.*

7.4.2 On Integration and Stability

Among the three-dimensional Lie–Poisson spaces admitting global Casimirs, most homogeneous quadratic Hamilton–Poisson systems are integrable by elementary functions. More precisely, among the normal forms presented in Theorem 16, all systems are integrable by elementary functions except for $((\mathfrak{g}^0_{3.5})^*_-, p_2^2 + p_3^2)$ which is integrable by Jacobi elliptic functions. A number of inhomogeneous systems are also integrable by Jacobi elliptic functions (cf. [7]).

It is generally not difficult to determine the (Lyapunov) stability nature of the equilibria of the quadratic systems on three-dimensional Lie–Poisson spaces. Instability usually follows from spectral instability; stability can usually be proved by using the (extended) energy Casimir method [85].

As an illustrative and typical example, we examine the family of inhomogeneous systems $H_\alpha(p) = p_1 + \tfrac{1}{2}\left(\alpha p_2^2 + p_3^2\right)$ on $\mathfrak{se}(2)^*_-$. The equations of motion are

$$\begin{cases} \dot{p}_1 = p_2 p_3 \\ \dot{p}_2 = -p_1 p_3 \\ \dot{p}_3 = (\alpha p_1 - 1)p_2. \end{cases}$$

The equilibrium states are

$$e_1^\mu = (\mu, 0, 0), \quad e_2^\nu = (\tfrac{1}{\alpha}, \nu, 0) \quad \text{and} \quad e_3^\nu = (0, 0, \nu)$$

where $\mu, \nu \in \mathbb{R}, \; \nu \neq 0$.

Theorem 18 (cf. [7]) *The equilibrium states have the following behaviour:*

1. *The states* e_1^μ, $0 < \mu < \frac{1}{\alpha}$ *are spectrally unstable.*
2. *The state* e_1^μ, $\mu = \frac{1}{\alpha}$ *is unstable.*
3. *The states* e_1^μ, $\mu \in (-\infty, 0] \cup (\frac{1}{\alpha}, \infty)$ *are stable.*
4. *The states* e_2^ν *are spectrally unstable*
5. *The states* e_3^ν *are stable.*

Proof

(1) The linearization

$$
D\vec{H}_\alpha(e_1^\mu) = \begin{bmatrix} 0 & 0 & 0 \\ 0 & 0 & -\mu \\ 0 & -1 + \alpha\mu & 0 \end{bmatrix}
$$

of $\vec{H}_\alpha$ at e_1^μ has eigenvalues $\lambda_1 = 0$, $\lambda_{2,3} = \pm\sqrt{\mu(1 - \alpha\mu)}$. Hence for $0 < \mu < \frac{1}{\alpha}$ the state e_1^μ is spectrally unstable. Item (4) follows similarly.

(2) The linearization of $\vec{H}_\alpha$ at $e_1^{1/\alpha}$ has eigenvalues all zero; hence, we resort to a direct approach to prove instability. We have that

$$
p(t) = \left(\frac{t^2 - \alpha}{\alpha(t^2 + \alpha)}, \frac{2t}{\sqrt{\alpha}(t^2 + \alpha)}, \frac{2\sqrt{\alpha}}{t^2 + \alpha} \right)
$$

is an integral curve of $\vec{H}_\alpha$ such that $\lim_{t \to -\infty} p(t) = e_1^{1/\alpha}$ and $p(0) = \left(-\frac{1}{\alpha}, 0, \frac{2}{\sqrt{\alpha}} \right)$. Let $U = \{p : \|p - e_1^{1/\alpha}\| < \frac{\sqrt{1+\alpha}}{\alpha}\}$. (Here $\|p\| = \sqrt{p_1^2 + p_2^2 + p_3^2}$.) Hence, for any neighbourhood $V \subset U$ of $e_1^{1/\alpha}$, there exists $t_0 < 0$ such that $p(t_0) \in V$ but $p(0) \notin V$. Consequently, the state $e_1^{1/\alpha}$ is unstable.

(3) Consider the energy function $G = \lambda_1 H + \lambda_2 C$. Let $\lambda_1 = 2\mu^2$ and $\lambda_2 = -\mu$. We have

$$
dG(e_1^\mu) = 0 \quad \text{and} \quad d^2G(e_1^\mu) = \begin{bmatrix} -2\mu & 0 & 0 \\ 0 & 2\mu(-1 + \alpha\mu) & 0 \\ 0 & 0 & 2\mu^2 \end{bmatrix}.
$$

The restriction of quadratic form $d^2G(e_1^\mu)$ to

$$
\ker dH(e_1^\mu) \cap \ker dC(e_1^\mu) = \{(0, p_2, p_3) : p_2, p_3 \in \mathbb{R}\}
$$

is positive definite for $\mu < 0$ and $\mu > \frac{1}{\alpha}$. Therefore, by the extended energy Casimir method, it follows that the states e_1^μ, $\mu \in (-\infty, 0) \cup (\frac{1}{\alpha}, \infty)$ are stable. On the other hand, we notice that the intersection $C^{-1}(0) \cap H_\alpha^{-1}(0)$

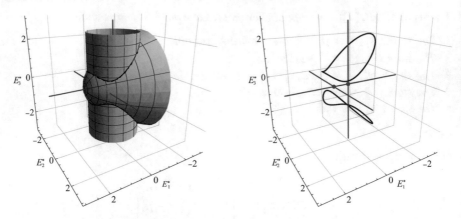

Fig. 7.1 Intersection of $C^{-1}(c_0)$ and $H_\alpha^{-1}(h_0)$ for some typical values

is a singleton, namely the origin e_1^0. Thus, by the continuous energy Casimir method, the origin is stable. Item (5) can likewise be proved by use of the extended energy Casimir method.   □

There are a number of qualitatively different types of integral curves of $\vec{H}_\alpha$, each type having a different explicit expression. These qualitative different types can be characterized in terms of the energy values (c_0, h_0) for the Casimir C and Hamiltonian H_α: we have a change in qualitative behaviour when the surfaces $C^{-1}(c_0)$ and $H_\alpha^{-1}(h_0)$ are tangent to one another. We consider here the case when $c_0 > \frac{1}{\alpha^2}$ and $h_0 > \frac{1+\alpha^2 c_0}{2\alpha}$. In Fig. 7.1 the cylinder $C^{-1}(c_0)$, the paraboloid $H_\alpha^{-1}(h_0)$, and their intersection are graphed some typical values of c_0, h_0, and α satisfying $c_0 > \frac{1}{\alpha^2}$ and $h_0 > \frac{1+\alpha^2 c_0}{2\alpha}$; the equilibria are also graphed (the stable equilibria in blue and the unstable equilibria in red).

Given a modulus $k \in [0, 1]$, the basic Jacobi elliptic functions $\mathrm{sn}(\cdot, k)$, $\mathrm{cn}(\cdot, k)$ and $\mathrm{dn}(\cdot, k)$ can be defined as (see, e.g., [11, 75, 101])

$$\mathrm{sn}(x, k) = \sin \mathrm{am}(x, k)$$

$$\mathrm{cn}(x, k) = \cos \mathrm{am}(x, k)$$

$$\mathrm{dn}(x, k) = \sqrt{1 - k^2 \sin^2 \mathrm{am}(x, k)}$$

where $\mathrm{am}(\cdot, k) = F(\cdot, k)^{-1}$ is the amplitude and $F(\varphi, k) = \int_0^\varphi \frac{dt}{1 - k^2 \sin^2 t}$. (For the degenerate case $k = 0$ and $k = 1$, we recover the circular functions and the hyperbolic functions, respectively.) The number K is defined as $K = F(\frac{\pi}{2}, k)$. The functions $\mathrm{sn}(\cdot, k)$ and $\mathrm{cn}(\cdot, k)$ are $4K$ periodic, whereas $\mathrm{dn}(\cdot, k)$ is $2K$ periodic. In terms of these functions, the integral curves of $\vec{H}_\alpha$ can be expressed as follows.

Theorem 19 ([7]) *Let* $p(\cdot) : (-\varepsilon, \varepsilon) \to \mathfrak{se}(2)^*$ *be an integral curve of* $\vec{H}_\alpha$ *and let* $h_0 = H(p(0))$, $c_0 = C(p(0))$. *If* $c_0 > \frac{1}{\alpha^2}$ *and* $h_0 > \frac{1+\alpha^2 c_0}{2\alpha}$, *then there exists* $t_0 \in \mathbb{R}$ *and* $\sigma \in \{-1, 1\}$ *such that* $p(t) = \bar{p}(t + t_0)$ *for* $t \in (-\varepsilon, \varepsilon)$, *where*

$$
\begin{cases}
\bar{p}_1(t) = \sqrt{c_0}\ \dfrac{\sqrt{h_0 - \delta} - \sqrt{h_0 + \delta}\ \operatorname{cn}(\Omega\, t, k)}{\sqrt{h_0 + \delta} - \sqrt{h_0 - \delta}\ \operatorname{cn}(\Omega\, t, k)} \\[4mm]
\bar{p}_2(t) = \sigma\sqrt{2c_0\delta}\ \dfrac{\operatorname{sn}(\Omega\, t, k)}{\sqrt{h_0 + \delta} - \sqrt{h_0 - \delta}\ \operatorname{cn}(\Omega\, t, k)} \\[4mm]
\bar{p}_3(t) = 2\sigma\delta\ \dfrac{\operatorname{dn}(\Omega\, t, k)}{\sqrt{h_0 + \delta} - \sqrt{h_0 - \delta}\ \operatorname{cn}(\Omega\, t, k)}.
\end{cases}
$$

Here $\delta = \sqrt{h_0^2 - c_0}$, $\Omega = \sqrt{2\delta}$ *and* $k = \frac{1}{\sqrt{2\delta}}\sqrt{(h_0 - \delta)(\alpha h_0 + \alpha\delta - 1)}$.

Proof (Sketch) We square the equation $\dot{p}_1 = p_2 p_3$ both sides to obtain $\dot{p}_1^2 = p_2^2 p_3^2$. By use of the constants of motion $c_0 = p_1^2 + p_2^2$ and $h_0 = p_1 + \frac{1}{2}\left(\alpha p_2^2 + p_3^2\right)$, we obtain the separable differential equation

$$
\frac{dp_1}{dt} = \pm\sqrt{(c_0 - p_1{}^2)(h_0 - 2p_1 - (c_0 - p_1{}^2)\alpha)}.
$$

The integral

$$
\int \frac{dp_1}{\sqrt{(c_0 - p_1{}^2)(h_0 - 2p_1 - (c_0 - p_1{}^2)\alpha)}}
$$

is transformed into a standard form and an elliptic integral formula is applied (see, e.g., [11, 75, 101]) to obtain

$$
\bar{p}_1(t) = \sqrt{c_0}\ \frac{\sqrt{h_0 - \delta} - \sqrt{h_0 + \delta}\ \operatorname{cn}(\Omega\, t, k)}{\sqrt{h_0 + \delta} - \sqrt{h_0 - \delta}\ \operatorname{cn}(\Omega\, t, k)}.
$$

The above step is quite involved computationally. Moreover, the transformation into standard form is dependent on the inequalities $c_0 > \frac{1}{\alpha^2}$ and $h_0 > \frac{1+\alpha^2 c_0}{2\alpha}$; in general, more inequalities may arise in the process of integration (than just those obtained by distinguishing qualitatively different cases) and other elliptic integral formulas may be applicable. We note that in this computation some translations in the independent variable t have been discarded.

Once a prospective expression has been found for $\bar{p}_1(t)$, the coordinates $\bar{p}_2(t)$ and $\bar{p}_3(t)$ are recovered by means of the identities $c_0 = p_1^2 + p_2^2$ and $h_0 = p_1 + \frac{1}{2}\left(\alpha p_2^2 + p_3^2\right)$, up to some possible changes in sign. By explicitly calculating $\dot{\bar{p}}(t) - \vec{H}_\alpha(\bar{p}(t))$, we identify for which choices of sign $\bar{p}(t)$ is an integral curve of $\vec{H}_\alpha$. (This step then also ensures that no mistakes were made in computing $\bar{p}(t)$.)

Finally, it remains to be shown that $p(\cdot)$ is identical to $\bar{p}(\cdot)$ up to a translation in the independent variable. We have that $\bar{p}_1(0) = -\sqrt{c_0}$ and $\bar{p}_1(\frac{2K}{\Omega}) = \sqrt{c_0}$. As $p_1^2 + p_2^2 = c_0$, we have that $-\sqrt{c_0} \leqslant p_1(0) \leqslant \sqrt{c_0}$. Thus there exists $t_1 \in [0, \frac{2K}{\Omega}]$ such that $\bar{p}_1(t_1) = p_1(0)$. By choosing $\sigma \in \{-1, 1\}$ appropriately and using the constant of motions, we get that $p(0) = \bar{p}(t_0)$ for $t_0 = t_1$ or $t_0 = -t_1$. Consequently the integral curves $t \mapsto p(t)$ and $t \mapsto \bar{p}(t + t_0)$ solve the same Cauchy problem and therefore are identical. $\qquad\square$

Note 5 Inhomogeneous systems on $\mathfrak{se}(2)^*_-$ were studied in [7] (see also [4]) whereas homogeneous systems were treated in [6]. An extensive investigation (covering equivalence, stability, and some integration) of systems on $\mathfrak{so}(3)^*_-$ and $\mathfrak{se}(1,1)^*_-$ is carried out in [1, 6] and [16, 20], respectively.

7.5 Conclusion

A central theme of this survey is the equivalence and classification of invariant control systems, cost-extended systems, and (the associated) Hamilton–Poisson systems. In Sect. 7.2.3, we presented a complete classification of the invariant control affine systems in three dimensions. There is no complete classification of the cost-extended systems in three dimensions. However, there are classifications of the invariant sub-Riemannian structures up to isometry [9] and invariant Riemannian structures up to $\mathfrak{L}$-isometry [55]. It turns out that two sub-Riemannian (but not Riemannian) structures on the same three-dimensional simply connected Lie group are isometric up to rescaling if and only if the associated cost-extended systems are equivalent (cf. [9, 99]). The extent to which this holds true in higher dimensions is an interesting open problem. Classifications of cost-extended systems in four dimensions (and beyond) are also topics for future research. Another avenue of investigation would be to interpret the inhomogeneous cost-extended systems as invariant affine distributions together with quadratic forms and consider diffeomorphisms relating such structures (cf. [45, 46]).

In order to find the extremal trajectories for a cost-extended system, one needs to integrate the associated Hamilton–Poisson system (see Theorem 10). In the last decade or so several authors have considered quadratic Hamilton–Poisson systems on low-dimensional Lie–Poisson spaces (see, e.g., [4, 7, 13, 23, 49, 97]). In Sect. 7.4.1, we presented a classification of a subclass of such systems in three dimensions. As several systems may be equivalently realized on more than one (non-isomorphic) Lie–Poisson spaces, it is desirable to have a complete classification. A systematic treatment (involving classification, stability, and integration) of the homogeneous positive semidefinite quadratic systems is currently in preparation (for some partial results, see [6, 8, 20, 32]). So far, inhomogeneous systems have only been studied, in this manner, on some specific spaces [1, 4, 7, 16].

Acknowledgements The research leading to these results has received funding from the European Union's Seventh Framework Programme (FP7/2007–2013) under grant agreement no. 317721. Also, Rory Biggs would like to acknowledge the financial assistance of the Claude Leon Foundation.

Appendix: Three-Dimensional Lie Algebras and Groups

There are eleven types of three-dimensional real Lie algebras; in fact, nine algebras and two parametrized infinite families of algebras (see, e.g., [71, 78, 82]). In terms of an (appropriate) ordered basis (E_1, E_2, E_3), the commutation operation is given by

$$[E_2, E_3] = n_1 E_1 - a E_2$$
$$[E_3, E_1] = a E_1 + n_2 E_2$$
$$[E_1, E_2] = n_3 E_3.$$

The structure parameters a, n_1, n_2, n_3 for each type are given in Table 7.1.

A classification of the three-dimensional (real connected) Lie groups can be found in [84]. Let G be a three-dimensional (real connected) Lie group with Lie algebra $\mathfrak{g}$.

1. If $\mathfrak{g}$ is Abelian, i.e., $\mathfrak{g} \cong 3\mathfrak{g}_1$, then G is isomorphic to $\mathbb{R}^3$, $\mathbb{R}^2 \times \mathbb{T}$, $\mathbb{R} \times \mathbb{T}$, or $\mathbb{T}^3$.
2. If $\mathfrak{g} \cong \mathfrak{g}_{2.1} \oplus \mathfrak{g}_1$, then G is isomorphic to $\mathsf{Aff}(\mathbb{R})_0 \times \mathbb{R}$ or $\mathsf{Aff}(\mathbb{R})_0 \times \mathbb{T}$.
3. If $\mathfrak{g} \cong \mathfrak{g}_{3.1}$, then G is isomorphic to the Heisenberg group H_3 or the Lie group $\mathsf{H}_3^* = \mathsf{H}_3 / \mathsf{Z}(\mathsf{H}_3(\mathbb{Z}))$, where $\mathsf{Z}(\mathsf{H}_3(\mathbb{Z}))$ is the group of integer points in the centre $\mathsf{Z}(\mathsf{H}_3) \cong \mathbb{R}$ of H_3.
4. If $\mathfrak{g} \cong \mathfrak{g}_{3.2}$, $\mathfrak{g}_{3.3}$, $\mathfrak{g}_{3.4}^0$, $\mathfrak{g}_{3.4}^a$, or $\mathfrak{g}_{3.5}^a$, then G is isomorphic to the simply connected Lie group $\mathsf{G}_{3.2}$, $\mathsf{G}_{3.3}$, $\mathsf{G}_{3.4}^0 = \mathsf{SE}(1,1)$, $\mathsf{G}_{3.4}^a$, or $\mathsf{G}_{3.5}^a$, respectively. (The centres of these groups are trivial.)
5. If $\mathfrak{g} \cong \mathfrak{g}_{3.5}^0$, then G is isomorphic to the Euclidean group $\mathsf{SE}(2)$, the n-fold covering $\mathsf{SE}_n(2)$ of $\mathsf{SE}_1(2) = \mathsf{SE}(2)$, or the universal covering group $\widetilde{\mathsf{SE}}(2)$.
6. If $\mathfrak{g} \cong \mathfrak{g}_{3.6}$, then G is isomorphic to the pseudo-orthogonal group $\mathsf{SO}(2,1)_0$, the n-fold covering A_n of $\mathsf{SO}(2,1)_0$, or the universal covering group $\widetilde{\mathsf{A}}$. Here $\mathsf{A}_2 \cong \mathsf{SL}(2, \mathbb{R})$.
7. If $\mathfrak{g} \cong \mathfrak{g}_{3.7}$, then G is isomorphic to either the unitary group $\mathsf{SU}(2)$ or the orthogonal group $\mathsf{SO}(3)$.

Among these Lie groups, only H_3^*, A_n, $n \geqslant 3$, and $\widetilde{\mathsf{A}}$ are not matrix Lie groups.

Table 7.1 Three-dimensional Lie algebras

	a	n_1	n_2	n_3	Unimodular	Nilpotent	Compl. solv.	Exponential	Solvable	Simple	Connected groups
$3\mathfrak{g}_1$	0	0	0	0	●	●	●	●	●		$\mathbb{R}^3,\ \mathbb{R}^2\times\mathbb{T},\ \mathbb{R}\times\mathbb{T}^2,\ \mathbb{T}^3$
$\mathfrak{g}_{2.1}\oplus\mathfrak{g}_1$	1	1	−1	0			●	●	●		$\mathrm{Aff}(\mathbb{R})_0\times\mathbb{R},\ \mathrm{Aff}(\mathbb{R})_0\times\mathbb{T}$
$\mathfrak{g}_{3.1}$	0	1	0	0	●	●	●	●	●		$\mathrm{H}_3,\ \mathrm{H}_3^*$
$\mathfrak{g}_{3.2}$	1	1	0	0			●	●	●		$\mathrm{G}_{3.2}$
$\mathfrak{g}_{3.3}$	1	0	0	0			●	●	●		$\mathrm{G}_{3.3}$
$\mathfrak{g}_{3.4}^{0}$	0	1	−1	0	●		●	●	●		$\mathrm{SE}(1,1)$
$\mathfrak{g}_{3.4}^{a}$	$a>0$ $a\neq 1$	1	−1	0			●	●	●		$\mathrm{G}_{3.4}^{a}$
$\mathfrak{g}_{3.5}^{0}$	0	1	1	0	●				●		$\widetilde{\mathrm{SE}}(2),\ \mathrm{SE}_n(2),\ \mathrm{SE}(2)$
$\mathfrak{g}_{3.5}^{a}$	$a>0$	1	1	0				●	●		$\mathrm{G}_{3.5}^{a}$
$\mathfrak{g}_{3.6}$	0	1	1	−1	●					●	$\widetilde{\mathrm{A}},\ \mathrm{A}_n,\ \mathrm{SL}(2,\mathbb{R}),\ \mathrm{SO}(2,1)_0$
$\mathfrak{g}_{3.7}$	0	1	1	1	●					●	$\mathrm{SU}(2),\ \mathrm{SO}(3)$

Matrix Representations for Solvable Groups

We have the following parametrizations of the solvable three-dimensional matrix Lie groups and their Lie algebras. (We omit the Abelian groups.)

$$\mathsf{Aff}\,(\mathbb{R})_0 \times \mathbb{R}\;:\;\begin{bmatrix} 1 & 0 & 0 \\ x-y\,e^{2z} & 0 \\ 0 & 0 & e^{x+y} \end{bmatrix} \qquad \mathfrak{g}_{2.1} \oplus \mathfrak{g}_1\;:\;\begin{bmatrix} 0 & 0 & 0 \\ x-y & 2z & 0 \\ 0 & 0 & x+y \end{bmatrix}$$

$$\mathsf{Aff}\,(\mathbb{R})_0 \times \mathbb{T}\;:\;\begin{bmatrix} 1 & 0 & 0 \\ x-y\,e^{2z} & 0 \\ 0 & 0 & e^{i(x+y)} \end{bmatrix} \qquad \mathfrak{g}_{2.1} \oplus \mathfrak{g}_1\;:\;\begin{bmatrix} 0 & 0 & 0 \\ x-y & 2z & 0 \\ 0 & 0 & i(x+y) \end{bmatrix}$$

$$\mathsf{H}_3\;:\;\begin{bmatrix} 1 & y & x \\ 0 & 1 & z \\ 0 & 0 & 1 \end{bmatrix} \qquad \mathfrak{h}_3\;:\;\begin{bmatrix} 0 & y & x \\ 0 & 0 & z \\ 0 & 0 & 0 \end{bmatrix}$$

$$\mathsf{G}_{3.2}\;:\;\begin{bmatrix} 1 & 0 & 0 \\ y & e^z & 0 \\ x & -ze^z & e^z \end{bmatrix} \qquad \mathfrak{g}_{3.2}\;:\;\begin{bmatrix} 0 & 0 & 0 \\ y & z & 0 \\ x & -z & z \end{bmatrix}$$

$$\mathsf{G}_{3.3}\;:\;\begin{bmatrix} 1 & 0 & 0 \\ y & e^z & 0 \\ x & 0 & e^z \end{bmatrix} \qquad \mathfrak{g}_{3.3}\;:\;\begin{bmatrix} 0 & 0 & 0 \\ y & z & 0 \\ x & 0 & z \end{bmatrix}$$

$$\mathsf{SE}\,(1,1)\;:\;\begin{bmatrix} 1 & 0 & 0 \\ x & \cosh z & -\sinh z \\ y & -\sinh z & \cosh z \end{bmatrix} \qquad \mathfrak{se}\,(1,1)\;:\;\begin{bmatrix} 0 & 0 & 0 \\ x & 0 & -z \\ y & -z & 0 \end{bmatrix}$$

$$\mathsf{G}^a_{3.4}\;:\;\begin{bmatrix} 1 & 0 & 0 \\ x & e^{az}\cosh z & -e^{az}\sinh z \\ y & -e^{az}\sinh z & e^{az}\cosh z \end{bmatrix} \qquad \mathfrak{g}^a_{3.4}\;:\;\begin{bmatrix} 0 & 0 & 0 \\ x & az & -z \\ y & -z & az \end{bmatrix}$$
$$a>0 \qquad a\neq 1 \qquad\qquad\qquad a>0 \qquad a\neq 1$$

$$\widetilde{\mathsf{SE}}\,(2)\;:\;\begin{bmatrix} 1 & 0 & 0 & 0 \\ x & \cos z & -\sin z & 0 \\ y & \sin z & \cos z & 0 \\ 0 & 0 & 0 & e^z \end{bmatrix} \qquad \widetilde{\mathfrak{se}}\,(2)\;:\;\begin{bmatrix} 0 & 0 & 0 & 0 \\ x & 0 & -z & 0 \\ y & z & 0 & 0 \\ 0 & 0 & 0 & z \end{bmatrix}$$

$$\mathsf{SE}_n(2)\;:\;\begin{bmatrix} 1 & 0 & 0 & 0 \\ x & \cos z & -\sin z & 0 \\ y & \sin z & \cos z & 0 \\ 0 & 0 & 0 & e^{\frac{iz}{n}} \end{bmatrix} \qquad \mathfrak{se}_n(2)\;:\;\begin{bmatrix} 0 & 0 & 0 & 0 \\ x & 0 & -z & 0 \\ y & z & 0 & 0 \\ 0 & 0 & 0 & \frac{iz}{n} \end{bmatrix}$$

$$\mathsf{SE}(2) \; : \; \begin{bmatrix} 1 & 0 & 0 \\ x & \cos z & -\sin z \\ y & \sin z & \cos z \end{bmatrix} \qquad \mathfrak{se}(2) \; : \; \begin{bmatrix} 0 & 0 & 0 \\ x & 0 & -z \\ y & z & 0 \end{bmatrix}$$

$$\mathsf{G}_{3.5}^a \; : \; \begin{bmatrix} 1 & 0 & 0 \\ x & e^{az}\cos z & -e^{az}\sin z \\ y & e^{az}\sin z & e^{az}\cos z \end{bmatrix} \qquad \mathfrak{g}_{3.5}^a \; : \; \begin{bmatrix} 0 & 0 & 0 \\ x & az & -z \\ y & z & az \end{bmatrix}$$

An appropriate ordered basis for the Lie algebra in each case is given by setting $(x, y, z) = (1, 0, 0)$ for E_1, $(x, y, z) = (0, 1, 0)$ for E_2, and $(x, y, z) = (0, 0, 1)$ for E_3.

Matrix Representations for Semisimple Groups

Matrix Lie Groups with Algebra $\mathfrak{g}_{3.6}$

There are only two connected matrix Lie groups with Lie algebra $\mathfrak{g}_{3.6}$, namely the pseudo-orthogonal group $\mathsf{SO}(2, 1)_0$ and the special linear group $\mathsf{SL}(2, \mathbb{R})$; $\mathsf{SL}(2, \mathbb{R})$ is a double cover of $\mathsf{SO}(2, 1)_0$.

The pseudo-orthogonal group

$$\mathsf{SO}(2, 1) = \{g \in \mathbb{R}^{3\times3} \; : \; g^\top J g = J, \; \det g = 1\}$$

has two connected components. Here $J = \mathrm{diag}(1, 1, -1)$. The identity component of $\mathsf{SO}(2, 1)$ is $\mathsf{SO}(2, 1)_0 = \{g \in \mathsf{SO}(2, 1) \; : \; g_{33} > 0\}$ where $g = [g_{ij}]$ (for $g \in \mathsf{SO}(2, 1)$). Its Lie algebra is given by

$$\mathfrak{so}(2, 1) = \{A \in \mathbb{R}^{3\times3} \; : \; A^\top J + JA = 0\}$$

$$= \left\{ \begin{bmatrix} 0 & z & y \\ -z & 0 & x \\ y & x & 0 \end{bmatrix} \; : \; x, y, z \in \mathbb{R} \right\}.$$

On the other hand, the special linear group is given by

$$\mathsf{SL}(2, \mathbb{R}) = \{g \in \mathbb{R}^{2\times2} \; : \; \det g = 1\}.$$

Its Lie algebra is given by

$$\mathfrak{sl}(2, \mathbb{R}) = \left\{ \begin{bmatrix} \frac{x}{2} & \frac{y-z}{2} \\ \frac{y+z}{2} & -\frac{x}{2} \end{bmatrix} \; : \; x, y, z \in \mathbb{R} \right\}.$$

Matrix Lie Groups with Algebra $\mathfrak{g}_{3.7}$

There are exactly two connected Lie groups with Lie algebra $\mathfrak{g}_{3.7}$; both are matrix Lie groups. The special unitary group and its Lie algebra are given by

$$\mathsf{SU}\,(2) = \{g \in \mathbb{C}^{2\times 2} : gg^\dagger = \mathbf{1},\ \det g = 1\}$$

$$\mathfrak{su}\,(2) = \left\{ \begin{bmatrix} \frac{i}{2}x & \frac{1}{2}(iz+y) \\ \frac{1}{2}(iz-y) & -\frac{i}{2}x \end{bmatrix} : x, y, z \in \mathbb{R} \right\}$$

$\mathsf{SU}\,(2)$ is a double cover of the orthogonal group $\mathsf{SO}\,(3)$. The orthogonal group $\mathsf{SO}\,(3)$ and its Lie algebra are given by

$$\mathsf{SO}\,(3) = \{g \in \mathbb{R}^{3\times 3} : gg^\top = \mathbf{1},\ \det g = 1\}$$

$$\mathfrak{so}\,(3) = \left\{ \begin{bmatrix} 0 & -z & y \\ z & 0 & -x \\ -y & x & 0 \end{bmatrix} : x, y, z \in \mathbb{R} \right\}.$$

Note 6 Again, an appropriate ordered basis for the Lie algebra in each case is given by setting $(x, y, z) = (1, 0, 0)$ for E_1, $(x, y, z) = (0, 1, 0)$ for E_2, and $(x, y, z) = (0, 0, 1)$ for E_3.

Automorphism Groups

A standard computation yields the automorphism group for each three-dimensional Lie algebra (see, e.g., [57]). With respect to the given ordered basis (E_1, E_2, E_3), the automorphism group of each solvable Lie algebra has parametrization:

$$\mathsf{Aut}(\mathfrak{g}_{3.1}) : \begin{bmatrix} yw - vz & x & u \\ 0 & y & v \\ 0 & z & w \end{bmatrix} \qquad \mathsf{Aut}(\mathfrak{g}_{2.1} \oplus \mathfrak{g}_1) : \begin{bmatrix} x & y & u \\ y & x & v \\ 0 & 0 & 1 \end{bmatrix}$$

$$\mathsf{Aut}(\mathfrak{g}_{3.2}) : \begin{bmatrix} u & x & y \\ 0 & u & z \\ 0 & 0 & 1 \end{bmatrix} \qquad \mathsf{Aut}(\mathfrak{g}_{3.3}) : \begin{bmatrix} x & y & z \\ u & v & w \\ 0 & 0 & 1 \end{bmatrix}$$

$$\mathsf{Aut}(\mathfrak{g}_{3.4}^0) : \begin{bmatrix} x & y & u \\ y & x & v \\ 0 & 0 & 1 \end{bmatrix}, \begin{bmatrix} x & y & u \\ -y & -x & v \\ 0 & 0 & -1 \end{bmatrix} \qquad \mathsf{Aut}(\mathfrak{g}_{3.4}^a) : \begin{bmatrix} x & y & u \\ y & x & v \\ 0 & 0 & 1 \end{bmatrix}$$

$$\mathsf{Aut}(\mathfrak{g}_{3.5}^0)) : \begin{bmatrix} x & y & u \\ -y & x & v \\ 0 & 0 & 1 \end{bmatrix}, \begin{bmatrix} x & y & u \\ y & -x & v \\ 0 & 0 & -1 \end{bmatrix} \qquad \mathsf{Aut}(\mathfrak{g}_{3.5}^a) : \begin{bmatrix} x & y & u \\ -y & x & v \\ 0 & 0 & 1 \end{bmatrix}$$

For the semisimple Lie algebras, we have

$$\text{Aut}(\mathfrak{g}_{3.6}) = \text{SO}(2, 1) \qquad \text{and} \qquad \text{Aut}(\mathfrak{g}_{3.7}) = \text{SO}(3).$$

In several cases, for a connected Lie group G, we wish to find the subgroup of linearized group automorphisms $d\,\text{Aut}(G) = \{T_1\phi \,:\, \phi \in \text{Aut}(G)\} \leqslant \text{Aut}(\mathfrak{g})$. If G is simply connected, then $d\,\text{Aut}(G) = \text{Aut}(\mathfrak{g})$. If G is not simply connected, then the subgroup $d\,\text{Aut}(G)$ may be determined by use of the following result.

Proposition 5 (cf. [58]) *Let G be a connected Lie group with Lie algebra $\mathfrak{g}$ and let $q : \widetilde{G} \to G$ be a universal covering. If $\psi \in \text{Aut}(\mathfrak{g})$, then $\psi \in d\,\text{Aut}(G)$ if and only if $\phi(\ker q) = \ker q$, where $\phi \in \text{Aut}(\widetilde{G})$ is the unique automorphism such that $T_1\phi = (T_1q)^{-1} \cdot \psi \cdot T_1q$.*

Casimir Functions for Lie–Poisson Spaces

An exhaustive list of Casimir functions (not necessarily globally defined), for low-dimensional Lie algebras, was obtained by Patera et al. [86]. For each three-dimensional Lie–Poisson space $\mathfrak{g}_-^*$ (associated to the three-dimensional Lie algebra $\mathfrak{g}$) we exhibit its Casimir function:

$$\mathfrak{g}_{2.1} \oplus \mathfrak{g}_1 \;:\; C(p) = p_3 \qquad\qquad \mathfrak{g}_{3.1} \;:\; C(p) = p_1$$

$$\mathfrak{g}_{3.2} \;:\; C(p) = p_1 e^{\frac{p_2}{p_1}} \qquad\qquad \mathfrak{g}_{3.3} \;:\; C(p) = \frac{p_2}{p_1}$$

$$\mathfrak{g}_{3.4}^0 \;:\; C(p) = p_1^2 - p_2^2 \qquad\qquad \underset{1 \neq \alpha > 0}{\mathfrak{g}_{3.4}^\alpha} \;:\; C(p) = \frac{\frac{1}{2}p_1 + \frac{1}{2}p_2}{(\pm\frac{1}{2}p_1 \mp \frac{1}{2}p_2)^{\frac{\alpha-1}{\alpha+1}}}$$

$$\mathfrak{g}_{3.5}^0 \;:\; C(p) = p_1^2 + p_2^2 \qquad\qquad \underset{\alpha > 0}{\mathfrak{g}_{3.5}^\alpha} \;:\; C(p) = (p_1^2 + p_2^2)\left(\frac{p_1 - ip_2}{p_1 + ip_2}\right)^{i\alpha}$$

$$\mathfrak{g}_{3.6} \;:\; C(p) = p_1^2 + p_2^2 - p_3^2 \qquad \mathfrak{g}_{3.7} \;:\; C(p) = p_1^2 + p_2^2 + p_3^2.$$

On the trivial Lie–Poisson space $(3\mathfrak{g}_1)_-^*$, every function is a Casimir function. Here $p \in \mathfrak{g}^*$ is written in coordinates as $p = p_1 E_1^* + p_2 E_2^* + p_3 E_3^*$ where (E_1^*, E_2^*, E_3^*) is the dual of the ordered basis (E_1, E_2, E_3). Note that only $3\mathfrak{g}_1$, $\mathfrak{g}_{2.1} \oplus \mathfrak{g}_1$, $\mathfrak{g}_{3.1}$, $\mathfrak{g}_{3.4}^0$, $\mathfrak{g}_{3.5}^0$, $\mathfrak{g}_{3.6}$, and $\mathfrak{g}_{3.7}$ admit globally defined Casimir functions.

References

1. R.M. Adams, R. Biggs, W. Holderbaum, C.C. Remsing, Stability and integration of Hamilton-Poisson systems on so(3)*. Rend. Mat. Appl. **37**(1–2), 1–42 (2016)
2. R.M. Adams, R. Biggs, C.C. Remsing, Equivalence of control systems on the Euclidean group SE(2). Control Cybern. **41**(3), 513–524 (2012)
3. R.M. Adams, R. Biggs, C.C. Remsing, On the equivalence of control systems on the orthogonal group SO(4), in *Recent Researches in Automatic Control, Systems Science and Communications, Porto, 2012*, ed. by H.R. Karimi (WSEAS Press, 2012), pp. 54–59. ISBN: 978-1-61804-103-6
4. R.M. Adams, R. Biggs, C.C. Remsing, Single-input control systems on the Euclidean group SE (2). Eur. J. Pure Appl. Math. **5**(1), 1–15 (2012)
5. R.M. Adams, R. Biggs, C.C. Remsing, Control systems on the orthogonal group SO (4). Commun. Math. **21**(2), 107–128 (2013)
6. R.M. Adams, R. Biggs, C.C. Remsing, On some quadratic Hamilton-Poisson systems. Appl. Sci. **15**, 1–12 (2013)
7. R.M. Adams, R. Biggs, C.C. Remsing, Two-input control systems on the Euclidean group SE (2). ESAIM Control Optim. Calc. Var. **19**(4), 947–975 (2013)
8. R.M. Adams, R. Biggs, C.C. Remsing, Quadratic Hamilton-Poisson systems on $\mathfrak{so}^*(3)$: classifications and integration, in *Proceedings of the 15th International Conference on Geometry, Integrability and Quantization*, Varna, 2013, ed. by I.M. Mladenov, A. Ludu, A. Yoshioka (Bulgarian Academy of Sciences, Sofia, 2014), pp. 55–66
9. A. Agrachev, D. Barilari, Sub-Riemannian structures on 3D Lie groups. J. Dyn. Control Syst. **18**(1), 21–44 (2012)
10. A.A. Agrachev, Y.L. Sachkov, *Control Theory from the Geometric Viewpoint* (Springer, Berlin, 2004)
11. J.V. Armitage, W.F. Eberlein, *Elliptic Functions* (Cambridge University Press, Cambridge, 2006)
12. A. Aron, C. Pop, Quadratic and homogeneous Hamilton-Poisson systems on the 13th Lie algebra from Bianchi's classification, in *Proceedings of the International Conference of Differential Geometry and Dynamical Systems*, Bucharest, 2009, vol. 17 (Geometry Balkan Press, Bucharest 2010), pp. 12–20
13. A. Aron, C. Dăniasă, M. Puta, Quadratic and homogeneous Hamilton-Poisson system on (so(3)). Int. J. Geom. Methods Mod. Phys. **4**(7), 1173–1186 (2007)
14. A. Aron, C. Pop, M. Puta, Some remarks on (sl(2, ℝ)) and Kahan's integrator. An. Ştiinţ. Univ. Al. I. Cuza Iaşi. Mat. (N.S.) **53**(Suppl. 1), 49–60 (2007)
15. A. Aron, M. Craioveanu, C. Pop, M. Puta, Quadratic and homogeneous Hamilton-Poisson systems on $A^*_{3,6,-1}$. Balkan J. Geom. Appl. **15**(1), 1–7 (2010)
16. D.I. Barrett, R. Biggs, C.C. Remsing, Quadratic Hamilton-Poisson systems on $\mathfrak{se}(1,1)^*$: the inhomogeneous case. Preprint
17. D.I. Barrett, R. Biggs, C.C. Remsing, Affine subspaces of the Lie algebra $\mathfrak{se}(1,1)$. Eur. J. Pure Appl. Math. **7**(2), 140–155 (2014)
18. D.I. Barrett, R. Biggs, C.C. Remsing, Optimal control of drift-free invariant control systems on the group of motions of the Minkowski plane, in *Proceedings of the 13th European Control Conference*, Strasbourg, 2014 (European Control Association, 2014), pp. 2466–2471. doi:10.1109/ECC.2014.6862313
19. D.I. Barrett, R. Biggs, C.C. Remsing, Affine distributions on a four-dimensional extension of the semi-Euclidean group. Note Mat. **35**(2), 81–97 (2015)
20. D.I. Barrett, R. Biggs, C.C. Remsing, Quadratic Hamilton-Poisson systems on $\mathfrak{se}(1,1)^*$: the homogeneous case. Int. J. Geom. Methods Mod. Phys. **12**, 1550011 (17 pp.) (2015)
21. C.E. Bartlett, R. Biggs, C.C. Remsing, Control systems on the Heisenberg group: equivalence and classification. Publ. Math. Debr. **88**(1–2), 217–234 (2016)

22. J. Biggs, W. Holderbaum, Planning rigid body motions using elastic curves. Math. Control Signals Syst. **20**(4), 351–367 (2008)
23. J. Biggs, W. Holderbaum, Integrable quadratic Hamiltonians on the Euclidean group of motions. J. Dyn. Control Syst. **16**(3), 301–317 (2010)
24. R. Biggs, P.T. Nagy, A classification of sub-Riemannian structures on the Heisenberg groups. Acta Polytech. Hungar. **10**(7), 41–52 (2013)
25. R. Biggs, C.C. Remsing, A category of control systems. An. Ştiinţ. Univ. "Ovidius" Constanţa Ser. Mat. **20**(1), 355–367 (2012)
26. R. Biggs, C.C. Remsing, A note on the affine subspaces of three-dimensional Lie algebras. Bul. Acad. Ştiinţe Repub. Mold. Mat. **2012**(3), 45–52 (2012)
27. R. Biggs, C.C. Remsing, On the equivalence of cost-extended control systems on Lie groups, in *Recent Researches in Automatic Control, Systems Science and Communications*, Porto, 2012, ed. by H.R. Karimi (WSEAS Press, 2012), pp. 60–65. ISBN: 978-1-61804-103-6
28. R. Biggs, C.C. Remsing, Control affine systems on solvable three-dimensional Lie groups, I. Arch. Math. (Brno) **49**(3), 187–197 (2013)
29. R. Biggs, C.C. Remsing, Control affine systems on solvable three-dimensional Lie groups, II. Note Mat. **33**(2), 19–31 (2013)
30. R. Biggs, C.C. Remsing, Control affine systems on semisimple three-dimensional Lie groups. An. Ştiinţ. Univ. Al. I. Cuza Iaşi. Mat. (N.S.) **59**(2), 399–414 (2013)
31. R. Biggs, C.C. Remsing, Feedback classification of invariant control systems on three-dimensional Lie groups, in *Proceedings of the 9th IFAC Symposium on Nonlinear Control Systems*, Toulouse (2013), pp. 506–511
32. R. Biggs, C.C. Remsing, A classification of quadratic Hamilton-Poisson systems in three dimensions, in *Proceedings of the 15th International Conference on Geometry, Integrability and Quantization*, Varna, 2013, ed. by I.M. Mladenov, A. Ludu, A. Yoshioka (Bulgarian Academy of Sciences, Sofia, 2014), pp. 67–78
33. R. Biggs, C.C. Remsing, Control systems on three-dimensional Lie groups: equivalence and controllability. J. Dyn. Control Syst. **20**(3), 307–339 (2014)
34. R. Biggs, C.C. Remsing, Control systems on three-dimensional Lie groups, in *Proceedings of the 13th European Control Conference*, Strasbourg, 2014 (European Control Association, 2014), pp. 2442–2447. doi:10.1109/ECC.2014.6862312
35. R. Biggs, C.C. Remsing, Cost-extended control systems on Lie groups. Mediterr. J. Math. **11**(1), 193–215 (2014)
36. R. Biggs, C.C. Remsing, Some remarks on the oscillator group. Differ. Geom. Appl. **35**(Suppl.), 199–209 (2014)
37. R. Biggs, C.C. Remsing, Subspaces of the real four-dimensional Lie algebras: a classification of subalgebras, ideals, and full-rank subspaces. Extracta Math. **31**(1), 41–93 (2015)
38. R. Biggs, C.C. Remsing, On the equivalence of control systems on Lie groups. Commun. Math. **23**(2), 119–129 (2015)
39. R. Biggs, C.C. Remsing, Equivalence of control systems on the pseudo-orthogonal group SO (2, 1). An. Ştiinţ. Univ. "Ovidius" Constanţa Ser. Mat. **24**(2), 45–65 (2016)
40. A.M. Bloch, *Nonholonomic Mechanics and Control* (Springer, New York, 2003)
41. A.M. Bloch, P.E. Crouch, N. Nordkvist, A.K. Sanyal, Embedded geodesic problems and optimal control for matrix Lie groups. J. Geom. Mech. **3**(2), 197–223 (2011)
42. B. Bonnard, V. Jurdjevic, I. Kupka, G. Sallet, Transitivity of families of invariant vector fields on the semidirect products of Lie groups. Trans. Am. Math. Soc. **271**(2), 525–535 (1982)
43. R.W. Brockett, System theory on group manifolds and coset spaces. SIAM J. Control **10**(2), 265–284 (1972)
44. L. Capogna, E. Le Donne, Smoothness of subRiemannian isometries. Am. J. Math. **138**(5), 1439–1454 (2016)
45. J.N. Clelland, C.G. Moseley, G.R. Wilkens, Geometry of control-affine systems. SIGMA Symmetry Integrability Geom. Methods Appl. **5**(Paper 095), 28 (2009)
46. J.N. Clelland, C.G. Moseley, G.R. Wilkens, Geometry of optimal control for control-affine systems. SIGMA Symmetry Integrability Geom. Methods Appl. **9**(Paper 034), 31 (2013)

47. M. Craioveanu, C. Pop, A. Aron, C. Petrişor, An optimal control problem on the special Euclidean group $SE(3, \mathbb{R})$, in *International Conference of Differential Geometry and Dynamical Systems*, Bucharest, 2009, vol. 17 (Geometry Balkan Press, Sofia, 2010), pp. 68–78

48. D. D'Alessandro, M. Dahleh, Optimal control of two-level quantum systems. IEEE Trans. Autom. Control **46**(6), 866–876 (2001)

49. C. Dăniasă, A. Gîrban, R.M. Tudoran, New aspects on the geometry and dynamics of quadratic Hamiltonian systems on $(\mathfrak{so}(3))^*$. Int. J. Geom. Methods Mod. Phys. **8**(8), 1695–1721 (2011)

50. V.I. Elkin, Affine control systems: their equivalence, classification, quotient systems, and subsystems. J. Math. Sci. **88**(5), 675–721 (1998)

51. R.V. Gamkrelidze, Discovery of the maximum principle. J. Dyn. Control Syst. **5**(4), 437–451 (1999)

52. R.B. Gardner, W.F. Shadwick, Feedback equivalence of control systems. Syst. Control Lett. **8**(5), 463–465 (1987)

53. V.V. Gorbatsevich, A.L. Onishchik, E.B. Vinberg, *Foundations of Lie Theory and Lie Transformation Groups* (Springer, Berlin, 1997)

54. S. Goyal, N.C. Perkins, C.L. Lee, Nonlinear dynamics and loop formation in Kirchhoff rods with implications to the mechanics of DNA and cables. J. Comput. Phys. **209**(1), 371–389 (2005)

55. K.Y. Ha, J.B. Lee, Left invariant metrics and curvatures on simply connected three-dimensional Lie groups. Math. Nachr. **282**(6), 868–898 (2009)

56. U. Hamenstädt, Some regularity theorems for Carnot-Carathéodory metrics. IEEE Trans. Autom. Control **32**(3), 819–850 (1990)

57. A. Harvey, Automorphisms of the Bianchi model Lie groups. J. Math. Phys. **20**(2), 251–253 (1979)

58. J. Hilgert, K.-H. Neeb, *Structure and Geometry of Lie Groups* (Springer, New York, 2012)

59. D.D. Holm, *Geometric Mechanics. Part I: Dynamics and Symmetry* (Imperial College Press, London, 2008)

60. D.D. Holm, *Geometric Mechanics. Part II: Rotating, Translating and Rolling*, 2nd edn. (Imperial College Press, London, 2011)

61. B. Jakubczyk, Equivalence and invariants of nonlinear control systems, in *Nonlinear Controllability and Optimal Control*, ed. by H.J. Sussmann (Dekker, New York, 1990), pp. 177–218

62. B. Jakubczyk, W. Respondek, On linearization of control systems. Bull. Acad. Polon. Sci. Sér. Sci. Math. **28**(9–10), 517–522 (1980)

63. V. Jurdjevic, The geometry of the plate-ball problem. Arch. Ration. Mech. Anal. **124**(4), 305–328 (1993)

64. V. Jurdjevic, *Geometric Control Theory* (Cambridge University Press, Cambridge, 1997)

65. V. Jurdjevic, Integrable Hamiltonian systems on Lie groups: Kowalewski type. Ann. Math. **150**(2), 605–644 (1999)

66. V. Jurdjevic, Optimal control on Lie groups and integrable Hamiltonian systems. Regul. Chaotic Dyn. **16**(5), 514–535 (2011)

67. V. Jurdjevic, F. Monroy-Pérez, Variational problems on Lie groups and their homogeneous spaces: elastic curves, tops, and constrained geodesic problems, in *Contemporary Trends in Nonlinear Geometric Control Theory and its Applications*, México City, 2000 (World Scientific, River Edge, NJ, 2002), pp. 3–51

68. V. Jurdjevic, H.J. Sussmann, Control systems on Lie groups. J. Differ. Equ. **12**, 313–329 (1972)

69. V. Jurdjevic, J. Zimmerman, Rolling sphere problems on spaces of constant curvature. Math. Proc. Camb. Philos. Soc. **144**(3), 729–747 (2008)

70. I. Kishimoto, Geodesics and isometries of Carnot groups. J. Math. Kyoto Univ. **43**(3), 509–522 (2003)

71. A. Krasiński, C.G. Behr, E. Schücking, F.B. Estabrook, H.D. Wahlquist, G.F.R. Ellis, R. Jantzen, W. Kundt, The Bianchi classification in the Schücking-Behr approach. Gen. Relativ. Gravit. **35**(3), 475–489 (2003)
72. A.J. Krener, On the equivalence of control systems and linearization of nonlinear systems. SIAM J. Control **11**, 670–676 (1973)
73. C. Laurent-Gengoux, A. Pichereau, P. Vanhaecke, *Poisson Structures* (Springer, Heidelberg, 2013)
74. J. Lauret, Modified H-type groups and symmetric-like Riemannian spaces. Differ. Geom. Appl. **10**(2), 121–143 (1999)
75. D.F. Lawden, *Elliptic Functions and Applications* (Springer, New York, 1989)
76. E. Le Donne, A. Ottazzi, Isometries of Carnot groups and subFinsler homogeneous manifolds. J. Geom. Anal. **26**(1), 330–345 (2016)
77. N.E. Leonard, P.S. Krishnaprasad, Motion control of drift-free, left-invariant systems on Lie groups. IEEE Trans. Automat. Control **40**, 1539–1554 (1995)
78. M.A.H. MacCallum, On the classification of the real four-dimensional Lie algebras, in *On Einstein's Path*, New York, 1996 (Springer, New York, 1999), pp. 299–317
79. J.E. Marsden, T.S. Ratiu, *Introduction to Mechanics and Symmetry*, 2nd edn. (Springer, New York, 1999)
80. C. Meyer, *Matrix Analysis and Applied Linear Algebra* (Society for Industrial and Applied Mathematics, Philadelphia, PA, 2000)
81. F. Monroy-Pérez, A. Anzaldo-Meneses, Optimal control on the Heisenberg group. J. Dyn. Control Syst. **5**(4), 473–499 (1999)
82. G.M. Mubarakzyanov, On solvable Lie algebras. Izv. Vysš. Učehn. Zaved. Matematika **1**(32), 114–123 (1963)
83. P.T. Nagy, M. Puta, Drift Lee-free left invariant control systems on SL(2, ℝ) with fewer controls than state variables. Bull. Math. Soc. Sci. Math. Roumanie (N.S.) **44**(92)(1), 3–11 (2001)
84. A.L. Onishchik, E.B. Vinberg, *Lie Groups and Lie Algebras, III* (Springer, Berlin, 1994)
85. J.-P. Ortega, V. Planas-Bielsa, T.S. Ratiu, Asymptotic and Lyapunov stability of constrained and Poisson equilibria. J. Differ. Equ. **214**(1), 92–127 (2005)
86. J. Patera, R.T. Sharp, P. Winternitz, H. Zassenhaus, Invariants of real low dimension Lie algebras. J. Math. Phys. **17**(6), 986–994 (1976)
87. R.O. Popovych, V.M. Boyko, M.O. Nesterenko, M.W. Lutfullin, Realizations of real low-dimensional Lie algebras. J. Phys. A **36**(26), 7337–7360 (2003)
88. M. Puta, Stability and control in spacecraft dynamics. J. Lie Theory **7**(2), 269–278 (1997)
89. M. Puta, Optimal control problems on matrix Lie groups, in *New Developments in Differential Geometry*, ed. by J. Szenthe (Kluwer, Dordrecht, 1999), pp. 361–373
90. W. Respondek, I.A. Tall, Feedback equivalence of nonlinear control systems: a survey on formal approach, in *Chaos in Automatic Control (Control Engineering)* (CRC, Taylor & Francis, Boca Raton, 2006), pp. 137–262
91. L. Saal, The automorphism group of a Lie algebra of Heisenberg type. Rend. Sem. Mat. Univ. Politec. Torino **54**(2), 101–113 (1996)
92. Y.L. Sachkov, Conjugate points in the Euler elastic problem. J. Dyn. Control Syst. **14**(3), 409–439 (2008)
93. Y.L. Sachkov, Maxwell strata in the Euler elastic problem. J. Dyn. Control Syst. **14**(2), 169–234 (2008)
94. Y.L. Sachkov, Control theory on Lie groups. J. Math. Sci. **156**(3), 381–439 (2009)
95. H.J. Sussmann, An extension of a theorem of Nagano on transitive Lie algebras. Proc. Am. Math. Soc. **45**(3), 349–356 (1974)
96. H.J. Sussmann, Lie brackets, real analyticity and geometric control, in *Differential Geometric Control Theory*, ed. by Brockett, R.W., Millman, R.S., Sussmann, H.J. (Birkhäuser, Boston, MA, 1983), pp. 1–116
97. R.M. Tudoran, The free rigid body dynamics: generalized versus classic. J. Math. Phys. **54**(7), 072704, 10 (2013)

98. R.M. Tudoran, R.A. Tudoran, On a large class of three-dimensional Hamiltonian systems. J. Math. Phys. **50**(1), 012703, 9 (2009)
99. A.M. Vershik, V.Ya. Gershkovich, Nonholonomic dynamical systems, geometry of distributions and variational problems, in *Dynamical Systems VII*, ed. by V.I. Arnol'd, S.P. Novikov (Springer, Berlin, 1994), pp. 1–81
100. G.C. Walsh, R. Montgomery, S.S. Sastry, Optimal path planning on matrix Lie groups, in *Proceedings of the 33rd Conference on Decision and Control*, Lake Buena Vista (1994), pp. 1258–1263
101. E.T. Whittaker, G.N. Watson, *A Course of Modern Analysis* (Cambridge University Press, Cambridge, 1927)
102. E.N. Wilson, Isometry groups on homogeneous nilmanifolds. Geom. Dedicata **12**(3), 337–346 (1982)

Chapter 8
An Optimal Control Problem for a Nonlocal Problem on the Plane

D. Devadze, Ts. Sarajishvili, and M. Abashidze

Abstract The present paper is dedicated to the investigation of optimal control problems whose behavior is described by quasilinear differential equations of first order on the plane with nonlocal Bitsadze–Samarski boundary conditions. A theorem of the existence and uniqueness of a generalized solution in the space $C_\mu(\overline{G})$ is proved for quasilinear differential equations; necessary conditions of optimality are obtained in terms of the principle of maximum; the Bitsadze–Samarski boundary value problem is proved for a linear differential equation of first order; the existence of a solution in the space $C_\mu^p(\overline{G})$ is proved and an a priori estimate is derived. A theorem on the necessary and sufficient condition of optimality is proved for a linear optimal control problem.

Nonlocal boundary value problems are quite an interesting generalization of classical problems and at the same time they are naturally obtained when constructing mathematical models in physics, engineering, sociology, ecology, and so on [1–5].

The investigation of nonlocal problems for differential equations originated in the last century. Here we should in the first place refer to the works of T. Carleman, R. Beals, F. Browder, and other works. The problems posed in [6–8] are the problems with nonlocal conditions, which are considered only on the boundary of the definition domain of a differential operator. In 1963, J. Cannon posed a nonlocal problem in his work [9] and thus gave an impetus to the development of a new trend in the investigation of nonlocal boundary value problems [10–13].

In 1969, the work of Bitsadze and Samarskiĭ [14] was published, which was dedicated to the investigation of a nonlocal problem of a new type. That problem arose in connection with the mathematical modeling of plasma processes . Intensive studies of Bitsadze–Samarskiĭ nonlocal problems and their various generalizations began in the 1980s of the last century (see the papers of D.G. Gordeziani, A.L. Skubachevski, V.P. Paneyakh, V.A. Ilyin, I. Moiseyev, G.V. Meladze, M.P. Sapagovas, D.V. Kapanadze, V.P. Mikhailov, A.K. Gushchin, G. Avalishvili,

D. Devadze (✉) • Ts. Sarajishvili • M. Abashidze
Batumi Shota Rustaveli State University, 35 Ninoshvili St., Batumi 6010, Georgia
e-mail: david.devadze@gmail.com; tsis55@yahoo.com; m.abashidze1@gmail.com

© Springer International Publishing AG 2017   183
G. Falcone (ed.), *Lie Groups, Differential Equations, and Geometry*,
UNIPA Springer Series, DOI 10.1007/978-3-319-62181-4_8

L. Gurevich [15–26], and other works). The papers of D.G. Sapagovas, G.K. Berikelashvili are certainly interesting from the standpoint of application and numerical methods (see, e.g., [11, 17]). The algorithm of reducing nonlocal problems of the Bitsadze–Samarskiĭ type to an iteration sequence of Dirichlet problems is investigated in Gordeziani's papers [15, 16].

Problems of the existence and uniqueness of a generalized analytic function for the Riemann–Hilbert problem are investigated in Vekua's monograph [27]. Problems of the existence and uniqueness of a generalized solution for quasilinear equations of first order on the plane with Riemann–Hilbert boundary conditions are considered in [28]. In [29, 30] the Bitsadze–Samarskiĭ nonlocal boundary value problem is considered for quasilinear differential equations.

For many optimization problems in elasticity theory, mechanics, diffusion processes, the kinetics of chemical reactions, and so on, the state of a system is described by partial differential equations [31–34]. Therefore control problems for systems with distributed parameters attract a great deal of attention [33, 35, 36].

The present paper is devoted to optimal control problems whose behavior is described by quasilinear differential equation of first order on the plane with nonlocal Bitsadze–Samarski boundary conditions. When dealing with optimization problems for systems with distributed parameters it is important to investigate the questions of existence of a generalized solution for discontinuous right-hand parts of an equation [27]. In the proposed paper, the necessary conditions of optimality are obtained using the approach elaborated in [37–39] for controlled systems of general form.

In the paper, the theorem on the existence and uniqueness theorem of a generalized solution in the space $C_\mu(\overline{G})$ is proved for quasilinear differential equations of first order with nonlocal boundary conditions; the Bitsadze–Samarskiĭ boundary value problem is considered for a linear differential equation of first order. The existence of a solution in the space $C_\mu^\rho(\overline{G})$ is proved and an a priori estimate is obtained. An optimal control problem is stated for a nonlocal boundary value problem with an integral quality criterion. Necessary conditions of optimality are obtained in terms of the maximum principle when the control domain is an arbitrary bounded set. An adjoint equation is constructed in the differential form. The optimal control problem for linear differential equations of first order is considered on the plane with nonlocal boundary conditions. The theorem on a necessary and sufficient condition of optimality is proved.

8.1   Existence of a Generalized Solution

Let G be a bounded domain of the complex plane E with boundary Γ which is a closed simple Lyapunov curve (i.e. the angle formed between the constant direction and the tangent to this curve is Hölder-continuous). Denote by γ the part of the boundary Γ, which is an open Lyapunov curve with the parametric equation $z = z(s)$, $0 \leqslant s \leqslant \delta$. Let γ_0 be the diffeomorphic image $z_0 = I(z)$ of γ, which lies in the domain G, with the parametric equation $z_0 = z_0(s)$, $0 \leqslant s \leqslant \delta$. Assume that γ_0 intersects with Γ, but not tangentially to it, $z = x + iy \in G$, $\partial_{\overline{z}} = \frac{1}{2}\left(\frac{\partial}{\partial x} + i\frac{\partial}{\partial y}\right)$ is the generalized Sobolev derivative [27, 40, 41].

Let us consider in the domain $\overline{G}$ the Bitsadze–Samarski boundary value problem [14] for quasilinear differential equations of first order

$$\partial_{\overline{z}} w = f(z, w, \overline{w}), \quad z \in G, \tag{8.1}$$

$$\mathrm{Re}[w(z)] = \varphi_1(z), \quad z \in \Gamma \setminus \gamma, \tag{8.2}$$

$$\mathrm{Re}\left[w(z(s))\right] = \sigma\,\mathrm{Re}\left[w(z_0(s))\right], \quad z(s) \in \gamma, \quad z_0(s) \in \gamma_0, \quad 0 \leqslant s \leqslant \delta, \tag{8.3}$$

$$\mathrm{Im}[w(z^*)] = c, \quad z^* \in \Gamma \setminus \gamma, \quad 0 < \sigma = \mathrm{const}. \tag{8.4}$$

It is assumed that the following conditions are fulfilled:

(A1) The function $f(z, w, \overline{w})$ is defined for $z \in G$, $|w| < R$, $f(z, 0, 0) \in L_p(\overline{G})$, $p > 2$ and

$$\left|f(z, w, \overline{w}) - f(z, w_0, \overline{w}_0)\right| \leqslant L\big(|w - w_0| + |w - \overline{w}_0|\big).$$

(A2) $\varphi_1(z) \in C_\mu(\Gamma \setminus \gamma)$, $\varphi_2(z) \in C_\mu(\gamma)$, $\mu > 1/2$, and the values of the function $\varphi_1(z)$, $\varphi_2(z)$ at the ends of the contours γ and γ_0 coincide.

In order to prove the existence of a solution of the problem (8.1)–(8.4), we consider the following iteration process

$$\partial_{\overline{z}} w_n = f(z, w_n, \overline{w}_n), \quad z \in G, \tag{8.5}$$

$$\mathrm{Re}[w_n(z)] = \varphi_1(z), \quad z \in \Gamma \setminus \gamma, \tag{8.6}$$

$$\mathrm{Re}\left[w_n(z(s))\right] = \sigma\,\mathrm{Re}\left[w_{n-1}(z_0(s))\right], \quad z(s) \in \gamma, \quad z_0(s) \in \gamma_0, \tag{8.7}$$

$$\mathrm{Im}[w_n(z^*)] = c, \quad z^* \in \Gamma \setminus \gamma, \quad 0 \leqslant s \leqslant \sigma, \quad n = 1, 2, \ldots, \tag{8.8}$$

where $w_0(z)$ is any function from $C_\mu(\gamma)$, $\mu > 1/2$, which continuously adjoins the values of the function $\varphi_1(z)$ at the ends of the contour γ.

For each $n \in N$ the problem (8.5)–(8.8) is the Riemann–Hilbert problem and its regular generalized solution belongs to the space $C_\mu(\overline{G})$ [27, 28].

Let us consider the function $v_n = w_{n+1} - w_n$. Then from (8.5)–(8.8) it follows that the function v_n is the solution of the following problem

$$\partial_{\overline{z}} v_n = f(z, w_{n+1}, \overline{w}_{n+1}) - f(z, w_n, \overline{w}_n)$$

$$\equiv F(z, w_n, \overline{w}_n, w_{n+1}, \overline{w}_{n+1}), \quad z \in G, \tag{8.9}$$

$$\mathrm{Re}[v_n(z)] = 0, \quad z \in \Gamma \setminus \gamma, \tag{8.10}$$

$$\mathrm{Re}\left[v_n(z(s))\right] = \sigma\,\mathrm{Re}\left[v_{n-1}(z_0(s))\right], \quad z(s) \in \gamma, \quad z_0(s) \in \gamma_0, \quad 0 \leqslant s \leqslant \delta, \tag{8.11}$$

$$\mathrm{Im}[v_n(z^*)] = c, \quad z^* \in \Gamma \setminus \gamma, \quad n = 1, 2, \ldots . \tag{8.12}$$

The solution of the problem (8.9)–(8.12) can be reduced to the following nonlinear integral equation [28]:

$$v^*(z) = \psi_n(z) + \phi_n(z)$$
$$- \frac{1}{\pi} \iint\limits_G \frac{F(\zeta, w_n(\zeta), \overline{w}_n(\zeta), w_{n+1}(\zeta), \overline{w}_{n+1}(\zeta))}{\zeta - z} \, d\xi \, d\eta, \qquad (8.13)$$

where $\zeta = \xi + i\eta$, $\psi_n(z)$ is a holomorphic function satisfying the conditions (8.10)–(8.12), and $\phi_n(z)$ is a holomorphic function such that the difference

$$\phi_n(z) - \frac{1}{\pi} \iint\limits_G \frac{F(\zeta, w_n, \overline{w}_n, w_{n+1}, \overline{w}_{n+1})}{\zeta - z} \, d\xi \, d\eta$$

satisfies the homogeneous boundary conditions; an a priori estimate has the form

$$\|\phi_n\|_{C_\mu(\overline{G})} \leq C_1 \|F\|_{L_p(\overline{G})}, \quad C_1 = \text{const} > 0.$$

Denote by T_G the integral operator in the right-hand side of Eq. (8.13). Note that the operator T_G maps the space $L_p(\overline{G})$ into $C_\beta(\overline{G})$, $\beta = (p-2)/p < \mu$ [27].
Consider the following conditions:

(A3) There exists a number $R_1 > 0$, $R_1 \leq R$ such that the following inequalities are fulfilled:

$$\|\psi_n\|_{C_\mu(\overline{G})} + \left(C_1 + \|T_G\|_{L_p(\overline{G}),C_\mu(\overline{G})}\right)\left(2L|G|^{1/p}R_1\right) \leq R_1.$$

(A4) $2|G|^{1/p}L\left(C_1 + \|T_G\|_{L_p(\overline{G}),C_\mu(\overline{G})}\right) < 1.$

Assume that the conditions (A1)–(A4) are fulfilled, then there exists a unique solution of the Riemann–Hilbert problem (8.9)–(8.12) in the ball $\|v_n\|_{C_\mu(\overline{G})} \leq R_1$ [28].
Let us estimate the function $v_n(z)$ from the equality (8.13) in the metric of the space $C(\overline{G})$:

$$\|v_n\|_{C(\overline{G})} \leq \|\psi_n\|_{C(\overline{G})} + \|\phi_n\|_{C(\overline{G})} + \|T_G[F]\|_{C(\overline{G})}. \qquad (8.14)$$

Using the previous estimates, from the inequality (8.14) we obtain

$$\|v_n\|_{C(\overline{G})} \leq \|\psi_n\|_{C(\overline{G})}\left(C_1 + \|T_G\|_{L_p(\overline{G}),C(\overline{G})}\right)\|F\|_{L_p(\overline{G})}. \qquad (8.15)$$

By virtue of the condition (A1) we have

$$\left|F(z, w_n, \overline{w}_n, w_{n+1}, \overline{w}_{n+1})\right|$$
$$= \left|f(z, w_{n+1}, \overline{w}_{n+1}) - f(z, w_n, \overline{w}_n)\right| \leqslant 2L|w_{n+1} - w_n| = 2L|v_n|.$$

Due to the condition (A1), from the last inequality it follows that the function $F(z, w_n, \overline{w}_n, w_{n+1}, \overline{w}_{n+1})$ belongs to the space $L_p(\overline{G})$. Then

$$\|F\|_{L_p(\overline{G})} \leqslant 2L\|v_n\|_{L_p(\overline{G})} \leqslant 2L|G|^{1/p}\|v_n\|_{C(\overline{G})}.$$

Thus, by the inequality (8.15) we can write that

$$\|v_n\|_{C(\overline{G})} \leqslant \|\psi_n\|_{C(\overline{G})} + 2L|G|^{1/p}\left(C_1 + \|T_G\|_{L_p(\overline{G}), C(\overline{G})}\right)\|v_n\|_{C(\overline{G})},$$

i.e., in view of the condition (A4), we finally obtain

$$\|v_n\|_{C(\overline{G})} \leqslant \frac{\|\psi_n\|_{C(\overline{G})}}{1 - 2L|G|^{1/p}\left(C_1 + \|T_G\|_{L_p(\overline{G}), C_\mu(\overline{G})}\right)}. \tag{8.16}$$

Note that the function $\psi_n(z)$ is the solution of the following problem

$$\partial_{\overline{z}} \psi_n(z) = 0, \quad z \in G,$$
$$\mathrm{Re}[\psi_n(z)] = 0, \quad z \in \Gamma \setminus \gamma,$$
$$\mathrm{Re}[\psi_n(z)] = \sigma \, \mathrm{Re}[\psi_{n-1}(z_0)], \quad z \in \gamma, \ z_0 \in \gamma_0,$$
$$\mathrm{Im}[\psi_n(z^*)] = 0, \quad n = 1, 2, \dots,$$
$$\psi_0(z) = w_1(z) - w_0(z).$$

Since $\mathrm{Re}[\psi_n(z)]$ is a harmonic function, all the conditions of Schwartz' lemma [42] are fulfilled for it and there exists $0 < q < 1$ not depending on ψ_n, such that the following inequality [16] holds for it:

$$\|\psi_n\|_{C(\overline{G})} \leqslant Mq^n,$$

where $M > 0$ is a constant depending only on $\varphi_1(z)$.

Taking this estimate from (8.16) we can write

$$\|v_n\|_{C(\overline{G})} \leqslant \frac{M}{1 - 2L|G|^{1/p}\left(C_1 + \|T_G\|_{L_p(\overline{G}), C(\overline{G})}\right)} q^n. \tag{8.17}$$

Then from (8.17) we can conclude that the series $\sum_{k=1}^{\infty} v_k$ converges uniformly to zero in the domain $\overline{G}$. Hence it follows that the sequence $\{w_n(z)\}$ is fundamental in $C(\overline{G})$ and has the limit $w(z) \in C(\overline{G})$.

Let us consider the integral representation for the function $w_n(z)$:

$$w_n(z) = \psi_n'(z) + \phi_n'(z) - \frac{1}{\pi} \iint\limits_G \frac{f(\zeta, w_n, \overline{w}_n)}{\zeta - z} \, d\xi \, d\eta, \tag{8.18}$$

where $\psi_n'(z)$ is a holomorphic function that satisfies the conditions (8.6)–(8.8), and $\phi_n'(z)$ is a holomorphic function such that the difference

$$\phi_n'(z) - \frac{1}{\pi} \iint\limits_G \frac{f(\zeta, w_n, \overline{w}_n)}{\zeta - z} \, d\xi \, d\eta$$

satisfies the homogeneous boundary conditions.

The representation (8.18) implies that $w(z)$ is the solution of the problem (8.1)–(8.4) and $w(z) \in C_\mu(\overline{G})$. From the uniqueness of the holomorphic solution and the integral representation (8.18) we can conclude that this solution is unique in the class $C_\mu(\overline{G})$.

We have thereby proved.

Theorem 1 *Let the conditions* (A1)–(A4) *be fulfilled, then the solution of the problem* (8.1)–(8.4) *exists in the space* $C_\mu(\overline{G})$ *and is unique.*

8.2 Linear Problem

Let us consider in the domain $\overline{G}$ the Bitsadze–Samarski boundary value problem for a linear differential equation of first order

$$\partial_{\overline{z}} w = A(z)w + B(z)\overline{w} + d(z), \quad z \in G,$$

$$\mathrm{Re}[w(z)] = 0, \quad z \in \Gamma \setminus \gamma,$$

$$\mathrm{Re}[w(z(s))] = \sigma \, \mathrm{Re}\left[w(z_0(s))\right], \quad z(s) \in \gamma, \ z_0(s) \in \gamma_0, \tag{8.19}$$

$$\mathrm{Im}[w(z^*)] = 0, \quad z^* \in \Gamma \setminus \gamma, \ 0 \leqslant s \leqslant \delta.$$

Assume that $A(z), B(z), d(z) \in L_p(\overline{G}), p > 2, |A|, |B| \leqslant N$.

Denote by $C_\mu^p(\overline{G})$ the set of functions $w(z) \in C_\mu(\overline{G})$ such that

$$\mathrm{Re}[w(z)] = 0, \quad z \in \Gamma \setminus \gamma,$$
$$\mathrm{Re}[w(z(s))] = \sigma \, \mathrm{Re}\left[w(z_0(s))\right], \quad z(s) \in \gamma, \ z_0(s) \in \gamma_0, \qquad (8.20)$$
$$\mathrm{Im}[w(z^*)] = 0, \quad z^* \in \Gamma \setminus \gamma, \ 0 \leqslant s \leqslant \delta$$

and having the finite norm

$$\|w\|_{C_\mu^p(\overline{G})} = \|w\|_{C_\mu(\overline{G})} + \|\partial_{\overline{z}} w\|_{L_p(\overline{G})} < \infty. \qquad (8.21)$$

It is easy to verify that the set $C_\mu^p(\overline{G})$ is a linear normed space over the real field with norm defined by means of the equality (8.21). If $p > q > 2$, then $C_\mu^p(\overline{G}) \subset C_\mu^q(\overline{G})$ and $\|w\|_{C_\mu^q(\overline{G})} \leqslant \alpha \|w\|_{C_\mu^p(\overline{G})}$, where α is a positive constant and w is any element from $C_\mu^p(\overline{G})$.

Lemma 8.2.1 *For any function $d(z) \in L_p(\overline{G})$, $p > 2$, the solution $w(z)$ of the problem* (8.19) *exists, belongs to the space $C_\mu^p(\overline{G})$ and the following a priori estimate holds for it*

$$\|w\|_{C_\mu^p(\overline{G})} \leqslant \alpha \|d\|_{L_p(\overline{G})}, \qquad (8.22)$$

where γ is a positive constant depending only on p, N and $|G| = \mathrm{mes}\, G$.

Proof The existence and uniqueness of the problem (8.19) immediately follow from Theorem 1. It remains for us to prove the validity of the a priori estimate (8.22).

Let us reduce the problem (8.19) to an integral equation. For this we apply the operator

$$T_G[z,f] = -\frac{1}{\pi} \iint\limits_G \frac{f(t)}{t-z} \, d\xi \, d\eta, \quad t = \xi + i\eta$$

and an operator $S_G[z,f]$ from $L_p(\overline{G})$ to the set of analytic functions which satisfy the conditions:

$$\mathrm{Re}\left\{T_G[z,f] + S_G[z,f]\right\} = 0, \quad z \in \Gamma \setminus \gamma,$$
$$\mathrm{Re}\left\{T_G[z,f] + S_G[z,f]\right\}$$
$$= \sigma \, \mathrm{Re}\left\{T_G[z_0,f] + S_G[z_0,f]\right\}, \quad z_0 \in \gamma_0, \ z \in \gamma, \qquad (8.23)$$
$$\mathrm{Im}\left\{T_G[z^*,f] + S_G[z^*,f]\right\} = 0,$$

where $z^* \in \Gamma \setminus \gamma$ is a fixed point.

Using these conditions, the operator $S_G[z,f]$ is defined uniquely. We define the operators

$$P(f) = T_G[z,f] + S_G[z,f],$$
$$P_{AB}(f) = P(Af) + P(B\bar{f}),$$

(8.24)

where the functions $A(z)$ and $B(z)$ are from the right-hand part of Eq. (8.19).

If we now take into account that $\partial_{\bar{z}} P(f) = f(z)$, then it can be easily verified that the solution of the problem (8.19) satisfies the integral equation

$$w(z) = P_{AB}(w) + P(d).$$

(8.25)

It is not difficult to show that the problems (8.19) and (8.25) are equivalent. Using the properties of the operators $T_G[z,f]$ and $P(f)$ [27], it can be shown that these operators are completely continuous over the field of real numbers. It is obvious that the operator $P_{AB}(f)$ is also completely continuous.

Since for $d(z) = 0$ Eq. (8.19) has only a trivial solution, the equation $w(z) = P_{AB}(w)$ will also have only a trivial solution. Hence, due to the complete continuity of the operator $P_{AB}(f)$, the existence and boundedness of the operator $(I - P_{AB})^{-1}$ follow, where is an identical operator.

Let us introduce the notations

$$\|I - P_A\|^{-1}_{C_\mu(\overline{G}), L_p(\overline{G})} = M, \quad \|P\|_{L_p(\overline{G}), C_\mu(\overline{G})} = M_p,$$

where M and M_p are positive constants. From Eq. (8.25) we immediately obtain

$$\|w(z)\|_{C_\mu(\overline{G})} \leqslant M M_p \|d\|_{L_p(\overline{G})}.$$

(8.26)

From Eq. (8.19) we can immediately obtain

$$\|\partial_{\bar{z}} w\|_{L_p(\overline{G})} \leqslant 2N \|w\|_{C_\mu(\overline{G})} + \|d\|_{L_p(\overline{G})}.$$

(8.27)

The inequalities (8.26), (8.27) imply the estimate

$$\|w\|_{C^p_\mu(\overline{G})} = \|w\|_{C_\mu(\overline{G})} + \|\partial_{\bar{z}} w\|_{L_p(\overline{G})} \leqslant \alpha \|d\|_{L_p(\overline{G})},$$

where $\alpha = M M_p (2N + 1) + 1$. The lemma is proved.

8.3   Statement of an Optimal Control Problem

Let U be some bounded subset from E. Each function $\omega(z) : G \to U$ will be called a control. The set U is called the control domain. We call the function $\omega(z)$ an admissible control if $\omega(z) \in L_p(G), p > 2$. The set of all admissible controls is denoted by Ω.

For each fixed $\omega \in \Omega$ in the domain $\overline{G}$ we consider the following Bitsadze–Samarski boundary value problem for a system of quasilinear differential equations of first order

$$
\begin{aligned}
\partial_{\overline{z}} w &= f(z, w, \overline{w}, \omega), \quad z \in G, \\
\mathrm{Re}[w(z)] &= \varphi_1(z), \quad z \in \Gamma \setminus \gamma, \\
\mathrm{Re}[w(z(s))] &= \sigma \, \mathrm{Re}\left[w(z_0(s))\right], \quad z(s) \in \gamma, \ z_0(s) \in \gamma_0, \\
\mathrm{Im}[w(z^*)] &= c, \quad z^* \in \Gamma \setminus \gamma, \ c = \mathrm{const}, \ 0 < \sigma = \mathrm{const},
\end{aligned}
\tag{8.28}
$$

and the functional

$$
I(u) = \iint\limits_{G} F(x, y, w_1, w_2, \omega_1, \omega_2) \, dx \, dy, \tag{8.29}
$$

where $z = x + iy$, $w = w_1 + iw_2$, $\omega = \omega_1 + i\omega_2$.

Let us pose the following optimal control problem: Find a function $\omega_0(z) \in \Omega$, for which the solution of the Bitsadze–Samarski boundary value problem (8.28) gives the functional (8.29) a minimal value. We call the function $\omega_0(z) \in \Omega$ an optimal control, and the corresponding generalized solution $w_0(z)$ of the problem (8.28) an optimal solution.

Assume that the conditions (A1)–(A4) are fulfilled. Then for each fixed $\omega \in \Omega$ the solution of the problem (8.1) exists in the space $C_\mu(\overline{\Omega})$ and is unique.

Furthermore, to obtain optimality conditions, it is additionally assumed that:

(A5) The function $f(z, w, q, \omega)$ has continuous partial derivatives which are also continuous with respect to the same arguments as f. The function F is continuous with respect to $w_1, w_2, \omega_1, \omega_2$, continuously differentiable with respect to w_1, w_2 and belongs to the space $L_p(G), p > 2$.

(A6) For any $(w, q) \in S_{wq}^{R_1} = \{(w, q) : |w|, |q| < R\}$, the following estimates hold in the domain G : $|f'_w|, |f'_q| \leqslant N_1(R) < +\infty$, $|f| \leqslant N_2(R) < +\infty$.

The following theorem is true:

Theorem 8.3.2 *Let the conditions* (A1)–(A6) *be fulfilled,* $w_0(z)$ *be an optimal control,* $w_0(z)$ *be the corresponding solution if the problem* (8.28), $\rho(x)$ *be a weight function,* $\psi(z)$ *be the solution of the adjoint problem*

$$\partial_{\bar{z}}(\rho(x)\psi(z)) + \frac{\partial f(\omega_0)}{\partial w}\rho(x)\psi(z) + \frac{\partial \bar{f}(\omega_0)}{\partial w}\rho(x)\bar{\psi}(z)$$

$$= -2\partial_w F(\omega_0), \quad z \in G, \tag{8.30}$$

$$\mathrm{Re}\left[\rho(x)\psi(z)\right] = 0, \quad z \in \Gamma,$$

$$\mathrm{Re}\left[\rho(x)\psi(z_0^+) - \rho(x)\psi(z_0^-)\right] = \sigma \,\mathrm{Re}\left[\rho(x)\psi(z)\right], \quad z_0 \in \gamma_0, \ z \in \gamma$$

then the relation

$$\mathrm{Re}\left[f(z, w_0, \bar{w}_0, \omega_0)\psi_0(z) + F(x, y, w_{01}, w_{02}, \omega_{01}, \omega_{02})\right]$$

$$= \inf_{\omega \in U} \mathrm{Re}\left[f(z, w_0, \bar{w}_0, \omega)\psi_0(z) + F(x, y, w_{01}, w_{02}, \omega_1, \omega_2)\right]$$

is fulfilled everywhere on G.

Proof Let us construct a variant of the impulse control $\omega_0(z)$ [43]. Choose any finite set of pairwise different Lebesgue points $\{z_i\}$ from the domain G. For each finite set of non-negative numbers $\gamma = \{\gamma_i^j\}$ there exists a positive number ε_0 such that for $0 \leqslant \varepsilon \leqslant \varepsilon_0$ the rectangles

$$\pi_{ij}^\varepsilon \equiv \left\{z: \ z = x + iy \in G,\right.$$

$$\left. x_i - \varepsilon \sum_{k=1}^{j} \gamma_i^k < x \leqslant x_i - \varepsilon \sum_{k=1}^{j-1} \gamma_i^k, \ y_i - j\varepsilon < y \leqslant y_i - (j-1)\varepsilon\right\}$$

do not intersect pairwise and all of them belong to the domain G. Let $u \equiv \{u_{i,j}\}$ be the finite set of points from U. Consider an arbitrary set

$$\{z_i\}, \ \{\omega_{i,j}\}, \ \{\gamma_i^j\}. \tag{8.31}$$

Define the variant $\omega_\varepsilon(z)$ as follows

$$\omega_\varepsilon(z) = \begin{cases} \omega_{ij}, & t \in \pi_{ij}^\varepsilon, \\ \omega_0(t), & t \in [a, b] \setminus \bigcup_{i,j} \pi_{ij}^\varepsilon. \end{cases} \tag{8.32}$$

Assume that the set of parameters (8.31) is fixed. For fixed ω_0 and ω_ε, $0 \leqslant \varepsilon \leqslant \varepsilon_0$, find an increment of the functional $I(\omega)$. Applying Lagrange's theorem, we find

$$
\Delta_\varepsilon \equiv I(\omega_\varepsilon) - I(\omega_0) = \mathrm{Re} \iint\limits_G \psi_\varepsilon(z) \Delta_\omega f(z, w_0, \omega_0, \omega_\varepsilon) \, dx \, dy
$$

$$
+ \iint\limits_G \Delta_\omega F(x, y, w_0, \omega_0, \omega_\varepsilon) \, dx \, dy, \tag{8.33}
$$

where

$$
\Delta_u F(x, y, w_0, \omega_0, \omega_\varepsilon) = F(x, y, w_{01}, w_{02}, \omega_{\varepsilon 1}, \omega_{\varepsilon 2}) - F(x, y, w_{01}, w_{02}, \omega_{01}, \omega_{02}),
$$

and the pulse variant $\omega_\varepsilon(z)$ is defined by means if the equality (8.28) for an arbitrary set of parameters

$$
\{z_i\}, \quad \omega = \{\omega_{i,j}\}, \quad \gamma = \{\gamma_i^j\}, \quad z_i \in G, \quad \omega_{ij} \in U, \quad \gamma_i^j \geqslant 0. \tag{8.34}
$$

For an increment of the functional $\Delta_\varepsilon I$, let us construct some integral representation. For this, in the space $C_\mu^p(\overline{G})$, for any ε, $0 \leqslant \varepsilon \leqslant \varepsilon_0$, we define the following linear operator

$$
S_\varepsilon(v) = \partial_{\overline{z}} v - \int_0^1 \frac{\partial f(\eta_\varepsilon)}{\partial w} \, dt \cdot v - \int_0^1 \frac{\partial f(\eta_\varepsilon)}{\partial \overline{w}} \, dt \cdot \overline{v}, \tag{8.35}
$$

where

$$
\eta_\varepsilon = (z, w_0 + t\Delta w, \overline{w}_0 + t\Delta \overline{w}, u_\varepsilon), \quad 0 \leqslant t \leqslant 1.
$$

The operator S_ε maps the space $C_\mu^p(\overline{G})$ onto $L_p(\overline{G})$. Indeed, on the one hand, the inclusion $S_\varepsilon(v(z)) \in L_p(\overline{G})$ is valid for any function $v(z) \in C_\mu^p(\overline{G})$ and, on the other hand, by virtue of Lemma 8.2.1 we can conclude that the equation

$$
S'_\varepsilon(v) = r(z) \tag{8.36}
$$

has a unique solution for any function $r(z) \in L_p(\overline{G})$. In this case the estimate

$$
\|v\|_{C_\mu^p(\overline{G})} \leqslant \alpha \|r\|_{L_p(\overline{G})}, \quad \alpha = \alpha\big(p, N_1, |G|\big) > 0 \tag{8.37}
$$

holds, from which it follows that for the operator S_ε there exists an inverse operator $S_\varepsilon^{-1} : L_p(\overline{G}) \to C_\mu^p(\overline{G})$ and $\|S_\varepsilon^{-1}\| \leqslant \alpha$.

While considering the increment $\Delta_\varepsilon I$, we observe that the second summand in the right-hand of equalities (8.33) defines uniquely some linear functional T_ε from the adjoint space $(C_\mu^p(\overline{G}))^*$:

$$\langle T_\varepsilon, \Delta w \rangle = 2 \iint\limits_G \text{Re} \left[\partial_{\overline{w}} F(\eta) \Delta w \right] dx\,dy, \tag{8.38}$$

where $\langle T_\varepsilon, \Delta w \rangle$ denotes a value of the functional T_ε on an element $\Delta w \in C_\mu^p(\overline{G})$, $\eta = (x, y, w_{01} + \theta \Delta w_1, w_{02} + \theta \Delta w_2, u_{01} + \theta \Delta u_1, u_{02} + \theta \Delta u_2), 0 \leqslant \theta \leqslant 1$.

Consider the adjoint operator $(S_\varepsilon^{-1})^*$ to the operator S_ε^{-1}. This operator maps the space $(C_\mu^p(\overline{G}))^*$ onto $L_p^*(G)$ since, by definition, $\langle T_\varepsilon, S_\varepsilon^{-1} f \rangle = \langle (S_\varepsilon^{-1})^*, f \rangle$ and

$$\langle T_\varepsilon, \Delta w \rangle = \langle T_\varepsilon, S_\varepsilon^{-1}(\Delta_\omega f) \rangle = \langle (S_\varepsilon^{-1})^* T_\varepsilon, \Delta_\omega f \rangle. \tag{8.39}$$

By Riesz' theorem [44] we can conclude that there exists a function $\psi_\varepsilon \in L_p(\overline{G})$, $1/p + 1/q = 1$ such that for $\Delta_\omega f \in L_p(\overline{G})$, $p > 2$, there hold the following representations

$$\langle (S_\varepsilon^{-1})^* T_\varepsilon, \Delta_\omega f \rangle = \text{Re} \iint\limits_G \psi_\varepsilon(z) \Delta_\omega f \, dx\,dy. \tag{8.40}$$

Taking into account the equalities (8.38)–(8.40), the increment of the functional $\Delta_\varepsilon I$ (8.23) can be written in the form

$$\Delta_\varepsilon I = \text{Re} \iint\limits_G \psi_\varepsilon(z) \Delta_\omega f \, dx\,dy + 2 \iint\limits_G \text{Re} \left[\partial_{\overline{\omega}} F(\eta) \Delta_\omega \right] dx\,dy, \tag{8.41}$$

where the function $\psi_\varepsilon \in L_p(\overline{G})$ is uniquely defined by the relation

$$\psi_\varepsilon = (S_\varepsilon^{-1})^* T_\varepsilon. \tag{8.42}$$

It can be easily shown that $\lim\limits_{\varepsilon \to 0} \| \psi_\varepsilon - \psi_0 \|_{L_q(\overline{G})} \longrightarrow 0$ [29]. Thus the function ψ_0 is defined by means of the equality

$$\psi_0 = (S_0^{-1})^* T_0. \tag{8.43}$$

Consider the first variation δI of the functional (8.29):

$$\delta I = \lim_{\varepsilon \to 0} \frac{1}{\varepsilon^2} \Delta_\varepsilon I = \lim_{\varepsilon \to 0} \frac{1}{\varepsilon^2} \left\{ \mathrm{Re} \iint\limits_{G} \Delta_\omega f(z, w_0, \omega_0, \omega_\varepsilon) \psi_\varepsilon(z) \, dx \, dy \right.$$

$$\left. + \iint\limits_{G} \Delta_\omega F(x, y, w_0, \omega_0, \omega_\varepsilon) \, dx \, dy \right\}$$

$$= \lim_{\varepsilon \to 0} \sum_{\substack{1 \le i \le s \\ 1 \le j \le q}} \left\{ \mathrm{Re} \iint\limits_{\pi_{ij}^\varepsilon(z_i)} \Delta_\omega f(z, w_0, \omega_0, \omega_\varepsilon) \psi_\varepsilon(z) \, dx \, dy \right.$$

$$\left. + \iint\limits_{\pi_{ij}^\varepsilon(z_i)} \Delta_\omega F(x, y, w_0, \omega_0, \omega_\varepsilon) \, dx \, dy \right\}.$$

So, on any variant $\omega_\varepsilon(z)$, for the finite set (8.41) the variation δI takes the form

$$\delta I = \sum_{i,j} \gamma_i^j \left\{ \mathrm{Re} \left[\Delta_\omega f(z_i, w_0(z_i), \omega_0(z_i), \omega_{i,j}) \psi_0(z_i) + \Delta_\omega F(z_i, w_0(z_i), \omega_0(z_i), \omega_{i,j}) \right] \right\}.$$

Therefore the relation

$$\mathrm{Re} \left[f(z, w_0, \overline{w}_0, \omega_0) \psi_0(z) + F(x, y, w_{01}, w_{02}, \omega_{01}, \omega_{02}) \right]$$

$$= \inf_{\omega \in U} \mathrm{Re} \left[f(z, w_0, \overline{w}_0, \omega_0) \psi_0(z) + F(x, y, w_{01}, w_{02}, \omega_1, \omega_2) \right]$$

holds almost everywhere on G [29, 43].

8.4   Construction of an Adjoint Equation in the Differential Form

From the relation (8.41) we obtain the equality

$$\langle \psi, S_0 w \rangle = \langle T_0, w \rangle, \quad \psi \in L_p^*(G) \tag{8.44}$$

for any $w \in C_\mu^p(\overline{G})$.

By Riesz' theorem [44], we can conclude that there exists a unique element from the space $L_q(G)$, which is identified with an element $\psi \in L_p^*(G)$, $1/p + 1/q = 1$ for which the equality (8.44) has the form

$$\mathrm{Re} \iint\limits_G \psi(z) S_0 w \, dx \, dy = \langle T_0, w \rangle. \tag{8.45}$$

Taking into account the form of the operator S_0 and also the form of the functional T_0, the equality (8.45) can be rewritten as follows

$$\mathrm{Re} \iint\limits_G \psi(z) \left[\partial_{\bar{z}} w - \frac{\partial f(\omega_0)}{\partial w} w - \frac{\partial f(\omega_0)}{\partial \overline{w}} \overline{w} \right] dx \, dy$$

$$= 2 \iint\limits_G \mathrm{Re} \left[\partial_w F(\omega_0) w \right] dx \, dy. \tag{8.46}$$

By Green's formula [27],

$$\mathrm{Re} \iint\limits_G \psi(z) \partial_{\bar{z}} w \, dx \, dy = \frac{1}{2i} \int\limits_\Gamma \psi w \, dz - \iint\limits_G w \partial_{\bar{z}} \psi \, dx \, dy,$$

from the relation (8.46), for any function $w \in C_\mu^p(\overline{G})$ we obtain the equality

$$\mathrm{Re} \left[\frac{1}{2i} \int\limits_\Gamma \psi w \, dz \right]$$

$$- \mathrm{Re} \iint\limits_G \left[\partial_{\bar{z}} \psi + \frac{\partial f(\omega_0)}{\partial w} \psi + \frac{\partial \overline{f}(\omega_0)}{\partial \overline{w}} \overline{\psi} + 2 \partial_w F(\omega_0) \right] w \, dx \, dy = 0.$$

$$\tag{8.47}$$

Let the domain $\overline{G}$ be the rectangle $\{z = x + iy: \ 0 \leqslant x \leqslant 1, \ 0 \leqslant y \leqslant 1\}$, Γ be the boundary of the domain G, and $\gamma_0 = \{z_0 = x_0 + iy: \ 0 \leqslant y \leqslant 1\}$, $\gamma_1 = \{z = 1 + iy: \ 0 \leqslant y \leqslant 1\}$, $\gamma_2 = \{z = x + i: \ 0 \leqslant x \leqslant 1\}$, $\gamma_3 = \{z = iy: \ 0 \leqslant y \leqslant 1\}$, $\gamma_4 = \{z = x: \ 0 \leqslant x \leqslant 1\}$, $0 < \mu < 1$, $\rho(x)$ be the real weight function of a real argument x defined as follows

$$\rho(x) = \begin{cases} \rho_1(x), & 0 \leqslant x \leqslant x_0, \\ \rho_2(x), & x_0 < x \leqslant 1, \end{cases}$$

$$\rho_1(x) \in C_\mu[0, x_0], \quad \rho_2(x) \in C_\mu[x_0, 1], \quad \lim_{x \to x_0^-} \rho_1(x) \neq \lim_{x \to x_0^+} \rho_2(x).$$

Then from the equality (8.47) we obtain for $\psi(z)$ the following problem

$$\partial_{\bar{z}}\psi + \frac{\partial f(\omega_0)}{\partial w}\psi + \frac{\partial \bar{f}(\omega_0)}{\partial \bar{w}}\bar{\psi} = -2\partial_w F(\omega_0), \quad z \in G,$$

$$\mathrm{Re}[\rho\psi] = 0, \quad z \in \gamma_2 \cup \gamma_4, \tag{8.48}$$

$$\mathrm{Re}[i\rho\psi] = 0, \quad z \in \gamma_3 \cup \gamma_1,$$

$$\mathrm{Re}\left[\rho\psi(z_0^+) - \rho\psi(z_0^-)\right] = \sigma\,\mathrm{Re}[\rho\psi(z)], \quad z_0 \in \gamma_0, \quad z \in \gamma_1.$$

8.5   Necessary and Sufficient Conditions of Optimality

Let the domain $\overline{G}$ be the rectangle $\{z = x + iy : \ 0 \leqslant x \leqslant 1, \ 0 \leqslant y \leqslant 1\}$, Γ be the boundary of the domain G, and $\gamma_0 = \{z_0 = x_0 + iy : \ 0 \leqslant y \leqslant 1\}$, $\gamma_1 = \{z = 1 + iy : \ 0 \leqslant y \leqslant 1\}$, $\gamma_2 = \{z = x + i : \ 0 \leqslant x \leqslant 1\}$, $\gamma_3 = \{z = iy : \ 0 \leqslant y \leqslant 1\}$, $\gamma_4 = \{z = x : \ 0 \leqslant x \leqslant 1\}$, $z^* \in \Gamma \setminus \gamma_1$, $c(z), d(z), B(z), f(z) \in L_p(\overline{G})$, $p > 2$, $g(z) \in C_\mu(\Gamma \setminus \gamma_1)$, $0 < \mu < 1$. For each fixed $\omega \in \Omega$ in the domain $\overline{G}$ we consider the following Bitsadze–Samarski boundary value problem for the system of linear differential equations of first order

$$\partial_{\bar{z}}w + B(z)\bar{w} = f(z)\omega, \quad z \in G,$$

$$\mathrm{Re}[w(z_1)] = g(z), \quad z \in \Gamma \setminus \gamma_1,$$

$$\mathrm{Re}[w(z_1)] = \sigma\,\mathrm{Re}[w(z_0)], \quad z_0 \in \gamma_0, \quad z_1 \in \gamma_1, \tag{8.49}$$

$$\mathrm{Im}[w(z^*)] = \mathrm{const}, \quad 0 < \sigma = \mathrm{const}.$$

Let us consider the functional

$$I(\omega) = \mathrm{Re}\iint\limits_{G} \left[c(z)w(z) + d(z)\omega(z)\right] dx\,dy \tag{8.50}$$

and state the following optimal control problem: Find a function $\omega_0(z) \in \Omega$, for which the solution of the Bitsadze–Samarski boundary value problem (8.49) gives the functional (8.50) a minimal value.

Theorem 8.5.1 *Let $\psi(z)$ be a solution of the adjoint problem*

$$\partial_{\bar{z}}(\rho(x)\psi(z)) - \bar{B}(z)\rho(x)\psi(z) = c(z),$$

$$\mathrm{Re}[\rho\psi] = 0, \quad z \in \gamma_2 \cup \gamma_4,$$

$$\mathrm{Re}[i\rho\psi] = 0, \quad z \in \gamma_3 \cup \gamma_1, \tag{8.51}$$

$$\mathrm{Re}\left[\rho\psi(z_0^+) - \rho\psi(z_0^-)\right] = \sigma\,\mathrm{Re}[\rho\psi(z)], \quad z_0 \in \gamma_0, \quad z \in \gamma_1,$$

then for the optimality of $\omega_0(z)$, $w_0(z)$ it is necessary and sufficient that the following equality be fulfilled almost everywhere on G

$$\text{Re}\left[(d(z) - \rho(x)\psi(z)f(z))\omega_0(z)\right] = \inf_{\omega \in U} \text{Re}\left[(d(z) - \rho(x)\psi(z)f(z))\omega(z)\right].$$

Proof Let $\omega_0(z) \in \Omega$ be an optimal control, $\omega_\varepsilon(z) \in \Omega$ be an arbitrary admissible control, $w_0(z)$, $w_\varepsilon(z)$ be the solutions of the problem (8.49), which correspond to the controls $\omega_0(z)$, $\omega_\varepsilon(z)$. Let us introduce the notations $\Delta w(z) = w_\varepsilon(z) - w_0(z)$, $\Delta\omega(z) = \omega_\varepsilon(z) - \omega_0(z)$. Then we obtain the following problem

$$\begin{aligned}
&\partial_{\bar{z}}\Delta w + B(z)\Delta\bar{w} = f(z)\omega, \quad z \in G, \\
&\text{Re}[\Delta w(z)] = g(z), \quad z \in \Gamma \setminus \gamma_1, \\
&\text{Re}[\Delta w(z_1)] = \sigma\,\text{Re}[\Delta w(z_0)], \quad z_0 \in \gamma_0, \quad z_1 \in \gamma_1, \\
&\text{Im}[\Delta w(z^*)] = \text{const}, \quad 0 < \sigma = \text{const}.
\end{aligned} \qquad (8.52)$$

Let $\psi(z) = \psi_1(x, y) + i\psi_2(x, y) \neq 0$ be an arbitrary integrable function. After multiplying (8.52) by $\rho(x)\psi(z)$, we obtain the equality

$$\text{Re}\iint_G \rho(x)\psi(z)[\partial_{\bar{z}}\Delta w + B(z)\Delta\bar{w}]\,dx\,dy = \text{Re}\iint_G \rho(x)\psi(z)f(z)\Delta\omega\,dx\,dy. \qquad (8.53)$$

For fixed $\omega_0(z)$ and $\omega_\varepsilon(z)$ let us find an increment of the functional (8.54):

$$\Delta I = I(\omega_\varepsilon) - I(\omega_0) = \text{Re}\iint_G \left[c(z)\Delta w(z) + d(z)\Delta\omega(z)\right]dx\,dy.$$

In view of (8.53), we obtain the equality

$$\Delta I = \text{Re}\iint_G \left[\rho(x)\psi(z)\partial_{\bar{z}}\Delta w + \left(c(z) + \rho(x)\bar{B}(z)\bar{\psi}(z)\right)\Delta w\right]dx\,dy$$

$$+ \text{Re}\iint_G \Delta\omega(z)\left(d(z) - \rho(x)\psi(z)f(z)\right)dx\,dy. \qquad (8.54)$$

Let $\Delta w = u + iv$ and consider the first summand in the right-hand part of the equality (8.54):

$$\operatorname{Re} \iint_G \rho(x)\psi(z)\partial_{\bar{z}} \, \Delta w \, dx \, dy$$

$$= \int_0^1 \int_0^1 \left[\rho(x)\psi_1(x,y)\left(\frac{\partial u}{\partial x} - \frac{\partial v}{\partial y}\right) - \rho(x)\psi_2(x,y)\left(\frac{\partial u}{\partial y} + \frac{\partial v}{\partial x}\right) \right] dx \, dy.$$

$$(8.55)$$

Furthermore, taking into account the boundary conditions from (8.52), we have

$$\int_0^1 \int_0^1 \left[\rho(x)\psi_1(x,y)\frac{\partial u}{\partial x} - \rho(x)\psi_2(x,y)\frac{\partial v}{\partial x} \right] dx \, dy$$

$$= \int_0^1 \left[\int_0^{x_0} \left(\rho_1(x)\psi_1(x,y)\frac{\partial u}{\partial x} - \rho_1(x)\psi_2(x,y)\frac{\partial v}{\partial x} \right) dx \right.$$

$$\left. + \int_{x_0}^1 \left(\rho_2(x)\psi_1(x,y)\frac{\partial u}{\partial x} - \rho_2(x)\psi_2(x,y)\frac{\partial v}{\partial x} \right) dx \right] dy$$

$$= \int_0^1 \left\{ \rho_1(x_0^-)\psi_1(x_0^-,y)u(x_0,y) - \rho_1(0)\psi_1(0,y)u(0,y) \right.$$

$$- \rho_1(x_0^-)\psi_2(x_0^-,y)v(x_0,y) - \rho_1(0)\psi_2(0,y)v(0,1)$$

$$- \int_0^{x_0} \left[\frac{\partial}{\partial x}(\rho_1\psi_1)u - \frac{\partial}{\partial x}(\rho_1\psi_2)v \right] dx$$

$$+ \rho_2(1)\psi_1(1,y)u(1,y) - \rho_2(x_0^+)\psi_1(x_0^+,y)u(x_0,y)$$

$$- \rho_2(1)\psi_2(1,y)v(1,y) + \rho_2(x_0^+)\psi_2(x_0^+,y)v(x_0,y)$$

$$\left. - \int_{x_0}^1 \left[\frac{\partial}{\partial x}(\rho_2\psi_1)u - \frac{\partial}{\partial x}(\rho_2\psi_2)v \right] dx \right\} dy$$

$$= \int_0^1 \left\{ \left[\rho_1(x_0^-)\psi_1(x_0^-,y) - \rho_2(x_0^+)\psi_1(x_0^+,y) + \sigma\rho_2(1)\psi_1(1,y) \right] u(x_0,y) \right.$$

$$+ \left[\rho_2(x_0^+)\psi_2(x_0^+, y) - \rho_1(x_0^-)\psi_2(x_0^-, y) \right] v(x_0, y)$$

$$- \rho_1(0)\psi_2(0, y)v(0, y) - \rho_2(1)\psi_2(1, y)v(1, y)$$

$$- \int_0^{x_0} \left[\frac{\partial}{\partial x}(\rho\psi_1)u - \frac{\partial}{\partial x}(\rho\psi_2)v \right] dx - \int_{x_0}^1 \left[\frac{\partial}{\partial x}(\rho\psi_1)u - \frac{\partial}{\partial x}(\rho\psi_2)v \right] dx \bigg\} \, dy.$$

Analogously, we obtain the equality

$$\int_0^1 \int_0^1 \left[\rho(x)\psi_2(x, y) \frac{\partial u}{\partial y} + \rho(x)\psi_1(x, y) \frac{\partial v}{\partial y} \right] dy \, dx$$

$$= \int_0^{x_0} \bigg\{ -\rho(x)\psi_1(x, 0)v(x, 0) + \rho(x)\psi_1(x, 1)v(x, 1)$$

$$\int_0^1 \left[\frac{\partial}{\partial y}(\rho\psi_2)u + \frac{\partial}{\partial y}(\rho\psi_1)v \right] dy \bigg\} \, dx$$

$$+ \int_{x_0}^1 \bigg\{ -\rho(x)\psi_1(x, 0)v(x, 0) + \rho(x)\psi_1(x, 1)v(x, 1)$$

$$\int_0^1 \left[\frac{\partial}{\partial y}(\rho\psi_2)u + \frac{\partial}{\partial y}(\rho\psi_1)v \right] dy \bigg\} \, dx.$$

Let us assume that the following conditions are fulfilled:

$$\begin{aligned}
&\rho_1(x_0^-)\psi_1(x_0^-, y) - \rho_2(x_0^+)\psi_1(x_0^+, y) + \sigma\rho_2(1)\psi_1(1, y) = 0, \\
&\rho_2(x_0^+)\psi_2(x_0^+, y) - \rho_1(x_0^-)\psi_2(x_0^-, y) = 0, \\
&\rho_1(0)\psi_2(0, y) = 0, \quad \rho_2(1)\psi_2(1, y) = 0, \\
&\rho(x)\psi_1(x, 0) = 0, \quad \rho(x)\psi_1(x, 1) = 0.
\end{aligned} \tag{8.56}$$

By the above reasoning, from the equality (8.55) we obtain the relation

$$\text{Re} \iint_G \rho(x)\psi(z)\partial_{\bar{z}}\Delta w \, dx \, dy = \text{Re} \iint_{G_1} \partial_{\bar{z}}(\rho(x)\psi(z))\Delta w \, dx \, dy$$

$$- \text{Re} \iint_{G_2} \partial_{\bar{z}}(\rho(x)\psi(z))\Delta w \, dx \, dy, \tag{8.57}$$

where

$$\overline{G}_1 = \{z = x + iy : \ 0 \leqslant x \leqslant x_0, \ 0 \leqslant y \leqslant 1\},$$
$$\overline{G}_2 = \{z = x + iy : \ x_0 \leqslant x \leqslant 1, \ 0 \leqslant y \leqslant 1\}.$$

Using (8.57), for the increment of the functional ΔI from (8.54) we obtain the equality

$$\Delta I = \mathrm{Re} \iint\limits_{G_1} \left[-\partial_{\overline{z}}(\rho(x)\psi(z)) + c(z) + \rho(x)\overline{B}(z)\overline{\psi}(z) \right] \Delta w \, dx \, dy$$

$$+ \mathrm{Re} \iint\limits_{G} \Delta \omega(z)\big(d(z) - \rho(x)\psi(z)f(z)\big) \, dx \, dy$$

$$+ \mathrm{Re} \iint\limits_{G_2} \left[-\partial_{\overline{z}}(\rho(x)\psi(z)) + c(z) + \rho(x)\overline{B}(z)\overline{\psi}(z) \right] \Delta w \, dx \, dy. \qquad (8.58)$$

Let us consider the equation

$$\partial_{\overline{z}}\rho(x)\psi(z)) - \overline{B}(z)(\rho(x)\overline{\psi}(z)) = c(z) \qquad (8.59)$$

and the function

$$\psi_i(x, y) = \begin{cases} \psi_i^{(1)}(x, y), & (x, y) \in G_1, \\ \psi_i^{(2)}(x, y), & (x, y) \in G_2, \end{cases} \quad i = 1, 2.$$

where $\rho_1(\psi_1^{(1)} + i\psi_2^{(1)})$ is a solution of Eq. (8.59) in the domain G_1, and $\rho_2(\psi_1^{(2)} + i\psi_2^{(2)})$ is a solution in the domain G_2. It is assumed that the function $\rho(x)\psi(z)$ satisfies also the conditions (8.56). In that case, the increment of the functional ΔI (8.58) takes the form

$$\Delta I = I(\omega_\varepsilon) - I(\omega_0) = \mathrm{Re} \iint\limits_{G} \Delta \omega(z)\big(d(z) - \rho(x)\psi(z)f(z)\big) \, dx \, dy. \qquad (8.60)$$

From the representation (8.60) it obviously follows that for the optimality of $\omega_0(z)$, $w_0(z)$ it is necessary and sufficient that the following equality be fulfilled almost everywhere on G

$$\mathrm{Re}\left[\big(d(z) - \rho(x)\psi(z)f(z)\big)\omega_0(z)\right] = \inf_{\omega \in U} \mathrm{Re}\left[\big(d(z) - \rho(x)\psi(z)f(z)\big)\omega(z)\right].$$

The theorem is proved.

Acknowledgements The authors were supported by the European Union's Seventh Framework Programme (FP7/2007–2013) under grant agreement no. 317721.

References

1. V.V. Shelukhin, A non-local in time model for radionuclides propagation in Stokes fluid. Dinamika Sploshn. Sredy **107**, 180–193 (1993)
2. C.V. Pao, Reaction diffusion equations with nonlocal boundary and nonlocal initial conditions. J. Math. Anal. Appl. **195**(3), 702–718 (1995)
3. E. Obolashvili, Nonlocal problems for some partial differential equations. Appl. Anal. **45**(1–4), 269–280 (1992)
4. T.V. Aloyev, E.N. Aslanova, Nonlocal problems of conductive radial heat exchange, in *Abstracts of International Conference "Non-local Boundary Problems and Related Mathematical Biology, Informatic and Physic Problems"*, Nalchik, 1996
5. J.I. Diaz, J.-M. Rakotoson, On a nonlocal stationary free-boundary problem arising in the confinement of a plasma in a stellarator geometry. Arch. Rational Mech. Anal. **134**(1), 53–95 (1996)
6. T. Carleman, Sur la théorie des equations integrals et ses applications. in *Verh. Internat. Math. Kongr., Zurich* (Orell Fussli, Zurich, 1932/1933), pp. 138–151
7. R. Beals, Nonlocal elliptic boundary value problems. Bull. Am. Math. Soc. **70**, 693–696 (1964)
8. F.E. Browder, Non-local elliptic boundary value problems. Am. J. Math. **86**, 735–750 (1964)
9. J.R. Cannon, The solution of the heat equation subject to the specification of energy. Q. Appl. Math. **21**, 155–160 (1963)
10. N.I. Ionkin, The solution of a certain boundary value problem of the theory of heat conduction with a nonclassical boundary condition. Differ. Uravn. **13**(2), 294–304, 381 (1977, in Russian)
11. G. Berikelashvili, To a nonlocal generalization of the Dirichlet problem. J. Inequal. Appl. **2006**, Art. ID 93858, 6 pp. (2006)
12. F. Shakeri, M. Dehghan, The method of lines for solution of the one-dimensional wave equation subject to an integral conservation condition. Comput. Math. Appl. **56**(9), 2175–2188 (2008)
13. A. Ashyralyev, O. Gercek, Nonlocal boundary value problems for elliptic-parabolic differential and difference equations. Discret. Dyn. Nat. Soc. **2008**, Art. ID 904824, 16 pp. (2008)
14. A.V. Bitsadze, A.A. Samarskiĭ, On some simple generalizations of linear elliptic boundary problems. Dokl. Akad. Nauk SSSR **185**, 739–740 (1969, in Russian); English transl.: Sov. Math. Dokl. **10**, 398–400 (1969)
15. D.G. Gordeziani, T.Z. Dzhioev, The solvability of a certain boundary value problem for a nonlinear equation of elliptic type. Sakharth. SSR Mecn. Akad. Moambe **68**, 289–292 (1972, in Russian)
16. D.G. Gordeziani, *Methods for Solving a Class of Nonlocal Boundary Value Problems*. With Georgian and English summaries (Tbilis. Gos. Univ., Inst. Prikl. Mat., Tbilisi, 1981, in Russian), 32 pp.
17. M.P. Sapagovas, R.Yu. Chegis, Boundary-value problems with nonlocal conditions. Differ. Uravn. **23**(7), 1268–1274 (1987, in Russian); translation in Differ. Equ. **23**(7), 858–863 (1987)
18. B.P. Paneyakh, Some nonlocal boundary value problems for linear differential operators. Mat. Zametki **35**(3), 425–434 (1984, in Russian)
19. A.L. Skubachevski, On a spectrum of some nonlocal boundary value problems. Mat. Sb. **117**(7), 548–562 (1982, in Russian)
20. D. Devadze, V. Beridze, An optimal control problem for Helmholtz equations with Bitsadze–Samarskiĭ boundary conditions. Proc. A. Razmadze Math. Inst. **161**, 47–53 (2013)
21. D. Devadze, M. Dumbadze, An optimal control problem for a non-local boundary value problem. Bull. Georgian Natl. Acad. Sci. (N.S.) **7**(2), 71–74 (2013)

22. D.V. Kapanadze, On a nonlocal Bitsadze–Samarski boundary value problem. Differ. Uravn. **23**(3), 543–545, 552 (1987, in Russian)
23. V.A. Il'in, E.I. Moiseev, A two-dimensional nonlocal boundary value problem for the Poisson operator in the differential and the difference interpretation. Mat. Model. **2**(8), 139–156 (1990, in Russian)
24. D. Gordeziani, N. Gordeziani, G. Avalishvili, Non-local boundary value problems for some partial differential equations. Bull. Georgian Acad. Sci. **157**(3), 365–368 (1998)
25. A.K. Gushchin, V.P. Mikhailov, On the solvability of nonlocal problems for a second-order elliptic equation. Mat. Sb. **185**(1), 121–160 (1994, in Russian); translation in Russian Acad. Sci. Sb. Math. **81**(1), 101–136 (1995)
26. P.L. Gurevich, Asymptotics of solutions for nonlocal elliptic problems in plane bounded domains. Funct. Differ. Equ. **10**(1–2), 175–214 (2003). Functional differential equations and applications (Beer-Sheva, 2002), No. 4, 773–775
27. I.N. Vekua, *Generalized Analytic Functions*, 2nd edn. Edited and with a preface by O.A. Oleinik and B.V. Shabat (Nauka, Moscow,1988, in Russian)
28. G.F. Mandzhavidze, V. Tuchke, Some boundary value problems for first-order nonlinear differential systems on the plane, in *Boundary Value Problems of the Theory of Generalized Analytic Functions and Their Applications* (Tbilis. Gos. Univ., Tbilisi, 1983, in Russian), pp. 79–124
29. G.V. Meladze, T.S. Tsutsunava, D.Sh. Devadze, in *The Optimal Control Problem for Quasilinear Differential Equations of First Order on the Plane with Nonlocal Boundary Conditions* (Tbilisi State University Press, Tbilisi, 1987). Deposited at Georgian Res. Inst. Sci. Eng. Inform., 25.12.87, No. 372, 87, 61 pp.
30. D.Sh. Devadze, V.Sh. Beridze, Optimality conditions for quasilinear differential equations with nonlocal boundary conditions. Uspekhi Mat. Nauk **68**(4(412)), 179–180 (2013, in Russian); translation in Russian Math. Surv. **68**(4), 773–775 (2013)
31. Z.-L. Lions, The optimal control of distributed systems. Uspehi Mat. Nauk **28**(4(172)), 15–46 (1973, in Russian); Translated from the French: Enseignement Math. (2) **19**, 125–166 (1973)
32. V.A. Troitski, L.V. Petukhov, *Optimization of the Form of Elastic Bodies* (Nauka, Moscow, 1982, in Russian)
33. A.G. Butkovski, *Theory of Optimal Control of Systems with Distributed Parameters* (Nauka, Moscow, 1977)
34. R. Glovinski, Zh.L. Lions, R. Tremol'er, *Numerical Analysis of Variational Inequalities*. Translated from the French by A.S. Kravchuk (Mir, Moscow, 1979, in Russian)
35. A.G. Butkovskiy, *Metody upravleniya sistemami s raspredelennymi parametrami (Distributed-Parameter Systems: Methods of Control)* (Nauka, Moscow, 1975)
36. D. Devadze, V. Beridze, An algorithm of the solution of an optimal control problem for a nonlocal problem. Bull. Georgian Natl. Acad. Sci. (N.S.) **7**(1), 44–48 (2013)
37. L.S. Pontryagin, V.G. Boltyanski, R.V. Gamkrelidze, E.F. Mishchenko, *The Mathematical Theory of Optimal Processes*, 4th edn. (Nauka, Moscow, 1983, in Russian)
38. Z.-L. Lions, in *Optimal Control of Systems That Are Governed by Partial Differential Equations*, ed. by R.V. Gamkrelidze. Translated from the French by N.H. Rozov (Izdat. Mir, Moscow, 1972, in Russian)
39. V.I. Plotnikov, Necessary and sufficient conditions for optimality and conditions for uniqueness of the optimizing functions for control systems of general form. Izv. Akad. Nauk SSSR Ser. Mat. **36**, 652–679 (1972, in Russian)
40. O.A. Ladyzhenskaya, N.N. Uraltseva, *Linear and Quasilinear Equations of Elliptic Type*, 2nd rev. edn. (Izdat. Nauka, Moscow, 1973, in Russian)
41. V.S. Vladimirov, *The Equations of Mathematical Physics*, 4th edn. (Nauka, Moscow, 1981, in Russian)
42. R. Courant, D. Hilbert, *Methods of Mathematical Physics*, vol. I (Interscience, New York, 1953)

43. V.I. Plotnikov, V.I. Sumin, Optimization of objects with distributed parameters that can be described by Goursat–Darboux systems. Z. Vycisl. Mat. i Mat. Fiz. **12**, 61–77 (1972, in Russian)
44. N. Danford, Dz. Shvarc, *Linear Operators. Part I: General Theory* (Izdat. Inostran. Lit., Moscow, 1962, in Russian)

Chapter 9
On the Geometry of the Domain of the Solution of Nonlinear Cauchy Problem

Á. Figula and M.Z. Menteshashvili

Abstract We consider the Cauchy problem for a second order quasi-linear partial differential equation with an admissible parabolic degeneration such that the given functions described the initial conditions are defined on a closed interval. We study also a variant of the inverse problem of the Cauchy problem and prove that the considered inverse problem has a solution under certain regularity condition. We illustrate the Cauchy and the inverse problems in some interesting examples such that the families of the characteristic curves have either common envelopes or singular points. In these cases the definition domain of the solution of the differential equation contains a gap.

9.1 Introduction

Let us consider a Cauchy problem for second order non-strictly quasi-linear hyperbolic partial differential equations: find a solution $u(x, y)$ of the equation by the initial conditions $u|_{y=0} = \tau(x)$, $u_y|_{y=0} = \nu(x)$, where $\nu(x) \in C^1(R)$, $\tau(x) \in C^2(R)$ are given functions such that $\nu(x)$ is once-, and $\tau(x)$ is twice-continuously differentiable. To solve this problem one can use the method of characteristics (cf. [18, pp. 164–166]). The characteristic roots of the partial differential equation are the solutions of the corresponding characteristic equation. The class of hyperbolic equations is defined through the characteristic roots by the inequality $\lambda_1 \neq \lambda_2$.

Á. Figula (✉)
Institute of Mathematics, University of Debrecen, P.O. Box 400, 4002 Debrecen, Hungary
e-mail: figula@science.unideb.hu

M.Z. Menteshashvili
Muskhelishvili Institute of Computational Mathematics of the Georgian Technical University, 4, Grigol Peradze Str., 0131 Tbilisi, Georgia

Sokhumi State University, 0186 Tbilisi, Georgia
e-mail: m.menteshashvili@gtu.ge

© Springer International Publishing AG 2017
G. Falcone (ed.), *Lie Groups, Differential Equations, and Geometry*,
UNIPA Springer Series, DOI 10.1007/978-3-319-62181-4_9

Characteristic directions are defined at every point by the relations

$$\frac{dy}{dx} = \lambda_1(x, y, u, u_x, u_y), \quad \frac{dy}{dx} = \lambda_2(x, y, u, u_x, u_y). \tag{9.1}$$

Solving the differential equations (9.1) we obtain the characteristic invariants (cf. [17]).

Among the papers based on the application of the method of characteristics to nonlinear hyperbolic problems we refer to [2, 6, 7, 9–13, 15, 16], where the structure of the definition domains of the solutions and that of influence domains of initial and characteristic perturbations are investigated in singular cases.

The aim of our investigation is to discuss some questions stating an initial problem on the data support $[a, b] \in R$ for the following second order non-strictly hyperbolic equation

$$(u_y^2 - 1)u_{xx} - 2u_x u_y u_{xy} + u_x^2 u_{yy} = 0 \tag{9.2}$$

(cf. [5, p. 442]). We consider also a variant of the inverse problem of the Cauchy problem for Eq. (9.2) on some open and closed data supports. Equation (9.2) is interesting for physical applications since the two-dimensional flow of an inviscid incompressible fluid in gravitational field can be described by it (cf. [19, p. 535]).

The characteristic roots of (9.2), i.e. the solution of the equation $(u_y^2 - 1)\lambda^2 + 2u_x u_y \lambda + u_x^2 = 0$, are

$$-\frac{u_x}{u_y + 1}, \quad -\frac{u_x}{u_y - 1}.$$

Therefore Eq. (9.2) is hyperbolic everywhere, except for the points at which the derivative u_x of the sought solution $u(x, y)$ has zero values. At these points, Eq. (9.2) parabolically degenerates. Hence Eq. (9.2) has mixed type (cf. [4]). It is one of the most important problems of mathematical physics to study the properties of solutions of equations of mixed type (cf. [3, 20, 22, 23]). The first fundamental results in this direction were obtained by the Italian mathematicians Francesco Tricomi in the twenties of the nineteenth century.

The combinations $u_x - u_y$, $u_x + u_y$ of the first derivatives of the solution which are constant for the string vibration equation $u_{xx} - u_{yy} = 0$ differ from those for non-linear equation (9.2). In the latter case these combinations depend on an unknown solution and its first derivatives. Hence it is rather difficult to use the characteristic invariants for the solution of problems in the case of quasi-linear equations. In aerodynamics, these combinations are called Riemann invariants (cf. [4, p. 23]). Solving the equations $\frac{dy}{dx} = -\frac{u_x}{u_y+1}, \frac{dy}{dx} = -\frac{u_x}{u_y-1}$ we obtain that the characteristic invariants for Eq. (9.2) are $u + y = \text{const}$ and $u - y = \text{const}$.

According to Hadamard (cf. [14]) the Cauchy problem can appear to be well-posed in some cases involving the closed initial support. Such problems were studied by Aleksandryan [1], Sobolev [21], Vakhania [24], Wolfersdorf [25], and others for linear equations.

The Cauchy problem of Eq. (9.2) such that the given functions τ, ν are defined in the circumference of the unit circle, i.e. the initial problem on the closed data support γ : $\rho = 1, x = \rho \cos \vartheta$, $y = \rho \sin \vartheta, 0 \leqslant \vartheta \leqslant 2\pi$, was investigated in [16].

In this paper we study the Cauchy problem of Eq. (9.2) on the data support $[a, b]$ (see Problem 1 in Sect. 9.2). In Theorem 1 we give the solution of the Cauchy problem of (9.2) in integral form. To establish this result we use the representation formula of the general integral for (9.2). After this we find a domain where the solution can be completely defined. For this we have to write equations for the characteristic curves of both families. Then we consider the set of intersection points for both families of characteristic curves. This set creates the domain within which the initial support is located. In Sect. 9.3 we formulate a variant of the inverse problem of the Cauchy problem for Eq. (9.2) (see Problem 2). We prove that the considered inverse problem has a solution under certain regularity condition (cf. Theorem 2). We illustrate the Cauchy and the inverse problems for Eq. (9.2) in Examples 1–3. In Examples 2, 3 we deal with the interesting cases that the families of the characteristic curves have either common envelopes or singular points.

9.2 Cauchy Problem on Eq. (9.2)

Problem 1 *Let $\tau(x)$, respectively $\nu(x)$ be real functions, which are twice, respectively once continuously differentiable. Find a function $u(x, y)$, which satisfies Eq. (9.2) and the conditions*

$$u(x, 0) = \tau(x), \quad u_y(x, 0) = \nu(x), \quad a \leqslant x \leqslant b. \tag{9.3}$$

It is also required to find a domain where the solution can be completely defined.

Theorem 1 *If $\tau'(x) \neq 0$ for all $x \in [a, b]$, then the integral of problem (9.2), (9.3) can be written into the form*

$$x = a + \frac{1}{2} \int_a^{T(u+y)} (1 - \nu(t))dt + \frac{1}{2} \int_a^{T(u-y)} (1 + \nu(t))dt, \tag{9.4}$$

where $x = T(z)$ denotes the inverse function of $z = \tau(x)$.

Proof We start the investigation of the problem by considering the form of the general integral of Eq. (9.2). Using a straightforward calculation it can be checked that the solution $u(x, y)$ of Eq. (9.2) satisfies the following functional equation

$$f(u + y) + g(u - y) = x, \tag{9.5}$$

where f, g are arbitrary functions (see [8]). In order to provide the required smoothness of the solution $u(x, y)$, here it is assumed that the arbitrary functions f, g belong to the class $C^2(R)$. Taking the derivations of (9.5) with respect to the variables x and y and putting into these $y = 0$ we obtain

$$f'(u(x, 0) + 0)u_x(x, 0) + g'(u(x, 0) - 0)u_x(x, 0) = 1 \qquad (9.6)$$

$$f'(u(x, 0) + 0)(u_y(x, 0) + 1) + g'(u(x, 0) - 0)(u_y(x, 0) - 1) = 0. \qquad (9.7)$$

Hence we come to a system of two linear algebraic equations (9.6), (9.7) with respect to the variables $f'(u(x, 0))$, $g'(u(x, 0))$. Solving this system of linear equations one gets

$$f'(u(x, 0)) = \frac{1 - v(x)}{2\tau'(x)} := F(x) \qquad (9.8)$$

$$g'(u(x, 0)) = \frac{1 + v(x)}{2\tau'(x)} := G(x). \qquad (9.9)$$

If the condition $\tau'(x) \neq 0$ is fulfilled for all $x \in [a, b]$, then on the closed interval $[a, b]$ the equation $\tau(x) = z$ is uniquely solvable in the class of real solutions. We denote this solution by $x = T(z)$. Integrating relations (9.8), (9.9) between $\tau(a)$ and an arbitrary value $z \in [\tau(a), \tau(b)]$ we obtain

$$f(z) - f(\tau(a)) = \int_{\tau(a)}^{z} f'(v)dv = \int_{a}^{x} F(t)\tau'(t)dt,$$

$$g(z) - g(\tau(a)) = \int_{\tau(a)}^{z} g'(v)dv = \int_{a}^{x} G(t)\tau'(t)dt,$$

or equivalently

$$f(z) = f(\tau(a)) + \int_{a}^{T(z)} F(t)\tau'(t)dt, \qquad (9.10)$$

$$g(z) = g(\tau(a)) + \int_{a}^{T(z)} G(t)\tau'(t)dt. \qquad (9.11)$$

Their sum already yields an implicit solution of problem (9.2), (9.3) which contains the undefined free functions for $x = a$. They can be identified by normalization. In particular, if for the functions f, g defined by (9.10), (9.11), relation (9.5) is fulfilled at the point $(a, 0)$, then we have

$$f(\tau(a)) + g(\tau(a)) + \int_{a}^{T[u(a,0)+0]} F(t)\tau'(t)dt + \int_{a}^{T[u(a,0)-0]} G(t)\tau'(t)dt = a. \qquad (9.12)$$

Note that the identity $T(\tau(x)) = x$ is valid for all values of $x \in [a, b]$, including $x = a$, and therefore the upper bounds of both integrals in (9.12) are a. Hence one has $f(\tau(a)) + g(\tau(a)) = a$. Finally, the implicit solution of problem (9.2), (9.3) can be written as

$$x - a = \int_a^{T(u+y)} F(t)\tau'(t)dt + \int_a^{T(u-y)} G(t)\tau'(t)dt, \tag{9.13}$$

where the functions $F(t)$ and $G(t)$ are defined by (9.8), (9.9). This representation is obtained if condition $\tau'(x) \neq 0$ is fulfilled for all $x \in [a, b]$ and the expressions under the integral sign are integrable.   □

In order to establish the definition domain of the solution $u(x, y)$ of (9.2) given by (9.13), it is necessary to investigate the structure of the characteristic curves. Keeping in mind representation (9.13) and that the relation $u + y = \text{const}$ is fulfilled along the characteristic curves of the first family, we can obtain an equation for each characteristic of this family that passes through an arbitrary point x of the data support $[a, b]$. At this arbitrarily chosen point, with which we associate the argument x_0, we have

$$u(x_0, 0) = \tau(x_0).$$

Hence the relation

$$u(x, y) + y = \tau(x_0) \tag{9.14}$$

is fulfilled along the characteristic curve satisfying the relation $u + y = \text{const}$ and passing through the point $(x_0, 0)$. Since the right-hand side of this equality is a completely defined expression, we conclude that

$$u(x, y) - y = \tau(x_0) - 2y. \tag{9.15}$$

Substituting (9.14) and (9.15) into (9.13), we obtain the equation for a characteristic curve of the first family in the form

$$x - a = \int_a^{T(\tau(x_0))} F(t)\tau'(t)dt + \int_a^{T(\tau(x_0))-2y} G(t)\tau'(t)dt. \tag{9.16}$$

Keeping in mind that the identity $T(\tau(x_0)) = x_0$ is valid, the upper bound of the first integral in (9.16) will be x_0. If we introduce the notation $x_0 = c$, then this expression becomes the parameter which takes values from the interval $[a, b]$. Therefore, all characteristic curves of the family $u + y = \text{const}$ which pass through the points of the data support have the representation

$$x = a + \int_a^c F_1(t)dt + \int_a^{T(\tau(c))-2y} G_1(t)dt, \tag{9.17}$$

where

$$F_1(t) = F(t)\tau'(t) = \frac{1 - v(t)}{2}$$

$$G_1(t) = G(t)\tau'(t) = \frac{1 + v(t)}{2}.$$

Analogously, for all characteristic curves of the family $u - y = \text{const}$ we obtain

$$x = a + \int_a^{T(\tau(c)+2y)} F_1(t)dt + \int_a^c G_1(t)dt. \tag{9.18}$$

Here the parameter c takes its values from the interval $[a, b]$. For both families, in (9.17) and (9.18) there is a one-to-one correspondence between the equation for a characteristic curve and the parameter value. This fact is stipulated by the condition $\tau'(x) \neq 0$ for all $x \in [a, b]$.

9.3 Inverse Problem

Now we consider certain inverse problems of the Cauchy problem for Eq. (9.2). Below we will treat a variant of inverse problem which requires the construction of a given equation under the condition that the characteristic curves of both families are known a priori. The cases that the families of characteristic curves have either common envelopes or singular points are particularly interesting to investigate. In these situations, gaps may appear in the definition domain of a solution of the problem.

As has been mentioned above, Eq. (9.2) is of hyperbolic type and the set of parabolic degeneration is not defined a priori because it depends on the behavior of an arbitrary solution $u(x, y)$ and its derivative with respect to the variable x. The characteristic invariants (cf. [4, p. 23]) corresponding to Eq. (9.2) are given as follows

$$u + y = \text{const}, \quad \text{for} \quad \lambda_1 = -\frac{u_x}{u_y + 1}, \tag{9.19}$$

$$u - y = \text{const}, \quad \text{for} \quad \lambda_2 = -\frac{u_x}{u_y - 1}. \tag{9.20}$$

From these equations we conclude that the families of characteristic curves of the equation are not defined a priori because they depend on the sought solution and on its first order derivatives.

In the case of Eq. (9.2), we can admit in the role of characteristics the families of characteristic curves along which relations (9.19) and (9.20) are fulfilled.

Characteristic families can therefore be given a priori in an arbitrary manner. The statement of the inverse problem considered here rests on this fact.

Let us consider two one-parameter families of plane curves which are given explicitly by the equations

$$y = \varphi_1(x, c), \tag{9.21}$$

$$y = \varphi_2(x, c), \tag{9.22}$$

where φ_1, φ_2 are given, twice differentiable functions with respect to the variable x. Let φ_1, φ_2 be defined for any value of the real parameter c. Assume that any curve of these families necessarily intersects once the straight line $y = 0$. We denote by D_1 the domain of the plane (x, y) which is completely covered by the family of characteristic curves given by Eq. (9.21) if the parameter c runs continuously through all real values. Analogously, we denote by D_2 the domain of the plane (x, y) which is completely covered by the characteristic lines given by Eq. (9.22). Also, we introduce the notation $D = D_1 \cap D_2, I = D \cap \{y = 0\}$.

Problem 2 (Inverse Problem)   *Find the initial conditions of a regular solution $u(x, y)$ of Eq. (9.2) and its derivative with respect to the normal direction on the interval I of the straight line $y = 0$ if the families of plane curves given by Eq. (9.21) are the characteristics corresponding to invariants (9.19), while (9.22) is the family of characteristics corresponding to invariants (9.20).*

Theorem 2   *If the condition $\varphi_1'(x, c') \neq \varphi_2'(x, c'')$ is fulfilled for all $c', c'' \in R$, then there exists a solution of Problem 2.*

Proof To investigate the posed problem, we need a structural analysis of the characteristic curves of Eq. (9.2). The characteristic roots corresponding to Eq. (9.2) define at every point two characteristic directions

$$\frac{dy}{dx} = -\frac{u_x}{u_y + 1} \equiv \lambda_1(x, y), \tag{9.23}$$

$$\frac{dy}{dx} = -\frac{u_x}{u_y - 1} \equiv \lambda_2(x, y). \tag{9.24}$$

It follows from the posed problem that the family of characteristic curves defined by Eq. (9.21) corresponds to the root λ_1 given by (9.23), while that given by Eq. (9.22) corresponds to the root λ_2 given by (9.24). Therefore we can calculate the values of first derivatives u_x, u_y of the unknown solution $u(x, y)$ along any characteristic curve. Note that the parameters contained in families (9.21) and (9.22) can be defined by the abscissa of the intersection point of the concrete curve and the axis $y = 0$. Indeed, let the curve corresponding to the parameter c^*, intersecting the straight line $y = 0$ at a point x_0 be

$$\varphi_1(x_0, c^*) = 0.$$

Solving the equation for the parameter $c^*(x_0) = c$, family (9.21) takes the form

$$y = \varphi_1(x, x_0),$$

which is equal to zero for $x = x_0$. If family (9.21) corresponds to the characteristic root λ_1, then the relation

$$\frac{d\varphi_1(x, x_0)}{dx} = -\frac{u_x(x, \varphi_1(x, x_0))}{u_y(x, \varphi_1(x, x_0)) + 1}. \tag{9.25}$$

is fulfilled. Analogously, along the second family of the characteristic curves the following equality is fulfilled

$$\frac{d\varphi_2(x, x_0)}{dx} = -\frac{u_x(x, \varphi_2(x, x_0))}{u_y(x, \varphi_2(x, x_0)) - 1}. \tag{9.26}$$

Equalities (9.25), (9.26) have the following at the points of the interval I:

$$\frac{d\varphi_1(x, x_0)}{dx}\Big|_{x=x_0} = -\frac{\tau'(x_0)}{v(x_0) + 1}, \quad x_0 \in I, \tag{9.27}$$

$$\frac{d\varphi_2(x, x_0)}{dx}\Big|_{x=x_0} = -\frac{\tau'(x_0)}{v(x_0) - 1}, \quad x_0 \in I. \tag{9.28}$$

Since the point $(x_0, 0)$ is arbitrarily chosen from the interval I, x_0 in Eqs. (9.27), (9.28) can be replaced with x. Hence we can easily define the unknown functions τ' and v. Integrating the function τ', we finally obtain the solutions $\tau(x)$, $v(x)$ of the problem. Naturally, the function τ is defined up to an integration constant which will be defined uniquely if we give the value of the function $u(x, y)$ at one arbitrary point of the interval I. The theorem is thereby proved.   □

From the above general argumentation we can draw a conclusion. If families (9.21) and (9.22) do not have common characteristic directions at anyone of the points, i.e. if Eq. (9.2) is of strictly hyperbolic type, then the solution of the inverse problem presents no difficulty (see Example 1). However, if the equation parabolically degenerates on certain set of points, then the situation drastically changes. We illustrate this in Examples 2, 3.

Example 1 We solve the Cauchy and inverse problems in the following concrete example. Cauchy problem: Let $u(x, 0) = \tau(x)$ be the function x and $u_y(x, 0) = v(x)$ be the function $1 - e^x$. According to (9.4) the solution of the Cauchy problem in implicit form is given by equation

$$\frac{1}{2}e^{u+y} + u - y - \frac{1}{2}e^{u-y} - x = 0.$$

The characteristic curves (9.17) of the family $u + y = $ const, as parametric curve $f(x, y, c)$ with respect to the parameter c are given by

$$0 = \frac{1}{2}e^c + c - 2y - \frac{1}{2}e^{c-2y} - x = f_1(x, y, c). \tag{9.29}$$

The characteristic curves (9.18) of the family $u - y = $ const, as parametric curve $f(x, y, c)$ with respect to the parameter c are given by

$$0 = \frac{1}{2}e^{c+2y} + c - \frac{1}{2}e^c - x = f_2(x, y, c). \tag{9.30}$$

Differentiating the function $f_1(x, y, c)$ in (9.29) with respect to the parameter c we get

$$0 = \frac{1}{2}e^c(1 - e^{-2y}) + 1 = f_{1,c}(x, y, c). \tag{9.31}$$

Expressing from (9.31) the parameter c we have

$$c = \ln\left(\frac{2e^{2y}}{1 - e^{2y}}\right). \tag{9.32}$$

Putting expression (9.32) into Eq. (9.29) we obtain

$$e^{1+x} = \frac{2}{1 - e^{2y}}. \tag{9.33}$$

Since the derivative $\frac{\partial f_{1,c}(x,y,c)}{\partial y} = e^{c-2y}$ is non-zero and the derivative $\frac{\partial f_{1,c}(x,y,c)}{\partial c} = \frac{1}{2}(1 - e^{-2y})e^c$ is non-zero for all $y \neq 0$ we obtain that the envelope of the characteristic curves (9.29) of the first family is given by (9.33). An analogous computation shows that this envelope is also the envelope of characteristic curves (9.30) of the second family. In the plane (x, y) for the domain which is below of envelope (9.33) there does not exist solution of the Cauchy problem (see Fig. 9.1 for parameter $c = -1$).

Inverse problem: Assume that the families of characteristic curves are given by:

$$x = \varphi_1(y, c) \equiv \frac{1}{2}e^c - \frac{1}{2}e^{c-2y} + c - 2y, \quad \text{for} \quad \lambda_1, \tag{9.34}$$

$$x = \varphi_2(y, c) \equiv -\frac{1}{2}e^c + \frac{1}{2}e^{c+2y} + c, \quad \text{for} \quad \lambda_2. \tag{9.35}$$

From (9.27), (9.28) we get

$$-\frac{u_x(x_0, 0)}{u_y(x_0, 0) + 1} = \frac{1}{e^{x_0} - 2}, \tag{9.36}$$

$$c = -1.$$

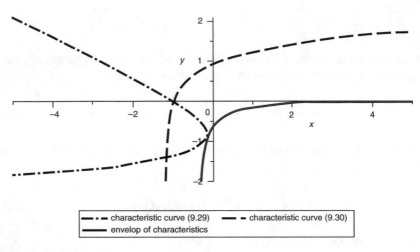

Characteristic curves legend:
- — · — characteristic curve (9.29)   — — characteristic curve (9.30)
- ——— envelop of characteristics

Fig. 9.1 The characteristic curves (9.29), (9.30) and their envelopes

for family (9.34) and

$$-\frac{u_x(x_0,0)}{u_y(x_0,0)-1} = \frac{1}{e^{x_0}}, \tag{9.37}$$

for family (9.35).

The right-hand sides of the equalities (9.36), (9.37) depend only on x_0. From the relations (9.36) and (9.37) for the point $(x_0, 0)$ we find the values of the functions τ' and v:

$$\tau'(x) = 1, \tag{9.38}$$

$$v(x) = 1 - e^x.$$

From equality (9.38) we obtain by integration the value of the solution $u(x, y)$ on the axis $y = 0$:

$$\tau(x) = x + c, \quad c = \text{const.} \tag{9.39}$$

The constant c in (9.39) can be obtained if we know the value of $\tau(x)$ in one arbitrary point.

Example 2 Now we consider the following inverse problem. Assume that the families of characteristic curves have common envelopes and are given in the form

$$x = \varphi_1(y, c) \equiv c - a\sqrt{\frac{a}{y+b} - 1}, \quad -b \leqslant y \leqslant a - b, \tag{9.40}$$

$$x = \varphi_2(y, c) \equiv c + a\sqrt{\frac{a}{y+b} - 1}, \quad -b \leqslant y \leqslant a - b, \tag{9.41}$$

$$a > b > 0, \quad c = \text{const}.$$

Each curve of family (9.40) defined on an interval $x \in (-\infty, c]$ monotonically increases, has the asymptote $y = -b$. For $x = c$ each curve of family (9.40) is tangent to the straight line $y = a - b$ and therefore is tangent to one of the curves of family (9.40). One of the curves of family (9.40) passes at every point of the straight line $y = a - b$ and smoothly continues from the same point to the completely defined curve of family (9.41). Therefore both characteristic directions at the points of this straight line coincide. Hence the straight line $y = a - b$ is the line of parabolic degeneration for Eq. (9.2). The straight line $y = -b$ is also the line of parabolic degeneration because all characteristic curves of both families are tangent to the straight line at infinity (see Fig. 9.2).

By direct calculations we establish that

$$\frac{u_x(x, y)}{u_y(x, y) + 1} = \frac{2a^3(x - c_1)}{[(x - c_1)^2 + a^2]^2}, \quad c_1 = x_0 + a\sqrt{\frac{a-b}{b}}, \tag{9.42}$$

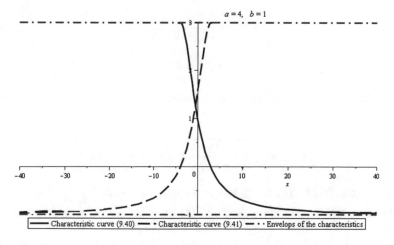

$$a = 4, \quad b = 1$$

Characteristic curve (9.40) ──── • Characteristic curve (9.41) ── • • Envelops of the characteristics

Fig. 9.2 The characteristic curves (9.40), (9.41) and their envelopes

holds for family (9.40) and that

$$\frac{u_x(x,y)}{u_y(x,y)-1} = \frac{2a^3(x-c_2)}{[(x-c_2)^2+a^2]^2}, \quad c_2 = x_0 - a\sqrt{\frac{a-b}{b}}, \tag{9.43}$$

is satisfied for family (9.41). Note that the right-hand sides of the equalities (9.42), (9.43) are represented by expressions depending solely on the variable x. Relations (9.42) and (9.43) for the point $(x_0, 0)$ yield the equalities

$$\frac{\tau'(x_0)}{\nu(x_0)+1} = \frac{2a^3(x_0-c_1)}{[(x_0-c_1)^2+a^2]^2},$$

for the first family of the characteristic curves, and

$$\frac{\tau'(x_0)}{\nu(x_0)-1} = \frac{2a^3(x_0-c_2)}{[(x_0-c_2)^2+a^2]^2},$$

for the second family. Let us assume that the point $x_0 \in (-\infty, +\infty)$ is arbitrary, then for any $(x, 0)$ we obtain

$$\frac{\tau'(x)}{\nu(x)+1} = -\frac{2b^2}{a^2}\sqrt{\frac{a-b}{b}}, \tag{9.44}$$

$$\frac{\tau'(x)}{\nu(x)-1} = \frac{2b^2}{a^2}\sqrt{\frac{a-b}{b}}. \tag{9.45}$$

Hence we find the values of the functions τ' and ν:

$$\tau'(x) = -\frac{2b^2}{a^2}\sqrt{\frac{a-b}{b}}, \tag{9.46}$$

$$\nu(x) = 0.$$

From equality (9.46) we find by integration the values of the solution $u(x, y)$ on the axis $y = 0$:

$$\tau(x) = -\frac{2b^2}{a^2}\sqrt{\frac{a-b}{b}}\, x + c, \quad c = \text{const}.$$

As relations (9.44), (9.45) show the characteristic roots $-\frac{u_x}{u_y+1}$, $-\frac{u_x}{u_y-1}$ of both families take a constant value. Despite this, we cannot assert that the straight line $y = 0$ is the line of parabolic degeneration. To make such an assertion it is necessary that the invariants $u + y$ and $u - y$ are constant along $y = 0$. The constancy of the invariants is due to the fact that families (9.40), (9.41) are sets of curves obtained by a parallel translation along the axis of abscissas. Therefore every curve of both families has one and the same slope with respect to the axis of abscissas. This indicates that the values of the derivative $\frac{dy}{dx}$ along this axis preserve constancy.

Example 3 The situation is more difficult if the families of the characteristic curves have, besides the lines of parabolic degeneration, also common singular points. As an example let us consider the case that families (9.21), (9.22) have the common node. Each contour of the one-parameter family of curves

$$F(r, \vartheta, \vartheta_0) = \frac{(1 - r\cos\vartheta)}{r^2 - 2r\cos\vartheta + 1} - \left\{ 1 + \left[\frac{r\sin\vartheta}{r^2 - 2r\cos\vartheta + 1} + \vartheta_0 \right]^2 \right\}^{-1} = 0,$$

$$(9.47)$$

$0 \leqslant \vartheta \leqslant 2\pi$ is closed. The whole family lies in the half plane to the left of the straight line $x = 1$. The circumference

$$(2x - 1)^2 + 4y^2 = 1 \tag{9.48}$$

is a common envelope of a family of curves. The point $(1, 0)$ is the node of the considered family. The unit circumference completely lies in the definition domain of family (9.47). Every curve of this family intersects twice the unit circumference. Every contour of family (9.47) can be represented as the union of two arcs. The first arc is considered from the point $(1, 0)$ to the point of tangency to circumference (9.48) if the movement occurs in the positive direction. The remaining part is considered to be the arc of the second family. If the polar angle of the point of tangency of the concrete curve to circumference (9.48) is denoted by ϑ^*, then family (9.47) can be divided into two parts:

$$F(r, \vartheta, \vartheta_0) = 0, \quad 0 \leqslant \vartheta \leqslant \vartheta^*, \tag{9.49}$$

$$F(r, \vartheta, \vartheta_0) = 0, \quad \vartheta^* \leqslant \vartheta \leqslant 2\pi. \tag{9.50}$$

Problem 3 (Inverse Problem)   *Find the values of a regular solution of Eq. (9.2) and its derivative with respect to the normal direction on the circumference $x^2 + y^2 = 1$, i.e. find the functions*

$$\lim_{r \to 1} u(r, \vartheta) = \tau(\vartheta), \qquad \lim_{r \to 1} u_r(r, \vartheta) = \nu(\vartheta), \tag{9.51}$$

if the family of plane curves (9.49) represents the characteristic curves corresponding to invariants (9.19), while family (9.50) corresponds to the characteristic invariants (9.20).

Let $(1, \varphi_0)$, $(\varphi_0 \neq 2\pi k, k \in Z)$ be a point of the unit circumference. Then the following curves of families (9.21) and (9.22) pass, respectively, through this point:

$$\frac{(1 - r\cos\vartheta)}{r^2 - 2r\cos\vartheta + 1} - \left\{1 + \left[\frac{r\sin\vartheta}{r^2 - 2r\cos\vartheta + 1} + \frac{\sin\varphi_0}{2(1 - \cos\varphi_0)} + 1\right]^2\right\}^{-1} = 0,$$

(9.52)

where $0 \leq \vartheta \leq \vartheta_1^*$,

$$\frac{(1 - r\cos\vartheta)}{r^2 - 2r\cos\vartheta + 1} - \left\{1 + \left[\frac{r\sin\vartheta}{r^2 - 2r\cos\vartheta + 1} - \frac{\sin\varphi_0}{2(1 - \cos\varphi_0)} - 1\right]^2\right\}^{-1} = 0,$$

(9.53)

where $\vartheta_2^* \leq \vartheta \leq 2\pi$.

Here $\vartheta_1^*, \vartheta_2^*$ are the polar angles of points of tangency of these curves with the circumference (9.48), respectively.

The relation

$$\frac{u_x(r, \vartheta, \varphi_0)}{u_y(r, \vartheta, \varphi_0) + 1} = \frac{f_1(\vartheta, \varphi_0)}{g_1(\vartheta, \varphi_0)}$$

(9.54)

is fulfilled for the characteristic curve (9.52), and the relation

$$\frac{u_x(r, \vartheta, \varphi_0)}{u_y(r, \vartheta, \varphi_0) - 1} = \frac{f_2(\vartheta, \varphi_0)}{g_2(\vartheta, \varphi_0)}$$

(9.55)

is fulfilled along curve (9.53), where

$$f_i(\vartheta, \varphi_0) = \frac{4h_i r \sin\vartheta (1 - \cos\vartheta)}{\left(1 + h_i^2\right)^2} + (r^2 \cos 2\vartheta - 2r\cos\vartheta + 1),$$

$$g_i(\vartheta, \varphi_0) = \frac{2h_i(r^2 \cos 2\vartheta - 2r\cos\vartheta + 1)}{\left(1 + h_i^2\right)^2} - 2r\sin\vartheta (1 - \cos\vartheta),$$

$$h_i = \frac{r\sin\vartheta}{r^2 - 2r\cos\vartheta + 1} - \frac{\sin\varphi_0}{2(1 - \cos\varphi_0)} - (-1)^i, \quad i = 1, 2.$$

From (9.54) and (9.55) we conclude that the following equalities hold at the point $(1, \varphi_0)$:

$$\frac{u_x(1, \varphi_0)}{u_y(1, \varphi_0) + 1} = \frac{f_1^0}{g_1^0},$$

(9.56)

$$\frac{u_x(1, \varphi_0)}{u_y(1, \varphi_0) - 1} = \frac{f_2^0}{g_2^0},$$

(9.57)

where

$$f_i^0 = (-1)^i \frac{1}{2} \sin \varphi_0 + \cos \varphi_0,$$

$$g_i^0 = (-1)^{i+1} \frac{1}{2} \cos \varphi_0 + \sin \varphi_0.$$

Using relations (9.56) and (9.57) we obtain for the values of the derivatives u_x and u_y at the point $(1, \varphi_0)$

$$u_x(1, \varphi_0) = \frac{2 f_1^0 f_2^0}{f_2^0 g_1^0 - f_1^0 g_2^0},$$

$$u_y(1, \varphi_0) = \frac{f_1^0 g_2^0 + f_2^0 g_1^0}{f_2^0 g_1^0 - f_1^0 g_2^0}.$$

Assuming that φ_0 takes all values on the circumference $x^2 + y^2 = 1$, $0 \leqslant \varphi_0 \leqslant 2\pi$, we have

$$u_x(1, \vartheta) = 2 \cos^2 \vartheta - \frac{1}{2} \sin^2 \vartheta,$$

$$u_y(1, \vartheta) = \frac{5}{4} \sin 2\vartheta.$$

Since the equalities

$$u_\vartheta(1, \vartheta) = -u_x \sin \vartheta + u_y \cos \vartheta,$$

$$u_r(1, \vartheta) = u_x \cos \vartheta + u_y \sin \vartheta$$

are fulfilled, we obtain

$$u_\vartheta(1, \vartheta) = \frac{1}{2} \sin \vartheta, \tag{9.58}$$

$$u_r(1, \vartheta) = 2 \cos \vartheta. \tag{9.59}$$

Finally, integrating the value of (9.58) on the circumference $x^2 + y^2 = 1$, $0 \leqslant \varphi_0 \leqslant 2\pi$, we define also the function

$$u(1, \vartheta) = -\frac{1}{2} \cos \vartheta + c. \tag{9.60}$$

If we consider the initial problem: find a function $u(x, y)$, which satisfies Eq. (9.2) with initial data defined by functions (9.59), (9.60) on the unit circumference γ: $r = 1$, $x = r\cos\vartheta$, $y = r\sin\vartheta$, then the definition domain of a solution of this problem is constructed by the set of characteristic curves represented by formulas (9.49), (9.50). By the structure of these characteristic curves we conclude that for $a > b$ the definition domain of a solution of the initial problem is the domain of the plane (x, y) which lies to the left of the straight line $x = 1$, with a gap given the circumference (9.48).

Acknowledgements The authors were supported by the European Union's Seventh Framework Programme (FP7/2007–2013) under grant agreements no. 317721, no. 318202.

References

1. R.A. Aleksandryan, The boundary problem of Dirichlet type for degenerative hyperbolic equations. Reports of Symposium in Continuum Mechanics and related problems of Analysis. Tbilisi (1971, in Russian)
2. G. Baghaturia, Nonlinear versions of hyperbolic problems for one quasi-linear equation of mixed type. J. Math. Sci. **208**, 621–634 (2015)
3. L. Bers, *Mathematical Aspects of Subsonic and Transonic Gas Dynamics*. Surveys in Applied Mathematics, vol. 3 (Wiley, New York; Chapman and Hall, London, 1958)
4. A.V. Bitsadze, *Equations of Mixed Type* (Pergamon Press, Oxford, 1964)
5. A.V. Bitsadze, *Some Classes of Partial Differential Equations* (Gordon and Breach Science, New York, 1988)
6. R. Bitsadze, On one version of the initial-characteristic Darboux problem for one equation of nonlinear oscillation. Reports of an Enlarged Session of the Seminar of I. Vekua Inst. Appl. Math. **8**, 4–6 (1993)
7. R.G. Bitsadze, M. Menteshashvili, On one nonlinear analogue of the Darboux problem. Proc. A. Razmadze Math. Inst. **169**, 9–21 (2015)
8. J.K. Gvazava, *On Some Classes of Quasi-Linear Equations of Mixed Type* (Metsniereba, Tbilisi, 1981, in Russian)
9. J.K. Gvazava, Second-order nonlinear equations with complete characteristic systems and characteristic problems for them. Trudy Tbiliss. Mat. Inst. Razmadze Akad. Nauk Gruzin. SSR **87**, 45–53 (1987, in Russian)
10. J.K. Gvazava, Nonlocal and initial problems for quasi-linear, non-strictly hyperbolic equations with general solutions represented by superposition of arbitrary functions. Georgian Math. J. **10**, 687–707 (2003)
11. J.K. Gvazava, The mean value property for nonstrictly hyperbolic second order quasilinear equations and the nonlocal problems. Proc. A. Razmadze Math. Inst. **135**, 79–96 (2004)
12. J.K. Gvazava, On one nonlinear version of the characteristic problem with a free support of data. Proc. A. Razmadze Math. Inst. **140**, 91–107 (2006)
13. J. Gvazava, M. Menteshashvili, G. Baghaturia, Cauchy problem for a quasi-linear hyperbolic equation with closed support of data. J. Math. Sci. **193**, 364–368 (2013)
14. J. Hadamard, *Leçons sur la propagation dés ondes et les équations de l'hydrodinamique* (Hermann, Paris, 1903)
15. M. Klebanskaya, Some nonlinear versions of Darboux and Goursat problems for a hyperbolic equation with parabolic degeneracy, in *International Symposium on Differential Equations and Mathematical Physics dedicated to the 90th Birthday Anniversary of Academician I. Vekua*, Tbilisi, 21–25 June 1997

16. M.Z. Menteshashvili, On the Cauchy problem on the unit circumference for a degenerating quasi-linear equation. Soobsh. Akad. Nauk Gruzii **148**, 190–193 (1993, in Russian)
17. G. Monge, Second mémoire sur le calcul integral de quelques équations aux differences partilelles. Mem. R. Accad. Sci. Turin, années **5**, 79–122 (1770–1773)
18. P.J. Olver, *Applications of Lie Groups to Differential Equations*. Graduate Texts in Mathematics, vol. 107 (Springer, Berlin, 1993)
19. I.Yu. Popov, Stratified flow in an electric field, the Schrödinger Equation, and the operator extension theory model. Theor. Math. Phys. **103**, 535–542 (1995)
20. L.I. Sedov, *Problems in the Plane in Hydrodynamics and Aerodynamics* (Gastekhizdat, Moscow-Leningrad, 1950)
21. S.L. Sobolev, On mixed problems for partial differential equations with two independent variables. Dokl. Akad. Nauk SSSR, **122**, 555–558 (1958, in Russian)
22. F. Tricomi, Beispiel einer Strömung mit Durchgang durch die Schallgeschwindigkeit. Monatshefte Math. **58**, 160–171 (1954)
23. F. Tricomi, Lectures on Partial Equations, Moscow, 1957
24. N. Vakhania, On a boundary problem with the prescription on the whole boundary for the hyperbolic system equivalent to the vibrating string equation. Dokl. Akad. Nauk SSSR, **116**, 906–909 (1957, in Russian)
25. L. Wolfersdorf, Zum Problem der Richtungsableitung für die Tricomi-Gleichung. Math. Nachr. **25**, 111–127 (1963)

Chapter 10
Reduction of Some Semi-discrete Schemes for an Evolutionary Equation to Two-Layer Schemes and Estimates for the Approximate Solution Error

Jemal Rogava, David Gulua, and Romeo Galdava

Abstract In the paper, using the perturbation algorithm, purely implicit three-layer and four-layer semi-discrete schemes for an abstract evolutionary equation are reduced to two-layer schemes. The solutions of these two-layer schemes are used to construct an approximate solution of the initial problem. By using the associated polynomials the estimates for the approximate solution error are proved.

10.1 Introduction

Various initial boundary value problems for evolutionary equations with partial derivatives can be reduced to the Cauchy problem for an abstract parabolic equation. One of the methods used to solve these problems is the semi-discretization method (this is the method based on the discretization of a derivative with respect to a time variable). The semi-discretization method for a parabolic equation is also known as the Rothe method [1]. The advantage of this method is that it enables us to solve the obtained system by the finite-difference method with subsequent discretization of derivatives with respect to spatial derivatives or to apply other methods (including analytical ones) which are easy to realize. Among these methods we want to mention the projective-network, variational and finite-element methods (see, e.g., Marchuk and Agoshkov [2], Mikhlin [3], Streng and Fiks [4]).

J. Rogava
I. Vekua Institute of Applied Mathematics, Iv. Javakhishvili Tbilisi State University, 2 University St., Tbilisi 0186, Georgia
e-mail: jemal.rogava@tsu.ge

D. Gulua (✉)
Georgian Technical University, 77 M. Kostava St., Tbilisi 0175, Georgia
e-mail: d_gulua@gtu.ge

R. Galdava
Sokhumi State University, 9, Anna Politkovskaia St., Tbilisi 0186, Georgia
e-mail: romeogaldava@gmail.com

© Springer International Publishing AG 2017
G. Falcone (ed.), *Lie Groups, Differential Equations, and Geometry*,
UNIPA Springer Series, DOI 10.1007/978-3-319-62181-4_10

Questions connected with the construction and investigation of approximate solution algorithms of evolutionary problems are considered, for example in the well-known books by Godunov and Ryabenki [5], Marchuk [6], Richtmayer and Morton [7], Samarskiĭ [8], Ianenko [9]. We also refer to the works by Crouzeix [10], Crouzeix and Raviart [11], and Le Rouxe [12].

The questions connected with construction and research of algorithms of the approximate solution of evolutionary problems, including research and realization of multi-layer schemes for these problems are important. The main difficulty which arises at realization of multi-layer schemes (especially for multidimensional problems) consists in use of large random access memory, which increases in proportion with growth of number of layers. One of the opportunities of overcoming this problem is decomposition of multi-layer schemes. In [13] the perturbation algorithm was considered for decomposition of three-layer scheme. The perturbation algorithm is widely used when solving problems of mathematical physics (see, e.g., Marchuk et al. [14]).

We would note [15, 16] where a purely implicit three-layer semi-discrete scheme for evolutionary equation is reduced to two-layer schemes and explicit estimates for the approximate solution at rather general assumptions about data the tasks are proved in Banach space. Furthermore, in these works, by reducing with the aid of the perturbation algorithm the four-layer scheme to two-layer schemes we demonstrate the generality of the algorithm when it is applied to difference schemes.

In the present paper, purely implicit multi-layer semi-discrete schemes are considered for an approximate solution of the Cauchy problem for an evolutionary equation with the self-adjoint positively defined operator in Hilbert space. Using the perturbation algorithm, the considered scheme is reduced to two-layer schemes. An approximate solution of the original problem is constructed by means of the solutions of these schemes. Note that the first two-layer scheme gives an approximate solution to an accuracy of first order, whereas the solution of each subsequent scheme is the refinement of the preceding solution by one order.

In the present work, for an estimate of the error of the approximate solution, we applied the approach offered in [17], where the stability of linear many-step methods is investigated by means of the properties of the class of polynomials of many variables (which are called associated polynomials). They are a natural generalization of classical Chebyshev polynomials of second kind.

The present article consists of two paragraphs. In the first paragraph, creation of algorithm and technique of proving estimate of approximate solution error is applied for three-layer scheme. In the second paragraph, same technique is applied for four-layer scheme. In our opinion, such presentation of the material makes it possible to demonstrate the generality of the proposed method for decomposition and investigation of multi-layer schemes.

In the first paragraph we also consider the results of the numerical experiments, which visually confirm the theoretical estimates received in this paragraph.

It should be noted that the algorithm proposed in this paper is close to the methods considered in the works by Marchuk and Shaĭdurov [18] and Pereyra [19, 20]. The paper is a generalization of the results obtained in [13].

10.2   The Case of Three-Layer Scheme

10.2.1   Reducing a Purely Implicit Three-Layer Scheme for an Evolutionary Problem to Two-Layer Schemes

Let us consider the following evolutionary problem in the Hilbert space H:

$$\frac{du(t)}{dt} + Au\,(t) = 0, \quad t \in]0, T], \tag{10.1}$$

$$u\,(0) = u_0, \tag{10.2}$$

where A is the self-adjoint positively defined operator in H with domain of definition $D(A)$; u_0 is a given vector from H; $u(t)$ is the sought function.

On $[0, T]$ we introduce a grid $t_k = k\tau$, $k = 0, 1, \ldots, n$, with the step $\tau = T/n$. Using the approximation of the first derivative

$$\left. \frac{du}{dt} \right|_{t=t_k} = \frac{\frac{3}{2}u(t_k) - 2u(t_{k-1}) + \frac{1}{2}u(t_{k-2})}{\tau} + \tau^2 R_k(\tau, u), \quad R_k(\tau, u) \in H,$$

Eq. (10.1) can be represented at the point $t = t_k$ as

$$\frac{\frac{3}{2}u(t_k) - 2u(t_{k-1}) + \frac{1}{2}u(t_{k-2})}{\tau} + Au\,(t_k) = -\tau^2 R_k(\tau, u), \quad k = 2, \ldots, n, \tag{10.3}$$

We write system (10.3) in the form

$$\frac{u(t_k) - u(t_{k-1})}{\tau} + Au\,(t_k) + \frac{\tau}{2}\left(\frac{u(t_k) - 2u(t_{k-1}) + u(t_{k-2})}{\tau^2}\right) = -\tau^2 R_k(\tau, u).$$

It is obvious that expression in brackets in case of $\frac{\tau}{2}$ is $u''(t_{k-1}) + \tau^2 R_{1,k-1}$, $R_{1,k} \in H$.

By analogy with the obtained system let us consider in the Hilbert space H the one-parametric family of equations

$$\frac{u_k - u_{k-1}}{\tau} + Au_k + \frac{\varepsilon}{2}\left(\frac{u_k - 2u_{k-1} + u_{k-2}}{\tau^2}\right) = \varepsilon^2 R_k, \quad R_k \in H. \tag{10.4}$$

Assume that u_k is analytic with respect to ε,

$$u_k = \sum_{j=0}^{\infty} \varepsilon^j u_k^{(j)}.$$ (10.5)

Substituting (10.5) into (10.4) and equating the members of identical powers ε, we obtain the following system of equations

$$\frac{u_k^{(0)} - u_{k-1}^{(0)}}{\tau} + A u_k^{(0)} = 0, \qquad u_0^{(0)} = u_0, \quad k = 1, \ldots, n,$$ (10.6)

$$\frac{u_k^{(1)} - u_{k-1}^{(1)}}{\tau} + A u_k^{(1)} = -\frac{1}{2} \frac{\Delta^2 u_{k-2}^{(0)}}{\tau^2}, \qquad k = 2, \ldots, n,$$ (10.7)

$$\frac{u_k^{(2)} - u_{k-1}^{(2)}}{\tau} + A u_k^{(2)} = -\frac{1}{2} \frac{\Delta^2 u_{k-2}^{(1)}}{\tau^2} + R_k,$$

$$\cdots\cdots\cdots\cdots\cdots\cdots\cdots\cdots\cdots\cdots\cdots\cdots\cdots\cdots\cdots\cdots\cdots\cdots$$

where $\Delta u_{k-1} = u_k - u_{k-1}$.

We introduce the notation

$$v_k = u_k^{(0)} + \tau u_k^{(1)}, \quad k = 2, \ldots, n.$$ (10.8)

Let the vector v_k be an approximate value of the exact solution of problem (10.1)–(10.2) for $t = t_k$, $u(t_k) \approx v_k$.

Note that in scheme (10.7) the starting vector $u_1^{(1)}$ is defined from the equality $v_1 = u_1^{(0)} + \tau u_1^{(1)}$, where $u_1^{(0)}$ is found by scheme (10.6), and v_1 is an approximate value of $u(t_1)$ with accuracy of $O(\tau^2)$. According to this assumption, we can write

$$\|v_1 - u(t_1)\| = O(\tau^2).$$

Then, taking into account that for the solution of the problem (10.1), (10.2) the formula (see, e.g., Kato [21]) $u(t) = \exp(-tA) u_0$ is true, we can write

$$v_1 = (I - \tau A) u_0.$$ (10.9)

From representation $v_1 = u_1^{(0)} + \tau u_1^{(1)}$, taking into account a formula (10.9) and equality $u_1^{(0)} = S u_0$, where $S = (I + \tau A)^{-1}$, for a start vector $u_1^{(1)}$ of the scheme (10.7), we can take

$$u_1^{(1)} = \frac{v_1 - u_1^{(0)}}{\tau} = \frac{1}{\tau} (I - \tau A - S) u_0 = \frac{1}{\tau} ((I - S) - \tau A) u_0$$

$$= \frac{1}{\tau} (\tau A S - \tau A) u_0 = A(S - I) u_0 = -\tau A^2 S u_0.$$ (10.10)

10.2.2 Estimate of Residual for Solution Obtained by Perturbation Algorithm

Let estimate residual of purely implicit three-layer scheme for solutions of scheme (10.6)–(10.8).

Multiply Eq. (10.7) by τ and add result with (10.6). We get that v_k is a solution of system of equations:

$$\frac{v_k - v_{k-1}}{\tau} + Av_k = -\frac{\tau}{2}\frac{\Delta^2 u_{k-2}^{(0)}}{\tau^2}, \qquad k = 2,\ldots,n. \tag{10.11}$$

Rewrite this system in the form

$$\frac{v_k - v_{k-1}}{\tau} + Av_k + \frac{\tau}{2}\frac{\Delta^2 v_{k-2}}{\tau^2} = \widetilde{R}_k(\tau), \tag{10.12}$$

where $k \geqslant 3$,

$$\widetilde{R}_k(\tau) = \frac{\tau}{2}\frac{\Delta^2 v_{k-2}}{\tau^2} - \frac{\tau}{2}\frac{\Delta^2 u_{k-2}^{(0)}}{\tau^2}.$$

It is obvious that (10.12) can be represented in the following form:

$$\frac{\frac{3}{2}v_k - 2v_{k-1} + \frac{1}{2}v_{k-2}}{\tau} + Av_k = \widetilde{R}_k(\tau), \qquad k = 3,\ldots,n. \tag{10.13}$$

Therefore $\widetilde{R}_k(\tau)$ is a residual of purely implicit three-layer scheme for solutions of scheme (10.6)–(10.8) (see (10.3)).

It is obviously the representation

$$\widetilde{R}_k(\tau) = \frac{\tau}{2}\frac{\Delta^2 v_{k-2}}{\tau^2} - \frac{\tau}{2}\frac{\Delta^2 u_{k-2}^{(0)}}{\tau^2} = \frac{\tau^2}{2}\frac{\Delta^2 u_{k-2}^{(1)}}{\tau^2}. \tag{10.14}$$

Note that representation (10.14) is true for $k > 2$.

We estimate the difference relation (10.14).

From (10.6) it is obviously $u_k^{(0)} = S^k u_0$, where $S = (I + \tau A)^{-1}$. Then

$$\Delta^2 u_{k-2}^{(0)} = u_k^{(0)} - 2u_{k-1}^{(0)} + u_{k-2}^{(0)}$$
$$= (S^k - 2S^{k-1} + S^{k-2})u_0 = \tau^2 A^2 S^k u_0. \tag{10.15}$$

Taking into account (10.15) from (10.7) we have:

$$
u_k^{(1)} = S\left(u_{k-1}^{(1)} - \frac{\tau}{2}\frac{\Delta^2 u_{k-2}^{(0)}}{\tau^2}\right)
$$

$$
= S u_{k-1}^{(1)} - \frac{\tau}{2}A^2 S^{k+1} u_0 = S^{k-1} u_1^{(1)} - \frac{\tau}{2}(k-1)A^2 S^{k+1} u_0. \tag{10.16}
$$

From here it follows:

$$
\Delta^2 u_{k-2}^{(1)} = u_k^{(1)} - 2u_{k-1}^{(1)} + u_{k-2}^{(1)}
$$

$$
= \left(\tau^2 A^3 S^{k+1} - \frac{\tau^3}{2}(k-3)A^4 S^{k+1}\right) u_0 + \tau^2 A^2 S^{k-1} u_1^{(1)}. \tag{10.17}
$$

From here, taking into account that $u_1^{(1)}$ uniformly bounded by τ (see (10.10)), for enough smooth initial data we receive the estimation

$$
\left\|\frac{\Delta^2 u_{k-2}^{(1)}}{\tau^2}\right\| \leqslant c, \qquad c = \text{const.} > 0. \tag{10.18}
$$

Therefore, taking into account (10.18), from (10.14) for residual $\widetilde{R}_k(\tau)$, we have the estimation

$$
\left\|\widetilde{R}_k(\tau)\right\| \leqslant c\tau^2, \qquad c = \text{const.} > 0. \tag{10.19}
$$

10.2.3 A Priori Estimate for the Approximate Solution Error

From (10.3) and (10.13), for error $z_k = u(t_k) - v_k$ we have:

$$
\frac{\frac{3}{2}z_k - 2z_{k-1} + \frac{1}{2}z_{k-2}}{\tau} + Az_k = r_k(\tau), \qquad k = 3,\ldots,n, \tag{10.20}
$$

where $r_k(\tau) = -\left(\tau^2 R_k(\tau, u) + \widetilde{R}_k(\tau)\right)$.

Remark 10.2.1 Taking into account (10.19) we conclude that if the solution of problem (10.1)–(10.2) is enough smooth function, then $\|r_k(\tau)\| = O(\tau^2)$.

The following theorem is valid.

Theorem 10.2.2 *Let A be a self-adjoint positively defined operator in H. Then*

$$
\|z_{k+1}\| \leqslant \frac{1}{2}\|z_1\| + \frac{3}{2}\|z_2\| + \tau \sum_{i=2}^{k} \|r_{i+1}(\tau)\|, \tag{10.21}
$$

where $k = 2,\ldots,n-1$.

Proof From (10.20), we have

$$z_{k+1} = L_1 z_k - L_2 z_{k-1} + g_{k+1}, \quad k = 2, \ldots, n-1.$$

where

$$L_1 = \frac{4}{3}L, \quad L_2 = \frac{1}{3}L, \quad g_{k+1} = \frac{2}{3}\tau L r_{k+1}(\tau),$$

$$L = \left(I + \frac{2}{3}\tau A \right)^{-1}.$$

Then (see [17], ch.1. §3):

$$z_{k+1} = U_{k-1} z_2 - L_2 U_{k-2} z_1 + \sum_{i=2}^{k} U_{k-i} g_{i+1}, \tag{10.22}$$

where $U_k(L_1, L_2)$ are the operator polynomials, which are defined by the following recurrence relation:

$$U_k(L_1, L_2) = L_1 U_{k-1} + L_2 U_{k-2},$$

$$U_0 = I, \quad U_{-1} = 0.$$

If in equality (10.22) we take the norms, we will have

$$\|z_{k+1}\| \leq \|U_{k-1}\|\|z_2\| + \|L_2\|\|U_{k-2}\|\|z_1\| + \frac{2}{3}\tau\|L\|\sum_{i=2}^{k}\|U_{k-i}\|\|r_{i+1}(\tau)\|.$$

$$\tag{10.23}$$

Be using the well-known fact that the norm of the operator polynomial is equal to C norm of corresponding scalar polynomial on the spectrum of this operator when the argument is self-adjoint bounded operator (see, e.g., [22], ch. IX, §5), we have

$$\|U_k(L_1, L_2)\| = \max_{x \in Sp(L_1)} \left| U_k\left(x, \frac{1}{4}x \right) \right|. \tag{10.24}$$

It is obvious that $Sp(L_1) \subset [0, \frac{4}{3}]$. Then we have

$$\max_{x \in [0, \frac{4}{3}]} \left| U_k\left(x, \frac{1}{4}x \right) \right| \leq \frac{3}{2}.$$

Taking into account this estimate from (10.24) it follows

$$\|U_k(L_1, L_2)\| \leqslant \frac{3}{2}.$$

Substituting this estimate in (10.23), we obtain the sought estimate (10.21).   □

Note that since the representation (10.14) is true only if $k > 2$, proof of the estimate (10.21) does not apply in the case of $k = 1$, i.e. for a vector v_2. So it is necessary to separately estimate the error $u(t_2) - v_2$.

From (10.11) follows

$$v_2 = Sv_1 + \tau S\varphi_1, \tag{10.25}$$

where $\varphi_1 = -\frac{\tau}{2}\frac{\Delta^2 u_0^{(0)}}{\tau^2}$. Substituting (10.9) and (10.15) in (10.25) and taking into account the expansion $S = I - \tau A + \tau^2 A^2 S$, we have

$$v_2 = (I - \tau A + \tau^2 A^2 S)(I - \tau A) u_0 + \widetilde{r}_1(\tau)u_0,   \|\widetilde{r}_1(\tau)u_0\| = O(\tau^2).$$

It is obvious that from here follows

$$v_2 = (I - 2\tau A) u_0 + \widetilde{r}_2(\tau)u_0,   \|\widetilde{r}(\tau)u_0\| = O(\tau^2).$$

Therefore for error $u(t_2) - v_2$ we have

$$\|u(t_2) - v_2\| = \|\exp(-2\tau A)u_0 - v_2\| = O(\tau^2). \tag{10.26}$$

Remark 10.2.3 Taking into account (10.9), (10.26) and Remark 10.2.1 from inequality (10.21), it follows

$$\|u(t_k) - v_k\| = O(\tau^2),   k = 1, \ldots, n.$$

It is obvious we imply that the solution of problem (10.1), (10.2) is a smooth enough function.

10.2.4   Results of Numerical Experiments

Now we will consider results of numerical experiments of solving some *model problems* by the algorithm studied in this chapter.

For the beginning we consider the following simple problem

$$\frac{du(t)}{dt} + \alpha u(t) = f(t), \quad t \in \,]0, 1],$$

$$u(0) = u_0.$$

We will look for approximate solution v_k $(k = 1, \ldots, 10)$ of this problem by the algorithm (10.6)–(10.8). Errors

$$\Delta = \max_k |u(t_k) - v_k|$$

for some test solutions $u(t)$ are given below. Let s be the order of the considered correction ($s = 0$ corresponds to the solution $u_k^{(0)}$ and $s = 1$ corresponds to the specified solution $v_k = u_k^{(0)} + \tau u_k^{(1)}$).

1. $u(t) = 5t^2 - 4t$

$$\alpha = 2,$$
$$s = 0: \quad \Delta = 0.1045$$
$$s = 1: \quad \Delta = 0.0051.$$
$$\alpha = 0.001,$$
$$s = 0: \quad \Delta = 0.2373$$
$$s = 1: \quad \Delta = 0.0062.$$

2. $u(t) = 3\sin(2t)$

$$\alpha = 2,$$
$$s = 0: \quad \Delta = 0.1075$$
$$s = 1: \quad \Delta = 0.0032.$$
$$\alpha = 0.001,$$
$$s = 0: \quad \Delta = 0.2136$$
$$s = 1: \quad \Delta = 0.0017.$$

3. $u(t) = e^{3t}$

$$\alpha = 2,$$
$$s = 0: \quad \Delta = 0.8902$$
$$s = 1: \quad \Delta = 0.0473.$$

$$\alpha = 0.001,$$
$$s = 0: \quad \Delta = 1.4543$$
$$s = 1: \quad \Delta = 0.0292.$$

4. $u(t) = 10\sin(t)$

$$\alpha = 2,$$
$$s = 0: \quad \Delta = 0.0629$$
$$s = 1: \quad \Delta = 0.0021.$$
$$\alpha = 0.001,$$
$$s = 0: \quad \Delta = 0.1162$$
$$s = 1: \quad \Delta = 0.0014.$$

5. $u(t) = \sin(10t)$

$$\alpha = 2,$$
$$s = 0: \quad \Delta = 0.3379$$
$$s = 1: \quad \Delta = 0.0389.$$
$$\alpha = 0.001,$$
$$s = 0: \quad \Delta = 0.4597$$
$$s = 1: \quad \Delta = 0.0507.$$

Now consider the following problem

$$\frac{\partial u(x,t)}{\partial t} = \alpha \frac{\partial^2 u(x,t)}{\partial x^2} + f(x,t), \quad (x,t) \in]a, b[\times]0, T],$$
$$u(x,0) = \varphi(x),$$
$$u(a,t) = \alpha_0(t), \quad u(b,t) = \alpha_1(t).$$

We will also look for approximate solution $v_{i,k}$ of this problem by the algorithm (10.6)–(10.8). Errors

$$\Delta = \max_{i,k} |u(x_i, t_k) - v_{i,k}|$$

for some test solutions $u(x,t)$ (s -order of the considered correction, n-number of steps on x, m -number of steps on t) are given below.

1. $u(x,t) = 5x^2t - x^2t^2$, $\alpha = 0.1$, $T = 2.1$, $a = 1, b = 5$, $m = 10, n = 20$

$$s = 0: \quad \Delta = 0.0636$$

$$s = 1: \quad \Delta = 0.00043.$$

2. $u(x,t) = 3\sin(2xt^2)$, $\alpha = 2$, $T = 1.2$, $a = 1.2, b = 2.4$, $m = 10, n = 30$

$$s = 0: \quad \Delta = 0.2851$$

$$s = 1: \quad \Delta = 0.0475.$$

3. $u(x,t) = e^{xt}$, $\alpha = 2$, $T = 1.2$, $a = 1.2, b = 2.4$, $m = 10, n = 30$

$$s = 0: \quad \Delta = 0.1331$$

$$s = 1: \quad \Delta = 0.0102.$$

As one would expect, the numerical experiments showed that the second step of correction ($s = 1$) reduces an error by one order, that is $\|u(t_k) - (u_k^{(0)} + \tau u_k^{(1)})\| = O(\tau^2)$, while $\|u(t_k) - u_k^{(0)}\| = O(\tau)$.

The numerical experiments also showed that the further correction $u_k^{(2)}$ (see (10.6)–(10.8)) cannot practically reduce an approximate solution error. The reason is that the equation for looking for $u_k^{(2)}$ contains unknown R_k and finding of this unknown R_k is equivalent to definition of the solution of an initial problem. This is the reason why the algorithm is limited to two Eqs. (10.6)–(10.7).

10.3 The Case of Four-Layer Scheme

10.3.1 Reducing a Purely Implicit Four-Layer Scheme for an Evolutionary Problem to Two-Layer Schemes

On $[0, T]$ we introduce a grid $t_k = k\tau$, $k = 0, 1, \ldots, n$, with the step $\tau = T/n$. Using the approximation of the first derivative

$$\left.\frac{du}{dt}\right|_{t=t_k} = \frac{\frac{11}{6}u(t_k) - 3u(t_{k-1}) + \frac{3}{2}u(t_{k-2}) - \frac{1}{3}u(t_{k-3})}{\tau} + \tau^3 R_k(\tau, u), \quad R_k(\tau, u) \in H.$$

Equation (10.1) can be represented at the point $t = t_k$ as

$$\frac{\frac{11}{6}u(t_k) - 3u(t_{k-1}) + \frac{3}{2}u(t_{k-2}) - \frac{1}{3}u(t_{k-3})}{\tau} + Au(t_k) = -\tau^3 R_k(\tau, u), \quad k = 3, \ldots, n.$$

$$(10.27)$$

We write system (10.27) in the form

$$\frac{u(t_k) - u(t_{k-1})}{\tau} + Au(t_k)$$

$$+ \frac{\tau}{2} \left(\frac{u(t_k) - 2u(t_{k-1}) + u(t_{k-2})}{\tau^2} \right)$$

$$+ \frac{\tau^2}{3} \left(\frac{u(t_k) - 3u(t_{k-1}) + 3u(t_{k-2}) - u(t_{k-3})}{\tau^3} \right) = -\tau^3 R_k(\tau, u). \qquad (10.28)$$

It is obvious that expression in brackets in case of $\frac{\tau}{2}$ is $u''(t_{k-1}) + \tau^2 R_{1,k-1}$ and expression in brackets in case of $\frac{\tau^2}{3}$ is $u'''(t_k) + \tau R_{2,k}$, $R_{1,k}, R_{2,k} \in H$.

By analogy with the obtained system let us consider in the Hilbert space H the one-parametric family of equations

$$\frac{u_k - u_{k-1}}{\tau} + Au_k + \frac{\varepsilon}{2} \left(\frac{u_k - 2u_{k-1} + u_{k-2}}{\tau^2} \right)$$

$$+ \frac{\varepsilon^2}{3} \left(\frac{u_k - 3u_{k-1} + 3u_{k-2} - u_{k-3}}{\tau^3} \right) = \varepsilon^3 R_k, \quad R_k \in H. \qquad (10.29)$$

Assume that u_k is analytic with respect to ε,

$$u_k = \sum_{j=0}^{\infty} \varepsilon^j u_k^{(j)}. \qquad (10.30)$$

Substituting (10.30) into (10.29) and equating the members of identical powers ε, we obtain the following system of equations

$$\frac{u_k^{(0)} - u_{k-1}^{(0)}}{\tau} + Au_k^{(0)} = 0, \qquad u_0^{(0)} = u_0, \quad k = 1, \ldots, n, \qquad (10.31)$$

$$\frac{u_k^{(1)} - u_{k-1}^{(1)}}{\tau} + Au_k^{(1)} = -\frac{1}{2} \frac{\Delta^2 u_{k-2}^{(0)}}{\tau^2}, \qquad k = 2, \ldots, n, \qquad (10.32)$$

$$\frac{u_k^{(2)} - u_{k-1}^{(2)}}{\tau} + Au_k^{(2)} = -\frac{1}{2} \frac{\Delta^2 u_{k-2}^{(1)}}{\tau^2} - \frac{1}{3} \frac{\Delta^3 u_{k-3}^{(0)}}{\tau^3}, \qquad k = 3, \ldots, n, \qquad (10.33)$$

$$\frac{u_k^{(3)} - u_{k-1}^{(3)}}{\tau} + Au_k^{(3)} = -\frac{1}{2} \frac{\Delta^2 u_{k-2}^{(2)}}{\tau^2} - \frac{1}{3} \frac{\Delta^3 u_{k-3}^{(1)}}{\tau^3} + R_k,$$

. . . .

where $\Delta u_{k-1} = u_k - u_{k-1}$.

We introduce the notation

$$v_k = u_k^{(0)} + \tau u_k^{(1)} + \tau^2 u_k^{(2)}, \quad k = 3, \ldots, n. \tag{10.34}$$

Let the vector v_k be an approximate value of the exact solution of problem (10.1)–(10.2) for $t = t_k$, $u(t_k) \approx v_k$.

Note that in scheme (10.32) the starting vector $u_1^{(1)}$ is defined from the equality $v_1 = u_1^{(0)} + \tau u_1^{(1)}$, where $u_1^{(0)}$ is found by scheme (10.31), and v_1 is an approximate value of $u(t_1)$ with accuracy of $O(\tau^3)$. Similarly, the starting vector $u_2^{(2)}$ is defined from the equality $v_2 = u_2^{(0)} + \tau u_2^{(1)} + \tau^2 u_2^{(2)}$, where $u_2^{(0)}$ and $u_2^{(1)}$ are found by scheme (10.31), (10.32), respectively, and v_2 is an approximate value of $u(t_2)$ with accuracy of $O(\tau^3)$.

According to this assumption, we can write

$$\|v_1 - u(t_1)\| = O(\tau^3), \quad \|v_2 - u(t_2)\| = O(\tau^3).$$

Then, taking into account that for the solution of the problem (10.1), (10.2) the formula (see, e.g., Kato [21]) $u(t) = \exp(-tA)u_0$ is true, we can write

$$v_1 = \left(I - \tau A + \frac{\tau^2 A^2}{2} \right) u_0, \tag{10.35}$$

$$v_2 = \left(I - 2\tau A + 2\tau^2 A^2 \right) u_0. \tag{10.36}$$

From representation $v_1 = u_1^{(0)} + \tau u_1^{(1)}$, taking into account a formula (10.35) and equality $u_1^{(0)} = S u_0$, for the start vector $u_1^{(1)}$ we have

$$u_1^{(1)} = \frac{v_1 - u_1^{(0)}}{\tau} = \frac{1}{\tau} \left(I - \tau A + \frac{\tau^2 A^2}{2} - S \right) u_0 = \frac{1}{\tau} \left((I - S) - \tau A + \frac{\tau^2}{2} A^2 \right) u_0$$

$$= \frac{1}{\tau} \left(\tau A S - \tau A + \frac{\tau^2}{2} A^2 \right) u_0 = \frac{1}{\tau} \left(\tau A (S - I) + \frac{\tau^2}{2} A^2 \right) u_0$$

$$= \frac{1}{\tau} \left(-\tau^2 A^2 S + \frac{\tau^2}{2} A^2 \right) u_0 = \left(\frac{\tau}{2} A^2 - \tau A^2 S \right) u_0. \tag{10.37}$$

From equality $v_2 = u_2^{(0)} + \tau u_2^{(1)} + \tau^2 u_2^{(2)}$, taking into account the formulas (10.16) and (10.37), it follows

$$u_2^{(2)} = \frac{1}{\tau^2} \left(v_2 - u_2^{(0)} - \tau u_2^{(1)} \right) = \frac{1}{\tau^2} \left(v_2 - u_2^{(0)} - \tau \left(S u_1^{(1)} - \frac{\tau}{2} A^2 S^3 u_0 \right) \right)$$

$$= \frac{1}{\tau^2} \left(v_2 - u_2^{(0)} - \tau^2 \left(\frac{1}{2} A^2 S - A^2 S^2 - \frac{1}{2} A^2 S^3 \right) u_0 \right). \tag{10.38}$$

Transform the difference $v_2 - u_2^{(0)}$. Taking into account (10.36) and equation $u_2^{(0)} = S^2 u_0$, we have

$$
\begin{aligned}
v_2 - u_2^{(0)} &= \left((I - S^2) - 2\tau A + 2\tau^2 A^2 \right) u_0 = \left(\tau A S(I + S) - 2\tau A + 2\tau^2 A^2 \right) u_0 \\
&= \left(\tau A(I + S) - 2\tau A S^{-1} + 2\tau^2 A^2 S^{-1} \right) S u_0 \\
&= \left(\tau A(I + S) - 2\tau A \left(I - \tau^2 A^2 \right) \right) S u_0 \\
&= \left(\tau A(S - I) + 2\tau^3 A^3 \right) S u_0 \\
&= \left(2\tau^3 A^3 - \tau^2 A^2 S \right) S u_0.
\end{aligned}
\tag{10.39}
$$

Substituting (10.39) into (10.38), for the start vector $u_2^{(2)}$, we have

$$
\begin{aligned}
u_2^{(2)} &= A^2 \left(2\tau A - \frac{1}{2} I + \frac{1}{2} S^2 \right) S u_0 = \frac{1}{2} A^2 \left(4\tau A + (S - I)(S + I) \right) S u_0 \\
&= \frac{\tau}{2} A^3 \left(4I - S(S + I) \right) S u_0.
\end{aligned}
\tag{10.40}
$$

10.3.2 Estimate of Residual for Solution Obtained by Perturbation Algorithm

Let us estimate residual of purely implicit four-layer scheme for solutions of schemes (10.31)–(10.34).

Multiply Eqs. (10.32) and (10.33) by τ and τ^2, respectively, and add result to (10.31). We get that v_k is a solution of system of equations:

$$
\frac{v_k - v_{k-1}}{\tau} + A v_k = -\frac{\tau}{2} \frac{\Delta^2 u_{k-2}^{(0)}}{\tau^2} - \frac{\tau^2}{2} \frac{\Delta^2 u_{k-2}^{(1)}}{\tau^2} - \frac{\tau^2}{3} \frac{\Delta^3 u_{k-3}^{(0)}}{\tau^3}, \qquad k = 3, \ldots, n.
\tag{10.41}
$$

Rewrite this system in the form

$$
\frac{v_k - v_{k-1}}{\tau} + A v_k + \frac{\tau}{2} \frac{\Delta^2 v_{k-2}}{\tau^2} + \frac{\tau^2}{3} \frac{\Delta^3 v_{k-3}}{\tau^3} = \widetilde{R}_k(\tau),
\tag{10.42}
$$

where $k \geqslant 5$,

$$
\widetilde{R}_k(\tau) = \frac{\tau}{2} \frac{\Delta^2 v_{k-2}}{\tau^2} + \frac{\tau^2}{3} \frac{\Delta^3 v_{k-3}}{\tau^3} - \frac{\tau}{2} \frac{\Delta^2 u_{k-2}^{(0)}}{\tau^2} - \frac{\tau^2}{2} \frac{\Delta^2 u_{k-2}^{(1)}}{\tau^2} - \frac{\tau^2}{3} \frac{\Delta^3 u_{k-3}^{(0)}}{\tau^3}.
$$

It is obvious that (10.42) can be represented in the following form:

$$
\frac{\frac{11}{6}v_k - 3v_{k-1} + \frac{3}{2}v_{k-2} - \frac{1}{3}v_{k-3}}{\tau} + Av_k = \widetilde{R}_k(\tau), \quad k = 5, \dots, n. \tag{10.43}
$$

Therefore $\widetilde{R}_k(\tau)$ is a residual of purely implicit four-layer scheme for solutions of scheme (10.31)–(10.34) (see (10.27)).

It is obviously the representation

$$
\widetilde{R}_k(\tau) = \left(\frac{\tau}{2}\frac{\Delta^2 u_{k-2}^{(0)}}{\tau^2} + \frac{\tau^2}{2}\frac{\Delta^2 u_{k-2}^{(1)}}{\tau^2} + \frac{\tau^3}{2}\frac{\Delta^2 u_{k-2}^{(2)}}{\tau^2} \right)
$$

$$
+ \left(\frac{\tau^2}{3}\frac{\Delta^3 u_{k-3}^{(0)}}{\tau^3} + \frac{\tau^3}{3}\frac{\Delta^3 u_{k-3}^{(1)}}{\tau^3} + \frac{\tau^4}{3}\frac{\Delta^3 u_{k-3}^{(2)}}{\tau^3} \right)
$$

$$
- \frac{\tau}{2}\frac{\Delta^2 u_{k-2}^{(0)}}{\tau^2} - \frac{\tau^2}{2}\frac{\Delta^2 u_{k-2}^{(1)}}{\tau^2} - \frac{\tau^2}{3}\frac{\Delta^3 u_{k-3}^{(0)}}{\tau^3}
$$

$$
= \tau^3 \left(\frac{1}{2}\frac{\Delta^2 u_{k-2}^{(2)}}{\tau^2} + \frac{1}{3}\frac{\Delta^3 u_{k-3}^{(1)}}{\tau^3} + \frac{\tau}{3}\frac{\Delta^3 u_{k-3}^{(2)}}{\tau^3} \right). \tag{10.44}
$$

Note that representation (10.44) is true for $k > 4$.

We estimate the difference relations included in $\widetilde{R}_k(\tau)$.

Taking into account (10.16), we have

$$
\Delta^2 u_{k-2}^{(1)} = u_k^{(1)} - 2u_{k-1}^{(1)} + u_{k-2}^{(1)}
$$

$$
= \left(\tau^2 A^3 S^{k+1} - \frac{\tau^3}{2}(k-3)A^4 S^{k+1} \right) u_0 + \tau^2 A^2 S^{k-1} u_1^{(1)}. \tag{10.45}
$$

As $\Delta^3 u_{k-3}^{(1)} = \Delta^2 u_{k-2}^{(1)} - \Delta^2 u_{k-3}^{(1)}$, taking into account (10.45) we have

$$
\Delta^3 u_{k-3}^{(1)} = -\tau^3 \left(\left(A^4 S^{k+1} - \frac{k-3}{2}\tau A^5 S^{k+1} + \frac{1}{2}A^4 S^k \right) u_0 + A^3 S^{k-1} u_1^{(1)} \right).
$$

From here, taking into account that $u_1^{(1)}$ is uniformly bounded by τ (see (10.37)), for smooth enough initial data we obtain the estimation

$$
\left\| \frac{\Delta^3 u_{k-3}^{(1)}}{\tau^3} \right\| \leq c, \quad c = \text{const.} > 0. \tag{10.46}
$$

Estimate the difference relation $\frac{\Delta^2 u_{k-2}^{(2)}}{\tau^2}$. From (10.33) we have

$$u_k^{(2)} = S^{k-2} u_2^{(2)} + \tau \sum_{i=1}^{k-2} S^i w_{k-i+1}, \qquad (10.47)$$

where

$$w_k = -\frac{1}{2} \frac{\Delta^2 u_{k-2}^{(1)}}{\tau^2} - \frac{1}{3} \frac{\Delta^3 u_{k-3}^{(0)}}{\tau^3}. \qquad (10.48)$$

Taking into account (10.47) we have

$$
\begin{aligned}
\Delta^2 u_{k-2}^{(2)} &= u_k^{(2)} - 2u_{k-1}^{(2)} + u_{k-2}^{(2)} = \left(S^{k-2} - 2S^{k-3} + S^{k-4} \right) u_2^{(2)} \\
&\quad + \tau \left[S(w_k - w_{k-1}) + (S^2 - S)w_{k-1} + (S^3 - 2S^2 + S)w_{k-2} + \cdots \right. \\
&\quad \left. + \left(S^{k-2} - 2S^{k-3} + S^{k-4} \right) w_3 \right] \\
&= \tau^2 A^2 S^{k-2} u_2^{(2)} + \tau S(w_k - w_{k-1}) - \tau^2 A S^2 w_{k-1} \\
&\quad + \tau^3 \left(A^2 S^3 w_{k-2} + \cdots + A^2 S^{k-2} w_3 \right)
\end{aligned}
\qquad (10.49)
$$

Taking into account (10.15) we have

$$\Delta^3 u_{k-3}^{(0)} = \Delta^2 u_{k-2}^{(0)} - \Delta^2 u_{k-3}^{(0)} = \tau^2 A^2 S^k u_0 - \tau^2 A^2 S^{k-1} u_0 = -\tau^3 A^3 S^k u_0. \qquad (10.50)$$

From (10.48), taking into account (10.15), it follows

$$
\begin{aligned}
w_k - w_{k-1} &= -\frac{1}{2\tau^2} \left(\Delta^2 u_{k-2}^{(1)} - \Delta^2 u_{k-3}^{(1)} \right) - \frac{1}{3\tau^3} \left(\Delta^3 u_{k-3}^{(0)} - \Delta^3 u_{k-4}^{(0)} \right) \\
&= -\frac{\tau}{2} \frac{\Delta^3 u_{k-3}^{(1)}}{\tau^3} - \frac{1}{3\tau^3} \Delta^4 u_{k-4}^{(0)} = -\frac{\tau}{2} \frac{\Delta^3 u_{k-3}^{(1)}}{\tau^3} - \frac{1}{3\tau^3} \tau^2 A^2 \Delta^2 (S^{k-2} u_0) \\
&= -\frac{\tau}{2} \frac{\Delta^3 u_{k-3}^{(1)}}{\tau^3} - \frac{1}{3\tau^3} \tau^2 A^2 \Delta^2 u_{k-2}^{(0)} = -\frac{\tau}{2} \frac{\Delta^3 u_{k-3}^{(1)}}{\tau^3} - \frac{\tau}{3} A^4 S^k u_0.
\end{aligned}
\qquad (10.51)
$$

from (10.49), taking into account (10.51), we have

$$
\begin{aligned}
\frac{\Delta^2 u_{k-2}^{(2)}}{\tau^2} &= A^2 S^{k-2} u_2^{(2)} - \frac{1}{2} S \frac{\Delta^3 u_{k-3}^{(1)}}{\tau^3} - \frac{1}{3} A^4 S^{k+1} u_0 - AS^2 w_{k-1} \\
&\quad + \tau A^2 S^3 \left(w_{k-2} + S w_{k-3} + \cdots + S^{k-5} w_3 \right).
\end{aligned}
\qquad (10.52)
$$

Note that from (10.48), taking into account (10.45) and (10.50), follows bounded-ness w_k,

$$\|w_k\| \leqslant c, \qquad c = \text{const.} > 0. \tag{10.53}$$

From (10.52), taking into account that, $u_2^{(2)}$ is uniformly bounded by τ (see (10.40)), and taking into account (10.46) and (10.53), for smooth enough initial data we obtain the estimation

$$\left\| \frac{\Delta^2 u_{k-2}^{(2)}}{\tau^2} \right\| \leqslant c, \qquad c = \text{const.} > 0. \tag{10.54}$$

Furthermore, since

$$\Delta^3 u_{k-3}^{(2)} = \Delta^2 u_{k-2}^{(2)} - \Delta^2 u_{k-3}^{(2)},$$

then taking into account (10.54) we have

$$\tau \left\| \frac{\Delta^3 u_{k-3}^{(2)}}{\tau^3} \right\| \leqslant c, \qquad c = \text{const.} > 0. \tag{10.55}$$

Finally taking into account (10.46), (10.54), and (10.55), from (10.44) for residual $\widetilde{R}_k(\tau)$, follows the estimate

$$\left\| \widetilde{R}_k(\tau) \right\| \leqslant c\tau^3, \qquad c = \text{const.} > 0. \tag{10.56}$$

10.3.3   A Priori Estimate for the Approximate Solution Error

From (10.27) and (10.43), for error $z_k = u(t_k) - v_k$ we have:

$$\frac{\frac{11}{6} z_k - 3z_{k-1} + \frac{3}{2} z_{k-2} - \frac{1}{3} z_{k-3}}{\tau} + A z_k = r_k(\tau), \qquad k = 5, \ldots, n, \tag{10.57}$$

where $r_k(\tau) = -\left(\tau^3 R_k(\tau, u) + \widetilde{R}_k(\tau) \right)$.

Remark 10.3.1 Taking into account (10.56) we conclude that if the solution of problem (10.1)–(10.2) is smooth enough function, then $\|r_k(\tau)\| = O(\tau^3)$.

The following theorem is valid.

Theorem 10.3.2 *Let A be a self-adjoint positively defined operator in H. Then*

$$\|z_{k+2}\| \leqslant c \left(\|z_2\| + \|z_3\| + \|z_4\| + \tau \sum_{i=3}^{k} \|r_{i+2}(\tau)\| \right), \qquad c = \text{const.} > 0.$$

$$(10.58)$$

where $k = 3, \ldots, n-2$.

Proof From (10.57), we have

$$z_{k+1} = L_1 z_k + L_2 z_{k-1} + L_3 z_{k-2} + g_{k+1}. \qquad (10.59)$$

where

$$L_1 = \frac{18}{11}L, \quad L_2 = -\frac{9}{11}L, \quad L_3 = \frac{2}{11}L, \quad g_{k+1} = \frac{6}{11}\tau L r_{k+1}(\tau),$$

$$L = \left(I + \frac{6}{11}\tau A \right)^{-1}.$$

From here, by induction, we have (see [17, p. 59])

$$z_{k+2} = U_{k-2}z_4 + (L_2 U_{k-3} + L_3 U_{k-4})z_3 + L_3 U_{k-3}z_2 + \sum_{i=3}^{k} U_{k-i}g_{i+2}, \qquad (10.60)$$

where $U_k(L_1, L_2, L_3)$ are the operator polynomials, which are defined by the following recurrence relation:

$$U_k(L_1, L_2, L_3) = L_1 U_{k-1} + L_2 U_{k-2} + L_3 U_{k-3},$$

$$U_0 = I, \quad U_{-1} = U_{-2} = 0.$$

Consider the characteristic equation associated with the difference equation (10.59)

$$Q_1(\lambda) = \lambda^3 - \frac{18}{11}x\lambda^2 + \frac{9}{11}x\lambda - \frac{2}{11}x = 0, \qquad (10.61)$$

where $x \in Sp(L) \subset [0, 1]$.

If the real root of Eq. (10.61) is in the unit circle and complex roots are inside the unit circle then operator polynomials $U_k(L_1, L_2, L_3)$ are bounded in the aggregate (see [17], I, §3). Note that proving the above-mentioned result uses the well-known fact that the norm of the operator polynomial is equal to C norm of corresponding scalar polynomial on the spectrum of this operator when the argument is self-adjoint bounded operator (see, e.g., [22], ch. IX, §5).

We will show that there exists $\alpha > 1$ such that for any $x \in [0, 1)$, the real root of a characteristic equation (10.61) belongs to an interval $[\frac{2}{11}\alpha x, 1]$.

There exists $\alpha \in (1, \frac{11}{9})$ such that

$$Q_1\left(\frac{2}{11}\alpha x\right) < 0$$

for any $x \in (0, 1]$.

As

$$Q_1(1) = 1 - x > 0, \quad x \in [0, 1),$$

one of the roots of the characteristic equation (10.61) belongs to an interval $[\frac{2}{11}\alpha x, 1]$. Furthermore, we will show that remaining roots of this equation are complex.

Assume that λ_1 is a real root of the characteristic equation (10.61) and λ_2 and λ_3 are complex roots, $\lambda_3 = \overline{\lambda_2}$. By theorem of Vieta

$$|\lambda_1 \cdot \lambda_2 \cdot \overline{\lambda_2}| = \frac{2}{11}x,$$

from here

$$|\lambda_2|^2 = \frac{2x}{11\lambda_1} < \frac{1}{\alpha} < 1.$$

Obviously, when $x = 1$, a real root of the characteristic equation (10.61) is $\lambda_1 = 1$, and complex roots are λ_2 and λ_3 then

$$|\lambda_2|^2 = |\lambda_3|^2 = \frac{2}{11}.$$

We will show that the roots λ_2 and λ_3 of the characteristic equation (10.61) are complex.

It is known that if discriminant

$$D = -108\left(\frac{q^2}{4} + \frac{p^3}{27}\right),$$

where

$$q = \frac{2a^3}{27} - \frac{ab}{3} + c, \quad p = b - \frac{a^2}{3},$$

of cubic equation

$$\lambda^3 + a\lambda^2 + b\lambda + c = 0,$$

is negative, then the cubic equation has complex roots.

In our case we have:

$$p = \frac{9}{11}x\left(1 - \frac{12}{11}x\right),$$

$$q = -\frac{2^4 \cdot 3^3}{11^3}x^3 + \frac{2 \cdot 3^3}{11^2}x^2 - \frac{2}{11}x.$$

Obviously, for any $x \in [0, 1]$ inequalities:

$$p(x) = \frac{9}{11}x(1 - x) - \frac{9}{11^2}x^2 \geq -\frac{9}{11^2}x^2,$$

$$q(x) \leq -\frac{5}{11 \cdot 2^4}x.$$

are true.
From here:

$$p^3(x) \geq -\frac{9^3}{11^6}x^6,$$

$$q^2(x) \geq \frac{25}{11^2 \cdot 2^8}x^2, \quad x \in [0, 1].$$

According to these inequalities we have

$$D = -108\left(\frac{q^2}{4} + \frac{p^3}{27}\right) \leq -108\left(\frac{25}{11^2 \cdot 2^{10}}x^2 - \frac{9^3}{27 \cdot 11^6}x^6\right) < 0.$$

So one of the roots of the characteristic equation (10.61) is real and two other roots are complex. Thus the real root is on the unit circle and complex roots are inside the unit circle. From this follows that operators $U_k(L_1, L_2, L_3)$ are bounded in the aggregate (see [17], I, §3). If we consider this fact and we will pass to norms in (10.60) then we will have the estimate (10.3.2). □

10.3.4 Error Estimation for Vectors v_3, v_4 and Consequences of Theorem 10.3.2

Note that since the representation (10.44) is true only if $k > 4$, proof of the estimate (10.58) does not apply in the case of $k = 1, 2$, i.e. for a vector v_3, v_4. So it is necessary to separately estimate the error for vectors v_3, v_4.
At first we estimate the error $u(t_3) - v_3$.

From (10.41) follows

$$v_3 = Sv_2 + \tau S\varphi_1,$$ (10.62)

where

$$\varphi_1 = -\frac{\tau}{2}\frac{\Delta^2 u_1^{(0)}}{\tau^2} - \frac{\tau^2}{2}\frac{\Delta^2 u_1^{(1)}}{\tau^2} - \frac{\tau^2}{3}\frac{\Delta^3 u_0^{(0)}}{\tau^3}.$$ (10.63)

Taking into account (10.15), (10.45), (10.50), and (10.37) we have:

$$\Delta^2 u_1^{(0)} = \tau^2 A^2 S^3 u_0,$$

$$\Delta^2 u_1^{(1)} = \tau^2 A^3 S^4 u_0 + \tau^2 A^2 S^2 u_1^{(1)} = \tau^2 \left(A^3 S^4 + \frac{\tau}{2}A^4 S^2 - \tau A^4 S^3\right) u_0,$$

$$\Delta^3 u_0^{(0)} = -\tau^3 A^3 S^3 u_0.$$

Substituting these values in (10.63) we have

$$\varphi_1 = \left(-\frac{\tau}{2}A^2 S^3 + \frac{\tau^2}{3}A^3 S^3 - \frac{\tau^2}{2}A^3 S^4 - \frac{\tau^3}{2}\left(\frac{1}{2}A^4 S^2 - A^4 S^3\right)\right) u_0.$$ (10.64)

Substituting (10.36) in (10.64) and (10.62) we have

$$v_3 = S\left(I - 2\tau A + 2\tau^2 A^2 - \frac{\tau^2}{2}A^2 S^3\right) u_0 + \widetilde{r}_1(\tau)u_0, \quad \|\widetilde{r}_1(\tau)u_0\| = O(\tau^3).$$

As $S = I - \tau AS$, that

$$\frac{\tau^2}{2}A^2 S^3 u_0 = \frac{\tau^2}{2}A^2 u_0 + \widetilde{r}_2(\tau)u_0, \quad \|\widetilde{r}_2(\tau)u_0\| = O(\tau^3).$$

Therefore for v_3 we have

$$v_3 = S\left(I - 2\tau A + \frac{3\tau^2}{2}A^2\right) u_0 + \widetilde{r}_3(\tau)u_0, \quad \|\widetilde{r}_3(\tau)u_0\| = O(\tau^3).$$ (10.65)

Taking into account the expansion

$$S = I - \tau A + \tau^2 A^2 - \tau^3 A^3 S,$$

from (10.65) follows

$$v_3 = \left(I - \tau A + \tau^2 A^2\right)\left(I - 2\tau A + \frac{3\tau^2}{2}A^2\right) u_0 + \widetilde{r}_4(\tau)u_0, \quad \|\widetilde{r}_4(\tau)u_0\| = O(\tau^3).$$

Obviously, from here it follows

$$v_3 = \left(I - 3\tau A + \frac{9\tau^2}{2}A^2\right)u_0 + \widetilde{r}_5(\tau)u_0, \quad \|\widetilde{r}_5(\tau)u_0\| = O(\tau^3).$$

Therefore for error $u(t_3) - v_3$ we have

$$\|u(t_3) - v_3\| = \|\exp(-3\tau A)u_0 - v_3\| = O(\tau^3). \tag{10.66}$$

We will pass to an error assessment $u(t_4) - v_4$.
From (10.41) it follows

$$v_4 = Sv_3 + \tau S\varphi_2, \tag{10.67}$$

where

$$\varphi_2 = -\frac{\tau}{2}\frac{\Delta^2 u_2^{(0)}}{\tau^2} - \frac{\tau^2}{2}\frac{\Delta^2 u_2^{(1)}}{\tau^2} - \frac{\tau^2}{3}\frac{\Delta^3 u_1^{(0)}}{\tau^3}. \tag{10.68}$$

Taking into account (10.15), (10.45), (10.50), and (10.37) we have:

$$\Delta^2 u_2^{(0)} = \tau^2 A^2 S^4 u_0,$$

$$\Delta^2 u_2^{(1)} = \left(\tau^2 A^3 S^5 - \frac{\tau^3}{4}A^4 S^5\right)u_0 + \tau^3 A^2 S^3 \left(\frac{1}{2}A^2 - A^2 S\right)u_0,$$

$$\Delta^3 u_1^{(0)} = -\tau^3 A^3 S^4 u_0.$$

Substituting these values in (10.68) we have

$$\varphi_2 = -\frac{\tau}{2}A^2 S^4 u_0 + \widetilde{r}_6(\tau)u_0, \quad \|\widetilde{r}_6(\tau)u_0\| = O(\tau^2).$$

Substituting values of v_3 and φ_2 in (10.67) we have

$$v_4 = S\left(I - 3\tau A + \frac{9\tau^2}{2}A^2 - \frac{\tau^2}{2}A^2 S^4\right)u_0 + \widetilde{r}_7(\tau)u_0, \quad \|\widetilde{r}_7(\tau)u_0\| = O(\tau^3).$$

Furthermore, arguing as in the previous case, we obtain

$$v_4 = \left(I - 4\tau A + 8\tau^2 A^2\right)u_0 + \widetilde{r}_8(\tau)u_0, \quad \|\widetilde{r}_8(\tau)u_0\| = O(\tau^3).$$

Therefore for error $u(t_4) - v_4$ we have

$$\|u(t_4) - v_4\| = \|\exp(-4\tau A) - v_4\| = O(\tau^3). \tag{10.69}$$

Theorem 10.3.3 *Let A be a self-adjoint positively defined operator in H and let the solution of problem* (10.1)–(10.2) *be a smooth enough function. If* $\|u(t_k) - v_k\| = O(\tau^3)$, $k = 1, 2$, *then the estimate is true*

$$\|u(t_k) - v_k\| = O(\tau^3), \qquad k = 3, \dots, n. \tag{10.70}$$

At this point, for $k = 3, 4$ we already proved the estimate (10.70) provided that the solution of problem (10.1)–(10.2) is smooth enough function and $\|u(t_k) - v_k\| = O(\tau^3)$, $k = 1, 2$. The estimate (10.70) for $k > 4$ follows from a priori estimate (10.58).

Acknowledgements J. Rogava and D. Gulua were supported by the European Union's Seventh Framework Programme (FP7/2007–2013) under grant agreement no. 317721.

References

1. E. Rothe, Über die Wärmeleitungsgleichung mit nichtkonstanten Koeffizienten im räumlichen Falle. Math. Ann. **104**(1), 340–354 (1931)
2. G.I. Marchuk, V.I. Agoshkov, *Introduction to Projection-Grid Methods* (Nauka, Moscow, 1981)
3. S.G. Mikhlin, *Numerical Realization of Variational Methods*. With an appendix by T.N. Smirnova (Izdat. Nauka, Moscow, 1966)
4. G. Streng, J. Fiks, in *Theory of the Finite Element Method*, ed. by G.I. Marchuk. Translated from the English by V.I. Agoshkov, V.A. Vasilenko, and V.V. Shaĭdurov (Izdat. Mir, Moscow, 1977)
5. S.K. Godunov, V.S. Ryabenki, *Difference Schemes* (Nauka, Moscow, 1973)
6. G.I. Marchuk, *Methods of Computational Mathematics* (Nauka, Moscow, 1977)
7. R. Richtmayer, K. Morton, *Difference Methods of Solution of Initial-Value Problems* (Mir, Moscow, 1972)
8. A.A. Samarskiĭ, *Theory of Difference Schemes* (Nauka, Moscow, 1977)
9. N.N. Ianenko, *A Method of Fractional Steps of Solution of Multi-Dimensional Problems of Mathematical Physics* (Nauka, Novosibirsk, 1967)
10. M. Crouzeix, Une methode multipas implicite-explicite pour l'approximation des équations d'évolution paraboliques. Numer. Math. **35**(3), 257–276 (1980)
11. M. Crouzeix, P.-A. Raviart, Approximation des équations d'évolution linéaires par des méthodes à pas multiples. C. R. Acad. Sci. Paris **283**, A367–A370 (1976)
12. M.-N. Le Roux, Semidiscretization in time for parabolic problems. Math. Comput. **33**(147), 919–931 (1979)
13. V.I. Agoshkov, D.V. Gulua, A perturbation algorithm for realization of finite-dimensional realization problems. Comput. Math. Dept. USSR Acad. Sci., Moscow, 35 pp. (1990). Preprint 253
14. G.I. Marchuk, V.I. Agoshkov, V.P. Shutyaev, *Adjoint Equations and Perturbation Methods in Nonlinear Problems of Mathematical Physics* (VO Nauka, Moscow, 1993)
15. J.L. Rogava, D.V. Gulua, Perturbation algorithm for implementing a finite-difference approximation to an abstract evolutionary problem and explicit error estimation of its solution. Dokl. Math. **89**(3), 335–337 (2014)
16. J.L. Rogava, D.V. Gulua, Reduction of a three-laver semi-discrete scheme for an abstract parabolic equation to two-layer schemes. Explicit estimates for the approximate solution error. J. Math. Sci. **206**(4), 424–444 (2015)

17. J.L. Rogava, *Semidiscrete Schemes for Operator Differential Equations* (Technical University Press, Tbilisi, 1995)
18. G.I. Marchuk, V.V. Shaĭdurov, *Difference Methods and Their Extrapolations* (Springer, Berlin, 1983)
19. V. Pereyra, On improving an approximate solution of a functional equation by deferred corrections. Numer. Math. **8**(3), 376–391 (1966)
20. V. Pereyra, Accelerating the convergence of discretization algorithms. SIAM J. Numer. Anal. **4**, 508–533 (1967)
21. T. Kato, in *Perturbation Theory for Linear Operators*, ed. by V.P. Maslov. Translated from the English by G.A. Voropaeva, A.M. Stepin and I.A. Sismarev (Izdat. Mir, Moscow, 1972)
22. L.V. Kantorovich, G.P. Akilov, *Functional Analysis* (VO Nauka, Moscow, 1977)

Chapter 11
Hilbert's Fourth Problem and Projectively Flat Finsler Metrics

Xinyue Cheng, Xiaoyu Ma, Yuling Shen, and Shuhua Liu

Abstract This survey article mainly introduces some important research progresses on the smooth solutions of Hilbert's fourth problem in the regular case, which we call projectively flat Finsler metrics. We characterize and classify projectively flat Finsler metrics of constant flag curvature. We also discuss and classify projectively flat Finsler metrics with isotropic S-curvature. In particular, we study and characterize projectively flat Randers metrics, square metrics, (α, β)-metrics, and their curvature properties.

11.1 Hilbert's Fourth Problem

In 1900, David Hilbert gave his famous speech about mathematical problems [18]. Among Hilbert's 23 problems, the fourth problem is "Problem of the straight line as the shortest distance between two points." Concretely, Hilbert's fourth problem asks to construct and study all metrics on an open subset in $\mathbb{R}^n$ such that straight line segments are the shortest paths joining any two points. Busemann called these metrics *projective metrics* and proposed an integral-geometric construction which is inspiringly simple (see [8]). Further, Pogorelov proved that every projective metric that is sufficiently smooth is of the metric form defined by Busemann's construction [21]. However, Álvarez Paiva pointed out in [1] that "a large number of projective metrics cannot be constructed by taking positive measures on the space of hyperplanes and it is not clear how to construct all quasi-positive measures" and "Busemann's construction, by itself, does not shed much light on the properties of projective metric spaces." On the other hand, all of the researches in [8] and [1, 21] are just for the absolutely homogeneous metrics, which means that the distance functions determined by the metrics are symmetric. Therefore, it is necessary to complement the results of Busemann and Pogorelov with other characterizations of

X. Cheng (✉) • X. Ma • Y. Shen • S. Liu
School of Mathematics and Statistics, Chongqing University of Technology, Chongqing 400054, People's Republic of China
e-mail: chengxy@cqut.edu.cn; 1066719732@qq.com; sylyiyi@163.com; 12612954@qq.com

© Springer International Publishing AG 2017
G. Falcone (ed.), *Lie Groups, Differential Equations, and Geometry*,
UNIPA Springer Series, DOI 10.1007/978-3-319-62181-4_11

projective metric spaces that either yield new explicit examples or shed more light on the general properties of these spaces.

Because any intrinsic quasimetric induces a Finsler metric and projective metrics that are sufficiently smooth are Finsler metrics, Hilbert's fourth problem in the regular case is to study the Finsler metrics with straight lines as their geodesics. We call the smooth solutions of Hilbert's fourth problem in the regular case *projectively flat Finsler metrics*.

It is easy to see that a Finsler metric $F = F(x, y)$ on an open subset $\mathcal{U} \subset \mathbb{R}^n$ is projectively flat if and only if the spray coefficients are in the following form

$$G^i = Py^i,$$

where $P = P(x, y)$ satisfies $P(x, \lambda y) = \lambda P(x, y), \forall \lambda > 0$ and G^i are defined by

$$G^i = \frac{1}{4} g^{il} \left\{ [F^2]_{x^m y^l} y^m - [F^2]_{x^l} \right\}. \tag{11.1}$$

We call G^i the *geodesic coefficients* of F. In 1903, G. Hamel found a system of partial differential equations that characterize projectively flat metrics $F = F(x, y)$ on an open subset $\mathcal{U} \subset \mathbb{R}^n$, that is,

$$F_{x^m y^i} y^m = F_{x^i}, \tag{11.2}$$

see [17]. A natural problem is to find projectively flat metrics by solving (11.2). According to the Beltrami Theorem, a Riemannian metric $F = \sqrt{g_{ij}(x) y^i y^j}$ is projectively flat if and only if it is of constant sectional curvature. Thus, Hilbert's fourth problem has been solved in Riemannian geometry. However, this problem is far from being solved for Finsler metrics.

11.2   Preliminaries

A *Finsler metric* on a C^∞ manifold M is a function $F : TM \rightarrow [0, \infty)$ with the following properties:

(1) Smoothness: $F(x, y)$ is C^∞ on $TM \backslash \{0\}$;
(2) Homogeneity: $F(x, \lambda y) = \lambda F(x, y), \quad \forall \lambda > 0$;
(3) Regularity/Convexity: $\left(g_{ij}(x, y) \right)$ is positive definite, where

$$g_{ij}(x, y) := \frac{1}{2} [F^2]_{y^i y^j}(x, y).$$

For a Finsler manifold (M, F) and for each $y \in T_x M \setminus \{0\}$, we can define an inner product $g_y : T_x M \times T_x M \rightarrow \mathbf{R}$:

$$g_y(u, v) = g_{ij}(x, y) u^i v^j,$$

where $u = u^i \frac{\partial}{\partial x^i}|_x$, $v = v^j \frac{\partial}{\partial x^j}|_x$. By the homogeneity of F, we have

$$F(x, y) = \sqrt{g_{ij}(x, y)y^i y^j}.$$

Remark 11.2.1 The following are some special Finsler metrics.

(1) Minkowski metric: $F(x, y) = \sqrt{g_{ij}(y)y^i y^j}$. In this case, $F(x, y)$ is independent of the position $x \in M$.
(2) Riemann metric: $\alpha(x, y) = \sqrt{a_{ij}(x)y^i y^j}$, where a_{ij} are independent of the direction $y \in T_x M$.
(3) Randers metric [22]: $F = \alpha + \beta$, where $\alpha = \sqrt{a_{ij}(x)y^i y^j}$ denotes a Riemannian metric and $\beta = b_i(x)y^i$ denotes a 1-form with $\|\beta\|_\alpha(x) := \sqrt{a^{ij}(x)b_i(x)b_j(x)} < 1$, $\forall x \in M$. Randers metrics were first introduced by physicist G. Randers in 1941 from the standpoint of general relativity. Randers metrics can also be naturally characterized as the solution of the Zermelo navigation problem. It is easy to show that a Finsler metric F is a Randers metric if and only if it is the solution of Zermelo navigation problem on a Riemann space (M, h) under the influence of a force field W with $|W|_h < 1$, where $|W|_h$ denotes the length of W with respect to Riemannian metric h. We call the couple (h, W) the *navigation data* of Randers metric F. This fact shows that it is unavoidable to meet non-Riemannian Finsler metrics in studying natural sciences. For more details about Randers metrics, see [13].
(4) (α, β)-metrics: $F = \alpha\phi(s)$, $s = \beta/\alpha$, where β satisfies $\|\beta_x\|_\alpha < b_0$ (or $\|\beta_x\|_\alpha \leq b_0$), $\forall x \in M$ and $\phi(s)$ is a C^∞ function on $(-b_0, b_0)$ satisfying

$$\phi(s) > 0, \quad \phi(s) - s\phi'(s) + (b^2 - s^2)\phi''(s) > 0, \quad (|s| \leq b < b_0).$$

Such metrics are called *regular* (or *almost regular*) (α, β)-metrics [20]. An (α, β)-space can be considered as a "perturbed Riemannian space" by an external force.

(α, β)-Metrics form a very important class of Finsler metrics which contains many important and interesting metrics. When $\phi = 1$, we get the Riemannian metric $F = \alpha$. If $\phi = 1 + s$, the (α, β)-metric $F = \alpha + \beta$ is just a Randers metric. More generally, if $\phi = \sqrt{1 + ks^2} + \epsilon s$, the (α, β)-metrics $F = \sqrt{\alpha^2 + k\beta^2} + \epsilon\beta$ are called (α, β)-metrics of Randers type. When $\phi = \frac{1}{1-s}$, we obtain the Matsumoto metric $F = \frac{\alpha^2}{\alpha-\beta}$. The square metric in the form $F = \frac{(\alpha+\beta)^2}{\alpha}$ is defined by $\phi = (1 + s)^2$. The famous Berwald's metric defined by L. Berwald in 1929 is an important class of square metrics which is positively complete and projectively flat Finsler metric with $\mathbf{K} = 0$ (see Sect. 11.3 or [4, 15]).

The notion of projective flatness is closely connected with the curvature properties of Finsler metrics. Let $\sigma = \sigma(t)$ $(a \leq t \leq b)$ be a geodesic on a Finsler manifold (M, F). Let $H(t, s)$ be a variation of σ such that each curve $\sigma_s(t) := H(t, s)$ $(a \leq t \leq b)$ is a geodesic. Let

$$J(t) := \frac{\partial H}{\partial s}(t, 0).$$

Then the vector field $J(t)$ is a *Jacobi field* along σ satisfying the Jacobi equation:

$$D_{\dot{\sigma}} D_{\dot{\sigma}} J(t) + \mathbf{R}_{\dot{\sigma}}\big(J(t)\big) = 0.$$

Here, $\mathbf{R}$ denotes the Riemann curvature of F. Locally, for any $x \in M$ and $y \in T_x M \backslash \{0\}$, the *Riemann curvature* $\mathbf{R}_y = R^i{}_k \frac{\partial}{\partial x^i} \otimes dx^k$ is defined by

$$R^i{}_k = 2\frac{\partial G^i}{\partial x^k} - \frac{\partial^2 G^i}{\partial x^m \partial y^k} y^m + 2G^m \frac{\partial^2 G^i}{\partial y^m \partial y^k} - \frac{\partial G^i}{\partial y^m}\frac{\partial G^m}{\partial y^k}. \tag{11.3}$$

The flag curvature of (M, F) is the function $\mathbf{K} = \mathbf{K}(x, y, \Pi)$ of a two-dimensional plane called "flag" $\Pi \subset T_x M$ and a "flagpole" $y \in \Pi \backslash \{0\}$ defined by

$$\mathbf{K}(x, y, \Pi) := \frac{g_y(\mathbf{R}_y(u), u)}{g_y(y, y)g_y(u, u) - \big[g_y(u, y)\big]^2},$$

where $\Pi = \text{span}\{y, u\}$. The flag curvature is the natural generalization of the sectional curvature in Riemannian geometry. When F is a Riemannian metric, $\mathbf{K}(x, y, \Pi) = \mathbf{K}(x, \Pi)$ is independent of the flagpole y and is just the sectional curvature. A Finsler metric F is said to be of *scalar flag curvature* if the flag curvature is independent of the flag Π, $\mathbf{K} = \mathbf{K}(x, y)$. F is said to be of *weakly isotropic flag curvature* if

$$\mathbf{K} = \frac{3\theta}{F} + \sigma(x),$$

where $\sigma = \sigma(x)$ is a scalar function and θ is a 1-form on M. When $\mathbf{K} = $ constant, F is said to be of *constant flag curvature*. The flag curvature governs the Jacobi equation and the second variation of length.

A fundamental fact is that projectively flat Finsler metrics with $G^i = P(x, y)y^i$ must be of scalar flag curvature,

$$\mathbf{K} = \frac{P^2 - P_{x^k}y^k}{F^2}.$$

Another fundamental fact is that Beltrami Theorem is no longer true in Finsler geometry. Let us look at an example.

Example 11.2.1 ([24] Shen's Fish Tank) Let $\Omega \subset R^3$ be an open domain given by

$$\Omega := \{(x^1, x^2, x^3)|(x^1)^2 + (x^2)^2 < 1\}.$$

For each $x = (x^1, x^2, x^3) \in \Omega$ and $y = \{y^1, y^2, y^3\} \in T_p\Omega$, define a Randers metric on Ω as

$$F = \alpha + \beta,$$

where

$$\alpha := \frac{\sqrt{[-x^2y^1 + x^1y^2]^2 + [(y^1)^2 + (y^2)^2 + (y^3)^2][1 - (x^1)^2 - (x^2)^2]}}{1 - (x^1)^2 - (x^2)^2}, \quad (11.4)$$

$$\beta := -\frac{-x^2y^1 + x^1y^2}{1 - (x^1)^2 - (x^2)^2}. \quad (11.5)$$

It has been proved that $\mathbf{K} = 0$. However, by Bácsó–Matsumoto theorem in [2], F is not projectively flat Finsler metric because β is not closed.

In Finsler geometry, there are some important quantities which all vanish for Riemannian metrics. Hence they are said to be *non-Riemannian*. Let F be a Finsler metric on an n-dimensional manifold M. Let $\{\mathbf{b}_i\}$ be a basis for T_xM and $\{\omega^i\}$ be the basis for T_x^*M dual to $\{\mathbf{b}_i\}$. Define the Busemann–Hausdorff volume form by

$$dV_{\mathrm{BH}} := \sigma_{\mathrm{BH}}(x)\omega^1 \wedge \cdots \wedge \omega^n,$$

where

$$\sigma_{\mathrm{BH}}(x) := \frac{\mathrm{Vol}\big(\mathbf{B}^n(1)\big)}{\mathrm{Vol}\big\{(y^i) \in R^n | F(x, y) < 1\big\}}.$$

Here $\mathrm{Vol}\{\cdot\}$ denotes the Euclidean volume function on subsets in $\mathbb{R}^n$ and $\mathbf{B}^n(1)$ denotes the unit ball in $\mathbb{R}^n$.

If $F = \sqrt{g_{ij}(x)y^iy^j}$ is a Riemannian metric, then

$$\sigma_{\mathrm{BH}}(x) = \sqrt{\det(g_{ij}(x))}.$$

However, in general, for a Finsler metric F, $\sigma_{\mathrm{BH}}(x) \neq \sqrt{\det(g_{ij}(x, y))}$. Define

$$\tau(x, y) := \ln\left[\frac{\sqrt{\det(g_{ij}(x, y))}}{\sigma_{\mathrm{BH}}(x)}\right].$$

$\tau = \tau(x, y)$ is well defined, which is called the *distortion* of F. The distortion τ characterizes the geometry of tangent space $(T_x M, F_x)$. It is well known that a Finsler metric F is Riemannian if and only if $\tau(x, y) = 0$.

It is natural to study the rate of change of the distortion along geodesics. For a vector $y \in T_x M \setminus \{0\}$, let $\sigma = \sigma(x)$ be the geodesic with $\sigma(0) = x$ and $\dot{\sigma}(0) = y$. Put

$$\mathbf{S}(x, y) := \frac{d}{dt} [\tau(\sigma(t), \dot{\sigma}(t))]\,|_{t=0}.$$

Equivalently,

$$\mathbf{S}(x, y) := \tau_{;m}(x, y) y^m,$$

where ";" denotes the horizontal covariant derivative with respect to F [15]. $\mathbf{S} = \mathbf{S}(x, y)$ is called the *S-curvature* of Finsler metric F [23].

The S-curvature $\mathbf{S}(x, y)$ measures the rate of change of $(T_x M, F_x)$ in the direction $y \in T_x M$. As we know, for any piecewise C^∞ curve $c = c(t)$ from p to q in any Berwald manifold (M, F), the parallel translation P_c is a linear isometry between $(T_p M, F_p)$ and $(T_q M, F_q)$. Hence, it is easy to see that, for any Berwald metrics, $\mathbf{S} = 0$. In particular, $\mathbf{S} = 0$ for Riemannian metrics [15, 23]. Hence, S-curvature is a non-Riemannian quantity.

We say that F is of *isotropic S-curvature* if there exists a scalar function $c(x)$ on M such that $\mathbf{S}(x, y) = (n + 1)c(x)F(x, y)$, equivalently,

$$\frac{\tau_{;m}(x, y) y^m}{F(x, y)} = (n + 1)c(x). \tag{11.6}$$

Equation (11.6) means that the rate of change of the tangent space $(T_x M, F_x)$ along the direction $y \in T_x M$ at each $x \in M$ is independent of the direction y but just dependent on the point x. If $c(x) =$constant, we say that F has *constant S-curvature*.

11.3 Projectively Flat Finsler Metrics of Constant Flag Curvature

Firstly, let us recall three important kinds of projectively flat Finsler metrics.

1. **Funk metric** [16]. Let $\Theta = \Theta(x, y)$ be a Finsler metric on a strongly convex domain $\Omega \subset \mathbb{R}^n$ satisfying

$$x + \frac{y}{\Theta(x, y)} \in \partial\Omega.$$

Θ is called a *Funk metric*. Funk metric has the following important property:

$$\Theta_{x^k} = \Theta\Theta_{y^k}. \tag{11.7}$$

From this, we know that Θ is a projectively flat Finsler metric with constant flag curvature $\mathbf{K} = -\frac{1}{4}$. Further, Θ is of constant S-curvature, $\mathbf{S} = \frac{n+1}{2}\Theta$. When $\Omega = \mathbf{B}^n(1)$, $\Theta(x, y) = \bar{\alpha} + \bar{\beta}$ is actually a Randers metric given by

$$\Theta(x, y) = \frac{\sqrt{(1 - |x|^2)|y|^2 + <x, y>^2}}{1 - |x|^2} + \frac{<x, y>}{1 - |x|^2}. \tag{11.8}$$

More generally, a Finsler metric $\Theta = \Theta(x, y)$ satisfying (11.7) on an open subset $\Omega \subset \mathbb{R}^n$ is called a Funk metric.

2. **Randers metric** [2]. In 1997, S. Bácsó and M. Matsumoto proved that a Randers metric $F = \alpha + \beta$ is projectively flat if and only if α is of constant sectional curvature (that is, α is projectively flat) and β is closed.

3. **Berwald's metric** [4]. Let $\mathbf{B}^n(1) \subset \mathbb{R}^n$ be the standard unit ball. Define

$$F = \frac{(\alpha + \beta)^2}{\alpha}, \quad y \in T_x\mathbf{B}^n(1), \tag{11.9}$$

where $\alpha = \lambda\bar{\alpha}$, $\beta = \lambda\bar{\beta}$ and

$$\bar{\alpha} = \frac{\sqrt{(1 - |x|^2)|y|^2 + <x, y>^2}}{1 - |x|^2}, \quad \bar{\beta} = \frac{<x, y>}{1 - |x|^2},$$

$$\lambda = \frac{1}{1 - |x|^2}.$$

The Finsler metric in (11.9) is constructed by Berwald [4]. It is easy to verify that $F = F(x, y)$ satisfies the following equations

$$F_{x^k y^l} y^k = F_{x^l}$$

and $F_{x^k} y^k = 2\Theta F$, where $\Theta = \Theta(x, y)$ is the Funk metric on $\mathbf{B}^n(1)$ defined by (11.8). Therefore, we can conclude that F is projectively flat and has zero flag curvature, $\mathbf{K} = 0$ (see [15]). The Finsler metrics in the form (11.9) are called *square metrics*.

11.3.1 Projectively Flat Randers Metrics of Constant Flag Curvature

As we mentioned before, a Randers metric $F = \alpha + \beta$ is projectively flat if and only if α is of constant sectional curvature and β is closed. In this case, α is locally isometric to the following metric defined on a ball $\mathbf{B}^n(r_\mu) \subset \mathbb{R}^n$,

$$\alpha_\mu(x, y) = \frac{\sqrt{|y|^2 + \mu(|x|^2|y|^2 - \langle x, y \rangle^2)}}{1 + \mu|x|^2}.$$

Moreover, if F is of constant flag curvature, β can be completely determined.

Theorem 11.3.1 ([25]) *Let $F = \alpha + \beta$ be an n-dimensional Randers metric of constant Ricci curvature* **Ric** $= (n-1)\sigma F^2$ *with* $\beta \not\equiv 0$. *Suppose that F is locally projectively flat. Then* $\sigma \leqslant 0$. *Further, if* $\sigma = 0$, *F is locally Minkowskian. If* $\sigma = -1/4$, *F can be expressed in the following form*

$$F = \frac{\sqrt{|y|^2 - (|x|^2|y|^2 - \langle x, y\rangle^2)}}{1 - |x|^2} \pm \frac{\langle x, y\rangle}{1 - |x|^2} \pm \frac{\langle a, y\rangle}{1 + \langle a, x\rangle}, \quad y \in T_x\mathbb{R}^n, \quad (11.10)$$

where $a \in \mathbb{R}^n$ is a constant vector with $|a| < 1$. The Randers metric in (11.10) has the following properties:

(a) $\mathbf{K} = -1/4$;
(b) $\mathbf{S} = \pm\frac{1}{2}(n+1)F$;
(c) *all geodesics of F are straight lines.*

The Finsler metrics defined by (11.10) can be expressed as follows

$$\Theta_{\pm a} := \Theta(x, y) \pm \frac{<a, y>}{1 + <a, x>}, \quad (11.11)$$

where $\Theta = \Theta(x, y)$ is Funk metric on $\mathbf{B}^n(1)$ given by (11.8). Theorem 11.3.1 is the first classification theorem in Finsler geometry without the restriction that the manifold is closed.

11.3.2 Projectively Flat Square Metrics of Constant Flag Curvature

Berwald's metric in (11.9) is the first projectively flat square metric of constant flag curvature. In 2008, by Hamel equation (11.2), Z. Shen and G.C. Yildirim first proved the following

Theorem 11.3.2 ([28]) *Let $F = (\alpha + \beta)^2/\alpha$ be a Finsler metric on a manifold M. F is projectively flat if and only if*

(i) $b_{i|j} = \tau\{(1 + 2b^2)a_{ij} - 3b_ib_j\}$,
(ii) the spray coefficients G^i_α of α are in the form: $G^i_\alpha = \theta y^i - \tau\alpha^2 b^i$,

where $b := \|\beta_x\|_\alpha$, $b_{i|j}$ denote the covariant derivatives of β with respect to α, $\tau = \tau(x)$ is a scalar function and $\theta = a_i(x)y^i$ is a 1-form on M.

There are plenty of non-trivial Finsler metrics satisfying the conditions (i) and (ii) in Theorem 11.3.2. By Theorem 11.3.2, we can completely determine the local structure of a projectively flat Finsler metric F in the form $F = (\alpha + \beta)^2/\alpha$ which is of constant flag curvature.

Theorem 11.3.3 ([28]) *Let* $F = (\alpha + \beta)^2/\alpha$ *be a square Finsler metric on a manifold M. Then F is locally projectively flat with constant flag curvature if and only if one of the following conditions holds*

(a) *α is flat and β is parallel with respect to α. In this case, F is locally Minkowskian;*

(b) *Up to a scaling, F is isometric to the following metric F_a for any constant vector $a \in \mathbb{R}^n$ with $|a| < 1$:*

$$F_a := \frac{(\lambda_a \bar{\alpha} + \lambda_a \bar{\beta}_a)^2}{\lambda_a \bar{\alpha}}, \qquad (11.12)$$

where

$$\bar{\alpha} = \frac{\sqrt{(1 - |x|^2)|y|^2 + <x, y>^2}}{1 - |x|^2}, \quad \bar{\beta}_a = \frac{<x, y>}{1 - |x|^2} + \frac{<a, y>}{1 + <a, x>}$$

and

$$\lambda_a := \frac{(1 + <a, x>)^2}{1 - |x|^2}.$$

In both cases (a) and (b), the flag curvature of F must be zero, $\mathbf{K} = 0$.

Recently, Zhou has proved that, if a square metric F has constant flag curvature $\mathbf{K}$, then F must be locally projectively flat Finsler metric [30]. In this case, according to Theorem 11.3.3, $\mathbf{K} = 0$.

11.3.3 Projectively Flat (α, β)-Metrics of Constant Flag Curvature

In Finsler geometry, it is very difficult in general to compute the curvatures of a Finsler metric. Some Finsler metrics are defined by some elementary functions, but their expressions of curvatures are extremely complicated so that one cannot easily determine their values. (α, β)-Metrics are relatively simple and "computable" Finsler metrics with interesting curvature properties. In recent years, various curvatures in Finsler geometry have been studied and their geometric meanings are better understood. This is partially due to the study of (α, β)-metrics.

In [27], Shen has characterized all projectively flat (α, β)-metrics.

Theorem 11.3.4 ([27]) *Let* $F = \alpha\phi(\beta/\alpha)$ *be an (α, β)-metric on an open subset $\mathcal{U}$ in the n-dimensional Euclidean space R^n ($n \geq 3$), where $\alpha = \sqrt{a_{ij}(x)y^i y^j}$ and $\beta = b_i(x)y^i \neq 0$. Suppose that the following conditions hold:*

(a) *β is not parallel with respect to α,*

(b) *$\phi \neq a_2\sqrt{1 + a_1 s^2} + a_3 s$ for some constants $a_1, a_2,$ and a_3 with $a_2 > 0$, that is, F is not (α, β)-metric of Randers type.*

Then F is projectively flat on $\mathscr{U}$ if and only if

(i) $\phi = \phi(s)$ *satisfies*

$$\left\{1 + (k_1 + k_2 s^2)s^2 + k_3 s^2\right\}\phi'' = (k_1 + k_2 s^2)\left\{\phi - s\phi'\right\}.$$

(ii) α *and* β *satisfy*

$$b_{i|j} = 2\tau\left\{(1 + k_1 b^2)a_{ij} + (k_2 b^2 + k_3)b_i b_j\right\},$$

$$G_\alpha^i = \xi y^i - \tau\left(k_1 \alpha^2 + k_2 \beta^2\right)b^i,$$

where $\tau = \tau(x)$ *is a scalar function on* $\mathscr{U}$ *and* k_1, k_2 *and* k_3 *are constants with* $(k_2, k_3) \neq (0, 0)$.

Using Theorem 11.3.4, one can classify locally projectively flat (α, β)-metrics $F = \alpha\phi(\beta/\alpha)$ of constant flag curvature. Roughly speaking, if F is not trivial, then $\phi = \sqrt{1 + ks^2} + \epsilon s$ or $\phi = (\sqrt{1 + ks^2} + \epsilon s)^2/\sqrt{1 + ks^2}$.

Theorem 11.3.5 ([19]) *Let* $F = \alpha\phi(\beta/\alpha)$ *be an* (α, β)-*metric on an open subset* $\mathscr{U} \subseteq R^n$ ($n \geqslant 3$), *where* $\alpha = \sqrt{a_{ij}(x)y^i y^j}$ *and* $\beta = b_i(x)y^i \neq 0$. *Then F is projectively flat Finsler metric with constant flag curvature* **K** *if and only if one of the following holds:*

(i) α *is projectively flat and* β *is parallel with respect to* α.
(ii) $\phi = \sqrt{1 + ks^2} + \epsilon s$, *where* $k, \epsilon(\neq 0)$ *are constants. In this case,* $F = \bar{\alpha} + \bar{\beta}$ *is of Randers type with constant flag curvature* **K** < 0, *where* $\bar{\alpha} = \sqrt{\alpha^2 + k\beta^2}$ *and* $\bar{\beta} = \epsilon\beta$.
(iii) $\phi = (\sqrt{1 + ks^2} + \epsilon s)^2/\sqrt{1 + ks^2}$, *where* $k, \epsilon(\neq 0)$ *are constants. In this case,* $F = (\bar{\alpha} + \bar{\beta})^2/\bar{\alpha}$ *is a square metric with constant flag curvature* **K** $= 0$, *where* $\bar{\alpha} = \sqrt{\alpha^2 + k\beta^2}$ *and* $\bar{\beta} = \epsilon\beta$

By Theorem 11.3.1, we can determine the structure of the metrics in Theorem 11.3.5(ii). By Theorem 11.3.3, we can determine the structure of the metrics in Theorem 11.3.5(iii).

Recently, Yu studied and characterized all projectively flat (α, β)-metrics. He classified projectively flat (α, β)-metrics in dimension $n \geqslant 3$ by a special class of deformations. His results show that the projective flatness of an (α, β)-metric always arises from the projective flatness of some Riemannian metrics when dimension $n \geqslant 3$. See [29].

11.3.4 Projectively Flat Finsler Metrics of Constant Flag Curvature

On every strongly convex domain $\mathcal{U}$ in $\mathbf{R}^n$, Hilbert constructed a complete reversible projectively flat metric $H = H(x, y)$ with negative constant flag curvature $\mathbf{K} = -1$. Then Funk constructed a positively complete projectively flat metric $\Theta = \Theta(x, y)$ with $\mathbf{K} = -1/4$ on $\mathcal{U}$ so that its symmetrization is just the Hilbert metric, $H(x, y) = \frac{1}{2}(\Theta(x, y) + \Theta(x, -y))$ [16]. When $\mathcal{U} = \mathbf{B}^n(1)$ is the unit ball in $\mathbf{R}^n$, the Funk metric is given by (11.8). On the other hand, Berwald [4] constructed a metric given by (11.9) on the unit ball $\mathbf{B}^n(1)$, which is projectively flat metric with zero flag curvature.

In [5–7], Bryant studied and characterized locally projectively flat Finsler metrics with constant flag curvature $\mathbf{K} = 1$ on S^2 and S^n. It is clear that Bryant's metrics cannot be expressed in terms of a Riemannian metric and a 1-form as Randers metrics and Berwald's metrics.

Further, Shen studied and characterized projectively flat Finsler metrics of constant flag curvature in [26]. Shen proved that such metrics can be described using algebraic equations or using Taylor expansions. Shen gave a formula for x-analytic projectively flat Finsler metrics with constant flag curvature using a power series with coefficients expressed in terms of $F(0, y)$ and $F_{x^k}(0, y)y^k$. Shen also gave a formula for general projectively flat Finsler metrics with constant flag curvature using some algebraic equations depending on $F(0, y)$ and $F_{x^k}(0, y)y^k$. By these formulas, he obtained several interesting projectively flat Finsler metrics of constant flag curvature which can be used as models in certain problems.

11.4 Projectively Flat Finsler Metrics with Isotropic S-Curvature

11.4.1 Projectively Flat Randers Metrics with Isotropic S-Curvature

By Theorem 11.3.1, Z. Shen has classified projectively flat Randers metrics of constant flag curvature. Later on, Bao and Robles proved the following result: If a Randers metric F is an Einstein metric with $\mathbf{Ric} = (n-1)\sigma(x)F^2$, then F is of constant S-curvature [3]. This naturally leads to the study of projectively flat Randers metrics with isotropic S-curvature.

Let $F = \alpha + \beta$ be a locally projectively flat Randers metric. Then α is locally projectively flat and β is closed. According to the Beltrami theorem in Riemann geometry, α is locally projectively flat if and only if it is of constant sectional curvature. Thus we may assume that α is of constant sectional curvature μ. It is locally isometric to the following standard metric α_μ on the unit ball $\mathbf{B}^n \subset \mathbf{R}^n$ or the whole $\mathbf{R}^n$ for $\mu = -1, 0, +1$:

$$\alpha_{-1}(x, y) = \frac{\sqrt{|y|^2 - (|x|^2|y|^2 - \langle x, y \rangle^2)}}{1 - |x|^2}, \qquad y \in T_x\mathbf{B}^n \cong \mathbf{R}^n, \qquad (11.13)$$

$$\alpha_0(x, y) = |y|, \qquad\qquad\qquad\qquad y \in T_x\mathbf{R}^n \cong \mathbf{R}^n, \qquad (11.14)$$

$$\alpha_{+1}(x, y) = \frac{\sqrt{|y|^2 + (|x|^2|y|^2 - \langle x, y \rangle^2)}}{1 + |x|^2}, \qquad y \in T_x\mathbf{R}^n \cong \mathbf{R}^n. \qquad (11.15)$$

Then we can determine β if F is of isotropic S-curvature and get the following classification theorem.

Theorem 11.4.1 ([14]) *Let $F = \alpha + \beta$ be a locally projectively flat Randers metric on an n-dimensional manifold M and μ denote the constant sectional curvature of α. Suppose that the S-curvature is isotropic, $\mathbf{S} = (n + 1)c(x)F$. Then F can be classified as follows.*

(A) If $\mu + 4c(x)^2 \equiv 0$, then $c(x) = $ constant and $\mathbf{K} = -c^2 \leqslant 0$.

  (A1) if $c = 0$, then F is locally Minkowskian with flag curvature $\mathbf{K} = 0$;
  (A2) if $c \neq 0$, then after a normalization, F is locally isometric to the following Randers metric on the unit ball $\mathbf{B}^n \subset \mathbb{R}^n$,

$$F(x, y) = \frac{\sqrt{|y|^2 - (|x|^2|y|^2 - \langle x, y \rangle^2)} \pm \langle x, y \rangle}{1 - |x|^2} \pm \frac{\langle a, y \rangle}{1 + \langle a, x \rangle}, \qquad (11.16)$$

  where $a \in \mathbb{R}^n$ with $|a| < 1$, and the flag curvature of F is negative constant, $\mathbf{K} = -\frac{1}{4}$.

(B) If $\mu + 4c(x)^2 \neq 0$, then F is given by

$$F(x, y) = \alpha(x, y) - \frac{2c_{x^k}(x)y^k}{\mu + 4c(x)^2} \qquad (11.17)$$

and the flag curvature of F is given by

$$\mathbf{K} = \frac{3c_{x^k}(x)y^k}{F(x, y)} + 3c(x)^2 + \mu. \qquad (11.18)$$

  (B1) when $\mu = -1$, $\alpha = \alpha_{-1}$ can be expressed in the form (11.13) on $\mathbf{B}^n$. In this case,

$$c(x) = \frac{\lambda + \langle a, x \rangle}{2\sqrt{(\lambda + \langle a, x \rangle)^2 \pm (1 - |x|^2)}}, \qquad (11.19)$$

  where $\lambda \in \mathbb{R}$ and $a \in \mathbb{R}^n$ with $|a|^2 < \lambda^2 \pm 1$.

(B2) *when* $\mu = 0$, $\alpha = \alpha_0$ *can be expressed in the form (11.14) on* $\mathbb{R}^n$. *In this case,*

$$c(x) = \frac{\pm 1}{2\sqrt{\kappa + \langle a, x \rangle + |x|^2}}, \tag{11.20}$$

where $\kappa > 0$ *and* $a \in \mathbb{R}^n$ *with* $|a|^2 < \kappa$.

(B3) *when* $\mu = 1$, $\alpha = \alpha_{+1}$ *can be expressed in the form (11.15) on* $\mathbb{R}^n$. *In this case,*

$$c(x) = \frac{\epsilon + \langle a, x \rangle}{2\sqrt{1 + |x|^2 - (\epsilon + \langle a, x \rangle)^2}}, \tag{11.21}$$

where $\epsilon \in \mathbb{R}$ *and* $a \in \mathbb{R}^n$ *with* $|\epsilon|^2 + |a|^2 < 1$.

Theorem 11.4.1 (A) follows from the classification theorem in [25] after we prove that the flag curvature is constant in this case (also see Theorem 11.3.1). From Theorem 11.4.1 we obtain some interesting projectively flat Randers metrics with isotropic S-curvature.

Example 11.4.1 Let

$$F_-(x, y) = \frac{\sqrt{(1 - |x|^2)|y|^2 + \langle x, y \rangle^2}\sqrt{(1 - |x|^2) + \lambda^2} + \lambda \langle x, y \rangle}{(1 - |x|^2)\sqrt{(1 - |x|^2) + \lambda^2}}, \quad y \in T_x \mathbf{B}^n, \tag{11.22}$$

where $\lambda \in \mathbb{R}$ is an arbitrary constant. The geodesics of F_- are straight lines in $\mathbf{B}^n$. One can easily verify that F_- is complete in the sense that every unit speed geodesic of F_- is defined on $(-\infty, \infty)$. Moreover F_- has strictly negative flag curvature $\mathbf{K} \leqslant -\frac{1}{4}$.

Example 11.4.2 Let

$$F_0(x, y) = \frac{|y|\sqrt{1 + |x|^2} + \langle x, y \rangle}{\sqrt{1 + |x|^2}}, \quad y \in T_x \mathbb{R}^n. \tag{11.23}$$

The geodesics of F_0 are straight lines in $\mathbb{R}^n$. One can easily verify that F_0 is positively complete in the sense that every unit speed geodesic of F_0 is defined on $(-a, \infty)$. Moreover F_0 has positive flag curvature $\mathbf{K} > 0$.

Theorem 11.4.1 is a local classification theorem. If we assume that the manifold is closed (compact without boundary), then the scalar function $c(x)$ takes much more special values [14]. In particular, we have the following

Theorem 11.4.2 ([14]) *Let* $S^n = (M, \alpha)$ *be the standard unit sphere and* $F = \alpha + \beta$ *be a projectively flat Randers metric on* S^n. *Suppose that S-curvature is isotropic,* $\mathbf{S} = (n + 1)c(x)F$. *Then*

$$c(x) = \frac{f(x)}{2\sqrt{1 - f(x)^2}}$$

and

$$F(x, y) = \alpha(x, y) - \frac{f_{x^k}(x)y^k}{\sqrt{1 - f(x)^2}},$$

where $f(x)$ is an eigenfunction of S^n corresponding to the first eigenvalue. Moreover,

(a) $\delta := \sqrt{|\nabla f|_\alpha^2(x) + f(x)^2} < 1$ *is a constant and we have the following estimates for flag curvature*

$$\frac{2 - \delta}{2(1 + \delta)} \leq \mathbf{K} \leq \frac{2 + \delta}{2(1 - \delta)}.$$

(b) *The geodesics of F are the great circles on S^n with F-length 2π.*

11.4.2 Projectively Flat (α, β)-Metrics with Isotropic S-Curvature

Note that projectively flat Finsler metrics must be of scalar flag curvature. We firstly recall the characterizations on (α, β)-metrics of scalar flag curvature with isotropic S-curvature. Based on the main result in [12] on (α, β)-metrics with isotropic S-curvature given by the first author and Z. Shen, the first author has proved the following theorem.

Theorem 11.4.3 ([10]) *A regular (α, β)-metric $F = \alpha\phi(\beta/\alpha)$ of non-Randers type on an n-dimensional manifold M is of isotropic S-curvature, $\mathbf{S} = (n + 1)cF$, if and only if β satisfies*

$$r_{ij} = 0, \quad s_j = 0. \tag{11.24}$$

In this case, $\mathbf{S} = 0$, regardless of the choice of a particular $\phi = \phi(s)$. Here,

$$r_{ij} := \frac{1}{2}(b_{i|j} + b_{j|i}), \quad s_{ij} := \frac{1}{2}(b_{i|j} - b_{j|i}),$$

$$s^i{}_j := a^{ih}s_{hj}, \quad s_j := b_i s^i{}_j = b^m s_{mj}$$

and "$|$" denotes the covariant derivative with respect to the Levi-Civita connection of α.

Further, the first author obtained the following classification theorem for regular (α, β)-metrics of non-Randers type of scalar flag curvature whose S-curvatures are isotropic.

Theorem 11.4.4 ([9, 10]) *Let* $F = \alpha\phi(\beta/\alpha)$ *be a regular* (α, β)*-metric of non-Randers type on an n-dimensional manifold M* $(n \geqslant 3)$*. Then F is of scalar flag curvature and of isotropic S-curvature if and only if the flag curvature* $\mathbf{K} = 0$ *and F is a Berwald metric. In this case, F is a locally Minkowski metric.*

If we regard locally Minkowski metrics as so-called trivial Finsler metrics, then, by Theorem 11.4.4, we have the following theorem.

Theorem 11.4.5 ([10]) *The non-trivial regular* (α, β)*-metrics of scalar flag curvature with isotropic S-curvature on an n-dimensional manifold M* $(n \geqslant 3)$ *must be Randers metrics. In this case, the metrics are of weakly isotropic flag curvature and are completely determined by navigation data* (h, W) *(for more details, see [13]).*

From Theorems 11.4.4 and 11.4.5, we can obtain the following theorem on projectively flat (α, β)-metrics with isotropic S-curvature.

Theorem 11.4.6 *Let* $F = \alpha\phi(\beta/\alpha)$ *be a regular* (α, β)*-metric on an n-dimensional manifold M* $(n \geqslant 3)$*. Then F is projectively flat metric with isotropic S-curvature if and only if F is a locally Minkowski metric of non-Randers type, or F is a projectively flat Randers metric with isotropic S-curvature.*

11.4.3 Projectively Flat Finsler Metrics with Isotropic S-Curvature

We have classified projectively flat Randers metrics and (α, β)-metrics with isotropic S-curvature. It is a natural problem to study and characterize projectively flat Finsler metrics with isotropic S-curvature. In [14], we have shown the following theorem.

Theorem 11.4.7 ([14]) *Let* (M, F) *be an n-dimensional Finsler manifold. Assume that F is of scalar curvature* $\mathbf{K} = \mathbf{K}(x, y)$*. If the S-curvature of F is isotropic, that is,* $\mathbf{S} = (n + 1)cF$*, where* $c = c(x)$ *is a scalar function, then the flag curvature* $\mathbf{K}$ *is given by*

$$\mathbf{K} = 3\frac{c_{x^m}y^m}{F} + \sigma, \tag{11.25}$$

where $\sigma = \sigma(x)$ *is a scalar function on M. In particular,* $c = $ *constant if and only if* $\mathbf{K} = \mathbf{K}(x)$ *is a scalar function on M.*

Shen and Yildirim have proved that the converse of Theorem 11.4.7 is also true for Randers metrics (see [28]). Because projectively flat Finsler metrics must be of scalar flag curvature, Theorem 11.4.7 plays a very important role in classifying projectively flat Randers metrics with isotropic S-curvature. Further, we can study and characterize locally projectively flat Finsler metrics with isotropic S-curvature by using Theorem 11.4.7.

There are many projectively flat Randers metrics with isotropic S-curvature [14]. Besides these Randers metrics, there is another special kind of projectively flat Finsler metrics with isotropic S-curvature—Funk metrics. More generally, we have the following classification theorem.

Theorem 11.4.8 ([11]) *Let $F = F(x, y)$ be a locally projectively flat Finsler metric on an open subset $\Omega \subset \mathbb{R}^n$. Suppose that F has isotropic S-curvature, $\mathbf{S} = (n + 1)c(x)F$. Then the flag curvature is in the form*

$$\mathbf{K} = 3\frac{c_{x^m}y^m}{F} + \sigma,$$

where $\sigma = \sigma(x)$ is a scalar function on Ω.

(a) *If $\mathbf{K} \neq -c^2 + \frac{c_{x^m}y^m}{F}$ on Ω, then $F = \alpha + \beta$ is a projectively flat Randers metric with isotropic S-curvature $\mathbf{S} = (n + 1)cF$*
(b) *If $\mathbf{K} \equiv -c^2 + \frac{c_{x^m}y^m}{F}$ on Ω, then $c = $ constant and F is either locally Minkowskian ($c = 0$) or, up to a scaling, locally isometric to the metric*

$$\Theta_a := \Theta(x, y) + \frac{<a, y>}{1+ <a, x>}, \qquad c = \frac{1}{2}$$

or its reverse

$$\bar{\Theta}_a := \Theta(x, -y) - \frac{<a, y>}{1+ <a, x>}, \qquad c = -\frac{1}{2},$$

where $a \in \mathbb{R}^n$ is a constant vector and $\Theta(x, y)$ is Funk metric on Ω.

In Theorem 11.4.8(a), the local structure of F has been completely determined in [14] (also see Theorem 11.4.1). Theorem 11.4.8 tells us that non-trivial projectively flat Finsler metrics with isotropic S-curvature are either projectively flat Randers metrics with isotropic S-curvature, or completely determined by Funk metrics.

Acknowledgements X. Cheng was supported by the National Natural Science Foundation of China (11371386) and the European Union's Seventh Framework Programme (FP7/2007–2013) under grant agreement no. 317721.

References

1. J.C. Álvarez Paiva, Symplectic geometry and Hilbert's fourth problem. J. Differ. Geom. **69**(2), 353–378 (2005)
2. S. Bácsó, M. Matsumoto, On Finsler spaces of Douglas type. A generalization of the notion of Berwald space. Publ. Math. Debr. **51**, 385–406 (1997)
3. D. Bao, C. Robles, On Randers metrics of constant curvature. Rep. Math. Phys. **51**, 9–42 (2003)
4. L. Berwald, Über die n-dimensionalen Geometrien konstanter Krümmung, in denen die Geraden die kürzesten sind. Math. Z. **30**, 499–469 (1929)

5. R. Bryant, Finsler structures on the 2-sphere satisfying $K = 1$, in *Finsler Geometry, Contemporary Mathematics*, vol. 196 (American Mathematical Society, Providence, 1996), pp. 27–42
6. R. Bryant, Projectively flat Finsler 2-spheres of constant curvature. Sel. Math. **3**, 161–204 (1997)
7. R. Bryant, Some remarks on Finsler manifolds with constant flag curvature. Houst. J. Math. **28**(2), 221–262 (2002)
8. H. Busemann, Problem IV: Desarguesian spaces in mathematical developments arising from Hilbert problems, in *Proceedings of Symposia in Pure Mathematics* (American Mathematical Society, Providence, 1976)
9. X. Cheng, On (α, β)-metrics of scalar flag curvature with constant S-curvature. Acta Math. Sin. **26**(9), 1701–1708 (2010)
10. X. Cheng, The (α, β)-metrics of scalar flag curvature. Differ. Geom. Appl. **35**, 361–369 (2014)
11. X. Cheng, Z. Shen, Projectively flat Finsler metrics with almost isotropic S-curvature. Acta Math. Sci. **26B**(2), 307–313 (2006)
12. X. Cheng, Z. Shen, A class of Finsler metrics with isotropic S-curvature. Isr. J. Math. **169**(1), 317–340 (2009)
13. X. Cheng, Z. Shen, *Finsler Geometry—An Approach via Randers Spaces* (Springer/Science Press, Berlin, 2012)
14. X. Cheng, X. Mo, Z. Shen, On the flag curvature of Finsler metrics of scalar curvature. J. Lond. Math. Soc. **68**(2), 762–780 (2003)
15. S.S. Chern, Z. Shen, *Riemann-Finsler Geometry*. Nankai Tracts in Mathematics, vol. 6 (World Scientific, Singapore, 2005)
16. P. Funk, Über Geometrien bei denen die Geraden die kürzesten sind. Math. Ann. **101**, 226–237 (1929)
17. G. Hamel, Über die Geometrien in denen die Geraden die kürzesten sind. Math. Ann. **57**, 231–264 (1903)
18. D. Hilbert, Mathematical problems. Bull. Am. Math. Soc. **8**, 437–479 (1902); Bull. Am. Math. Soc. **37**(4), 407–436 (2000)
19. B. Li, Z. Shen, On a class of projectively flat Finsler metrics with constant flag curvature. Int. J. Math. **18**(7), 1–12 (2007)
20. M. Matsumoto, On C-reducible Finsler spaces. Tensor N. S. **24**, 29–37 (1972)
21. A.V. Pogorelov, *Hilbert's Fourth Problem*. Scripta Series in Mathematics (Winston and Sons, Washington, 1979)
22. G. Randers, On an asymmetrical metric in the four-space of general relativity. Phys. Rev. **59**, 195–199 (1941)
23. Z. Shen, Volume comparison and its applications in Riemann-Finsler geometry. Adv. Math. **128**(2), 306–328 (1997)
24. Z. Shen, Finsler metrics with **K**=0 and **S**=0. Can. J. Math. **55**(1), 112–132 (2003)
25. Z. Shen, Projectively flat Randers metrics with constant flag curvature. Math. Ann. **325**, 19–30 (2003)
26. Z. Shen, Projectively flat Finsler metrics of constant flag curvature. Trans. Am. Math. Soc. **355**(4), 1713–1728 (2003)
27. Z. Shen, On projectively flat (α, β)-metrics. Can. Math. Bull. **52**(1), 132–144 (2009)
28. Z. Shen, G.C. Yildirim, On a class of projectively flat metrics with constant flag curvature. Can. J. Math. **60**(2), 443–456 (2008)
29. C. Yu, Deformations and Hilbert's fourth problem. Math. Ann. **365**, 1379–1408 (2016)
30. L. Zhou, A local classification of a class of (α, β) metrics with constant flag curvature. Differ. Geom. Appl. **28**, 170–193 (2010)

Chapter 12
Holonomy Theory of Finsler Manifolds

Zoltán Muzsnay and Péter T. Nagy

Abstract The holonomy group of a Riemannian or Finslerian manifold can be introduced in a very natural way: it is the group generated by parallel translations along loops with respect to the canonical connection. The Riemannian holonomy groups have been extensively studied and by now their complete classification is known. On the Finslerian holonomy, however, only few results are known and, as our results show, it can be essentially different from the Riemannian one.

In recent papers we have developed a method for the investigation of holonomy properties of non-Riemannian Finsler manifolds by constructing tangent Lie algebras to the holonomy group: the curvature algebra, the infinitesimal holonomy algebra, and the holonomy algebra. In this book chapter we present this method and give a unified treatment of our results. In particular we show that the dimension of these tangent algebras is usually greater than the possible dimensions of Riemannian holonomy groups and in many cases is infinite. We prove that the holonomy group of a locally projectively flat Finsler manifold of constant curvature is finite dimensional if and only if it is a Riemannian manifold or a flat Finsler manifold. We also show that the topological closure of the holonomy group of a certain class of simply connected, projectively flat Finsler 2-manifolds of constant curvature (spherically symmetric Finsler 2-manifolds) is not a finite dimensional Lie group, and we prove that its topological closure is the connected component of the full diffeomorphism group of the circle.

12.1 Introduction

The notion of the holonomy group of a Riemannian or Finslerian manifold can be introduced in a very natural way: it is the group generated by parallel translations along loops. In contrast to the Finslerian case, the Riemannian holonomy groups

Z. Muzsnay (✉)
Institute of Mathematics, University of Debrecen, P.O.B. 400, H-4010 Debrecen, Hungary
e-mail: muzsnay@science.unideb.hu

P.T. Nagy
Institute of Applied Mathematics, Óbuda University, Bécsi út 96/b, H-1034 Budapest, Hungary
e-mail: nagy.peter@nik.uni-obuda.hu

© Springer International Publishing AG 2017   265
G. Falcone (ed.), *Lie Groups, Differential Equations, and Geometry*,
UNIPA Springer Series, DOI 10.1007/978-3-319-62181-4_12

have been extensively studied. One of the earliest fundamental results is the theorem of Borel and Lichnerowicz [3] from 1952, claiming that the holonomy group of a simply connected Riemannian manifold is a closed Lie subgroup of the orthogonal group $O(n)$. By now, the complete classification of Riemannian holonomy groups is known. Similarly to the Riemannian case the holonomy group of a Finsler manifold is the subgroup of the diffeomorphism group of an indicatrix, generated by canonical homogeneous (nonlinear) parallel translations along closed loops. Before our investigation (c.f. [18–23]), the holonomy groups of non-Riemannian Finsler manifolds have been described only in special cases: for Berwald manifolds there exist Riemannian metrics with the same holonomy group (cf. Szabó, [31]), for positive definite Landsberg manifolds the holonomy groups are compact Lie groups consisting of isometries of the indicatrix with respect to an induced Riemannian metric (cf. Kozma [13, 14]). A thorough study of holonomy groups of homogeneous (nonlinear) connections was initiated by Barthel in his basic work [2] in 1963; he gave a construction for a holonomy algebra of vector fields on the tangent space. A general setting for the study of infinite dimensional holonomy groups and holonomy algebras of nonlinear connections was initiated by Michor in [17]. However the introduced holonomy algebras could not be used to estimate the dimension of the holonomy group since their tangential properties to the holonomy group were not clarified.

Now we give a unified treatment of our results on Finslerian holonomy theory. We construct and investigate tangent Lie algebras to the holonomy group and show that the dimension of these tangent algebras in many cases is infinite, particularly it is greater than the possible dimensions of Riemannian holonomy groups.

In the second section we collect the necessary definitions and constructions of spray and Finsler geometry. The third section is devoted to the investigation of tangential properties of subalgebras of the Lie algebra of vector fields on a manifold to the infinite dimensional diffeomorphism group of this manifold. Particularly we consider the case if the manifold is compact, in this case the diffeomorphism group is an infinite dimensional Lie group modeled on the Lie algebra of vector fields on the manifold.

In Sect. 12.4 we introduce the notion of curvature algebra of a Finsler manifold consisting of tangent vector fields on the indicatrix, which is a generalization of the matrix group generated by curvature operators of a Riemannian manifold. We show that the vector fields belonging to the curvature algebra are tangent to 1-parameter families of diffeomorphisms contained in the holonomy group. We prove that for a positive definite non-Riemannian Finsler manifold of non-zero constant curvature with dimension $n > 2$ the dimension of the curvature algebra is strictly greater than the dimension of the orthogonal group acting on the tangent space and hence it cannot be a compact Lie group. In addition, we provide an example of a left invariant singular (non y-global) Finsler metric of Berwald-Moór-type on the three-dimensional Heisenberg group which has infinite dimensional curvature algebra and hence its holonomy is not a (finite dimensional) Lie group. These results give a positive answer to the following problem formulated by Chern and Shen in [6]

(p. 85): *Is there a Finsler manifold whose holonomy group is not the holonomy group of any Riemannian manifold?*

Section 12.5 contains constructions of further tangent Lie algebras to the holonomy group consisting of tangent vector fields on the indicatrix, namely the infinitesimal holonomy algebra and the holonomy algebra of a Finsler manifold. Our goal is to make an attempt to find the right notion of the holonomy algebra of Finsler spaces. The holonomy algebra should be the largest Lie algebra such that all its elements are tangent to the holonomy group. In our attempt we are building successively Lie algebras having the tangent properties. We define the *infinitesimal holonomy algebra* by the smallest Lie algebra of vector fields on an indicatrix, containing the curvature vector fields and their horizontal covariant derivatives with respect to the Berwald connection and prove the tangential property of this Lie algebra to the holonomy group. At the end we introduce the notion of the *holonomy algebra* of a Finsler manifold by all conjugates of infinitesimal holonomy algebras by parallel translations with respect to the Berwald connection. We prove that this holonomy algebra is tangent to the holonomy group. A different treatment of the holonomy algebra is given by Crampin and Saunders in [7]. The question of whether the holonomy algebra introduced in this way is the largest Lie algebra, which is tangent to the holonomy group, is still open.

In Sect. 12.6 we construct for interesting classes of locally projectively flat Finsler surfaces and manifolds of non-zero constant curvature infinite dimensional subalgebras in the tangent infinitesimal holonomy algebras. From the viewpoint of non-Euclidean geometry the most important Riemann-Finsler manifolds are the projectively flat spaces of constant flag curvature. We will turn our attention to non-Riemannian projectively flat Finsler manifolds of non-zero constant flag curvature. We consider the following classes of locally projectively flat non-Riemannian Finsler manifolds of non-zero constant flag curvature:

1. Randers manifolds,
2. manifolds having a two-dimensional subspace in the tangent space at some point, on which the Finsler norm is an Euclidean norm,
3. manifolds having a two-dimensional subspace in the tangent space at some point, on which the Finsler norm and the projective factor are linearly dependent.

The first class consists of positively complete Finsler manifolds of negative curvature, the second class contains a large family of (not necessarily complete) Finsler manifolds of negative curvature, and the third class contains a large family of not necessarily complete Finsler manifolds of positive curvature. The metrics belonging to these classes can be considered as (local) generalizations of a 1-parameter family of complete Finsler manifolds of positive curvature defined on S^2 by Bryant in [4, 5] and on S^n by Shen in [30], Example 7.1. It follows from these constructions that the holonomy group of Finsler manifolds belonging to the above classes and satisfying some additional technical assumption is infinite dimensional.

In Sect. 12.7 we study the holonomy group of an arbitrary locally projectively flat Finsler manifolds of constant curvature. Our aim is to characterize all locally projectively flat Finsler manifolds with finite dimensional holonomy group. To obtain such a characterization, we investigate the dimension of the infinitesimal holonomy algebra. We obtain that if $(M, \mathscr{F})$ is a non-Riemannian locally projectively flat Finsler manifolds of non-zero constant curvature, then its infinitesimal holonomy algebra is infinite dimensional. Using this general result and the tangent property of the infinitesimal holonomy algebra we obtain the characterization: *The holonomy group of a locally projectively flat Finsler manifold of constant curvature is finite dimensional if and only if it is a Riemannian manifold or a flat Finsler manifold.*

Section 12.8 is devoted to show that the topological closure of the holonomy group of a certain class of simply connected, projectively flat Finsler 2-manifolds of constant curvature is not a finite dimensional Lie group, and we prove that its topological closure is the connected component of the full diffeomorphism group of the circle. Before our investigation, perhaps because of technical difficulties, not a single infinite dimensional Finsler holonomy group has been described. Now, we provide the first such a description. This class of Finsler 2-manifolds contains the positively complete standard Funk plane of constant negative curvature (positively complete standard Funk plane), and the complete irreversible Bryant-Shen-spheres of constant positive curvature [5, 30]. We obtain that for every simply connected Finsler 2-manifold the topological closure of the holonomy group is a subgroup of $\mathrm{Diff}_+^\infty(\mathbb{S}^1)$. That means that in the examples mentioned above, the closed holonomy group is maximal. In the proof we use our constructive method developed in Sect. 12.6 for the study of infinite dimensional Lie subalgebras of the infinitesimal holonomy algebra.

12.2 Preliminaries

12.2.1 Spray Manifolds

A *spray* on a manifold M is a smooth vector field $\mathscr{S}$ on the slit tangent bundle $\hat{T}M := TM \setminus \{0\}$ expressed in a standard coordinate system (x^i, y^i) on TM as

$$\mathscr{S} = y^i \frac{\partial}{\partial x^i} - 2G^i(x, y) \frac{\partial}{\partial y^i}, \tag{12.1}$$

where the functions $G^i(x, y)$ of local coordinates (x^i, y^i) on TM satisfy

$$G^i(x, \lambda y) = \lambda^2 G^i(x, y), \quad \lambda > 0. \tag{12.2}$$

A manifold M with a spray $\mathscr{S}$ is called a *spray manifold* $(M, \mathscr{S})$, cf. [27], Chapter 5.

A curve $c(t)$ is called *geodesic* of the spray manifold $(M, \mathscr{S})$ if its coordinate functions $c^i(t)$ satisfy the system of 2nd order ordinary differential equations

$$\ddot{c}^i(t) + 2G^i(c(t), \dot{c}(t)) = 0, \tag{12.3}$$

where the functions $G^i(x, y)$ are the *geodesic coefficients* the spray manifold $(M, \mathscr{S})$.

12.2.1.1   Horizontal Distribution, Covariant Derivative and Curvature

Let (TM, π, M) and (TTM, τ, TM) be the first and the second tangent bundle of the manifold M, respectively, and let $\mathscr{V}TM \subset TTM$ be the (integrable) vertical distribution on TM given by $\mathscr{V}TM := \operatorname{Ker} \pi_*$. The *horizontal distribution* $\mathscr{H}TM \subset TTM$ associated to the spray manifold $(M, \mathscr{S})$ is the image of the horizontal lift which is the vector space isomorphism $l_y \colon T_x M \to \mathscr{H}_y TM$ for $x \in M$ and $y \in T_x M$ defined by

$$l_y\left(\frac{\partial}{\partial x^i}\right) = \frac{\partial}{\partial x^i} - G_i^k(x, y)\frac{\partial}{\partial y^k}, \quad \text{where} \quad G_j^i = \frac{\partial G^i}{\partial y^j} \tag{12.4}$$

in the coordinate system (x^i, y^i) of TM. The horizontal distribution is complementary to the vertical distribution, hence we have the decomposition $T_y TM = \mathscr{H}_y TM \oplus \mathscr{V}_y TM$. The projectors corresponding to this decomposition will be denoted by $h : TTM \to \mathscr{H}TM$ and $v : TTM \to \mathscr{V}TM$.

The vertical distribution over the slit tangent bundle $\hat{T}M = TM \setminus \{0\}$ will be denoted by $(\hat{\mathscr{V}}TM, \tau, \hat{T}M)$ and the pull-back bundle of $(\hat{T}M, \pi, M)$ corresponding to the map $\pi : TM \to M$ by $(\pi^*TM, \bar{\pi}, \hat{T}M)$. Clearly, the mapping

$$\left(x, y, \xi^i \frac{\partial}{\partial y^i}\right) \mapsto \left(x, y, \xi^i \frac{\partial}{\partial x^i}\right) : \hat{\mathscr{V}}TM \to \pi^*TM \tag{12.5}$$

is a canonical bundle isomorphism. In the following we will use the isomorphism (12.5) for the identification of these bundles.

Let $\mathfrak{X}^\infty(M)$ be the vector space of smooth vector fields on the manifold M and $\hat{\mathfrak{X}}^\infty(TM)$ the vector space of smooth sections of the bundle $(\hat{\mathscr{V}}TM, \tau, \hat{T}M)$. The *horizontal covariant derivative* of a section $\xi \in \hat{\mathfrak{X}}^\infty(TM)$ by a vector field $X \in \mathfrak{X}^\infty(M)$ is given by

$$\nabla_X \xi := [l(X), \xi].$$

If $\xi(x, y) = \xi^i(x, y)\frac{\partial}{\partial y^i}$ and $X(x) = X^i(x)\frac{\partial}{\partial x^i}$, then $\nabla_X \xi$ can be expressed as

$$\nabla_X \xi = \left(\frac{\partial \xi^i(x, y)}{\partial x^j} - G_j^k(x, y)\frac{\partial \xi^i(x, y)}{\partial y^k} + G_{jk}^i(x, y)\xi^k(x, y)\right) X^j \frac{\partial}{\partial y^i}, \tag{12.6}$$

where $G_{jk}^i := \frac{\partial G_j^i}{\partial y^k}$.

Defining the horizontal covariant derivative

$$\nabla_X \phi = l(X)\phi = \left(\frac{\partial \phi}{\partial x^j} - G_j^k(x, y)\frac{\partial \phi(x, y)}{\partial y^k}\right) X^j$$

of a smooth function $\phi : \hat{T}M \to \mathbb{R}$, the horizontal covariant derivation (12.6) can be extended to sections of the tensor bundle over $(\pi^*TM, \pi, \hat{T}M)$, using the canonical bundle isomorphism (12.5).

The *curvature tensor* field

$$K_{(x,y)}(X, Y) := v[X^h, Y^h], \qquad X, Y \in T_xM. \tag{12.7}$$

on the pull-back bundle $(\pi^*TM, \pi, \hat{T}M)$ of the spray manifold $(M, \mathscr{S})$ in a local coordinate system (x^i, y^i) of TM is given by

$$K_{(x,y)} = K^i_{jk}(x, y)dx^j \otimes dx^k \otimes \frac{\partial}{\partial x^i},$$

where

$$K^i_{jk}(x, y) = \frac{\partial G^i_j(x, y)}{\partial x^k} - \frac{\partial G^i_k(x, y)}{\partial x^j} + G^m_j(x, y)G^i_{km}(x, y) - G^m_k(x, y)G^i_{jm}(x, y). \tag{12.8}$$

The curvature tensor field characterizes the integrability of the horizontal distribution. Namely, if the horizontal distribution $\mathscr{H}TM$ is integrable, then the curvature is identically zero.

12.2.1.2 Parallel Translation

For a spray manifold $(M, \mathscr{S})$ the *parallel vector fields* $X(t) = X^i(t)\frac{\partial}{\partial x^i}$ along a curve $c(t)$ are defined by the solutions of the differential equation

$$D_{\dot{c}}X(t) := \left(\frac{dX^i(t)}{dt} + G^i_j(c(t), X(t))\dot{c}^j(t) \right) \frac{\partial}{\partial x^i} = 0. \tag{12.9}$$

Using the relations (12.2) and Euler theorem on homogeneous functions we see that the functions $G^i_j(x, y)$ are positive homogeneous of first order with respect to the variable y, and hence $D_{\dot{c}}(\lambda X(t)) = \lambda D_{\dot{c}}X(t)$ for any $\lambda \geq 0$. The differential equation (12.9) can be expressed by the horizontal covariant derivative (12.6) using the bundle isomorphism (12.5) as follows: a vector field $X(t) = X^i(t)\frac{\partial}{\partial x^i}$ along a curve $c(t)$ is parallel if it satisfies the equation

$$\nabla_{\dot{c}}X(t) = \left(\frac{dX^i(t)}{dt} + G^i_j(c(t), X(t))\dot{c}^j(t) \right) \frac{\partial}{\partial x^i} = 0. \tag{12.10}$$

Clearly, for any $X_0 \in T_{c(0)}M$ there is a unique parallel vector field $X(t)$ along the curve c such that $X_0 = X(0)$. Moreover, if $X(t)$ is a parallel vector field along c, then

$\lambda X(t)$ is also parallel along c for any $\lambda \geq 0$. Then the *homogeneous (nonlinear) parallel translation*

$$\tau_c : T_{c(0)}M \to T_{c(1)}M \quad \text{along a curve} \quad c(t)$$

of the spray manifold $(M, \mathscr{S})$ is defined by the positive homogeneous map $\tau_c :$ $X_0 \mapsto X_1$ given by the value $X_1 = X(1)$ at $t = 1$ of the parallel vector field with initial value $X(0) = X_0$ (Fig. 12.1).

Since the parallel translation of a spray manifold $(M, \mathscr{S})$ is determined by its horizontal distribution $\mathscr{H}TM \subset TTM$, a spray manifold can be considered as a particular case of a fibered manifold equipped with an Ehresmann connection (cf. [8]). An Ehresmann connection of a fibered manifold is given by a horizontal distribution, which is complement to the vertical distribution consisting of the tangent spaces of the fibers. For a spray manifold the fibered manifold is the tangent bundle of M and the horizontal distribution determined by the horizontal lift $l_y: T_xM \to \mathscr{H}_yTM$ expressed by Eq. (12.4).

The parallel translation can be introduced in a very nice geometrical way with the help of the notion of horizontal distribution. Namely, we call a curve in TM horizontal if the tangent vectors of this curve are contained in the horizontal distribution $\mathscr{H}TM \subset TTM$. Let now $c(t)$ be a curve in the manifold M joining the points p and q. The *horizontal lift* $c^h(t) = (c(t), X^i(t)\frac{\partial}{\partial x^i})$ of $c(t)$ is the curve $c^h(t)$ in TM defined by the properties that $c^h(t)$ projects on $c(t)$ and $c^h(t)$ is horizontal that is $\dot{c}^h(t) \in H_{c(t)}$. This means according to Eq. (12.4) that

$$\dot{c}^i(t)\frac{\partial}{\partial x^i} + \frac{d}{dt}X^i(t)\frac{\partial}{\partial y^i} = \left(\frac{\partial}{\partial x^i} - G_i^k(x, y)\frac{\partial}{\partial y^k}\right)\dot{c}^i(t),$$

i.e., the tangent vector of the lifted curve $c^h(t)$ is the horizontal lift of the tangent vector $\dot{c}^i(t)\frac{\partial}{\partial x^i}$ of $c(t)$. It follows that a vector field $X(t)$ along a curve $c(t)$ is parallel if and only if it is a solution of the differential equation

$$\frac{d}{dt}\left(c(t), X^i(t)\frac{\partial}{\partial x^i}\right) = l_{X(t)}(\dot{c}(t)), \tag{12.11}$$

or equivalently $X(t)$ satisfies the differential equation (12.9). Hence the parallel translation along a curve $c(t)$ joining the points p and q is the map $\tau_c : T_pM \to T_qM$

Fig. 12.1 Parallel translation

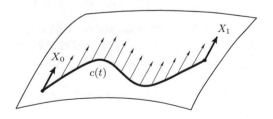

Fig. 12.2 Geometric
construction of the parallel
translation

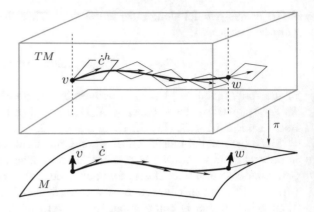

determined by the intersection points of the horizontal lifts of the curve $c(t)$ with the tangent spaces T_p and T_q. The construction can be illustrated by Fig. 12.2.

12.2.1.3 Totally Geodesic and Auto-Parallel Submanifolds

A submanifold $\bar{M}$ in a spray manifold $(M, \mathscr{S})$ is called *totally geodesic* if any geodesic of $(M, \mathscr{S})$ which is tangent to $\bar{M}$ at some point is contained in $\bar{M}$.

A totally geodesic submanifold $\bar{M}$ of $(M, \mathscr{S})$ is called *auto-parallel* if the homogeneous (nonlinear) parallel translations $\tau_c : T_{c(0)}M \to T_{c(1)}M$ along curves in the submanifold $\bar{M}$ leave invariant the tangent bundle $T\bar{M}$ and for every $\xi \in \hat{\mathfrak{X}}^\infty(T\bar{M})$ the horizontal Berwald covariant derivative $\nabla_X \xi$ belongs to $\hat{\mathfrak{X}}^\infty(T\bar{M})$.

Lemma 1 *Let $\bar{M}$ be a totally geodesic submanifold in a spray manifold $(M, \mathscr{S})$. The following assertions hold:*

(a) *the spray $\mathscr{S}$ induces a spray $\bar{\mathscr{S}}$ on the submanifold $\bar{M}$,*
(b) *$\bar{M}$ is an auto-parallel submanifold.*

Proof Assume that the manifolds $\bar{M}$ and M are k, respectively, $n = k + p$ dimensional. Let $(x^1, \ldots, x^k, x^{k+1}, \ldots, x^n)$ be an adapted coordinate system, i. e. the submanifold $\bar{M}$ is locally given by the equations $x^{k+1} = \cdots = x^n = 0$. We denote the indices running on the values $\{1, \ldots, k\}$ or $\{k + 1, \ldots, n\}$ by α, β, γ or σ, τ, respectively. The differential equation (12.3) of geodesics yields that the geodesic coefficients $G^\sigma(x, y)$ satisfy

$$G^\sigma(x^1, \ldots, x^k, 0, \ldots, 0; y^1, \ldots, y^k, 0, \ldots, 0) = 0$$

identically, hence their derivatives with respect to $y^1, \ldots, y^k$ are also vanishing. It follows that $G^\sigma_\alpha = 0$ and $G^\sigma_{\alpha\,\beta} = 0$ at any $(x^1, \ldots, x^k, 0, \ldots, 0; y^1, \ldots, y^k, 0, \ldots, 0)$. Hence the induced spray $\bar{\mathscr{S}}$ on $\bar{M}$ is defined by the geodesic coefficients

$$\bar{G}^\beta(x^1, \ldots, x^k; y^1, \ldots, y^k) = G^\beta(x^1, \ldots, x^k, 0, \ldots, 0; y^1, \ldots, y^k, 0, \ldots, 0).$$

$$(12.12)$$

The homogeneous (nonlinear) parallel translation $\tau_c : T_{c(0)}M \rightarrow T_{c(1)}M$ along curves in the submanifold $\bar{M}$ and the horizontal covariant derivative on $\bar{M}$ with respect to the spray $\mathscr{S}$ coincide with the translation and the horizontal covariant derivative on $\bar{M}$ with respect to the spray $\bar{\mathscr{S}}$. Hence the assertions are true.

12.2.1.4   Holonomy

The holonomy group of an Ehresmann connection, or particularly of a spray manifold is the group generated by parallel translations along loops with respect to the associated connection, cf. [12, 33], pp. 82–86. The holonomy properties of a spray manifold depend essentially on its curvature properties. This can be easily understood by considering the geometric construction of the parallel translation.

Let $(M, \mathscr{S})$ be a spray manifold, and let us assume that M is connected. We choose a fixed base point $p \in M$. For each closed piecewise smooth curve $c : [0, 1] \rightarrow M$ through p the parallel translation $\tau_c : T_pM \rightarrow T_pM$ along the curve $c : [0, 1] \rightarrow M$ is a diffeomorphism of the tangent space T_pM. All these diffeomorphisms form together the holonomy group at the point p, a subgroup of the diffeomorphism group of T_pM. Clearly, the holonomy group depends on the base point p only up to conjugation, and therefore the holonomy groups at different points of M are isomorphic.

- Case $K \equiv 0$. Let us consider a point $p \in M$, a tangent vector $v \in T_pM$ at p, and an arbitrary closed curve c on M starting from p.

If the curvature identically vanishes, then the horizontal distribution is integrable. Therefore TTM has a horizontal foliation, that is for every $v \in T_pM$ there is a unique n-dimensional submanifold $\mathscr{H}_v$ in TM such that $\mathscr{H}_v \cap T_pM = \{v\}$ and the tangent spaces are horizontal. In Fig. 12.3 the level surface corresponding to $v \in T_pM$ is represented by dashed lines in tangent manifold TM. The horizontal lift

Fig. 12.3 Trivial holonomy: $R \equiv 0$

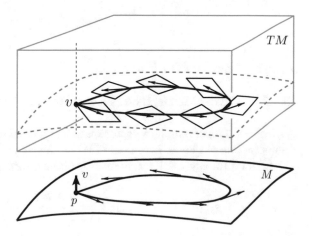

Fig. 12.4 Nontrivial
holonomy: $R \not\equiv 0$

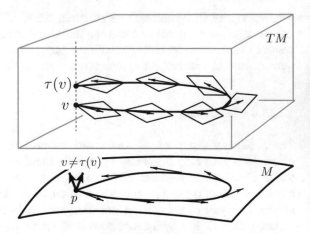

$c^h = l \circ c$ of the curve c starts at v and (because the integrability condition) stays
on the horizontal submanifold $\mathscr{H}_v$. Its endpoint is an element of $\mathscr{H}_v$ and also an
element of T_pM that is v. Consequently we can obtain that $c^h(0) = c^h(1) = v$, and
the holonomy is trivial.

- Case $K \not\equiv 0$. Although Fig. 12.4 describing this case looks similar to that of
  Fig. 12.3, the situation is quite different. The horizontal distribution is non-integ-
  rable and therefore there is no horizontal foliation. Let us consider a point $p \in M$
  and a vector $v \in T_pM$. The smallest integrable distribution $\mathscr{N}_v$ at v containing
  the horizontal distribution is at least $(n+1)$-dimensional. The reachable sets in
  T_pM is at least one-dimensional. In particular $\{v\} \subsetneq \mathscr{N}_v \cap T_pM$ and there are
  other elements T_pM reachable from v. That is, there are elements $w \in T_pM$ and
  a horizontal curve $c^h \in \mathscr{N}_v$ such that $c^h(0) = v$, $c^h(1) = w$. Considering the
  projection of c^h we can obtain a curve $c := \pi \circ c^h$ such that $c(0) = c(1) = p$ and
  $\tau_c(v) \neq v$. Consequently we obtained that the holonomy is nontrivial (cf. [27],
  Remark 8.1.3).

12.2.2 Finsler Manifolds

12.2.2.1 Finsler Metric, Its Associated Spray and Parallel Translation

A *Minkowski functional* on a vector space V is a continuous function $\mathscr{F}$, positively
homogeneous of degree two, i.e. $\mathscr{F}(\lambda y) = \lambda^2 \mathscr{F}(y)$ if $\lambda > 0$, smooth on $\hat{V} :=
V \setminus \{0\}$, and for any $y \in \hat{V}$ the symmetric bilinear form $g_y: V \times V \to \mathbb{R}$ defined by

$$g_y: (u, v) \mapsto g_{ij}(y)u^iv^j = \frac{1}{2}\frac{\partial^2 \mathscr{F}(y + su + tv)}{\partial s\, \partial t}\bigg|_{t=s=0}$$

is non-degenerate. If g_y is positive definite for any $y \in \hat{V}$, then $\mathscr{F}$ is said positive definite and $(V, \mathscr{F})$ is called *positive definite Minkowski space*. A Minkowski functional $\mathscr{F}$ is called *semi-Euclidean* if there exists a symmetric bilinear form $\langle\, ,\, \rangle$ on V such that $g_y(u, v) = \langle u, v \rangle$ for any $y \in \hat{V}$ and $u, v \in V$. A semi-Euclidean positive definite Minkowski functional is called *Euclidean*.

A *Finsler manifold* is a pair $(M, \mathscr{F})$ of an n-manifold M and a function $\mathscr{F} : TM \to \mathbb{R}$ (called *Finsler metric*, cf. [27]) defined on the tangent bundle of M, which is smooth on $\hat{T}M := TM \setminus \{0\}$ and its restriction $\mathscr{F}_x = \mathscr{F}|_{T_xM}$ is a Minkowski functional on T_xM for all $x \in M$. If the Minkowski functional $\mathscr{F}_x$ is positive definite on T_xM for all $x \in M$, then $(M, \mathscr{F})$ is called *positive definite* Finsler manifold. A point $x \in M$ is called *(semi-)Riemannian* if the Minkowski functional $\mathscr{F}_x$ is (semi-)Euclidean.

We remark that in many applications the metric $\mathscr{F}$ is defined and smooth only on an open cone $\mathscr{C}M \subset TM \setminus \{0\}$, where $\mathscr{C}M = \cup_{x \in M} \mathscr{C}_x M$ is a fiber bundle over M such that each $\mathscr{C}_x M$ is an open cone in $T_xM \setminus \{0\}$. In such case $(M, \mathscr{F})$ is called *singular* (or *non y-global*) Finsler space (cf. [27]).

The symmetric bilinear form

$$g_{x,y} : (u, v) \mapsto g_{ij}(x, y) u^i v^j = \frac{1}{2} \left. \frac{\partial^2 \mathscr{F}_x^2(y + su + tv)}{\partial s\, \partial t} \right|_{t=s=0}, \quad u, v \in T_xM$$

is called the metric tensor of the Finsler manifold $(M, \mathscr{F})$. The Finsler function is called *absolutely homogeneous* at $x \in M$, if $\mathscr{F}_x(\lambda y) = |\lambda| \mathscr{F}_x(y)$ for all $\lambda \in \mathbb{R}$. If $\mathscr{F}$ is absolutely homogeneous at every $x \in M$, then the Finsler manifold $(M, \mathscr{F})$ is *reversible*.

The simplest non-Riemannian Finsler metrics are the Randers metrics firstly studied by Randers in [26]. A Finsler manifold $(M, \mathscr{F})$ is called *Randers manifold* if the Finsler metric $\mathscr{F}$ can be expressed in the form $F(x, y) = a_x(y) + b_x(y)$, where $a_x(y) = \sqrt{a_{ij}(x) y^i y^j}$ is a Riemannian metric and $b_x(y) = b_i(x) y^i$ is a nowhere zero 1-form.

The *canonical spray* of a Finsler manifold $(M, \mathscr{F})$ is locally given by $\mathscr{S} = y^i \frac{\partial}{\partial x^i} - 2G^i(x, y) \frac{\partial}{\partial y^i}$, where

$$G^i(x, y) := \frac{1}{4} g^{il}(x, y) \left(2 \frac{\partial g_{jl}}{\partial x^k}(x, y) - \frac{\partial g_{jk}}{\partial x^l}(x, y) \right) y^j y^k. \tag{12.13}$$

are the *geodesic coefficients*. The *geodesics* of a Finsler manifold $(M, \mathscr{F})$ are the geodesics of the canonical spray of $(M, \mathscr{F})$ determined by the differential equations (12.3). The horizontal covariant derivative with respect to the spray associated to the Finsler manifold $(M, \mathscr{F})$ is called the *horizontal Berwald covariant derivative*, cf. Eq. (12.6). The corresponding homogeneous (nonlinear) parallel translation $\tau_c : T_{c(0)}M \to T_{c(1)}M$ along a curve $c(t)$, given by Eqs. (12.9) and (12.10), is called the *canonical homogeneous (nonlinear) parallel translation* of the Finsler manifold $(M, \mathscr{F})$. Since the geodesic coefficients $G^i(x, y)$ are differentiable functions on the slit tangent bundle $\hat{T}M = TM \setminus \{0\}$ the canonical homogeneous (nonlinear) parallel translation $\tau_c : T_{c(0)}M \to T_{c(1)}M$ induces differentiable maps between the slit tangent spaces $T_{c(0)}M \setminus \{0\}$ and $T_{c(1)}M \setminus \{0\}$.

12.2.2.2 Curvature

The *Riemannian curvature tensor* field of the Finsler manifold $(M, \mathscr{F})$ is the *curvature tensor* field $v[X^h, Y^h]$ of the canonical spray manifold of $(M, \mathscr{F})$ is defined on the pull-back bundle $(\pi^*TM, \bar{\pi}, \hat{T}M)$, (cf. Eq. (12.7)). If $\mathscr{H}TM$ is integrable, then the Riemannian curvature is identically zero. According to Eq. (12.8) the expression of the Riemannian curvature tensor $R_{(x,y)} = R^i_{jk}(x, y)dx^j \otimes dx^k \otimes \frac{\partial}{\partial x^i}$ is

$$R^i_{jk}(x, y) = \frac{\partial G^i_j(x, y)}{\partial x^k} - \frac{\partial G^i_k(x, y)}{\partial x^j} + G^m_j(x, y)G^i_{km}(x, y) - G^m_k(x, y)G^i_{jm}(x, y)$$

in a local coordinate system. The manifold $(M, \mathscr{F})$ has constant flag curvature $\lambda \in \mathbb{R}$, if for any $x \in M$ the local expression of the Riemannian curvature is

$$R^i_{jk}(x, y) = \lambda \left(\delta^i_k g_{jm}(x, y)y^m - \delta^i_j g_{km}(x, y)y^m \right). \tag{12.14}$$

In this case the flag curvature of the Finsler manifold (cf. [6], Sect. 2.1 pp. 43–46) does not depend either on the point or on the 2-flag.

The *Berwald curvature tensor* field $B_{(x,y)} = B^i_{jkl}(x, y)dx^j \otimes dx^k \otimes dx^l \otimes \frac{\partial}{\partial x^i}$ is

$$B^i_{jkl}(x, y) = \frac{\partial G^i_{jk}(x, y)}{\partial y^l} = \frac{\partial^3 G^i(x, y)}{\partial y^j \partial y^k \partial y^l}. \tag{12.15}$$

The *mean Berwald curvature tensor* field $E_{(x,y)} = E_{jk}(x, y)dx^j \otimes dx^k$ is the trace

$$E_{jk}(x, y) = B^l_{jkl}(x, y) = \frac{\partial^3 G^l(x, y)}{\partial y^j \partial y^k \partial y^l}. \tag{12.16}$$

The *Landsberg curvature tensor* field $L_{(x,y)} = L^i_{jkl}(x, y)dx^j \otimes dx^k \otimes dx^l \otimes \frac{\partial}{\partial x^i}$ is

$$L_{(x,y)}(u, v, w) = g_{(x,y)} \left(\nabla_w B_{(x,y)}(u, v, w), y \right), \quad u, v, w \in T_xM.$$

According to Lemma 6.2.2, Eq. (6.30), p. 85 in [27], one has for $u, v, w \in T_xM$

$$\nabla_w g_{(x,y)}(u, v) = -2L_{(x,y)}(u, v, w).$$

Lemma 2 *The horizontal Berwald covariant derivative of the tensor field*

$$Q_{(x,y)} = \left(\delta^i_j g_{km}(x,y)y^m - \delta^i_k g_{jm}(x,y)y^m\right) dx^j \otimes dx^k \otimes dx^l \otimes \frac{\partial}{\partial x^i}$$

vanishes.

Proof For any vector field $W \in \mathfrak{X}^\infty(M)$ we have $\nabla_W y = 0$ and $\nabla_W \mathsf{Id}_{TM} = 0$. Moreover, since $L_{(x,y)}(y, v, w) = 0$ (cf. Eq. (6.28), p. 85 in [27]) we get the assertion.

12.2.2.3 Projectively Flat Finsler Manifold

A Finsler manifold $(D, \mathscr{F})$ on an open subset $D \subset \mathbb{R}^n$ is said to be *projectively flat*, if all geodesics of $(D, \mathscr{F})$ are contained in straight lines of the affine space associated to $\mathbb{R}^n$. A Finsler manifold $(M, \mathscr{F})$ is said to be *locally projectively flat*, if for any point in $p \in M$ there exists a local coordinate map $x : U \to \mathbb{R}^n$ of a neighborhood $U \subset M$ of p such that the Finsler manifold induced by the Finsler function $\mathscr{F}$ on the image $x(U) = D$ is projectively flat. The space $\mathbb{R}^n$ containing D is called to be *projectively related to* $(M, \mathscr{F})$.

Let $(M, \mathscr{F})$ be a locally projectively flat Finsler manifold and $(x^1, \ldots, x^n) : U \to D$ a local coordinate map corresponding to canonical coordinates of the space $\mathbb{R}^n$ which is projectively related to $(M, \mathscr{F})$. Then the geodesic coefficients (12.13) are of the form

$$G^i(x,y) = \mathscr{P}(x,y)y^i, \quad G^i_k = \frac{\partial \mathscr{P}}{\partial y^k}y^i + \mathscr{P}\delta^i_k, \quad G^i_{kl} = \frac{\partial^2 \mathscr{P}}{\partial y^k \partial y^l}y^i + \frac{\partial \mathscr{P}}{\partial y^k}\delta^i_l + \frac{\partial \mathscr{P}}{\partial y^l}\delta^i_k,$$

$$\tag{12.17}$$

where $\mathscr{P}$ is a 1-homogeneous function in y, called the *projective factor* of $(M, \mathscr{F})$, (cf. [6], p. 63). Clearly, the intersections of 2-planes of $\mathbb{R}^n$ with the image D of the coordinate map $(x^1, \ldots, x^n) : U \to D$ are images of totally geodesic submanifolds of $(M, \mathscr{F})$.

Remark 1 The canonical homogeneous parallel translation $\tau_c : T_{c(0)}M \to T_{c(1)}M$ in a locally projectively flat Finsler manifold $(M, \mathscr{F})$ along curves $c(t)$ contained in the domain of the coordinate system $(x^1, \ldots, x^n)$ are linear maps if and only if the projective factor $\mathscr{P}(x, y)$ is a linear function in y. Hence the nonlinearity in y of the projective factor implies that the locally projectively flat Finsler manifold is non-Riemannian.

Locally projectively flat Randers manifolds with constant flag curvature were classified by Shen in [29]. He proved that any locally projectively flat non-Riemannian Randers manifold $(M, \mathscr{F})$ with non-zero constant flag curvature has negative curvature. These metrics can be normalized by a constant factor so that the curvature is $-\frac{1}{4}$. In this case $(M, \mathscr{F})$ is isometric to the Finsler manifold defined by the metric function

$$\mathscr{F}(x, y) = \frac{\sqrt{|y|^2 - (|x|^2|y|^2 - \langle x, y \rangle^2)}}{1 - |x|^2} \pm \left(\frac{\langle x, y \rangle}{1 - |x|^2} + \frac{\langle a, y \rangle}{1 + \langle a, x \rangle} \right) \qquad (12.18)$$

on the unit ball $\mathbb{D}^n \subset \mathbb{R}^n$, where $a \in \mathbb{R}^n$ is any constant vector with $|a| < 1$. According to Lemma 8.2.1 in [6], p.155, the projective factor $\mathscr{P}(x, y)$ can be computed by the formula

$$\mathscr{P}(x, y) = \frac{1}{2\mathscr{F}} \frac{\partial \mathscr{F}}{\partial x^i} y^i.$$

An easy calculation yields

$$\pm \frac{\partial \mathscr{F}}{\partial x^i} y^i = \left(\frac{\sqrt{|y|^2 - (|x|^2|y|^2 - \langle x, y \rangle^2)} \pm \langle x, y \rangle}{1 - |x|^2} \right)^2 - \left(\frac{\langle a, y \rangle}{1 + \langle a, x \rangle} \right)^2,$$

hence

$$\mathscr{P}(x, y) = \frac{1}{2} \left(\frac{\pm \sqrt{|y|^2 - (|x|^2|y|^2 - \langle x, y \rangle^2)} + \langle x, y \rangle}{1 - |x|^2} - \frac{\langle a, y \rangle}{1 + \langle a, x \rangle} \right). \qquad (12.19)$$

12.2.3 Finsler Holonomy Represented on the Indicatrix Bundle

The notion of the holonomy group of a Riemannian or Finslerian manifolds is an adaptation of the corresponding notion of spray manifolds: it is the group generated by parallel translations along loops with respect to the canonical associated connection. However, the parallel translation leaves invariant the indicatrix bundle, the holonomy group can be identified by its action on the indicatrix at the initial point. Hence the holonomy group can be considered as a subgroup of the diffeomorphism group of the indicatrix (Fig. 12.5a, b).

The holonomy properties of Finsler spaces are essentially different from the Riemannian one, and it is far from being well understood. The main difficulty comes from the fact that in the general case the canonical connection of a Finsler manifold is neither linear nor metrical (that is, the parallel translation is not necessarily preserving the metric). Only much weaker properties are fulfilled: instead of the linearity it is only 1-homogeneous, and instead of the metrical property it is preserving only the norm function. Nonetheless these properties allow us to consider the parallel translations as maps between the indicatrices and therefore the holonomy group as a subgroup of the diffeomorphism group of the indicatrix.

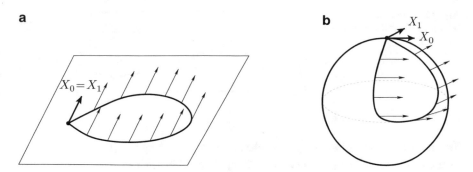

Fig. 12.5 (a) Trivial holonomy: the plane. (b) Nontrivial holonomy: sphere

Let $(M, \mathscr{F})$ be an n-dimensional Finsler manifold. The *indicatrix* $\mathfrak{I}_x M$ at $x \in M$ is a hypersurface of $T_x M$ defined by

$$\mathfrak{I}_x M := \{y \in T_x M; \ \mathscr{F}(y) = \pm 1\}.$$

If the Finsler manifold $(M, \mathscr{F})$ is positive definite, then the indicatrix $\mathfrak{I}_x M$ is a compact hypersurface in the tangent space $T_x M$, diffeomorphic to the standard $(n-1)$-dimensional sphere.

In the sequel $(\mathfrak{I}M, \pi, M)$ will denote the *indicatrix bundle* of $(M, \mathscr{F})$ and $i :$ $\mathfrak{I}M \hookrightarrow TM$ the natural embedding of the indicatrix bundle into the tangent bundle (TM, π, M).

The *parallel translation* $\tau_c : T_{c(0)}M \rightarrow T_{c(1)}M$ along a curve $c : [0, 1] \rightarrow \mathbb{R}$ on a Finsler manifold $(M, \mathscr{F})$ is defined by the parallel translation of the associated spray manifold (cf. Sect. 12.2.1.2). It is determined by vector fields $X(t)$ along $c(t)$ which are solutions of the differential equation (12.10). Since $\tau_c : T_{c(0)}M \rightarrow T_{c(1)}M$ is a differentiable map between the slit tangent spaces $\hat{T}_{c(0)}M$ and $\hat{T}_{c(1)}M$ preserves the value of the Finsler function, it induces a map

$$\tau_c^{\mathfrak{I}} : \mathfrak{I}_{c(0)}M \longrightarrow \mathfrak{I}_{c(1)}M \tag{12.20}$$

between the indicatrices. Since the parallel translation is 1-homogeneous, the parallel translation τ_c is entirely characterized by the map $\tau_c^{\mathfrak{I}}$: we have $\tau_C(0) = 0$ and for every non-zero vector $v \in T_{c(0)}M$ we have

$$\tau_c(v) = |v| \cdot \tau_c^{\mathfrak{I}} \left(\frac{v}{|v|} \right).$$

It follows from these observations that the holonomy group $\mathsf{Hol}(x)$ of the spray manifold associated to a Finsler manifold $(M, \mathscr{F})$ (cf. Sect. 12.2.1.4) is uniquely determined by its action on the indicatrix in the tangent space $T_x M$ at the point x. Hence we can formulate

Fig. 12.6 Holonomy
transformation induced on the
indicatrix

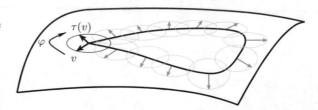

Definition 1 The *holonomy group* $\mathsf{Hol}(x)$ of a Finsler space $(M, \mathscr{F})$ at $x \in M$ is the subgroup of the group of diffeomorphisms $\mathsf{Diff}(\mathfrak{I}_x M)$ of the indicatrix $\mathfrak{I}_x M$ determined by parallel translation of $\mathfrak{I}_x M$ along piece-wise differentiable closed curves initiated at the point $x \in M$ (Fig. 12.6).

We note that the holonomy group $\mathsf{Hol}(x)$ is a topological subgroup of the regular infinite dimensional Lie group $\mathsf{Diff}(\mathfrak{I}_x M)$, but its differentiable structure is not known in general. ($\mathsf{Diff}(\mathfrak{I}_x M)$ denotes the group of all C^∞-diffeomorphism of $\mathfrak{I}_x M$ with the C^∞-topology.)

12.3 Diffeomorphism Groups and Their Tangent Algebras

The group $\mathsf{Diff}(M)$ of all smooth diffeomorphisms of a differentiable manifold M is a regular infinite dimensional Lie group modeled on the vector space $\mathfrak{X}_c(M)$ of smooth vector fields on M with compact support. The Lie algebra of the infinite dimensional Lie group $\mathsf{Diff}(M)$ is the vector space $\mathfrak{X}_c(M)$, equipped with the negative of the usual Lie bracket (c.f. Kriegl and Michor [15], Sect. 43.1, pp. 454–456).

Here we discuss the tangential properties of Lie algebras of vector fields to an abstract subgroup of the diffeomorphism group of a manifold. The results of this section will be applied in the following to the investigation of tangent Lie algebras of the holonomy subgroup of the diffeomorphism group of an indicatrix $\mathfrak{I}_x M$ and to the fibered holonomy subgroup of the diffeomorphism group of the indicatrix bundle $\mathfrak{I}(M)$.

Let M be a C^∞ manifold, let $\mathscr{G}$ be a (not necessarily differentiable) subgroup of the diffeomorphism group $\mathsf{Diff}^\infty(M)$ and let $\mathfrak{X}^\infty(M)$ be the Lie algebra of smooth vector fields on M.

Definition 2 A vector field $X \in \mathfrak{X}^\infty(M)$ is called *tangent* to the subgroup $\mathscr{G}$ of $\mathsf{Diff}^\infty(M)$, if there exists a $\mathscr{C}^1$-differentiable 1-parameter family $\{\phi_t \in \mathscr{G}\}_{t \in (-\varepsilon, \varepsilon)}$ of diffeomorphisms of M such that $\phi_0 = \mathsf{Id}$ and $\frac{\partial \phi_t}{\partial t}\big|_{t=0} = X$. A Lie subalgebra $\mathfrak{g}$ of $\mathfrak{X}^\infty(M)$ is called *tangent* to $\mathscr{G}$, if all elements of $\mathfrak{g}$ are tangent vector fields to $\mathscr{G}$.

Unfortunately, it is not true in general that tangent vector fields to a group $\mathscr{G}$ generate a tangent Lie algebra to $\mathscr{G}$. This is why we have to introduce a stronger tangential property in Definition 4.

Definition 3 A $\mathscr{C}^\infty$-differentiable k-parameter family $\{\phi_{(t_1,...,t_k)} \in \mathsf{Diff}^\infty(M)\}_{t_i \in (-\varepsilon,\varepsilon)}$ of diffeomorphisms of M is called a *commutator-like family* if it satisfies the equations

$$\phi_{(t_1,...,t_k)} = \mathsf{Id}, \quad \text{whenever} \quad t_j = 0 \quad \text{for some} \quad 1 \leq j \leq k.$$

We remark that any $\mathscr{C}^1$-differentiable 1-parameter family $\{\phi_t \in \mathscr{G}\}_{t \in (-\varepsilon,\varepsilon)}$ of diffeomorphisms of M with $\phi_0 = \mathsf{Id}$ is a commutator-like family. Moreover, the commutator of commutator-like families is commutator-like, and the inverse of a commutator-like family is commutator-like.

Definition 4 A vector field $X \in \mathfrak{X}^\infty(M)$ is called *strongly tangent* to the subgroup $\mathscr{G}$ of $\mathsf{Diff}^\infty(M)$, if there exists a commutator-like family $\{\phi_{(t_1,...,t_k)} \in \mathsf{Diff}^\infty(M)\}_{t_i \in (-\varepsilon,\varepsilon)}$ of diffeomorphisms satisfying the conditions

(A) $\phi_{(t_1,...,t_k)} \in \mathscr{G}$ for all $t_i \in (-\varepsilon, \varepsilon)$, $1 \leq i \leq k$,

(B) $\left. \dfrac{\partial^k \phi_{(t_1,...,t_k)}}{\partial t_1 \cdots \partial t_k} \right|_{(0,...,0)} = X.$

It follows from the commutator-like property that $\left. \dfrac{\partial^k \phi_{(t_1,...,t_k)}}{\partial t_1 \cdots \partial t_k} \right|_{(0,...,0)}$ is the first non-necessarily vanishing derivative of the diffeomorphism family $\{\phi_{(t_1,...,t_k)}\}$ at any point $x \in M$, and therefore it determines a vector field. On the other hand, by parameterizing the commutator like family of diffeomorphism, it can be shown that if a vector field is strongly tangent to a group $\mathscr{G}$, then it is also tangent to $\mathscr{G}$. Moreover, we have the following

Theorem 1 *Let $\mathscr{V}$ be a set of vector fields strongly tangent to the subgroup $\mathscr{G}$ of $\mathsf{Diff}^\infty(M)$. The Lie subalgebra $\mathfrak{v}$ of $\mathfrak{X}^\infty(M)$ generated by $\mathscr{V}$ is tangent to $\mathscr{G}$.*

Proof First, we investigate some properties of vector fields strongly tangent to the group $\mathscr{G}$.

Lemma 1 *Let $\{\psi_{(t_1,...,t_h)} \in \mathsf{Diff}^\infty(U)\}_{t_i \in (-\varepsilon,\varepsilon)}$ be a $\mathscr{C}^\infty$-differentiable h-parameter commutator-like family of (local) diffeomorphisms on a neighborhood $U \subset \mathbb{R}^n$. Then*

(i) $\left. \dfrac{\partial^{i_1 + \cdots + i_h} \psi_{(t_1,...,t_h)}}{\partial t_1^{i_1} \cdots \partial t_h^{i_h}} \right|_{(0,...,0)} (x) = 0, \quad \text{if} \quad i_p = 0 \quad \text{for some} \quad 1 \leq p \leq h;$

(ii) $\left. \dfrac{\partial^h (\psi_{(t_1,...,t_h)})^{-1}}{\partial t_1 \cdots \partial t_h} \right|_{(0,...,0)} (x) = - \left. \dfrac{\partial^h \psi_{(t_1,...,t_h)}}{\partial t_1 \cdots \partial t_h} \right|_{(0,...,0)} (x);$

(iii) $\left. \dfrac{\partial^h \psi_{(t_1,...,t_h)}}{\partial t_1 \cdots \partial t_h} \right|_{(0,...,0)} (x) = \left. \dfrac{\partial \psi_{(\sqrt[h]{t},...,\sqrt[h]{t})}}{\partial t} \right|_{t=0} (x)$

at any point $x \in U$.

Proof Assertions (i) and (ii) can be obtained by direct computation. It follows from (i) that $\left.\frac{\partial^h \psi_{(t_1,\ldots,t_h)}}{\partial t_1 \ldots \partial t_h}\right|_{(0,\ldots,0)}(x)$ is the first non-necessarily vanishing derivative of the diffeomorphism family $\{\psi_{(t_1,\ldots,t_h)}\}$ at any point $x \in M$. Using

$$\psi_{(t_1,\ldots,t_k)}(x) = x + t_1 \cdots t_k \left(X(x) + \omega(x, t_1, \ldots, t_k) \right),$$

where $\lim_{t_i \to 0} \omega(x, t_1, \ldots, t_k) = 0$ we obtain that

$$\left.\frac{\partial}{\partial t}\right|_{t=0} \psi_{(\sqrt[k]{t},\ldots,\sqrt[k]{t})}(x) = \left.\frac{\partial}{\partial t}\right|_{t=0} \left(x + t \left(X(x) + \omega(x, \sqrt[k]{t}, \ldots, \sqrt[k]{t}) \right) \right) = X(x),$$

which proves (iii).

We remark that assertion (*iii*) means that any vector field strongly tangent to $\mathscr{G}$ is tangent to $\mathscr{G}$. Now, we generalize a well-known relation between the commutator of vector fields and the commutator of their induced flows.

Lemma 2 *Let* $\{\phi_{(s_1,\ldots,s_k)}\}$ *and* $\{\psi_{(t_1,\ldots,t_l)}\}$ *be* $\mathscr{C}^\infty$*-differentiable k-parameter, respectively, l-parameter families of (local) diffeomorphisms defined on a neighborhood* $U \subset \mathbb{R}^n$*. Assume that* $\phi_{(s_1,\ldots,s_k)} = \mathsf{Id}$*, respectively,* $\psi_{(t_1,\ldots,t_l)} = \mathsf{Id}$*, if some of their variables is 0. Then the family of (local) diffeomorphisms* $[\phi_{(s_1,\ldots,s_k)}, \psi_{(t_1,\ldots,t_l)}]$ *defined by the commutator of the group* $\mathsf{Diff}^\infty(U)$ *fulfills* $[\phi_{(s_1,\ldots,s_k)}, \psi_{(t_1,\ldots,t_l)}] = \mathsf{Id}$*, if some of its variables equals 0. Moreover*

$$\left.\frac{\partial^{k+l}[\phi_{(s_1\ldots s_k)}, \psi_{(t_1\ldots t_l)}]}{\partial s_1 \ldots \partial s_k \partial t_1 \ldots \partial t_l}\right|_{(0\ldots0;0\ldots0)}(x) = -\left[\left.\frac{\partial^k \phi_{(s_1\ldots s_k)}}{\partial s_1 \ldots \partial s_k}\right|_{(0\ldots0)}, \left.\frac{\partial^l \psi_{(t_1\ldots t_l)}}{\partial t_1 \ldots \partial t_l}\right|_{(0\ldots0)} \right](x)$$

at any point $x \in U$.

Proof The group theoretical commutator $[\phi_{(s_1,\ldots,s_k)}, \psi_{(t_1,\ldots,t_l)}]$ of the families of diffeomorphisms satisfies $[\phi_{(s_1,\ldots,s_k)}, \psi_{(t_1,\ldots,t_l)}] = \mathsf{Id}$, if some of its variables equals 0. Hence

$$\left.\frac{\partial^{i_1+\cdots+i_k+j_1+\cdots+j_l}[\phi_{(s_1,\ldots,s_k)}, \psi_{(t_1,\ldots,t_l)}]}{\partial s_1^{i_1} \ldots \partial s_k^{i_k} \partial t_1^{j_1} \ldots \partial t_l^{i_l}}\right|_{(0,\ldots,0;0,\ldots,0)} = 0,$$

if $i_p = 0$ or $j_q = 0$ for some index $1 \leqslant p \leqslant k$ or $1 \leqslant q \leqslant l$. The families of diffeomorphisms $\{\phi_{(s_1,\ldots,s_l)}\}$, $\{\psi_{(t_1,\ldots,t_l)}\}$, $\{\phi_{(s_1,\ldots,s_l)}^{-1}\}$ and $\{\psi_{(t_1,\ldots,t_l)}^{-1}\}$ are the constant family Id, if some of their variables equals 0. Hence one has

$$\frac{\partial^{k+l}[\phi_{(s_1...s_k)}, \psi_{(t_1...t_l)}]}{\partial s_1 \ \ldots \ \partial s_k \, \partial t_1 \ \ldots \ \partial t_l}\Bigg|_{(0,...,0; \, 0,...,0)}(x) \qquad\qquad (12.21)$$

$$= \frac{\partial^k}{\partial s_1 \ldots \partial s_k}\Bigg|_{(0...0)}\left(\frac{\partial^l\left(\phi_{(s_1...s_k)}^{-1} \circ \psi_{(t_1...t_l)}^{-1} \circ \phi_{(s_1...s_k)} \circ \psi_{(t_1...t_l)}(x)\right)}{\partial t_1 \ldots \partial t_l}\Bigg|_{(0...0)}\right)$$

$$= \frac{\partial^k}{\partial s_1 \ldots \partial s_k}\Bigg|_{(0...0)}\left(d(\phi_{(s_1...s_k)}^{-1})_{\phi_{(s_1...s_k)}(x)}\frac{\partial^l\psi_{(t_1...t_l)}^{-1}}{\partial t_1 \ldots \partial t_l}\Bigg|_{(0,...,0)}(\phi_{(s_1...s_k)}(x))\right),$$

where $d\big(\phi_{(s_1,...,s_k)}^{-1}\big)_{\phi_{(s_1,...,s_k)}(x)}$ denotes the Jacobi operator of the map $\phi_{(s_1,...,s_k)}^{-1}$ at the point $\phi_{(s_1,...,s_k)}(x)$. Using the fact, that $\{\phi_{(s_1,...,s_k)}\}$ is the constant family Id, if some of its variables equals 0, and the relation $d(\phi_{(0,...,0)}^{-1})_{\phi_{(s_1,...,s_k)}(x)} = \text{Id}$, we obtain that (12.21) can be written as

$$d\left(\frac{\partial^k\phi_{(s_1...s_k)}^{-1}}{\partial s_1 \ldots \partial s_k}\Bigg|_{(0...0)}\right)_x\frac{\partial^l\psi_{(t_1...t_l)}^{-1}(x)}{\partial t_1 \ldots \partial t_l}\Bigg|_{(0...0)} + d\left(\frac{\partial^l\psi_{(t_1...t_l)}^{-1}}{\partial t_1 \ldots \partial t_l}\Bigg|_{(0...0)}\right)_x\frac{\partial^k\phi_{(s_1...s_k)}(x)}{\partial s_1 \ldots \partial s_k}\Bigg|_{(0,...,0)}.$$

According to assertion (ii) of Lemma 1 the last formula gives

$$d\left(\frac{\partial^k\phi_{(s_1...s_k)}}{\partial s_1 \ \ldots \ \partial s_k}\Bigg|_{(0...0)}\right)_x\frac{\partial^l\psi_{(t_1...t_l)}(x)}{\partial t_1 \ \ldots \ \partial t_l}\Bigg|_{(0...0)} - d\left(\frac{\partial^l\psi_{(t_1...t_l)}}{\partial t_1 \ \ldots \ \partial t_l}\Bigg|_{(0...0)}\right)_x\frac{\partial^k\phi_{(s_1...s_k)}(x)}{\partial s_1 \ \ldots \ \partial s_k}\Bigg|_{(0...0)},$$

which is the Lie bracket of vector fields

$$\left[\frac{\partial^l\psi_{(t_1,...,t_l)}}{\partial t_1 \ \ldots \ \partial t_l}\Bigg|_{(0,...,0)}, \frac{\partial^k\phi_{(s_1,...,s_k)}}{\partial s_1 \ \ldots \ \partial s_k}\Bigg|_{(0,...,0)}\right] : U \to \mathbb{R}^n.$$

The previous lemma gives for 1-parameter families $\{\phi_t \in \mathcal{G}\}_{t\in(-\varepsilon,\varepsilon)}$ and $\{\psi_t \in \mathcal{G}\}_{t\in(-\varepsilon,\varepsilon)}$ of diffeomorphisms of M with $\phi_0 = \psi_0 = \text{Id}$ the relation

$$\frac{\partial^2}{\partial s\partial t}\Bigg|_{s=0,t=0}[\phi_s, \psi_t] = -\left[\frac{\partial\phi_s}{\partial s}\Bigg|_{s=0}, \frac{\partial^l\psi_t}{\partial t}\Bigg|_{t=0}\right]. \qquad (12.22)$$

Lemma 3 *Any Lie subalgebra of $\mathfrak{X}^\infty(M)$ algebraically generated by strongly tangent vector fields to the group $\mathcal{G}$ has a basis consisting of vector fields strongly tangent to the group $\mathcal{G}$.*

Proof Let $\mathcal{V}$ be a set of strongly tangent vector fields to the group $\mathcal{G}$ and $\mathfrak{v}$ the Lie algebra algebraically generated by $\mathcal{V}$. The iterated Lie brackets of vector fields belonging to $\mathcal{V}$ linearly generate the vector space $\mathfrak{v}$. It follows from Lemma 2 that these iterated Lie brackets of vector fields are strongly tangent to the group $\mathcal{G}$. Hence $\mathfrak{v}$ is linearly generated by vector fields strongly tangent to $\mathcal{G}$.

Lemma 4 *Linear combinations of vector fields tangent to $\mathcal{G}$ are tangent to $\mathcal{G}$.*

Proof If X and Y are vector fields tangent to $\mathcal{G}$ then there exist $\mathscr{C}^1$-differentiable 1-parameter families of diffeomorphisms $\{\phi_t \in \mathcal{G}\}$ and $\{\psi_t \in \mathcal{G}\}$ such that

$$\phi_0 = \psi_0 = \mathsf{Id}, \qquad \left.\frac{\partial}{\partial t}\right|_{t=0}\phi_t = X, \qquad \left.\frac{\partial}{\partial t}\right|_{t=0}\psi_t = Y.$$

Hence the 1-parameter families of diffeomorphisms $\{\phi_t \circ \psi_t\}$ and $\{\phi_{ct}\}$ satisfy

$$X + Y = \left.\frac{\partial}{\partial t}\right|_{t=0}(\phi_t \circ \psi_t), \qquad cX = \left.\frac{\partial}{\partial t}\right|_{t=0}\phi_{(ct)}, \quad \text{for all} \quad c \in \mathbb{R}^n,$$

which proves the assertion.
Lemmas 1–4 prove Theorem 1.

12.3.1 Diffeomorphism Group of Compact Manifolds

If K is a compact manifold, then the group $\mathsf{Diff}^\infty(K)$ of diffeomorphisms is an infinite dimensional Lie group belonging to the class of Fréchet Lie groups. The Lie algebra of $\mathsf{Diff}^\infty(K)$ is the Lie algebra $\mathfrak{X}^\infty(K)$ of smooth vector fields on K endowed with the negative of the usual Lie bracket of vector fields. The Fréchet Lie group $\mathsf{Diff}^\infty(K)$ is modeled on the locally convex topological Fréchet vector space $\mathfrak{X}^\infty(K)$. A sequence $\{f_j\}_{j\in\mathbb{N}} \subset \mathfrak{X}^\infty(K)$ converges to f in the topology of $\mathfrak{X}^\infty(K)$ if and only if the functions f_j and all their derivatives converge uniformly to f, respectively to the corresponding derivatives of f. We note that the difficulty of the theory of Fréchet manifolds comes from the fact that the inverse function theorem and the existence theorems of differential equations, which are well known for Banach manifolds, are not true in this category. These problems have led to the concept of regular Fréchet Lie groups (cf. [25] Chapter III, [15] Chapter VIII). The distinguishing properties of regular Fréchet Lie groups can be summarized as (*a*) the existence of the smooth exponential map from the Lie algebra of the Fréchet Lie groups to the group itself, (*b*) the existence of product integrals, which produces the convergence of some approximation methods for solving differential equations (cf. Sect. III.5. in [25], pp. 83–89). J. Teichmann gave a detailed discussion of these properties in [32].

Proposition 1 *If a Lie subalgebra $\mathfrak{g}$ of the Lie algebra $\mathfrak{X}^\infty(K)$ of smooth vector fields on a compact manifold K is tangent to a subgroup $\mathcal{G}$ of the diffeomorphism group $\mathsf{Diff}^\infty(K)$ of K, then the group generated by the exponential image $\exp(\mathfrak{g})$ of $\mathfrak{g}$ is contained in the topological closure $\overline{\mathcal{G}}$ of $\mathcal{G}$ in $\mathsf{Diff}^\infty(K)$.*

Proof Let us denote by $\langle \exp(\mathfrak{g}) \rangle$ the group generated by the exponential image of $\mathfrak{g}$. For any element $X \in \mathfrak{g}$ there exists a $\mathscr{C}^1$-differentiable 1-parameter family $\{\Phi(t) \in \mathscr{G}\}_{t \in \mathbb{R}}$ of diffeomorphisms of the manifold K such that

$$\Phi(0) = \mathrm{Id} \quad \text{and} \quad \left. \frac{\partial \Phi(s)}{\partial s} \right|_{s=0} = X.$$

Then, considering $\Phi(t)$ as "hair" and using the argument of Corollary 5.4. in [25], p. 85, we get that

$$\left\{ \Phi \left(\frac{t}{n} \right)^n \right\}_{t \in \mathbb{R}} = \left\{ \Phi \left(\frac{t}{n} \right) \circ \cdots \circ \Phi \left(\frac{t}{n} \right) \right\}_{t \in \mathbb{R}} \subset \mathscr{G}, \quad n = 1, 2 \ldots$$

as a sequence of $\mathrm{Diff}^\infty(K)$ converges uniformly in all derivatives to $\exp(tX)$. It follows that we have $\{\exp(tX); t \in \mathbb{R}\} \subset \overline{\mathscr{G}}$ for any $X \in \mathfrak{g}$) and therefore $\exp(\mathfrak{g}) \subset \overline{\mathscr{G}}$. Naturally, if for the generated group $\langle \exp(\mathfrak{g}) \rangle$, then the containing relation is preserved, that is $\langle \exp(\mathfrak{g}) \rangle \subset \overline{\mathscr{G}}$, which proves the proposition.

12.4 Curvature Algebra

12.4.1 Curvature Vector Fields at a Point

Definition 5 A vector field $\xi \in \mathfrak{X}^\infty(\mathfrak{I}M)$ on the indicatrix bundle $\mathfrak{I}M$ is a *curvature vector field* of the Finsler manifold $(M, \mathscr{F})$, if there exist vector fields $X, Y \in \mathfrak{X}^\infty(M)$ on the manifold M such that $\xi = r(X, Y)$, where for every $x \in M$ and $y \in \mathfrak{I}_x M$ we have

$$r(X, Y)(x, y) := R_{(x,y)}(X_x, Y_x). \tag{12.23}$$

If $x \in M$ is fixed and $X, Y \in T_x M$, then the vector field $y \to r(X, Y)(x, y)$ on $\mathfrak{I}_x M$ is a *curvature vector field at the point x*.

The Lie algebra $\mathfrak{R}(M)$ of vector fields generated by the curvature vector fields of $(M, \mathscr{F})$ is called the *curvature algebra* of the Finsler manifold $(M, \mathscr{F})$. For a fixed $x \in M$ the Lie algebra $\mathfrak{R}_x$ of vector fields generated by the curvature vector fields at x is called the *curvature algebra at the point x*.

In this section we investigate the properties of the curvature vector fields and of the curvature algebra at a fixed point $x \in M$. The curvature vector fields $\xi = r_x(X, Y)$ are tangent to the indicatrix

$$g_{(x,y)} \big(y, R_{(x,y)}(l_y(X), l_y(Y)) \big) = 0, \quad \text{for any} \quad y, X, Y \in T_x M$$

according to [27], Eq. (10.9). In the sequel we investigate the tangential properties of the curvature algebra to the holonomy group of the canonical connection ∇ of a Finsler manifold.

Proposition 2 *Any curvature vector field at $x \in M$ is strongly tangent to the holonomy group* $\mathsf{Hol}(x)$.

Proof Indeed, let us consider the curvature vector field $\xi := r_x(X, Y) \in \mathfrak{X}(\mathfrak{I}_x M)$ corresponding to the directions $X, Y \in T_x M$ and let $\hat{X}, \hat{Y} \in \mathfrak{X}(M)$ be commuting vector fields, i.e. $[\hat{X}, \hat{Y}] = 0$ such that $\hat{X}_x = X$, $\hat{Y}_x = Y$. By the geometric construction, the flows $\{\phi_t\}$ and $\{\psi_s\}$ of $\hat{X}$ and $\hat{Y}$ are commuting, that is $\phi_s \circ \psi_t \equiv \psi_t \circ \phi_s$. For any sufficiently small $s, t \in \mathbb{R}$ we can consider the curve $\mathscr{P}_{s,t}$ defined as follows:

$$
\mathscr{P}_{s,t}(u) = \begin{cases} \psi_u(x), & 0 \leq u \leq t, \\ \phi_{u-t}(\psi_t(x)), & t \leq u \leq t+s, \\ \psi_{u-(t+s)}^{-1}(\phi_s(\psi_t(x))), & t+s \leq u \leq 2t+s, \\ \phi_{u-(2t+s)}^{-1}(\psi_t^{-1}(\phi_s(\psi_t(x)))), & 2t+s \leq u \leq 2t+2s. \end{cases}
$$

Because of the commuting property of the flows $\{\phi_t\}$ and $\{\psi_s\}$ the curves $\mathscr{P}_{s,t}$ are closed parallelograms: their initial and final point at $u = 0$ and $u = 2t + 2s$ are the same $x \in M$. Consequently, the parallel translation $\tau_{s,t} \colon T_x M \to T_x M$ along the parallelogram $\mathscr{P}_{s,t}$ is a holonomy element for every small value of $t, s \in \mathbb{R}$ (see Fig. 12.7).

On the other hand, using the geometric construction of parallel translation presented in Sect. 12.2.1.2, we know that the flows $\{\phi_t^h\}$ and $\{\psi_s^h\}$ of the horizontal lifts $l(\hat{X})$ and $l(\hat{Y})$ can be considered as parallel translations along integral curves of $\hat{X}$ and $\hat{Y}$, respectively. They can be considered as fiber preserving diffeomorphisms of the bundle $\mathfrak{I}M$ for any $t, s \in \mathbb{R}$. Then the commutator

$$
\tau_{s,t} = [\phi_s^h, \psi_t^h] = \phi_{-s}^h \circ \psi_{-t}^h \circ \phi_s^h \circ \psi_t^h \colon \quad \mathfrak{I}M \to \mathfrak{I}M
$$

is also a fiber preserving diffeomorphism of the bundle $\mathfrak{I}M$ for any $t, s \in \mathbb{R}$. Therefore for $x \in M$ the restriction

$$
\tau_{s,t}(x) = \tau_{s,t}\big|_{\mathfrak{I}_x M} \colon \mathfrak{I}_x M \to \mathfrak{I}_x M
$$

Fig. 12.7 Parallel translation along a parallelogram

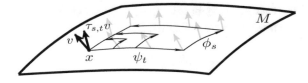

Fig. 12.8 Non-zero
curvature vector field

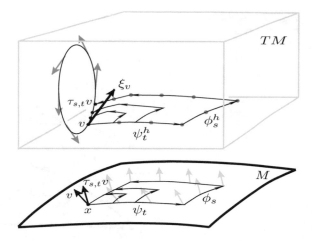

to the fiber $\mathfrak{J}_x M$ is a 2-parameter C^∞-differentiable family of diffeomorphisms
contained in the holonomy group $\mathsf{Hol}(x)$ such that

$$\tau_{s,0}(x) = \mathsf{Id}, \qquad \tau_{0,t}(x) = \mathsf{Id}, \qquad \text{and} \qquad \frac{\partial^2}{\partial t \partial s}\bigg|_{s=0,t=0} \tau_{s,t}(x) = r_x(X, Y),$$

which proves that the curvature vector field $\xi = r_x(X, Y)$ is strongly tangent to the
holonomy group $\mathsf{Hol}(x)$ and hence we obtain the assertion.

In order to enlighten the construction and the geometric meaning of the curvature
vector field $\xi = r_x(X, Y)$ we consider Fig. 12.8 which can be seen as the extension
of Fig. 12.7. Here we can present not only the geometric objects at the level of the
manifold M but at the level of the tangent manifold TM too. We remark that, as it
is usual, the tangent vectors of M are represented as "arrows" at the level of M, but
they are represented as "points" at the level of the tangent space TM. For example,
the vectors v and $\tau_{s,t}v$ are represented as arrows at x on M and points above x in TM.
The gray vectors at the level of M represent the elements of the parallel vector field
V along the parallelogram $\mathscr{P}_{s,t}$ with the initial condition $V_x = v$. The gray dots are
the points in TM corresponding to the elements of V. These dots lie on the curves
of the flows ϕ_t^h, ψ_s^h because V is a parallel field and these flows correspond to the
parallel translations along integral curves of $\hat{X}$ and $\hat{Y}$, respectively. As the picture
shows, the parallel translation along the parallelogram $P_{s,t}$ of a vector $v \in T_x M$
can be obtained by following in TM the flows ϕ_t^h, ψ_s^h, ϕ_{-t}^h, ψ_{-s}^h above $P_{s,t}$. The
indicatrix, or unite ball, at $x \in M$ is represented by the oval above x. Since the
parallel translation preserves the norm, if $v \in \mathfrak{J}_x M$, then $\tau_{s,t}v \in \mathfrak{J}_x M$. Therefore
$t \to \frac{1}{2}\tau_{t,t}(v)$ is a curve in $\mathfrak{J}_x M$. Its tangent vector at $t = 0$ is $\xi(v)$ which is therefore
a tangent vector of $\mathfrak{J}_x M$.

We remark that in the case, when the curvature is identically zero, the horizontal
lifts of commuting vector fields are also commuting vector fields. Therefore one
obtains $\phi_s^h \circ \psi_t^h \equiv \psi_t^h \circ \phi_s^h$ and $\tau_{s,t} = [\phi_s^h, \psi_t^h] : \mathfrak{J}M \to \mathfrak{J}M$ is the identity
transformation. In that case $\tau_{s,t}v \equiv v$ is a constant map and therefore its derivative is

Fig. 12.9 Vanishing
curvature vector field

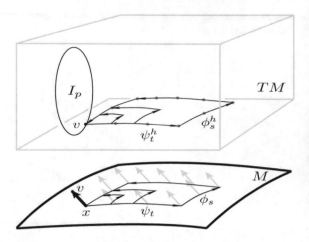

zero, that is $\xi_v = 0$. Geometrically that means that the horizontal lifts of the closed
parallelograms $\mathscr{P}_{s,t}$ are closed parallelograms. See Fig. 12.9.

Theorem 1 *The curvature algebra $\mathfrak{R}_x$ at a point $x \in M$ of a Finsler manifold
$(M, \mathscr{F})$ is tangent to the holonomy group $\mathsf{Hol}(x)$.*

*If $(M, \mathscr{F})$ is a positive definite Finsler manifold, then the group generated by the
exponential image $\exp(\mathfrak{R}_x)$ is a subgroup of the topological closure of the holonomy
group $\mathsf{Hol}(x)$.*

Proof Since by Proposition 2 the curvature vector fields are strongly tangent to
$\mathsf{Hol}(x)$ and the curvature algebra $\mathfrak{R}_x$ is algebraically generated by the curvature
vector fields, the first assertion follows from Theorem 1. The second assertion is a
consequence of Proposition 1.

Proposition 3 *The curvature algebra $\mathfrak{R}_x$ of a Riemannian manifold (M, g) at any
point $x \in M$ is isomorphic to the linear Lie algebra over the vector space T_xM
generated by the curvature operators of (M, g) at $x \in M$.*

Proof The curvature tensor field of a Riemannian manifold given by Eq. (12.7) is
linear with respect to $y \in T_xM$ and hence

$$R_{(x,y)}(\xi, \eta) = (R_x(\xi, \eta))^k_l y^l \frac{\partial}{\partial y^k},$$

where $R_x(\xi, \eta))^k_l$ is the matrix of the curvature operator $R_x(\xi, \eta): T_xM \to T_xM$
with respect to the natural basis $\left\{\frac{\partial}{\partial x^1}|_x, \ldots, \frac{\partial}{\partial x^n}|_x\right\}$. Hence any curvature vector field
$r_x(\xi, \eta)(y)$ with $\xi, \eta \in T_xM$ has the shape $r_x(\xi, \eta)(y) = (R_x(\xi, \eta))^k_l y^l \frac{\partial}{\partial y^k}$. It follows
that the flow of $r_x(\xi, \eta)(y)$ on the indicatrix $\mathfrak{I}_xM$ generated by the vector field
$r_x(\xi, \eta)(y)$ is induced by the action of the linear 1-parameter group $\exp t R_x(\xi, \eta))$
on T_xM, which implies the assertion.

Since for Finsler surfaces of non-vanishing curvature the curvature vector fields form a one-dimensional vector space and hence the generated Lie algebra is also one-dimensional, we have

Remark 2 The curvature algebra of Finsler surfaces is at most one-dimensional.

12.4.2 Constant Curvature

Now, we consider a Finsler manifold $(M, \mathscr{F})$ of non-zero constant curvature. In this case for any $x \in M$ the curvature vector field $r_x(X, Y)(y)$ has the shape (cf. (12.14))

$$r(X, Y)(y) = c \left(\delta_j^i g_{km}(y) y^m - \delta_k^i g_{jm}(y) y^m \right) X^j Y^k \frac{\partial}{\partial y^i}, \quad 0 \neq c \in \mathbb{R}.$$

Putting $y_j = g_{jm}(y) y^m$ we can write $r(X, Y)(y) = c \left(\delta_j^i y_k - \delta_k^i y_j \right) X^j Y^k \frac{\partial}{\partial y^i}$. Any linear combination of curvature vector fields has the form

$$r(A)(y) = A^{jk} \left(\delta_j^i y_k - \delta_k^i y_j \right) \frac{\partial}{\partial y^i},$$

where $A = A^{jk} \frac{\partial}{\partial x^j} \wedge \frac{\partial}{\partial x^k} \in T_x M \wedge T_x M$ is arbitrary bivector at $x \in M$.

Lemma 1 *Let $(M, \mathscr{F})$ be a Finsler manifold of non-zero constant curvature. The curvature algebra $\mathfrak{R}_x$ at any point $x \in M$ satisfies*

$$\dim \mathfrak{R}_x \geq \frac{n(n-1)}{2}, \tag{12.24}$$

where $n = \dim M$.

Proof Let us consider the curvature vector fields $r_{jk} = r_x(\frac{\partial}{\partial y^j}, \frac{\partial}{\partial y^k})(y)$ at a fixed point $x \in M$. If a linear combination

$$A^{jk} r_{jk} = A^{jk} (\delta_j^i y_k - \delta_k^i y_j) \frac{\partial}{\partial y^i} = (A^{ik} y_k - A^{ji} y_j) \frac{\partial}{\partial y^i} = 2A^{ik} y_k \frac{\partial}{\partial y^i}$$

of curvature vector fields r_{jk} with constant coefficients $A^{jk} = -A^{kj}$ satisfies $A^{jk} r_{jk} = 0$ for any $y \in T_x M$ then one has the linear equation $A^{ik} y_k = 0$ for any fixed index i. Since the covector fields $y_1, \ldots, y_n$ are linearly independent we obtain $A^{jk} = 0$ for all $j, k \in \{1, \ldots, n\}$. It follows that the curvature vector fields r_{jk} are linearly independent for any $j < k$ and hence $\dim \mathfrak{R}_x \geq \frac{n(n-1)}{2}$.

Corollary 1 *Let (M, g) be a Riemannian manifold of non-zero constant curvature with $n = \dim M$. The curvature algebra $\mathfrak{R}_x$ at any point $x \in M$ is isomorphic to the orthogonal Lie algebra $\mathfrak{o}(n)$.*

Proof The holonomy group of a Riemannian manifold is a subgroup of the orthogonal group $O(n)$ of the tangent space $T_x M$ and hence the curvature algebra $\mathfrak{R}_x$ is a subalgebra of the orthogonal Lie algebra $\mathfrak{o}(n)$. Hence the previous assertion implies the corollary.

Theorem 2 *Let $(M, \mathscr{F})$ be a Finsler manifold of non-zero constant curvature with $\dim M > 2$. If the point $x \in M$ is not (semi-)Riemannian, then the curvature algebra $\mathfrak{R}_x$ at $x \in M$ satisfies*

$$\dim \mathfrak{R}_x > \frac{n(n-1)}{2} \quad where \quad n = \dim M. \tag{12.25}$$

Proof We assume $\dim \mathfrak{R}_x = \frac{n(n-1)}{2}$. For any constant skew-symmetric matrices $\{A^{jk}\}$ and $\{B^{jk}\}$ the Lie bracket of vector fields $A^{ik} y_k \frac{\partial}{\partial y^i}$ and $B^{ik} y_k \frac{\partial}{\partial y^i}$ has the shape $C^{ik} y_k \frac{\partial}{\partial y^i}$, where $\{C^{ik}\}$ is a constant skew-symmetric matrix, too. Using the homogeneity of g_{hl} we obtain

$$\frac{\partial y_h}{\partial y^m} = \frac{\partial g_{hl}}{\partial y^m} y^l + g_{hm} = g_{hm} \tag{12.26}$$

and hence

$$\left[A^{mk} y_k \frac{\partial}{\partial y^m}, B^{ih} y_h \frac{\partial}{\partial y^i} \right] = \left(A^{mk} B^{ih} \frac{\partial y_h}{\partial y^m} - B^{mk} A^{ih} \frac{\partial y_h}{\partial y^m} \right) y_k \frac{\partial}{\partial y^i}$$

$$= \left(B^{ih} g_{hm} A^{mk} - A^{ih} g_{hm} B^{mk} \right) y_k \frac{\partial}{\partial y^i} = C^{ik} y_k \frac{\partial}{\partial y^i}.$$

Particularly, for the skew-symmetric matrices $E_{ab}^{ij} = \delta_a^i \delta_b^j - \delta_b^i \delta_a^j$, $a, b \in \{1, \ldots, n\}$, we have

$$\left[E_{ab}^{ij} y_j \frac{\partial}{\partial y^i}, E_{cd}^{kl} y_l \frac{\partial}{\partial y^k} \right] = \left(E_{cd}^{ih} g_{hm} E_{ab}^{mk} - E_{ab}^{ih} g_{hm} E_{cd}^{mk} \right) y_k \frac{\partial}{\partial y^i} = \Lambda_{ab,cd}^{im} y_m \frac{\partial}{\partial y^i},$$

where the constants $\Lambda_{ab,cd}^{ij}$ satisfy $\Lambda_{ab,cd}^{ij} = -\Lambda_{ab,cd}^{ji} = -\Lambda_{ba,cd}^{ij} = -\Lambda_{ab,dc}^{ij} = -\Lambda_{cd,ab}^{ij}$. Putting $i = a$ and computing the trace for these indices we obtain

$$(n-2)(g_{bd} y_c - g_{bc} y_d) = \Lambda_{b,cd}^l y_l, \tag{12.27}$$

where $\Lambda_{b,cd}^l := \Lambda_{ib,cd}^{il}$. The right-hand side is a linear form in variables $y_1, \ldots, y_n$. According to the identity (12.27) this linear form vanishes for $y_c = y_d = 0$, hence

$\Lambda^l_{b,cd} = 0$ for $l \neq c, d$. Denoting $\lambda^{(c)}_{bd} := \frac{1}{n-2}\Lambda^c_{b,cd}$ (no summation for the index c) we get the identities

$$g_{bd}\, y_c - g_{bc}\, y_d = \lambda^{(c)}_{bd}\, y_c - \lambda^{(d)}_{bc}\, y_d \quad \text{(no summation for } c \text{ and } d\text{).}$$

Putting $y_d = 0$ we obtain $g_{bd}\big|_{y_d=0} = \lambda^{(c)}_{bd}$ for any $c \neq d$. It follows $\lambda^{(c)}_{bd}$ is independent of the index c ($\neq d$). Defining $\lambda_{bd} := \lambda^{(c)}_{bd}$ with some c ($\neq d$) we obtain from (12.27) the identity

$$g_{bd}\, y_c - g_{bc}\, y_d = \lambda_{bd}\, y_c - \lambda_{bc}\, y_d \tag{12.28}$$

for any $b, c, d \in \{1, \dots, n\}$. We have

$$\lambda_{cd}\, y_b - \lambda_{cb}\, y_d = (g_{bd}\, y_c - g_{bc}\, y_d) - (g_{db}\, y_c - g_{dc}\, y_b) = (\lambda_{bd}\, y_c - \lambda_{bc}\, y_d) - (\lambda_{db}\, y_c - \lambda_{dc}\, y_b).$$

which implies the identity

$$(\lambda_{cd}\, y_b - \lambda_{cb}\, y_d) + (\lambda_{db}\, y_c - \lambda_{dc}\, y_b) + (\lambda_{bc}\, y_d - \lambda_{bd}\, y_c)$$
$$= (\lambda_{cd} - \lambda_{dc})\, y_b + (\lambda_{db} - \lambda_{bd})\, y_c + (\lambda_{bc} - \lambda_{cb})\, y_d = 0. \tag{12.29}$$

Since $\dim M > 2$, we can consider three different indices b, c, d and we obtain from the identity (12.29) that $\lambda_{bc} = \lambda_{cb}$ for any $b, c \in \{1, \dots, n\}$.

By derivation the identity (12.28) we get

$$\frac{\partial g_{bd}}{\partial y_a}\, y_c - \frac{\partial g_{bc}}{\partial y_a}\, y_d + g_{bd}\, \delta^a_c - g_{bc}\, \delta^a_d = \lambda_{bd}\, \delta^a_c - \lambda_{bc}\, \delta^a_d.$$

Using (12.26) we obtain

$$\frac{\partial y_a}{\partial y^q}\left(\frac{\partial g_{bd}}{\partial y_a}\, y_c - \frac{\partial g_{bc}}{\partial y_a}\, y_d\right) + g_{bd}\, g_{cq} - g_{bc}\, g_{dq}$$

$$= \frac{\partial g_{bd}}{\partial y^q}\, y_c - \frac{\partial g_{bc}}{\partial y^q}\, y_d + g_{bd}\, g_{cq} - g_{bc}\, g_{dq} = \lambda_{bd}\, g_{cq} - \lambda_{bc}\, g_{dq}.$$

Since

$$\left(\frac{\partial g_{bd}}{\partial y^q}\, y_c - \frac{\partial g_{bc}}{\partial y^q}\, y_d\right) y^b = 0$$

we get the identity

$$y_d\, g_{cq} - y_c\, g_{dq} = \lambda_{bd}\, y^b\, g_{cq} - \lambda_{bc}\, y^b\, g_{dq}.$$

Multiplying both sides of this identity by the inverse $\{g^{qr}\}$ of the matrix $\{g_{cq}\}$ and taking the trace with respect to the indices c, r we obtain the identity

$$(n-1)\,y_d = (n-1)\lambda_{bd}\,y^b.$$

Hence we obtain that $g_{bd}\,y^b = \lambda_{bd}\,y^b$ and hence $g_{bd} = \lambda_{bd}$, which means that the point $x \in M$ is (semi-)Riemannian. From this contradiction follows the assertion.

Corollary 2 *The curvature algebra $\mathfrak{R}_x$ at a point $x \in M$ of a Finsler manifold $(M, \mathscr{F})$ of non-zero constant curvature satisfies*

$$\dim \mathfrak{R}_x = \frac{n(n-1)}{2}, \quad \text{where} \quad n = \dim M, \tag{12.30}$$

if and only if $n = 2$ or the point $x \in M$ is (semi-)Riemannian.

Theorem 3 *Let $(M, \mathscr{F})$ be a positive definite Finsler manifold of non-zero constant curvature with $n = \dim M > 2$. The holonomy group of $(M, \mathscr{F})$ is a compact Lie group if and only if $(M, \mathscr{F})$ is a Riemannian manifold.*

Proof We assume that the holonomy group of a Finsler manifold $(M, \mathscr{F})$ of non-zero constant curvature with $\dim M \geq 3$ is a compact Lie transformation group on the indicatrix $\mathfrak{I}_x M$. The curvature algebra $\mathfrak{R}_x$ at a point $x \in M$ is tangent to the holonomy group $\mathsf{Hol}(x)$ and hence $\dim \mathsf{Hol}(x) \geq \dim \mathfrak{R}_x$. If there exists a not (semi-)Riemannian point $x \in M$, then $\dim \mathfrak{R}_x > \frac{n(n-1)}{2}$. The $(n-1)$-dimensional indicatrix $\mathfrak{I}_x M$ at x can be equipped with a Riemannian metric which is invariant with respect to the compact Lie transformation group $\mathsf{Hol}(x)$. Since the group of isometries of an $n-1$-dimensional Riemannian manifold is of dimension at most $\frac{n(n-1)}{2}$ (cf. Kobayashi [11], p. 46) we obtain a contradiction, which proves the assertion.

Since the holonomy group of a Landsberg manifold is a subgroup of the isometry group of the indicatrix, we obtain that any Landsberg manifold of non-zero constant curvature with dimension > 2 is Riemannian (c.f. Numata [24]).

We can summarize our results as follows:

Theorem 4 *The holonomy group of any non-Riemannian positive definite Finsler manifold of non-zero constant curvature with dimension > 2 does not occur as the holonomy group of any Riemannian manifold.*

12.4.3 Infinite Dimensional Curvature Algebra

Let us consider the singular (non y-global) Finsler manifold $(H_3, \mathscr{F})$, where H_3 is the three-dimensional Heisenberg group and $\mathscr{F}$ is a left-invariant Berwald-Moór metric (c.f. [27], Example 1.1.5, p. 8).

The group H_3 can be realized as the Lie group of matrices of the form $\begin{bmatrix} 1 & x^1 & x^2 \\ 0 & 1 & x^3 \\ 0 & 0 & 1 \end{bmatrix}$,

where $x = (x^1, x^2, x^3) \in \mathbb{R}^3$ and hence the multiplication can be written as

$$(x^1, x^2, x^3) \cdot (y^1, y^2, y^3) = (x^1 + y^1, x^2 + y^2 + x^1 y^3, x^3 + y^3).$$

The vector $0 = (0, 0, 0) \in \mathbb{R}^3$ gives the unit element of H_3. The Lie algebra $\mathfrak{h}_3 = T_0 H_3$ consists of matrices of the form $\begin{bmatrix} 0 & a^1 & a^2 \\ 0 & 0 & a^3 \\ 0 & 0 & 0 \end{bmatrix}$, corresponding to the tangent vector $a = a^1 \frac{\partial}{\partial x^1} + a^2 \frac{\partial}{\partial x^2} + a^3 \frac{\partial}{\partial x^3}$ at the unit element $0 \in H_3$. A left-invariant Berwald-Moór Finsler metric $\mathscr{F}$ is induced by the (singular) Minkowski functional $\mathscr{F}_0 \colon \mathfrak{h}_3 \to \mathbb{R}$:

$$\mathscr{F}_0(a) := \left(a^1 a^2 a^3\right)^{\frac{2}{3}}$$

of the Lie algebra in the following way: if $y = (y^1, y^2, y^3)$ is a tangent vector at $x \in H_3$, then

$$\mathscr{F}(x, y) := \mathscr{F}_0(x^{-1} y).$$

The coordinate expression of the singular (non y-global) Finsler metric $\mathscr{F}$ is

$$\mathscr{F}(x, y) = \left(y^1 \left(y^2 - x^1 y^3\right) y^3\right)^{\frac{2}{3}}.$$

Since $\mathscr{F}$ is left-invariant, the associated geometric structures (connection, geodesics, curvature) are also left-invariant and the curvature algebras at different points are isomorphic. Using the notation

$$r_x(i, j) = r_x\left(\frac{\partial}{\partial x^i}, \frac{\partial}{\partial x^j}\right), \quad i, j = 1, 2, 3,$$

for curvature vector fields, a direct computation yields

$$r_x(1, 2) = \frac{1}{4}\left(\frac{5y^{1^2}y^{3^2}}{(x^1 y^3 - y^2)^3}\frac{\partial}{\partial y^1} + \frac{y^1 y^{3^2}\left(3x^1 y^3 + y^2\right)}{(y^2 - x^1 y^3)^3}\frac{\partial}{\partial y^2} + \frac{4y^1 y^{3^3}}{(y^2 - x^1 y^3)^3}\frac{\partial}{\partial y^3}\right),$$

$$r_x(1, 3) = \frac{1}{4}\left(\frac{y^{1^2}y^3\left(6x^1 y^3 - 11 y^2\right)}{(x^1 y^3 - y^2)^3}\frac{\partial}{\partial y^1} + \frac{4y^1 y^{3^2}x^1\left(2x^1 y^3 - 3 y^2\right)}{(y^2 - x^1 y^3)^3}\frac{\partial}{\partial y^2}\right.$$

$$\left. + \frac{y^1 y^{3^2}\left(7x^1 y^3 - 11 y^2\right)}{(y^2 - x^1 y^3)^3}\frac{\partial}{\partial y^3}\right),$$

$$r_x(2, 3) = \frac{1}{4}\left(\frac{4y^{1^3}y_3}{(x^1 y^3 - y^2)^3}\frac{\partial}{\partial y^1} + \frac{y^{1^2}y^3\left(6x^1 y^3 - y^2\right)}{(y^2 - x^1 y^3)^3}\frac{\partial}{\partial y^2} + \frac{5y^{1^2}y^{3^2}}{(y^2 - x^1 y^3)^3}\frac{\partial}{\partial y^3}\right).$$

The curvature vector fields $r_0(i,j)$, $i,j = 1,2,3$, at the unit element $0 \in H_3$ generate the curvature algebra $\mathfrak{r}_0$. Let us denote $Y^{k,m} := \frac{y^{1k}y^{3m}}{y^{2k+m-1}}$, $k,m \in \mathbb{N}$, and consider the vector fields

$$A^{k,m}(a^1,a^2,a^3) = a^1 Y^{k+1,m} \frac{\partial}{\partial y^1}\Big|_0 + a^2 Y^{k,m} \frac{\partial}{\partial y^2}\Big|_0 + a^3 Y^{k,m+1} \frac{\partial}{\partial y^3}\Big|_0, \quad (12.31)$$

with $(a^1,a^2,a^3) \in \mathbb{R}^3$ and $k,m \in \mathbb{N}$. Then the curvature vector fields $r_0(i,j)$ at $0 \in H_3$ can be written in the form

$$r_0(1,2) = \frac{1}{4}A^{1,2}(-5,1,4), \quad r_0(1,3) = \frac{1}{4}A^{1,1}(11,0,-11), \quad r_0(2,3) = \frac{1}{4}A^{2,1}(-4,-1,5).$$

Proposition 4 *The curvature algebra $\mathfrak{r}_x$ at any point $x \in M$ is a Lie algebra of infinite dimension.*

Proof Since the Finsler metric is left-invariant, the curvature algebras at different points are isomorphic. Therefore it is enough to prove that the curvature algebra $\mathfrak{r}_0$ at $0 \in H_3$ has infinite dimension. We prove the statement by contradiction: let us suppose that $\mathfrak{r}_0$ is finite dimensional.

A direct computation shows that for any $(a^1,a^2,a^3), (b^1,b^2,b^3) \in \mathbb{R}^3$ one has

$$\left[A^{k,m}(a^1,a^2,a^3), A^{p,q}(b^1,b^2,b^3)\right] = A^{k+p,m+q}(c^1,c^2,c^3)$$

with some $(c^1,c^2,c^3) \in \mathbb{R}^3$. It follows that any iterated Lie bracket of curvature vector fields $r_0(i,j)$, $i,j = 1,2,3$, has the shape (12.31) and hence there exists a basis of the curvature algebra $\mathfrak{r}_0$ of the form $\{A^{k_i,m_i}(a_i^1,a_i^2,a_i^3)\}_{i=1}^N$, where $N \in \mathbb{N}$ is the dimension of $\mathfrak{r}_0$. We can assume that $\{(k_i,m_i)\}_{i=1}^N$ forms an increasing sequence, i.e. $(k_1,m_1) \leqslant (k_2,m_2) \leqslant \cdots \leqslant (k_N,m_N)$ holds with respect to the lexicographical ordering of $\mathbb{N} \times \mathbb{N}$. We can consider the vector fields

$$\frac{4}{11}r_0(1,3) = A^{1,1}(1,0,-1), \quad 4r_0(1,2) = A^{1,2}(-5,1,4), \quad 4r_0(2,3) = A^{2,1}(-4,-1,5)$$

as the first three members of this sequence. Hence $1 \leqslant k_N, m_N$ and

$$\left[A^{1,1}(1,0,-1), A^{k_N,m_N}(a_N^1,a_N^2,a_N^3)\right] = A^{1+k_N,1+m_N}(c^1,c^2,c^3)$$

belongs to $\mathfrak{r}_0$, too, where $c^1 = (k_N - m_N - 1)a_N^1 + 2a_N^2 - a_N^3$, $c^2 = (k_N - m_N)a_N^2$ and $c^3 = a_N^1 - 2a_N^2 + (k_N - m_N + 1)a_N^3$. Since $k_N < 1 + k_N$, $m_N < 1 + m_N$ we have $c^1 = c^2 = c^3 = 0$ and hence the homogeneous linear system

$$0 = (k_N - m_N - 1)a_N^1 + 2a_N^2 - a_N^3,$$

$$0 = (k_N - m_N)a_N^2,$$

$$0 = a_N^1 - 2a_N^2 + (k_N - m_N + 1)a_N^3$$

has a solution $(a_N^1, a_N^2, a_N^3) \neq (0,0,0)$. It follows that $k_N = m_N$.

Similarly, computing the Lie bracket

$$0 = \left[A^{1,2}(-5,1,4), A^{k_N,k_N}(a_N^1, a_N^2, a_N^3)\right] = A^{1+k_N,2+k_N}(d^1, d^2, d^3)$$

Since $k_N < 1 + k_N < 2 + k_N$ we have $d^1 = d^2 = d^3 = 0$ giving the homogeneous linear system

$$0 = (-3k_N + 5)a_N^1 - 15a_N^2 + 10a_N^3,$$

$$0 = -a_N^1 + (3 - 3k_N)a_N^2 - 2a_N^3,$$

$$0 = -4a_N^1 + 12a_N^2 - (3k_N + 8)a_N^3$$

for (a_N^1, a_N^2, a_N^3). The determinant of this system vanishes only for $k_N = 0$ which is a contradiction.

Corollary 3 *The holonomy group of the y-singular Finsler manifold* $(H_3, \mathscr{F})$ *has an infinite dimensional tangent Lie algebra.*

We remark here that it remains an interesting open question: *Is there a nonsingular (y-global) Finsler manifold whose curvature algebra is infinite dimensional?*

12.5   Holonomy Algebra

12.5.1   Fibered Holonomy Group

Now, we introduce the notion of the fibered holonomy group of a Finsler manifold $(M, \mathscr{F})$ as a subgroup of the diffeomorphism group of the total manifold $\mathfrak{J}M$ of the bundle $(\mathfrak{J}M, \pi, M)$ and apply our results on tangent vector fields to an abstract subgroup of the diffeomorphism group to the study of tangent Lie algebras to the fibered holonomy group.

Definition 6 The *fibered holonomy group* $\mathsf{Hol}_f(M)$ of $(M, \mathscr{F})$ consists of fiber preserving diffeomorphisms $\Phi \in \mathsf{Diff}^\infty(\mathfrak{J}M)$ of the indicatrix bundle $(\mathfrak{J}M, \pi, M)$ such that for any $p \in M$ the restriction $\Phi_p = \Phi|_{\mathfrak{J}_pM} \in \mathsf{Diff}^\infty(\mathfrak{J}_pM)$ belongs to the holonomy group $\mathsf{Hol}(p)$.
We note that the holonomy group $\mathsf{Hol}(p)$ and the fibered holonomy group $\mathsf{Hol}_f(M)$ are topological subgroups of the infinite dimensional Lie groups $\mathsf{Diff}^\infty(\mathfrak{J}_pM)$ and $\mathsf{Diff}^\infty(\mathfrak{J}M)$, respectively.
The definition of strongly tangent vector fields yields

Remark 3 A vector field $\xi \in \mathfrak{X}^\infty(\mathfrak{J}M)$ is strongly tangent to the fibered holonomy group $\mathsf{Hol}_f(M)$ if and only if there exists a family $\{\Phi_{(t_1,\ldots,t_k)}|_{\mathfrak{J}M}\}_{t_i \in (-\varepsilon,\varepsilon)}$ of fiber preserving diffeomorphisms of the bundle $(\mathfrak{J}M, \pi, M)$ such that for any indicatrix

$\mathfrak{I}_p$ the induced family $\left\{\Phi_{(t_1,\ldots,t_k)}\big|_{\mathfrak{I}_pM}\right\}_{t_i\in(-\varepsilon,\varepsilon)}$ of diffeomorphisms is contained in the holonomy group $\mathsf{Hol}(p)$ and $\xi\big|_{\mathfrak{I}_pM}$ is strongly tangent to $\mathsf{Hol}(p)$.

Since $\pi\left(\Phi_{(t_1,\ldots,t_k)}(p)\right)\equiv p$ and $\pi_*(\xi)=0$ for every $p\in U$, we get the

Corollary 4 *Strongly tangent vector fields to the fibered holonomy group* $\mathsf{Hol}_{\mathsf{f}}(M)$ *are vertical vector fields. If* $\xi\in\mathfrak{X}^\infty(\mathfrak{I}M)$ *is strongly tangent to* $\mathsf{Hol}_{\mathsf{f}}(M)$ *then its restriction* $\xi_p:=\xi\big|_{\mathfrak{I}_p}$ *to any indicatrix* $\mathfrak{I}_p$ *is strongly tangent to the holonomy group* $\mathsf{Hol}(p)$.

Let now $U\subset M$ be an open neighborhood in M diffeomorphic to $\mathbb{R}^n$ and consider the Finsler manifold $(U,\mathscr{F}|_U)$ induced on U. Now we prove that the first assertion of Theorem 1 on the tangential property to the holonomy group of curvature vector fields at a point can be extended to curvature vector fields defined on the full the indicatrix bundle of $(U,\mathscr{F}|_U)$.

Proposition 5 *Any curvature vector field* $\xi\in\mathfrak{X}^\infty(\mathfrak{I}U)$ *on the indicatrix bundle of the Finsler manifold* $(U,\mathscr{F}|_U)$ *is strongly tangent to the fibered holonomy group* $\mathsf{Hol}_{\mathsf{f}}(U)$.

Proof Since U is diffeomorphic to $\mathbb{R}^n$ we can identify the manifold U with the vector space $\mathbb{R}^n$. Let $\xi=r(X,Y)\in\mathfrak{X}^\infty(\mathfrak{I}\mathbb{R}^n)$ be a curvature vector field with $X,Y\in\mathfrak{X}^\infty(\mathbb{R}^n)$. According to Proposition 2 its restriction $\xi\big|_{\mathfrak{I}_p\mathbb{R}^n}$ to any indicatrix $\mathfrak{I}_p\mathbb{R}^n$ is strongly tangent to the holonomy groups $\mathsf{Hol}(p)$. We have to prove that there exists a family $\left\{\Phi_{(t_1,\ldots,t_k)}\big|_{\mathfrak{I}\mathbb{R}^n}\right\}_{t_i\in(-\varepsilon,\varepsilon)}$ of fiber preserving diffeomorphisms of the indicatrix bundle $(\mathfrak{I}\mathbb{R}^n,\pi,\mathbb{R}^n)$ such that for any $p\in\mathbb{R}^n$ the family of diffeomorphisms induced on the indicatrix $\mathfrak{I}_p$ is contained in $\mathsf{Hol}(p)$ and $\xi\big|_{\mathfrak{I}_p\mathbb{R}^n}$ is strongly tangent to $\mathsf{Hol}(p)$.

For any $p\in\mathbb{R}^n$ and $-1<s,t<1$ let $\Pi(sX_p,tY_p)$ be the parallelogram in $\mathbb{R}^n$ determined by the vertexes $p,p+sX_p,p+sX_p+tY_p,p+tY_p\in\mathbb{R}^n$ and let $\tau_{\Pi(sX_p,tY_p)}:\mathfrak{I}_p\to\mathfrak{I}_p$ denote the (nonlinear) parallel translation of the indicatrix $\mathfrak{I}_p$ along the parallelogram $\Pi(sX_p,tY_p)$ with respect to the associated homogeneous (nonlinear) parallel translation of the Finsler manifold $(\mathbb{R}^n,\mathscr{F})$. Clearly we have $\tau_{\Pi(sX_p,tY_p)}=\mathsf{Id}_{\mathfrak{I}\mathbb{R}^n}$, if $s=0$ or $t=0$ and

$$\frac{\partial^2\tau_{\Pi(sX_p,tY_p)}}{\partial s\partial t}\bigg|_{(s,t)=(0,0)}=\xi_p\quad\text{for every}\quad p\in\mathbb{R}^n.$$

Since $\Pi(sX_p,tY_p)$ is a differentiable field of parallelograms in $\mathbb{R}^n$, the maps $\tau_{\Pi(sX_p,tY_p)},p\in\mathbb{R}^n,0<s,t<1$, define a 2-parameter family of fiber preserving diffeomorphisms of the indicatrix bundle $\mathfrak{I}\mathbb{R}^n$. The diffeomorphisms induced by the family $\left\{\tau_{\Pi(sX_p,tY_p)}\right\}_{s,t\in(-1,1)}$ on any indicatrix $\mathfrak{I}_p$ are contained in $\mathsf{Hol}(p)$. Hence the vector field $\xi\in\mathfrak{X}^\infty(\mathbb{R}^n)$ is strongly tangent to the fibered holonomy group $\mathsf{Hol}_{\mathsf{f}}(U)$ and the assertion is proved.

Corollary 5 *If* M *is diffeomorphic to* $\mathbb{R}^n$, *then the curvature algebra* $\mathfrak{R}(M)$ *of* $(M,\mathscr{F})$ *is tangent to the fibered holonomy group* $\mathsf{Hol}_{\mathsf{f}}(M)$.

12.5.2 Infinitesimal Holonomy Algebra

The following assertion shows that similarly to the Riemannian case, the curvature algebra can be extended to a larger tangent Lie algebra containing all horizontal Berwald covariant derivatives of the curvature vector fields.

Proposition 6 *If $\xi \in \mathfrak{X}^\infty(\mathfrak{J}M)$ is strongly tangent to the fibered holonomy group* $\mathsf{Hol}_\mathsf{f}(M)$ *of $(M, \mathscr{F})$, then its horizontal Berwald covariant derivative $\nabla_X \xi$ along any vector field $X \in \mathfrak{X}^\infty(M)$ is also strongly tangent to* $\mathsf{Hol}_\mathsf{f}(M)$.

Proof Let τ be the (nonlinear) parallel translation along the flow φ of the vector field X, i.e. for every $p \in M$ and $t \in (-\varepsilon_p, \varepsilon_p)$ the map $\tau_t(p) \colon \mathfrak{J}_p M \to \mathfrak{J}_{\varphi_t(p)} M$ is the (nonlinear) parallel translation along the integral curve of X. If $\{\Phi_{(t_1,\ldots,t_k)}\}_{t_i \in (-\varepsilon,\varepsilon)}$ is a $\mathscr{C}^\infty$-differentiable k-parameter family $\{\Phi_{(t_1,\ldots,t_k)}\}_{t_i \in (-\varepsilon,\varepsilon)}$ of fiber preserving diffeomorphisms of the indicatrix bundle $(\mathfrak{J}M, \pi|_M, M)$ satisfying the conditions of Definition 4, then the commutator

$$[\Phi_{(t_1,\ldots,t_k)}, \tau_{t_{k+1}}] := \Phi_{(t_1,\ldots,t_k)}^{-1} \circ (\tau_{t_{k+1}})^{-1} \circ \Phi_{(t_1,\ldots,t_k)} \circ \tau_{t_{k+1}}$$

of the group $\mathsf{Diff}^\infty(\mathfrak{J}M)$ fulfills $[\Phi_{(t_1,\ldots,t_k)}, \tau_{t_{k+1}}] = \mathsf{Id}$, if some of its variables equals 0. Moreover

$$\left. \frac{\partial^{k+1}[\Phi_{(t_1\ldots t_k)}, \tau_{(t_{k+1})}]}{\partial t_1 \ \ldots \ \partial t_{k+1}} \right|_{(0\ldots0)} = -[\xi, X^h] \tag{12.32}$$

at any point of M, which shows that the vector field $[\xi, X^h]$ is strongly tangent to $\mathsf{Hol}_\mathsf{f}(M)$. Moreover, since the vector field ξ is vertical, we have $h[X^h, \xi] = 0$, and using $\nabla_X \xi := [X^h, \xi]$ we obtain

$$-[\xi, X^h] = [X^h, \xi] = v[X^h, \xi] = \nabla_X \xi$$

which yields the assertion.

Definition 7 Let $\mathfrak{hol}^*(M)$ be the smallest Lie algebra of vector fields on the indicatrix bundle $\mathfrak{J}M$ satisfying the properties

 (i) any curvature vector field ξ belongs to $\mathfrak{hol}^*(M)$,
 (ii) if $\xi, \eta \in \mathfrak{hol}^*(M)$, then $[\xi, \eta] \in \mathfrak{hol}^*(M)$,
 (iii) if $\xi \in \mathfrak{hol}^*(M)$ and $X \in \mathfrak{X}^\infty(M)$, then the horizontal Berwald covariant derivative $\nabla_X \xi$ also belongs to $\mathfrak{hol}^*(M)$.

The Lie algebra $\mathfrak{hol}^*(M) \subset \mathfrak{X}^\infty(\mathfrak{J}M)$ is called the *infinitesimal holonomy algebra* of the Finsler manifold $(M, \mathscr{F})$.

Remark 4 The infinitesimal holonomy algebra $\mathfrak{hol}^*(M)$ is invariant with respect to the horizontal Berwald covariant derivation, i.e.

$$\xi \in \mathfrak{hol}^*(M) \quad \text{and} \quad X \in \mathfrak{X}^\infty(M) \quad \Rightarrow \quad \nabla_X \xi \in \mathfrak{hol}^*(M). \tag{12.33}$$

Let $U \subset M$ be an open neighborhood in M diffeomorphic to $\mathbb{R}^n$ and let $(U, \mathscr{F}|_U)$ be the Finsler manifold $(U, \mathscr{F}|_U)$ induced on U. The results of this sections yield the following

Theorem 1 *The infinitesimal holonomy algebra* $\mathfrak{hol}^*(U)$ *of the Finsler manifold* $(U, \mathscr{F}|_U)$ *is tangent to the fibered holonomy group* $\mathsf{Hol}_f(U)$.

Let $\mathfrak{hol}^*(M) \subset \mathfrak{X}^\infty(\mathfrak{I}M)$ be the infinitesimal holonomy algebra of the Finsler manifold $(M, \mathscr{F})$ and let p be a given point in M.

Definition 8 The Lie algebra $\mathfrak{hol}^*(p) := \{\xi_p ; \xi \in \mathfrak{hol}^*(M)\} \subset \mathfrak{X}^\infty(\mathfrak{I}_pM)$ of vector fields on the indicatrix $\mathfrak{I}_pM$ is called the *infinitesimal holonomy algebra at the point* $p \in M$.

Clearly, for any $p \in M$ the curvature algebra $\mathfrak{R}_p$ at $p \in M$ is contained in the infinitesimal holonomy algebra $\mathfrak{hol}^*(p)$ at $p \in M$.

The following assertion is a direct consequence of the definition. It shows that the infinitesimal holonomy algebra at a point p of $(M, \mathscr{F})$ can be calculated in a neighborhood of p.

Remark 5 Let $(U, \mathscr{F}|_U)$ be an open submanifold of $(M, \mathscr{F})$ such that $U \subset M$ is diffeomorphic to $\mathbb{R}^n$ and let $p \in U$. The infinitesimal holonomy algebras at p of the Finsler manifolds $(M, \mathscr{F})$ and $(U, \mathscr{F}|_U)$ coincide.

Now, we can prove the following

Theorem 2 *The infinitesimal holonomy algebra* $\mathfrak{hol}^*(p)$ *at a point* $p \in M$ *of the Finsler manifold* $(M, \mathscr{F})$ *is tangent to the holonomy group* $\mathsf{Hol}(p)$.

If $(M, \mathscr{F})$ *is a positive definite Finsler manifold, then the group generated by the exponential image* $\exp(\mathfrak{hol}^*(p))$ *is a subgroup of the topological closure of the holonomy group* $\mathsf{Hol}(p)$.

Proof Let $U \subset M$ be an open submanifold of M, diffeomorphic to $\mathbb{R}^n$ and containing $p \in M$. According to the previous remark we have $\mathfrak{hol}^*(p) := \{\xi_p ; \xi \in \mathfrak{hol}_f(U)\}$. Since the fibered holonomy algebra $\mathfrak{hol}_f(U)$ is tangent to the fibered holonomy group $\mathsf{Hol}_f(U)$ we obtain the first assertion. The second assertion is a consequence of Proposition 1.

12.5.3 The Berwald Translate

Let $x(t)$, $0 \leqslant t \leqslant a$ be a smooth curve joining the points $q = x(0)$ and $p = x(a)$ in the Finsler manifold $(M, \mathscr{F})$. If $y(t) = \tau_t y(0) \in \mathfrak{I}_{x(t)}M$ is a parallel vector field along $x(t)$, $0 \leqslant t \leqslant a$, where $\tau_t : \mathfrak{I}_qM \to \mathfrak{I}_{x(t)}M$ denotes the homogeneous (nonlinear)

parallel translation, then we have $D_{\dot{x}}y(t) := \left(\frac{dy^i(t)}{dt} + G^i_j(x(t), y(t))\dot{x}^j(t)\right)\frac{\partial}{\partial x^i} = 0$.
Considering a vector field ξ on the indicatrix $\mathfrak{I}_qM$, the map $\tau_{a*}\xi \circ \tau_a^{-1} : (p, y) \mapsto \tau_{a*}\xi(y(a))$ gives a vector field on the indicatrix $\mathfrak{I}_pM$. Hence we can formulate

Lemma 1 *For any vector field* $\xi \in \mathfrak{hol}^*(q) \subset \mathfrak{X}^\infty(\mathfrak{I}_qM)$ *in the infinitesimal holonomy algebra at q the vector field* $\tau_{a*}\xi \circ \tau_a^{-1} \in \mathfrak{X}^\infty(\mathfrak{I}_pM)$ *is tangent to the holonomy group* $\mathsf{Hol}(p)$.

Proof Let $\{\phi_t \in \mathsf{Hol}(q)\}_{t\in(-\varepsilon,\varepsilon)}$ be a $\mathscr{C}^1$-differentiable 1-parameter family of diffeomorphisms of $\mathfrak{I}_qM$ belonging to the holonomy group $\mathsf{Hol}(q)$ and satisfying the conditions $\phi_0 = \mathsf{Id}$, $\frac{\partial\phi_t}{\partial t}\big|_{t=0} = \xi$. Since the 1-parameter family

$$\tau_a \circ \phi_t \circ \tau_a^{-1} \in \mathsf{Diff}^\infty(\mathfrak{I}_pM)\}_{t\in(-\varepsilon,\varepsilon)}$$

of diffeomorphisms consists of elements of the holonomy group $\mathsf{Hol}(p)$ and satisfies the conditions

$$\tau_a \circ \phi_0 \circ \tau_a^{-1} = \mathsf{Id}, \qquad \frac{\partial\left(\tau_a \circ \phi_t \circ \tau_a^{-1}\right)}{\partial t}\bigg|_{t=0} = \tau_{a*}\xi \circ \tau_a^{-1},$$

the assertion follows.

Definition 9 A vector field $\mathbf{B}_\gamma\xi \in \mathfrak{X}^\infty(\mathfrak{I}_pM)$ on the indicatrix $\mathfrak{I}_pM$ will be called *the Berwald translate* of the vector field $\xi \in \mathfrak{X}^\infty(\mathfrak{I}_qM)$ along the curve $\gamma = x(t)$ if

$$\mathbf{B}_\gamma\xi = \tau_{a*}\xi \circ (\tau_a)^{-1}.$$

Remark 6 Let $y(t) = \tau_t y(0) \in \mathfrak{I}_{x(t)}M$ be a parallel vector field along $\gamma = x(t)$, $0 \leqslant t \leqslant a$, started at $y(0) \in \mathfrak{I}_{x(0)}M$. Then, the vertical vector field $\xi_t = \xi(x(t), y(t))$ along $(x(t), y(t))$ is the Berwald translate $\xi_t = \tau_{t*}\xi_0 \circ \tau_t^{-1}$ if and only if

$$\nabla_{\dot{x}}\xi = \left(\frac{\partial\xi^i(x, y)}{\partial x^j} - G^k_j(x, y)\frac{\partial\xi^i(x, y)}{\partial y^k} + G^i_{jk}(x, y)\xi^k(x, y)\right)\dot{x}^j\frac{\partial}{\partial y^i} = 0.$$

Now, Lemma 1 yields the following

Corollary 6 *If* $\xi \in \mathfrak{hol}^*(q)$, *then its Berwald translate* $\mathbf{B}_\gamma\xi \in \mathfrak{X}^\infty(\mathfrak{I}_pM)$ *along any curve* $\gamma = x(t)$, $0 \leqslant t \leqslant a$, *joining* $q = x(0)$ *with* $p = x(a)$ *is tangent to the holonomy group* $\mathsf{Hol}(p)$.

This last statement motivates the following

Definition 10 The *holonomy algebra* $\mathfrak{hol}_p(M)$ of the Finsler manifold $(M, \mathscr{F})$ at the point $p \in M$ is defined by the smallest Lie algebra of vector fields on the indicatrix $\mathfrak{I}_pM$, containing the Berwald translates of all infinitesimal holonomy algebras along arbitrary curves $x(t)$, $0 \leqslant t \leqslant a$ joining any points $q = x(0)$ with the point $p = x(a)$.

Clearly, the holonomy algebras at different points of the Finsler manifold $(M, \mathscr{F})$ are isomorphic. Lemma 1, Corollary 6, and Proposition 1 yield the following

Theorem 3 *The holonomy algebra* $\mathfrak{hol}_p(M)$ *at a point* $p \in M$ *of a Finsler manifold* $(M, \mathscr{F})$ *is tangent to the holonomy group* $\mathsf{Hol}(p)$.

If $(M, \mathscr{F})$ *is a positive definite Finsler manifold, then the group generated by the exponential image* $\exp(\mathfrak{hol}_p(M))$ *is a subgroup of the topological closure of the holonomy group* $\mathsf{Hol}(p)$.

12.5.4 Finsler Surfaces with $\mathfrak{hol}^*(x) = \mathfrak{R}_x$

The relation between the infinitesimal holonomy algebra and the curvature algebra is enlightened by the following

Theorem 4 *Let* $(M, \mathscr{F})$ *be a Finsler surface with non-zero constant flag curvature. The infinitesimal holonomy algebra* $\mathfrak{hol}^*(x)$ *at a point* $x \in M$ *coincides with the curvature algebra* $\mathfrak{R}_x$ *at* x *if and only if the mean Berwald curvature* $E_{(x,y)}$ *of* $(M, \mathscr{F})$ *vanishes for any* $y \in \mathfrak{I}_x M$.

Proof Let $U \subset M$ be a neighborhood of $x \in M$ diffeomorphic to $\mathbb{R}^2$. Identifying U with $\mathbb{R}^2$ and considering a coordinate system (x_1, x_2) in $\mathbb{R}^2$ we can write

$$R^i_{jk}(x, y) = \lambda \left(\delta^i_j g_{km}(x, y)y^m - \delta^i_k g_{jm}(x, y)y^m \right), \quad \text{with} \quad \lambda \neq 0.$$

Since the curvature tensor field is skew-symmetric, $R_{(x,y)}$ acts on the one-dimensional wedge product $T_x M \wedge T_x M$. According to Lemma 2 the covariant derivative of the curvature vector field $\xi = R(X, Y) = \frac{1}{2}R(X \otimes Y - Y \otimes X) = R(X \wedge Y)$ can be written in the form

$$\nabla_Z \xi = \nabla_Z (r(X, Y)) = R (\nabla_Z(X \wedge Y)) = R(\nabla_Z X \wedge Y + X \wedge \nabla_Z Y),$$

where $X, Y, Z \in \mathfrak{X}(U)$. If $X = X^i \frac{\partial}{\partial x^i}$, $Y = Y^i \frac{\partial}{\partial x^i}$ and $Z = Z^i \frac{\partial}{\partial x^i}$, then we have $X \wedge Y = \frac{1}{2} \left(X^1 Y^2 - X^2 Y^1 \right) \frac{\partial}{\partial x^1} \wedge \frac{\partial}{\partial x^2}$ and hence we obtain

$$\nabla_Z \xi = R \left(\nabla_k \left((X^1 Y^2 - Y^1 X^2) \frac{\partial}{\partial x^1} \wedge \frac{\partial}{\partial x^2} \right) Z^k \right) \qquad (12.34)$$

$$= R \left(\frac{\partial (X^1 Y^2 - Y^1 X^2)}{\partial x^k} Z^k \frac{\partial}{\partial x^1} \wedge \frac{\partial}{\partial x^2} \right)$$

$$+ (X^1 Y^2 - Y^1 X^2) R \left(\nabla_k \left(\frac{\partial}{\partial x^1} \wedge \frac{\partial}{\partial x^2} \right) \right) Z^k,$$

where we denote the covariant derivative ∇_Z by ∇_k if $Z = \frac{\partial}{\partial x^k}$, $k = 1, 2$. For given vector fields $X, Y, Z \in \mathfrak{X}^\infty(U)$ the expression $\frac{\partial(X^1 Y^2 - Y^1 X^2)}{\partial x^k} Z^k$ is a function on U. Hence there exists a function ψ on U such that

$$R\left(\frac{\partial(X^j Y^h - Y^j X^h)}{\partial x^k} Z^k \frac{\partial}{\partial x^j} \wedge \frac{\partial}{\partial x^h} \right) = \psi R(X \wedge Y) = \psi R(X, Y),$$

and $\psi R(X, Y)$ is an element of the curvature algebra $\mathfrak{R}(U)$ of the submanifold $(U, \mathscr{F}|_U)$.

Now, we investigate the second term of the right-hand side of (12.34).

$$\nabla_k\left(\frac{\partial}{\partial x^1} \wedge \frac{\partial}{\partial x^2} \right) = \left(\nabla_k \frac{\partial}{\partial x^1} \right) \wedge \frac{\partial}{\partial x^2} + \frac{\partial}{\partial x^1} \wedge \left(\nabla_k \frac{\partial}{\partial x^2} \right) =$$

$$= G^l_{k1} \frac{\partial}{\partial x^l} \wedge \frac{\partial}{\partial x^2} + \frac{\partial}{\partial x^1} \wedge G^m_{k2} \frac{\partial}{\partial x^m} = \left(G^1_{k1} + G^2_{k2} \right) \frac{\partial}{\partial x^1} \wedge \frac{\partial}{\partial x^2}.$$

Hence

$$(X^1 Y^2 - Y^1 X^2) R\left(\nabla_k \left(\frac{\partial}{\partial x^1} \wedge \frac{\partial}{\partial x^2} \right) \right) Z^k = \left(G^1_{k1} + G^2_{k2} \right) Z^k R(X, Y) = \left(G^1_{k1} + G^2_{k2} \right) Z^k \xi$$

This expression belongs to the curvature algebra if and only if the function $G^1_{k1} + G^2_{k2}$ does not depend on the variable y, i.e. if and only if

$$E_{kh} = \frac{\partial \left(G^1_{k1} + G^2_{k2} \right)}{\partial y^h} = 0, \quad h, k = 1, 2,$$

identically.

Remark 7 Let $\xi = R(X, Y)$ be a curvature vector field. Assume that the vector fields $X, Y \in \mathfrak{X}^\infty(M)$ have constant coordinate functions in a local coordinate system $(x^1, \ldots, x^n)$ of the Finsler surface $(M, \mathscr{F})$. Then we have in this coordinate system

$$\nabla_Z \xi = \left(G^1_{k1} + G^2_{k2} \right) Z^k \xi.$$

Shen constructed in [28] families of Randers surfaces depending on the real parameter ϵ, which are of constant flag curvature 1 on the unit sphere $S^2 \subset \mathbb{R}^3$ and of constant flag curvature -1 on a disk $\mathbb{D}^2 \subset \mathbb{R}^2$. These Finsler surfaces are not projectively flat and have vanishing S-curvature (c.f. [28], Theorems 1.1 and 1.2). Their Finsler function is defined by

$$\alpha = \frac{\sqrt{\epsilon^2 h(v, y)^2 + h(y, y)(1 - \epsilon^2 h(v, v))}}{1 - \epsilon^2 h(v, v)}, \quad \beta = \frac{\epsilon h(v, y)}{1 - \epsilon^2 h(v, v)}, \tag{12.35}$$

where $h(v, y)$ is the standard metric of the sphere S^2, respectively $h(v, y)$ is the standard Klein metric on the unit disk $\mathbb{D}^2$ and v denotes the vector field defined by $(-x_2, x_1, 0)$ at $(x_1, x_2, x_3) \in S^2$, respectively by $(-x_2, x_1)$ at $(x_1, x_2) \in \mathbb{D}^2$.

Theorem 5 *For any Randers surface defined by (12.35) the infinitesimal holonomy algebra $\mathfrak{hol}^*(x)$ at a point $x \in M$ coincides with the curvature algebra $\mathfrak{R}_x$.*

Proof According to Theorems 1.1 and 1.2 in [28], the above classes of not locally projectively flat Randers surfaces with non-zero constant flag curvature have vanishing S-curvature. Moreover, Proposition 6.1.3 in [27], p. 80 yields that the mean Berwald curvature vanishes in this case. Hence the assertion follows from Corollary 4.

12.6 Infinite Dimensional Subalgebras of the Infinitesimal Holonomy Algebra

12.6.1 *Projective Finsler Surfaces of Constant Curvature*

A Finsler manifold $(M, \mathcal{F})$ of dimension 2 is called *Finsler surface*. In this case the indicatrix is one-dimensional at any point $x \in M$, hence the curvature vector fields at $x \in M$ are proportional to any given non-vanishing curvature vector field. It follows that the curvature algebra $\mathfrak{R}_x(M)$ has a simple structure: it is at most one-dimensional and commutative. Even in this case, the infinitesimal holonomy algebra $\mathfrak{hol}_x^*(M)$ can be higher dimensional, or potentially infinite dimensional. For the investigation of such examples we use a classical result of Lie claiming that the dimension of a finite-dimensional Lie algebra of vector fields on a connected one-dimensional manifold is less than 4 (cf. [1], Theorem 4.3.4). We obtain the following

Lemma 1 *If the infinitesimal holonomy algebra $\mathfrak{hol}_x^*(M)$ of a Finsler surface $(M, \mathcal{F})$ contains four simultaneously non-vanishing $\mathbb{R}$-linearly independent vector fields, then $\mathfrak{hol}_x^*(M)$ is infinite dimensional.*

Proof If the infinitesimal holonomy algebra is finite-dimensional, then the dimension of the corresponding Lie group acting locally effectively on the one-dimensional indicatrix would be at least 4, which is a contradiction.

Let $(M, \mathcal{F})$ be a locally projectively flat Finsler surface of non-zero constant curvature, let (x^1, x^2) be a local coordinate system centered at $x \in M$, corresponding to the canonical coordinates of the Euclidean space which is projectively related to $(M, \mathcal{F})$ and let (y^1, y^2) be the induced coordinate system in the tangent plane $T_x M$.

In the sequel we identify the tangent plane $T_x M$ with $\mathbb{R}^2$ with help of the coordinate system (y^1, y^2). We will use the euclidean norm $||(y^1, y^2)|| = \sqrt{(y^1)^2 + (y^2)^2}$ of $\mathbb{R}^2$ and the corresponding polar coordinate system (e^r, t), too.

Let $\varphi(y^1, y^2)$ be a positively 1-homogeneous function on $\mathbb{R}^2$ and let $r(t)$ be the 2π-periodic smooth function $r : \mathbb{R} \to \mathbb{R}$ determined by

$$\varphi(e^{r(t)}\cos t, e^{r(t)}\sin t) = 1 \quad \text{or} \quad \varphi(y^1, y^2) = e^{-r(t)}\sqrt{(y^1)^2 + (y^2)^2}, \qquad (12.36)$$

where

$$\cos t = \frac{y^1}{\sqrt{(y^1)^2 + (y^2)^2}}, \quad \sin t = \frac{y^2}{\sqrt{(y^1)^2 + (y^2)^2}}, \quad \tan t = \frac{y^2}{y^1},$$

i.e. the level set $\{\varphi(y^1, y^2) \equiv 1\}$ of the 1-homogeneous function φ in $\mathbb{R}^2$ is given by the parametrized curve $t \to (e^{r(t)}\cos t, e^{r(t)}\sin t)$.

Since the curvature κ of a smooth curve $t \to (e^{r(t)}\cos t, e^{r(t)}\sin t)$ in $\mathbb{R}^2$ is

$$\kappa = -\frac{e^r}{\sqrt{\dot{r}^2 + 1}}(\ddot{r} - \dot{r}^2 - 1), \qquad (12.37)$$

the vanishing of the expression $\ddot{r} - \dot{r}^2 - 1$ means the infinitesimal linearity of the corresponding positively homogeneous function in $\mathbb{R}^2$.

Definition 11 Let $\varphi(y^1, y^2)$ be a positively 1-homogeneous function on $\mathbb{R}^2$ and let $\kappa(t)$ be the curvature of the curve $t \to (e^{r(t)}\cos t, e^{r(t)}\sin t)$ defined by the Eq. (12.36). We say that $\varphi(y^1, y^2)$ is *strongly convex*, if $\kappa(t) \neq 0$ for all $t \in \mathbb{R}$.

Conditions (A)–(C) in the following theorem imply that the projective factor $\mathscr{P}$ at $x_0 \in M$ is a nonlinear function, and hence, according to Remark 1, $(M, \mathscr{F})$ is a non-Riemannian Finsler manifold.

Theorem 1 *Let $(M, \mathscr{F})$ be a projectively flat Finsler surface of non-zero constant curvature covered by a coordinate system (x^1, x^2). Assume that there exists a point $x_0 \in M$ such that one of the following conditions hold*

(A) *$\mathscr{F}$ induces a scalar product on $T_{x_0}M$ and the projective factor $\mathscr{P}$ at x_0 is a strongly convex positively 1-homogeneous function,*
(B) *$\mathscr{F}(x_0, y)$ is a strongly convex absolutely 1-homogeneous function on $T_{x_0}M$, and the projective factor $\mathscr{P}(x_0, y)$ on $T_{x_0}M$ satisfies $\mathscr{P}(x_0, y) = c \cdot \mathscr{F}(x_0, y)$ with $0 \neq c \in \mathbb{R}$,*
(C) *there is a projectively related Euclidean coordinate system of $(M, \mathscr{F})$ centered at x_0 and one has*

$$\mathscr{F}(0, y) = |y| \pm \langle a, y \rangle \quad \text{and} \quad \mathscr{P}(0, y) = \frac{1}{2}(\pm|y| - \langle a, y \rangle). \qquad (12.38)$$

Assume that the vector fields $U = U^i \frac{\partial}{\partial x^i}$, $V = V^i \frac{\partial}{\partial x^i} \in \mathfrak{X}^\infty(M)$ have constant coordinate functions and let $\xi = R(U, V)$ be the corresponding curvature vector field. Then the subalgebra of the infinitesimal holonomy algebra $\mathfrak{hol}_x^*(M)$ generated by the vector fields $\xi|_{x_0}$, $\nabla_1\xi|_{x_0}$, $\nabla_2\xi|_{x_0}$ and $\nabla_1(\nabla_2\xi)|_{x_0}$ is infinite dimensional.

Proof Since $(M, \mathscr{F})$ is of constant flag curvature, we can write

$$R^i_{jk}(x, y) = \lambda \left(\delta^i_j g_{km}(x, y) y^m - \delta^i_k g_{jm}(x, y) y^m \right), \quad \text{with} \quad \lambda = \text{const.}$$

According to Lemma 2 the horizontal Berwald covariant derivative $\nabla_W R$ of the tensor field $R = R^i_{jk}(x, y) dx^j \wedge dx^k \frac{\partial}{\partial x^i}$ vanishes and hence $\nabla_W R = 0$.

Since the curvature tensor field is skew-symmetric, $R_{(x,y)}$ acts on the one-dimensional wedge product $T_x M \wedge T_x M$. The covariant derivative $\nabla_W \xi$ of the curvature vector field $\xi = R(U, V) = \frac{1}{2} R(U \otimes V - V \otimes U) = R(U \wedge V)$ can be written in the form

$$\nabla_W \xi = R \left(\nabla_W (U \wedge V) \right) = R(\nabla_W U \wedge V + U \wedge \nabla_W V).$$

We have $U \wedge V = \frac{1}{2} \left(U^1 V^2 - U^2 V^1 \right) \frac{\partial}{\partial x^1} \wedge \frac{\partial}{\partial x^2}$ and hence

$$\nabla_W \xi = (U^1 V^2 - V^1 U^2) W^k R \left(\nabla_k \left(\frac{\partial}{\partial x^1} \wedge \frac{\partial}{\partial x^2} \right) \right), \tag{12.39}$$

where $\nabla_k \xi := \nabla_{\frac{\partial}{\partial x^k}} \xi$. Since

$$\nabla_k \left(\frac{\partial}{\partial x^1} \wedge \frac{\partial}{\partial x^2} \right) = \left(\nabla_k \frac{\partial}{\partial x^1} \right) \wedge \frac{\partial}{\partial x^2} + \frac{\partial}{\partial x^1} \wedge \left(\nabla_k \frac{\partial}{\partial x^2} \right)$$

$$= G^l_{k1} \frac{\partial}{\partial x^l} \wedge \frac{\partial}{\partial x^2} + \frac{\partial}{\partial x^1} \wedge G^m_{k2} \frac{\partial}{\partial x^m} = \left(G^1_{k1} + G^2_{k2} \right) \frac{\partial}{\partial x^1} \wedge \frac{\partial}{\partial x^2}$$

we obtain

$$\nabla_W \xi = \left(G^1_{k1} + G^2_{k2} \right) W^k R(U, V) = \left(G^1_{k1} + G^2_{k2} \right) W^k \xi.$$

Since the geodesic coefficients are given by (12.17) we have

$$\nabla_W \xi = G^m_{km} W^k \xi = 3 \frac{\partial \mathscr{P}}{\partial y^k} W^k \xi. \tag{12.40}$$

Hence

$$\nabla_Z (\nabla_W \xi) = 3 \nabla_Z \left(\frac{\partial \mathscr{P}}{\partial y^k} W^k \xi \right) = 3 \left\{ \nabla_Z \left(\frac{\partial \mathscr{P}}{\partial y^k} W^k \right) \xi + \left(\frac{\partial \mathscr{P}}{\partial y^k} W^k \right) \left(\frac{\partial \mathscr{P}}{\partial y^l} Z^l \right) \right\} \xi.$$

Let W be a vector field with constant coordinate functions. Then, using (12.17) we get

$$\nabla_Z \left(\frac{\partial \mathscr{P}}{\partial y^k} W^k \right) = \left(\frac{\partial^2 \mathscr{P}}{\partial x^j \partial y^k} - G^m_j \frac{\partial^2 \mathscr{P}}{\partial y^m \partial y^k} \right) W^k Z^j = \left(\frac{\partial^2 \mathscr{P}}{\partial x^j \partial y^k} - \mathscr{P} \frac{\partial^2 \mathscr{P}}{\partial y^k \partial y^j} \right) W^k Z^j,$$

and hence

$$\nabla_Z \left(\nabla_W \xi \right) = 3 \left\{ \frac{\partial^2 \mathscr{P}}{\partial x^j \partial y^k} - \mathscr{P} \frac{\partial^2 \mathscr{P}}{\partial y^k \partial y^j} + \frac{\partial \mathscr{P}}{\partial y^k} \frac{\partial \mathscr{P}}{\partial y^j} \right\} W^k Z^j \xi. \tag{12.41}$$

Let $x_0 \in M$ be the point with coordinates $(0,0)$ in the local coordinate system of $(M, \mathscr{F})$ corresponding to the canonical coordinates of the projectively related Euclidean plane. According to Lemma 8.2.1 in [6], p.155, we have

$$\frac{\partial^2 \mathscr{P}}{\partial x^1 \partial y^2} - \mathscr{P} \frac{\partial^2 \mathscr{P}}{\partial y^1 \partial y^2} + \frac{\partial \mathscr{P}}{\partial y^1} \frac{\partial \mathscr{P}}{\partial y^2} = 2 \frac{\partial \mathscr{P}}{\partial y^1} \frac{\partial \mathscr{P}}{\partial y^2} - \frac{1}{2} \frac{\partial^2 \mathscr{F}^2}{\partial y^1 \partial y^2} = 2 \frac{\partial \mathscr{P}}{\partial y^1} \frac{\partial \mathscr{P}}{\partial y^2} - \lambda \, g_{12}.$$

Hence the vector fields $\xi|_{x_0}$, $\nabla_1 \xi|_{x_0}$, $\nabla_2 \xi|_{x_0}$ and $\nabla_1 (\nabla_2 \xi)|_{x_0}$ are linearly independent if and only if the functions

$$1, \quad \frac{\partial \mathscr{P}}{\partial y^1}\Big|_{x_0}, \quad \frac{\partial \mathscr{P}}{\partial y^2}\Big|_{x_0}, \quad \left(2 \frac{\partial \mathscr{P}}{\partial y^1} \frac{\partial \mathscr{P}}{\partial y^2} - \lambda \, g_{12} \right)\Big|_{x_0} \tag{12.42}$$

are linearly independent, where $g_{12} = g_y(\frac{\partial}{\partial x^1}, \frac{\partial}{\partial x^2})$ is the component of the metric tensor of $(M, \mathscr{F})$.

Lemma 2 *The functions* $\frac{\partial \mathscr{P}(0,y)}{\partial y^1}$, $\frac{\partial \mathscr{P}(0,y)}{\partial y^2}$ *and* $\mathscr{P}(0,y) \frac{\partial^2 \mathscr{P}(0,y)}{\partial y^1 \partial y^2}$ *can be expressed in the polar coordinate system* (e^r, t) *by*

$$\frac{\partial \mathscr{P}(0,y)}{\partial y^1} = (\cos t + \dot{r} \sin t) e^{-r}, \quad \frac{\partial \mathscr{P}(0,y)}{\partial y^2} = (\sin t - \dot{r} \cos t) e^{-r},$$

$$\mathscr{P}(0,y) \frac{\partial^2 \mathscr{P}(0,y)}{\partial y^1 \partial y^2} = (\dot{r}^2 + 1 - \ddot{r}) e^{-2r} \sin t \cos t,$$

where the dot refers to differentiation with respect to the variable t.

Proof We obtain from $\frac{\partial e^{-r}}{\partial y^1} = -e^{-r} \dot{r} \frac{\partial t}{\partial y^1}$ and from $-\frac{y^2}{(y^1)^2} = \frac{\partial}{\partial y^1} (\frac{y^2}{y^1}) \frac{d \tan t}{dt} \frac{\partial t}{\partial y^1} = \frac{1}{\cos^2 t} \frac{\partial t}{\partial y^1}$ that $\frac{\partial e^{-r}}{\partial y^1} = e^{-r} \dot{r} \cos^2 t \frac{y^2}{(y^1)^2} = e^{-r} \dot{r} \frac{y^2}{(y^1)^2 + (y^2)^2}$. Hence

$$\frac{\partial \mathscr{P}(0,y)}{\partial y^1} = \frac{\partial \left(e^{-r} \sqrt{(y^1)^2 + (y^2)^2} \right)}{\partial y^1} = e^{-r} \left(\dot{r} \frac{y^2}{\sqrt{(y^1)^2 + (y^2)^2}} + \frac{y^1}{\sqrt{(y^1)^2 + (y^2)^2}} \right).$$

Similarly, we have $\frac{\partial e^{-r}}{\partial y^2} = -e^{-r} \dot{r} \cos^2 t \frac{1}{y^1} = -e^{-r} \dot{r} \frac{y^1}{(y^1)^2 + (y^2)^2}$. Hence

$$\frac{\partial \mathscr{P}(0,y)}{\partial y^2} = \frac{\partial \left(e^{-r} \sqrt{(y^1)^2 + (y^2)^2} \right)}{\partial y^2} = e^{-r} \left(-\dot{r} \frac{y^1}{\sqrt{(y^1)^2 + (y^2)^2}} + \frac{y^2}{\sqrt{(y^1)^2 + (y^2)^2}} \right).$$

Finally we have

$$\frac{\partial^2 \mathscr{P}(0,y)}{\partial y^1 \partial y^2} = \frac{\partial(\sin t - \dot{r}\cos t)e^{-r}}{\partial y^1} = (\ddot{r} - \dot{r}^2 - 1)e^{-r}\sin t \cos t \frac{1}{\sqrt{(y^1)^2 + (y^2)^2}}.$$

Replacing φ by the function $\mathscr{P}(0,y)$ in the expression (12.36) we get the assertion.

Lemma 3 *Let* $r : \mathbb{R} \to \mathbb{R}$ *be a* 2π-*periodic smooth function such that the inequality* $\ddot{r}(t) - \dot{r}^2(t) - 1 \neq 0$ *holds on a dense subset of* $\mathbb{R}$. *Then the functions*

$$1, \ (\cos t + \dot{r}\sin t)e^{-r}, \ (\sin t - \dot{r}\cos t)e^{-r}, \ (\cos t + \dot{r}\sin t)(\sin t - \dot{r}\cos t)e^{-2r} \tag{12.43}$$

are linearly independent.

Proof The derivative of $(\cos t + \dot{r}\sin t)e^{-r}$ and of $(\sin t - \dot{r}\cos t)e^{-r}$ are $(\ddot{r} - \dot{r}^2 - 1)e^{-r}\sin t$ and $(\ddot{r} - \dot{r}^2 - 1)e^{-r}\cos t$, respectively, hence the functions (12.43) do not vanish identically. Let us consider a linear combination

$$A + B(\cos t + \dot{r}\sin t)e^{-r} + C(\sin t - \dot{r}\cos t)e^{-r} + D(\cos t + \dot{r}\sin t)(\sin t - \dot{r}\cos t)e^{-2r} = 0$$

with constant coefficients A, B, C, D. We differentiate and divide by $e^{-t}(\ddot{r} - \dot{r}^2 - 1)$ and we have

$$B\sin t - C\cos t - D(\cos 2t + \dot{r}\sin 2t)e^{-r} = 0.$$

Putting $t = 0$ and $t = \pi$ we get $C = -De^{-r(0)} = De^{-r(\pi)}$. Since $e^{-r(0)}, \ e^{-r(\pi)} > 0$ we get $C = D = 0$ and hence $A = B = C = D = 0$.

Now, assume that condition (A) of Theorem 1 is fulfilled. According to Proposition 1 if the functions (12.42) are linearly independent, then the holonomy group $\mathsf{Hol}_{x_0}(M)$ is an infinite dimensional subgroup of $\mathsf{Diff}^\infty(\mathfrak{I}_{x_0}M)$. The function $\mathscr{F}(x_0,y)$ induces a scalar product on $T_{x_0}M$, consequently the component g_{12} of the metric tensor is constant on $T_{x_0}M$. Hence $\mathsf{Hol}_{x_0}(M)$ is infinite dimensional if the functions

$$1, \quad \left.\frac{\partial \mathscr{P}}{\partial y^1}\right|_{x_0}, \quad \left.\frac{\partial \mathscr{P}}{\partial y^2}\right|_{x_0}, \quad \left.\frac{\partial \mathscr{P}}{\partial y^1}\frac{\partial \mathscr{P}}{\partial y^2}\right|_{x_0} \tag{12.44}$$

are linearly independent. This follows from Lemma 3 and hence the assertion of the theorem is true.

Assume that condition (B) is satisfied. We denote $\varphi(y) = \mathscr{F}(x_0, y)$. Using the expressions (12.42) we obtain that the vector fields $\xi|_{x_0}$, $\nabla_1\xi|_{x_0}$, $\nabla_2\xi|_{x_0}$ and $\nabla_1(\nabla_2\xi)|_{x_0}$ are linearly independent if and only if the functions

$$1, \qquad \left.\frac{\partial \mathscr{P}}{\partial y^1}\right|_{x_0} = c\frac{\partial \varphi}{\partial y^1}, \qquad \left.\frac{\partial \mathscr{P}}{\partial y^2}\right|_{x_0} = c\frac{\partial \varphi}{\partial y^2}$$

$$\left.\left(2\frac{\partial \mathscr{P}}{\partial y^1}\frac{\partial \mathscr{P}}{\partial y^2} - \lambda\, g_{12}\right)\right|_{x_0} = (2c^2 - \lambda)\frac{\partial \varphi}{\partial y^1}\frac{\partial \varphi}{\partial y^2} - \lambda\, \varphi\frac{\partial^2 \varphi}{\partial y^1 \partial y^2}$$

are linearly independent. According to Lemma 2 this is equivalent to the linear independence of the functions

$$1, \qquad (\cos t + \dot{r}\sin t)\, e^{-r}, \quad (\sin t - \dot{r}\cos t)\, e^{-r},$$

$$(2c^2 - \lambda)(\cos t + \dot{r}\sin t)(\sin t - \dot{r}\cos t)\, e^{-2r} - \lambda(\ddot{r} - \dot{r}^2 - 1)\, e^{-2r}\sin t\cos t.$$

If $r = \text{const}$, then these functions are 1, $\cos t\, e^{-r}$, $\sin t\, e^{-r}$, $2c^2 \cos t \sin t\, e^{-2r}$, hence the assertion follows from Lemma 3. In the following we can assume that $r(t) \neq \text{const}$. Let $t_0 \in \mathbb{R}$ such that $\dot{r}(t_0) = 0$ and $\kappa(t_0) \neq 0$. We rotate the coordinate system at the angle $-t_0$ with respect to the euclidean norm $\sqrt{(y^1)^2 + (y^2)^2}$, then we get in the new polar coordinate system that $\dot{r}(0) = 0$ and $\kappa(0) \neq 0$. Consider the linear combination

$$A + B(\cos t + \dot{r}\sin t)e^{-r} + C(\sin t - \dot{r}\cos t)\, e^{-r}$$

$$+ D\big((2c^2 - \lambda)(\cos t + \dot{r}\sin t)(\sin t - \dot{r}\cos t)\, e^{-2r} - \lambda(\ddot{r} - \dot{r}^2 - 1)\, e^{-2r}\sin t\cos t\big) = 0 \tag{12.45}$$

with some constants A, B, C, D. Since the function φ is absolutely homogeneous, the function $r(t)$ is π-periodic. Putting $t + \pi$ into t, the value of

$$A + D(2c^2 - \lambda)(\cos t + \dot{r}\sin t)(\sin t - \dot{r}\cos t)\, e^{-2r} - \lambda(\ddot{r} - \dot{r}^2 - 1)\, e^{-2r}\sin t\cos t$$

does not change, but the value of

$$B(\cos t + \dot{r}\sin t)\, e^{-r} + C(\sin t - \dot{r}\cos t)\, e^{-r}$$

changes sign. Since Lemma 3 implies that $(\cos t + \dot{r}\sin t)\, e^{-r}$ and $(\sin t - \dot{r}\cos t)\, e^{-r}$ are linearly independent, we have $B = C = 0$ and (12.45) becomes

$$A\, e^{2r} + D\left((2c^2 - \lambda)\left[-\dot{r}\cos 2t + \frac{1}{2}(1 - \dot{r}^2)\sin 2t\right] - \frac{\lambda}{2}(\ddot{r} - \dot{r}^2 - 1)\sin 2t\right) = 0. \tag{12.46}$$

Since $\dot{r}(0) = 0$ at $t = 0$, we have $A = 0$. If $D \neq 0$, then (12.46) gives

$$(2c^2 - \lambda)\left[-\dot{r}\cos 2t + \frac{1}{2}(1 - \dot{r}^2)\sin 2t\right] - \frac{\lambda}{2}(\ddot{r} - \dot{r}^2 - 1)\sin 2t = 0.$$

By derivation and putting $t = 0$ we obtain

$$(2c^2 - \lambda)\left[-\ddot{r}(0) + 1 \right] - \lambda(\dot{r}(0) - 1) = 2c^2(1 - \ddot{r}(0)) = 0.$$

Using the relation (12.37) condition (B) gives $\kappa(0) = e^{r(0)}(1 - \ddot{r}(0)) \neq 0$, which is a contradiction. Hence $D = 0$ and the vector fields $\xi\big|_{x_0}$, $\nabla_1\xi\big|_{x_0}$, $\nabla_2\xi\big|_{x_0}$ and $\nabla_1(\nabla_2\xi)\big|_{x_0}$ are linearly independent. Using Proposition 1 we obtain the assertion.

Suppose now that the condition (C) holds. Hence we have

$$\frac{\partial \mathscr{F}}{\partial y^1}(0, y) = \frac{y^1}{|y|} \pm a^1, \qquad \frac{\partial \mathscr{F}}{\partial y^2}(0, y) = \frac{y^2}{|y|} \pm a^2, \qquad \frac{\partial^2 \mathscr{F}}{\partial y^1 \partial y^2}(0, y) = -\frac{y^1 y^2}{|y|^3},$$

and

$$g_{12} = \left(\frac{y^1}{|y|} \pm a^1 \right)\left(\frac{y^2}{|y|} \pm a^2 \right) - \left(1 \pm \left\langle a, \frac{y}{|y|} \right\rangle \right)\frac{y^1 y^2}{|y|^2}. \tag{12.47}$$

Similarly, we obtain from condition (C) that

$$\frac{\partial \mathscr{P}}{\partial y^1}(0, y) = \pm \frac{y^1}{|y|} - a^1, \qquad \frac{\partial \mathscr{P}}{\partial y^2}(0, y) = \pm \frac{y^2}{|y|} - a^2.$$

Using the expressions (12.42) we get that the vector fields $\xi\big|_{x_0}$, $\nabla_1\xi\big|_{x_0}$, $\nabla_2\xi\big|_{x_0}$, $\nabla_1(\nabla_2\xi)\big|_{x_0}$ are linearly independent if and only if the functions

$$1, \qquad \frac{\partial \mathscr{P}}{\partial y^1}\bigg|_{(0,y)} = \pm \frac{y^1}{|y|} - a^1, \qquad \frac{\partial \mathscr{P}}{\partial y^2}\bigg|_{(0,y)} = \pm \frac{y^2}{|y|} - a^2$$

and

$$2\frac{\partial \mathscr{P}}{\partial y^1}\frac{\partial \mathscr{P}}{\partial y^2} - \lambda g_{12}\bigg|_{(0,y)} =$$

$$= \mp \left\langle a, \frac{y}{|y|} \right\rangle \frac{y^1 y^2}{|y|^2} + (1 - \lambda)\frac{y^1 y^2}{|y|^2}$$

$$\mp (2 + \lambda)\left(a_2\frac{y^1}{|y|} + a_1\frac{y^2}{|y|} \right) + (2 - \lambda)a_1 a_2$$

are linearly independent. Putting

$$\cos t = \frac{y^1}{|y|}, \qquad \sin t = \frac{y^2}{|y|}$$

we obtain that this condition is true, since the trigonometric polynomials

$$1, \quad \cos t, \quad \sin t, \quad (1 - \lambda) \cos t \sin t \mp (a_1 \cos t + a_2 \sin t) \cos t \sin t$$

are linearly independent. Hence the vector fields $\xi|_{x_0}$, $\nabla_1 \xi|_{x_0}$, $\nabla_2 \xi|_{x_0}$, $\nabla_1 (\nabla_2 \xi)|_{x_0}$ generate an infinite dimensional subalgebra of $\mathsf{Hol}_{x_0}(M)$.

12.6.2 Projective Finsler Manifolds of Constant Curvature

Now we will prove that the infinitesimal holonomy algebra of a totally geodesic submanifold of a Finsler manifold can be embedded into the infinitesimal holonomy algebra of the entire manifold. This result yields a lower estimate for the dimension of the holonomy group.

12.6.2.1 Totally Geodesic and Auto-Parallel Submanifolds

Lemma 4 *Let $\bar{M}$ be a totally geodesic submanifold in a spray manifold $(M, \mathscr{S})$. The curvature vector fields at any point of $\bar{M}$ can be extended to a curvature vector field of M.*

Proof Assume that the manifolds $\bar{M}$ and M are k, respectively $n = k + p$ dimensional. Let $(x^1, \ldots, x^k, x^{k+1}, \ldots, x^n)$ be an adapted coordinate system, i.e. the submanifold $\bar{M}$ is locally given by the equations $x^{k+1} = \cdots = x^n = 0$. Using the notation of the proof of Lemma 1 we get from Eq. (12.12) that $G_\alpha^\sigma = 0$ and $G_{\alpha\,\beta}^\sigma = 0$ for any $(x^1, \ldots, x^k, 0, \ldots, 0; y^1, \ldots, y^k, 0, \ldots, 0)$ we have

$$\frac{\partial G_\alpha^\sigma}{\partial x^\beta} - \frac{\partial G_\beta^\sigma}{\partial x^\alpha} + G_\alpha^\tau G_{\beta\tau}^\sigma - G_\beta^\tau G_{\alpha\tau}^\sigma + G_\alpha^\gamma G_{\beta\gamma}^\sigma - G_\beta^\gamma G_{\alpha\gamma}^\sigma = 0$$

at $(x^1, \ldots, x^k, 0, \ldots, 0; y^1, \ldots, y^k, 0, \ldots, 0)$. Hence the curvature tensors $\bar{K}$ and K, corresponding to the spray $\mathscr{S}$, respectively to the spray $\mathscr{S}$ satisfy

$$\bar{K}(X, Y)(x, y) = K(X, Y)(x, y) \quad \text{if} \quad x \in \bar{M} \quad \text{and} \quad y, X, Y \in T_x\bar{M}.$$

It follows that for any given $X, Y \in T_x\bar{M}$ the curvature vector field $\bar{\xi}(y) = \bar{K}(X, Y)(x, y)$ at $x \in \bar{M}$ defined on $T_x\bar{M}$ can be extended to the curvature vector field $\xi(y) = K(X, Y)(x, y)$ at $x \in \bar{M}$ defined on T_xM.

Theorem 2 *Let $\bar{M}$ be a totally geodesic two-dimensional submanifold of a Finsler manifold $(M, \mathscr{F})$ such that the infinitesimal holonomy algebra $\mathfrak{hol}_x^*(\bar{M})$ of $\bar{M}$ is infinite dimensional. Then the infinitesimal holonomy algebra $\mathfrak{hol}_x^*(M)$ of M is infinite dimensional.*

Proof According to Lemma 1 any curvature vector field of $\bar{M}$ at $x \in \bar{M} \subset M$ defined on $\mathfrak{I}_x \bar{M}$ can be extended to a curvature vector field on the indicatrix $\mathfrak{I}_x M$. Hence the curvature algebra $\mathfrak{R}_x(\bar{M})$ of the submanifold $\bar{M}$ can be embedded into the curvature algebra $\mathfrak{R}_x(M)$ of the manifold $(M, \mathscr{F})$. Assume that $\bar{\xi}$ is a vector field belonging to the infinitesimal holonomy algebra $\mathfrak{hol}_x^*(\bar{M})$ which can be extended to the vector field ξ belonging to the infinitesimal holonomy algebra $\mathfrak{hol}_x^*(M)$. Any vector field $\bar{X} \in \mathfrak{X}^\infty(\bar{M})$ can be extended to a vector field $X \in \mathfrak{X}^\infty(M)$, hence the horizontal Berwald covariant derivative along $\bar{X} \in \mathfrak{X}^\infty(\bar{M})$ of $\bar{\xi}$ can be extended to the Berwald horizontal covariant derivative along $X \in \mathfrak{X}^\infty(M)$ of the vector field ξ. It follows that the infinitesimal holonomy algebra $\mathfrak{hol}_x^*(\bar{M})$ of the submanifold $\bar{M}$ can be embedded into the infinitesimal holonomy algebra $\mathfrak{hol}_x^*(M)$ of the Finsler manifold $(M, \mathscr{F})$. Consequently, $\mathfrak{hol}_x^*(M)$ is infinite dimensional and hence the holonomy group $\mathsf{Hol}_x(M)$ is an infinite dimensional subgroup of $\mathsf{Diff}^\infty(\mathfrak{I}_x M)$.

This result can be applied to locally projectively flat Finsler manifolds, as they have for each tangent 2-plane a totally geodesic submanifold which is tangent to this 2-plane.

Corollary 7 *If a locally projectively flat Finsler manifold has a two-dimensional totally geodesic submanifold satisfying one of the conditions of Theorem 1, then its infinitesimal holonomy algebra is infinite dimensional.*

According to Eqs. (12.18) and (12.19) the projectively flat Randers manifolds of non-zero constant curvature satisfy condition (C) of Theorem 1. We can apply Corollary 7 to these manifolds and we get the following

Theorem 3 *The infinitesimal holonomy algebra of any projectively flat Randers manifolds of non-zero constant flag curvature is infinite dimensional.*

Bryant in [4, 5] introduced and studied complete Finsler metrics of positive curvature on S^2. He proved that there exists exactly a 2-parameter family of Finsler metrics on S^2 with curvature $= 1$ with great circles as geodesics. Z. Shen generalized a 1-parameter family of complete Bryant metrics to S^n satisfying

$$\mathscr{F}(0, y) = |y| \cos \alpha, \quad \mathscr{P}(0, y) = |y| \sin \alpha \tag{12.48}$$

with $|\alpha| < \frac{\pi}{2}$ in a coordinate neighborhood centered at $0 \in \mathbb{R}^n$, (cf. Example 7.1. in [30] and Example 8.2.9 in [6]).

We investigate the holonomy groups of two families of metrics, containing the 1-parameter family of complete Bryant-Shen metrics (12.48). The first family in the following theorem is defined by condition (A), which is motivated by Theorem 8.2.3 in [6]. There is given the following construction:

If $\psi = \psi(y)$ is an arbitrary Minkowski norm on $\mathbb{R}^n$ and $\varphi = \varphi(y)$ is an arbitrary positively 1-homogeneous function on $\mathbb{R}^n$, then there exists a projectively flat Finsler metric $\mathscr{F}$ of constant flag curvature -1, defined on a neighborhood of the origin, such that $\mathscr{F}$ and its projective factor $\mathscr{P}$ satisfy $\mathscr{F}(0, y) = \psi(y)$ and $\mathscr{P}(0, y) = \varphi(y)$.

Condition (B) in the next theorem is confirmed by Example 7 in [30], p. 1726, where it is proved that for an arbitrary given Minkowski norm φ and $|\vartheta| < \frac{\pi}{2}$ there exists a projectively flat Finsler function $\mathscr{F}$ of constant curvature $= 1$ defined on a neighborhood of $0 \in \mathbb{R}^n$, such that

$$\mathscr{F}(0, y) = \varphi(y) \cos \vartheta \quad \text{and} \quad \mathscr{P}(0, y) = \varphi(y) \sin \vartheta.$$

Conditions (A) and (B) in Theorem 1 together with Corollary 7 yield the following

Theorem 4 *Let* $(M, \mathscr{F})$ *be a projectively flat Finsler manifold of non-zero constant curvature. Assume that there exists a point* $x_0 \in M$ *and a two-dimensional totally geodesic submanifold* $\bar{M}$ *through* x_0 *such that one of the following conditions holds*

(A) *$\mathscr{F}$ induces a scalar product on $T_{x_0}\bar{M}$, and the projective factor $\mathscr{P}$ on $T_{x_0}\bar{M}$ is a strongly convex positively 1-homogeneous function,*
(B) *$\mathscr{F}(x_0, y)$ on $T_{x_0}\bar{M}$ is a strongly convex absolutely 1-homogeneous function on $T_{x_0}M$, and the projective factor $\mathscr{P}(x_0, y)$ on $T_{x_0}\bar{M}$ satisfies $\mathscr{P}(x_0, y) = c \cdot \mathscr{F}(x_0, y)$ with $0 \neq c \in \mathbb{R}$.*

Then the infinitesimal holonomy $\mathfrak{hol}_x^*(M)$ *of* M *is infinite dimensional.*

12.7   Dimension of the Holonomy Group

Let (M, F) be a positive definite Finsler manifold and $x \in M$ an arbitrary point in M. According to Proposition 3 of [18], the infinitesimal holonomy algebra $\mathfrak{hol}_x^*(M)$ is tangent to the holonomy group $\mathsf{Hol}_x(M)$. Therefore the group generated by the exponential image of the infinitesimal holonomy algebra at $x \in M$ with respect to the exponential map $\exp_x \colon \mathfrak{X}^\infty(\mathfrak{I}_xM) \to \mathsf{Diff}^\infty(\mathfrak{I}_xM)$ is a subgroup of the closed holonomy group $\overline{\mathsf{Hol}_x(M)}$ (see Theorem 3.1 of [21]). Consequently, we have the following estimation on the dimensions:

$$\dim \mathfrak{hol}_x^*(M) \leqslant \dim \mathsf{Hol}_x(M). \tag{12.49}$$

Proposition 7 *The infinitesimal holonomy algebra* $\mathfrak{hol}_x^*(M)$ *of any locally projectively flat non-Riemannian Finsler surface* $(M, \mathscr{F})$ *of constant curvature* $\lambda \neq 0$ *is infinite dimensional.*

Proof We use for the proof Lemma 1 and the notations introduced in the proof of Theorem 1 in the previous section on projective Finsler surfaces of constant curvature. Using the assertion on the vector fields (12.42) we obtain

Lemma 1 *For any fixed* $1 \leqslant j, k \leqslant 2$

$$y \to \xi(x, y), \quad y \to \nabla_1 \xi(x, y), \quad y \to \nabla_2 \xi(x, y), \quad y \to \nabla_j(\nabla_k \xi)(x, y), \tag{12.50}$$

considered as vector fields on $\mathfrak{I}_x M$, are $\mathbb{R}$-linearly independent if and only if the

$$1, \quad \frac{\partial \mathscr{P}}{\partial y^1}, \quad \frac{\partial \mathscr{P}}{\partial y^2}, \quad \frac{\partial^2 \mathscr{P}}{\partial y^j \partial y^k} - \frac{\lambda}{4} g_{jk} \tag{12.51}$$

are linearly independent functions on $T_x M$.

Since we assumed that the Finsler function $\mathscr{F}$ is non-Riemannian at the point x, i.e. $\mathscr{F}^2(x, y)$ is non-quadratic in y, the function $\mathscr{P}(x, y)$ is nonlinear in y on $T_x M$ (cf. Eq. (12.17)). Let us choose a direction $y_0 = (y_0^1, y_0^2) \in T_x M$ with $y_0^1 \neq 0$, $y_0^2 \neq 0$ and having property that $\mathscr{P}$ is nonlinear 1-homogeneous function in a conic neighborhood U of y_0 in $T_x M$. By restricting U if it is necessary we can suppose that for any $y \in U$ we have $y^1 \neq 0, y^2 \neq 0$.

To avoid confusion between coordinate indexes and exponents, we rename the fiber coordinates of vectors belonging to U by $(u, v) = (y^1, y^2)$. Using the values of $\mathscr{P}$ on U we can define a 1-variable function $f = f(t)$ on an interval $I \subset \mathbb{R}$ by

$$f(t) := \frac{1}{v} \mathscr{P}(x_1, x_2, tv, v). \tag{12.52}$$

Then we can express $\mathscr{P}$ and its derivatives with f:

$$\mathscr{P} = v f(u/v), \qquad \frac{\partial \mathscr{P}}{\partial y^1} = f'(u/v), \qquad \frac{\partial \mathscr{P}}{\partial y^2} = f(u/v) - \frac{u}{v} f'(u/v),$$

$$\frac{\partial^2 \mathscr{P}}{\partial y^1 \partial y^1} = \frac{1}{v} f''(u/v), \qquad \frac{\partial^2 \mathscr{P}}{\partial y^1 \partial y^2} = -\frac{u}{v^2} f''(u/v), \qquad \frac{\partial^2 \mathscr{P}}{\partial y^2 \partial y^2} = \frac{u^2}{v^3} f''(u/v). \tag{12.53}$$

Lemma 2 *The functions $1, \frac{\partial \mathscr{P}}{\partial y^1}, \frac{\partial \mathscr{P}}{\partial y^2}$ are linearly independent.*

Proof A nontrivial relation $a + b \frac{\partial \mathscr{P}}{\partial y^1} + c \frac{\partial \mathscr{P}}{\partial y^2} = 0$ yields the differential equation $a + bf' + c(f - tf') = 0$. It is clear that both b and c cannot be zero. If $c \neq 0$, we get the differential equation

$$\frac{(a + cf)'}{a + cf} = \frac{1}{t - \frac{b}{c}}.$$

The solution is $f(t) = t - (a+b)/c$ and therefore the corresponding $\mathscr{P}(u, v) = u - v(a+b)/c$ is linear which is a contradiction. If $c = 0$, then $b \neq 0$ and $f = -\frac{a}{b}t + K$. The corresponding $\mathscr{P}(u, v) = -\frac{a}{b}u + Kv$ is again linear which is a contradiction. Let us assume now that the infinitesimal holonomy algebra is finite dimensional. We will show that this assumption leads to contradiction which will prove then that the infinitesimal holonomy algebra is actually infinite dimensional.

Since $\mathfrak{I}_x M$ is one-dimensional, according to Lemma 1, the four vector fields in (12.50) are linearly dependent for any $j, k \in \{1, 2\}$. Using Lemma 1 we get that the functions

$$1, \quad \mathscr{P}_1, \quad \mathscr{P}_2, \quad \mathscr{P}_j \mathscr{P}_k - \frac{\lambda}{4} g_{jk} \tag{12.54}$$

$(\mathscr{P}_i = \frac{\partial \mathscr{P}}{\partial y^i}, \mathscr{P}_{jk} = \frac{\partial^2 \mathscr{P}}{\partial y^j \partial y^k})$ are linearly dependent for any $j, k \in \{1, 2\}$. From Lemma 2 we know that the first three functions in (12.54) are linearly independent. Therefore by the assumption, the fourth function must be a linear combination of the first three, that is there exist constants $a_i, b_i, c_i \in \mathbb{R}$, $i = 1, 2, 3$, such that

$$\frac{\lambda}{4} g_{11} = \mathscr{P}_1 \mathscr{P}_1 + a_1 + b_1 \mathscr{P}_1 + c_1 \mathscr{P}_2,$$

$$\frac{\lambda}{4} g_{12} = \mathscr{P}_1 \mathscr{P}_2 + a_2 + b_2 \mathscr{P}_1 + c_2 \mathscr{P}_2, \tag{12.55}$$

$$\frac{\lambda}{4} g_{22} = \mathscr{P}_2 \mathscr{P}_2 + a_3 + b_3 \mathscr{P}_1 + c_3 \mathscr{P}_2.$$

Since $\partial_1 g_{21} - \partial_2 g_{11} = 0$ and $\partial_1 g_{22} - \partial_2 g_{12} = 0$ we have

$$\mathscr{P}_2 \mathscr{P}_{11} - \mathscr{P}_1 \mathscr{P}_{12} + b_2 \mathscr{P}_{11} + (c_2 - b_1) \mathscr{P}_{12} - c_1 \mathscr{P}_{22} = 0,$$
$$\mathscr{P}_1 \mathscr{P}_{22} - \mathscr{P}_2 \mathscr{P}_{12} - b_3 \mathscr{P}_{11} + (b_2 - c_3) \mathscr{P}_{12} + c_2 \mathscr{P}_{22} = 0. \tag{12.56}$$

Using the expressions (12.53) we obtain from (12.56) the equations

$$\left(f - \frac{u}{v}f'\right)\frac{1}{v}f'' + f'\frac{u}{v^2}f'' + b_2\frac{1}{v}f'' - (c_2 - b_1)\frac{u}{v^2}f'' - c_1\frac{u^2}{v^3}f'' = 0,$$

$$f'\frac{u^2}{v^3}f'' + \left(f - \frac{u}{v}f'\right)\frac{u}{v^2}f'' - b_3\frac{1}{v}f'' - (b_2 - c_3)\frac{u}{v^2}f'' + c_2\frac{u^2}{v^3}f'' = 0. \tag{12.57}$$

Since by the nonlinearity of $\mathscr{P}$ on U we have $f'' \neq 0$, Eq. (12.57) can divide by f''/v and we get

$$f + b_2 + (b_1 - c_2)\frac{u}{v} - \frac{c_1 u^2}{v^2} = 0$$

$$\frac{u}{v}f - b_3 + (c_3 - b_2)\frac{u}{v} + \frac{c_2 u^2}{v^2} = 0. \tag{12.58}$$

for any $t = u/v$ in an interval $I \subset \mathbb{R}$. The solution of this system of quadratic equations for the function f is $f(t) = -c_2 t - b_2$ with $c_1 = b_3 = 0$, $b_1 = 2c_2$, $c_3 = 2b_2$. But this is a contradiction, since we supposed that by the nonlinearity of P we have $f'' \neq 0$ on this interval. Hence the functions $1, \mathscr{P}_1, \mathscr{P}_2, \mathscr{P}_j \mathscr{P}_k - \frac{\lambda}{4}g_{jk}$ can not be linearly dependent for any $j, k \in \{1, 2\}$, from which follows the assertion.

Remark 8 From Proposition 7 we get that if $(M, \mathscr{F})$ is non-Riemannian and $\lambda \neq 0$, then the holonomy group has an infinite dimensional tangent algebra.
Indeed, according to Theorem 6.3 in [18] the infinitesimal holonomy algebra $\mathfrak{hol}_x^*(M)$ is tangent to the holonomy group $\mathsf{Hol}_x(M)$, from which follows the assertion.

Now, we can prove our main result:

Theorem 1 *The holonomy group of a locally projectively flat simply connected Finsler manifold $(M, \mathscr{F})$ of constant curvature λ is finite dimensional if and only if $(M, \mathscr{F})$ is Riemannian or $\lambda = 0$.*

Proof If $(M, \mathscr{F})$ is Riemannian, then its holonomy group is a Lie subgroup of the orthogonal group and therefore it is a finite dimensional compact Lie group. If $(M, \mathscr{F})$ has zero curvature, then the horizontal distribution associated to the canonical connection in the tangent bundle is integrable and hence the holonomy group is trivial.

If $(M, \mathscr{F})$ is non-Riemannian having non-zero curvature λ, then for each tangent 2-plane $S \subset T_x M$ the manifold M has a totally geodesic submanifold $\widetilde{M} \subset M$ such that $T_x \widetilde{M} = S$. This $\widetilde{M}$ with the induced metric is a locally projectively flat Finsler surface of constant curvature λ. Therefore from Proposition 7 we get that $\mathfrak{hol}_x^*(\widetilde{M})$ is infinite dimensional. Moreover, according to Theorem 4.3 in [22], if a Finsler manifold $(M, \mathscr{F})$ has a totally geodesic two-dimensional submanifold $\widetilde{M}$ such that the infinitesimal holonomy algebra of $\widetilde{M}$ is infinite dimensional, then the infinitesimal holonomy algebra $\mathfrak{hol}_x^*(M)$ of the containing manifold is also infinite dimensional. Using (12.49) we get that $\mathsf{Hol}_x(M)$ cannot be finite dimensional. Hence the assertion is true.

We note that there are examples of non-Riemannian type locally projectively flat Finsler manifolds with $\lambda = 0$ curvature, (cf. [16]).

Remark 9 In the discussion before the previous theorem, the key condition for the Finsler metric tensor was not the positive definiteness but its non-degenerate property. Therefore Theorem 1 can be generalized as follows.

A pair $(M, \mathscr{F})$ is called *semi-Finsler manifold* if in the definition of Finsler manifolds the positive definiteness of the Finsler metric tensor is replaced by the non-degenerate property. Then we have

Corollary 8 *The holonomy group of a locally projectively flat simply connected semi-Finsler manifold $(M, \mathscr{F})$ of constant curvature λ is finite dimensional if and only if $(M, \mathscr{F})$ is semi-Riemannian or $\lambda = 0$.*

12.8 Maximal Holonomy

12.8.1 The Group $\mathsf{Diff}_+^\infty(\mathbb{S}^1)$ and the Fourier Algebra

Let $(M, \mathscr{F})$ be a positive definite Finsler 2-manifold. In this case the indicatrix is diffeomorphic to the unit circle $\mathbb{S}^1$, at any point $x \in M$. Moreover, if there exists a non-vanishing curvature vector field at $x \in M$ then any other curvature vector field at $x \in M$ is proportional to it, which means that the curvature algebra is at most one-dimensional. The infinitesimal holonomy algebra, however, can be an infinite dimensional subalgebra of $\mathfrak{X}^\infty(\mathbb{S}^1)$, therefore the holonomy group can be an infinite dimensional subgroup of $\text{Diff}_+^\infty(\mathbb{S}^1)$, cf. [22].

Let $\mathbb{S}^1 = \mathbb{R}$ mod 2π be the unit circle with the standard counterclockwise orientation. The group $\text{Diff}_+^\infty(\mathbb{S}^1)$ of orientation preserving diffeomorphisms of the $\mathbb{S}^1$ is the connected component of $\text{Diff}^\infty(\mathbb{S}^1)$. The Lie algebra of $\text{Diff}_+^\infty(\mathbb{S}^1)$ is the Lie algebra $\mathfrak{X}^\infty(\mathbb{S}^1)$—denoted also by $\text{Vect}(\mathbb{S}^1)$ in the literature—can be written in the form $f(t)\frac{d}{dt}$, where f is a 2π-periodic smooth functions on the real line $\mathbb{R}$. A sequence $\{f_j\frac{d}{dt}\}_{j\in\mathbb{N}} \subset \text{Vect}(\mathbb{S}^1)$ converges to $f\frac{d}{dt}$ in the Fréchet topology of $\text{Vect}(\mathbb{S}^1)$ if and only if the functions f_j and all their derivatives converge uniformly to f, respectively to the corresponding derivatives of f. The Lie bracket on $\text{Vect}(\mathbb{S}^1)$ is given by

$$\left[f\frac{d}{dt}, g\frac{d}{dt}\right] = \left(g\frac{df}{dt} - \frac{dg}{dt}f\right)\frac{d}{dt}.$$

The *Fourier algebra* $\text{F}(\mathbb{S}^1)$ on $\mathbb{S}^1$ is the Lie subalgebra of $\text{Vect}(\mathbb{S}^1)$ consisting of vector fields $f\frac{d}{dt}$ such that $f(t)$ has finite Fourier series, i.e. $f(t)$ is a Fourier polynomial. The vector fields $\{\frac{d}{dt}, \cos nt\frac{d}{dt}, \sin nt\frac{d}{dt}\}_{n\in\mathbb{N}}$ provide a basis for $\text{F}(\mathbb{S}^1)$. A direct computation shows that the vector fields

$$\frac{d}{dt}, \quad \cos t\frac{d}{dt}, \quad \sin t\frac{d}{dt}, \quad \cos 2t\frac{d}{dt}, \quad \sin 2t\frac{d}{dt} \tag{12.59}$$

generate the Lie algebra $\text{F}(\mathbb{S}^1)$. The complexification $\text{F}(\mathbb{S}^1)\otimes_\mathbb{R}\mathbb{C}$ of $\text{F}(\mathbb{S}^1)$ is called the *Witt algebra* $\text{W}(\mathbb{S}^1)$ on $\mathbb{S}^1$ having the natural basis $\{ie^{int}\frac{d}{dt}\}_{n\in\mathbb{Z}}$, with the Lie bracket $[ie^{imt}\frac{d}{dt}, ie^{int}\frac{d}{dt}] = i(m-n)e^{i(n-m)t}\frac{d}{dt}$.

Lemma 1 *The group $\left\langle \overline{\exp(\text{F}(\mathbb{S}^1))} \right\rangle$ generated by the topological closure of the exponential image of the Fourier algebra $\text{F}(\mathbb{S}^1)$ is the orientation preserving diffeomorphism group $\text{Diff}_+^\infty(\mathbb{S}^1)$.*

Proof The Fourier algebra $\text{F}(\mathbb{S}^1)$ is a dense subalgebra of $\text{Vect}(\mathbb{S}^1)$ with respect to the Fréchet topology, i.e. $\overline{\text{F}(\mathbb{S}^1)} = \text{Vect}(\mathbb{S}^1)$. This assertion follows from the fact that any r-times continuously differentiable function can be approximated uniformly by the arithmetical means of the partial sums of its Fourier series (cf. [10], 2.12 Theorem). The exponential mapping is continuous (c.f. Lemma 4.1 in [25], p. 79), hence we have

$$\exp\left(\text{Vect}(\mathbb{S}^1)\right) = \exp\left(\overline{\text{F}(\mathbb{S}^1)}\right) \subset \overline{\exp\left(\text{F}(\mathbb{S}^1)\right)} \subset \text{Diff}_+^\infty(\mathbb{S}^1) \tag{12.60}$$

which gives for the generated groups the relations

$$\left\langle\exp\left(\text{Vect}(\mathbb{S}^1)\right)\right\rangle \subset \left\langle \overline{\exp\left(\text{F}(\mathbb{S}^1)\right)} \right\rangle \subset \text{Diff}_+^\infty(\mathbb{S}^1). \tag{12.61}$$

Moreover, the conjugation map $\mathsf{Ad} : \mathsf{Diff}_+^\infty(\mathbb{S}^1) \times \mathsf{Vect}(\mathbb{S}^1)$ satisfies the relation

$$h \exp s\xi\, h^{-1} = \exp s\mathsf{Ad}(h)\xi$$

for every $h \in \mathsf{Diff}_+^\infty(\mathbb{S}^1)$ and $\xi \in \mathsf{Vect}(\mathbb{S}^1)$. Clearly, the Lie algebra $\mathsf{Vect}(\mathbb{S}^1)$ is invariant under conjugation and hence the group $\langle \exp\left(\mathsf{Vect}(\mathbb{S}^1)\right)\rangle$ is also invariant under conjugation. Therefore $\langle \exp\left(\mathsf{Vect}(\mathbb{S}^1)\right)\rangle$ is a nontrivial normal subgroup of $\mathsf{Diff}_+^\infty(\mathbb{S}^1)$. On the other hand, $\mathsf{Diff}_+^\infty(\mathbb{S}^1)$ is a simple group (cf. [9]) which means that its only nontrivial normal subgroup is itself. Therefore, we have $\langle \exp\left(\mathsf{Vect}(\mathbb{S}^1)\right)\rangle = \mathsf{Diff}_+^\infty(\mathbb{S}^1)$, and using (12.61) we get

$$\left\langle \overline{\exp(\mathsf{F}(\mathbb{S}^1))} \right\rangle = \mathsf{Diff}_+^\infty(\mathbb{S}^1).$$

12.8.2 Holonomy of the Standard Funk Plane and the Bryant-Shen 2-Spheres

Using the results of the preceding chapter we can prove the following statement, which provides a useful tool for the investigation of the closed holonomy group of Finsler 2-manifolds.

Proposition 8 *If the infinitesimal holonomy algebra* $\mathfrak{hol}_x^*(M)$ *at a point* $x \in M$ *of a simply connected Finsler 2-manifold* $(M, \mathscr{F})$ *contains the Fourier algebra* $\mathsf{F}(\mathbb{S}^1)$ *on the indicatrix at* x, *then* $\overline{\mathsf{Hol}_x(M)}$ *is isomorphic to* $\mathsf{Diff}_+^\infty(\mathbb{S}^1)$.

Proof Since M is simply connected we have

$$\overline{\mathsf{Hol}_x(M)} \subset \mathsf{Diff}_+^\infty(\mathbb{S}^1). \tag{12.62}$$

On the other hand, using Proposition 1, we get

$$\exp\left(\mathsf{F}(\mathbb{S}^1)\right) \subset \overline{\mathsf{Hol}_x(M)} \;\Rightarrow\; \overline{\exp\left(\mathsf{F}(\mathbb{S}^1)\right)} \subset \overline{\mathsf{Hol}_x(M)} \;\Rightarrow\; \left\langle \overline{\exp\left(\mathsf{F}(\mathbb{S}^1)\right)} \right\rangle \subset \overline{\mathsf{Hol}_x(M)},$$

and from the last relation, using Lemma 1, we can obtain that

$$\mathsf{Diff}_+^\infty(\mathbb{S}^1) \subset \overline{\mathsf{Hol}_x(M)}. \tag{12.63}$$

Comparing (12.62) and (12.63) we get the assertion.

Using this proposition we can prove our main result:

Theorem 1 *Let* $(M, \mathscr{F})$ *be a simply connected projectively flat Finsler manifold of constant curvature* $\lambda \neq 0$. *Assume that there exists a point* $x_0 \in M$ *such that on* $T_{x_0}M$ *the induced Minkowski norm is an Euclidean norm, that is* $\mathscr{F}(x_0, y) = \|y\|$, *and the* *projective factor at x_0 satisfies* $\mathscr{P}(x_0, y) = c \cdot \|y\|$ *with* $c \in \mathbb{R}$, $c \neq 0$. *Then the closed holonomy group* $\overline{\mathrm{Hol}}_{x_0}(M)$ *at* x_0 *is isomorphic to* $\mathrm{Diff}_+^{\infty}(\mathbb{S}^1)$.

Proof Since $(M, \mathscr{F})$ is a locally projectively flat Finsler manifold of non-zero constant curvature, we can use an (x^1, x^2) local coordinate system centered at $x_0 \in M$, corresponding to the canonical coordinates of the Euclidean space which is projectively related to $(M, \mathscr{F})$. Let (y^1, y^2) be the induced coordinate system in the tangent plane $T_x M$. In the sequel we identify the tangent plane $T_{x_0}M$ with $\mathbb{R}^2$ by using the coordinate system (y^1, y^2). We will use the Euclidean norm $\|(y^1, y^2)\| = \sqrt{(y^1)^2 + (y^2)^2}$ of $\mathbb{R}^2$ and the corresponding polar coordinate system (e^r, t), too.

Let us consider the curvature vector field ξ at $x_0 = 0$ defined by

$$\xi = R\left(\frac{\partial}{\partial x_1}, \frac{\partial}{\partial x_2}\right)\bigg|_{x=0} = \lambda\left(\delta_2^i g_{1m}(0, y)y^m - \delta_1^i g_{2m}(0, y)y^m\right)\frac{\partial}{\partial x^i}$$

Since $(M, \mathscr{F})$ is of constant flag curvature, the horizontal Berwald covariant derivative $\nabla_W R$ of the tensor field R vanishes, c.f. Lemma 2. Therefore the covariant derivative of ξ can be written in the form

$$\nabla_W \xi = R\left(\nabla_k\left(\frac{\partial}{\partial x^1} \wedge \frac{\partial}{\partial x^2}\right)\right)W^k.$$

Since

$$\nabla_k\left(\frac{\partial}{\partial x^1} \wedge \frac{\partial}{\partial x^2}\right) = \left(G_{k1}^1 + G_{k2}^2\right)\frac{\partial}{\partial x^1} \wedge \frac{\partial}{\partial x^2}$$

we obtain $\nabla_W \xi = \left(G_{k1}^1 + G_{k2}^2\right)W^k\xi$. Using (12.17) we can express $G_{km}^m = 3\frac{\partial P}{\partial y^k} = 3c\frac{y^k}{\|y\|}$ and hence

$$\nabla_k \xi = 3\frac{\partial P}{\partial y^k}\xi = 3c\frac{y^k}{\|y\|}\xi,$$

where we use the notation $\nabla_k = \nabla_{\frac{\partial}{\partial x^k}}$. Moreover we have

$$\nabla_j\left(\frac{\partial \mathscr{P}}{\partial y^k}\right) = \frac{\partial^2 \mathscr{P}}{\partial x^j \partial y^k} - G_j^m\frac{\partial^2 \mathscr{P}}{\partial y^m \partial y^k} = \frac{\partial^2 \mathscr{P}}{\partial x^j \partial y^k} - \mathscr{P}\frac{\partial^2 \mathscr{P}}{\partial y^k \partial y^j},$$

and hence

$$\nabla_j\left(\nabla_k \xi\right) = 3\left\{\frac{\partial^2 \mathscr{P}}{\partial x^j \partial y^k} - \mathscr{P}\frac{\partial^2 \mathscr{P}}{\partial y^k \partial y^j} + 3\frac{\partial \mathscr{P}}{\partial y^k}\frac{\partial \mathscr{P}}{\partial y^j}\right\}\xi.$$

According to Lemma 8.2.1, Eq. (8.25) in [6], p. 155, we obtain

$$\frac{\partial^2 \mathscr{P}}{\partial x^j \partial y^k} = \frac{\partial \mathscr{P}}{\partial y^j} \frac{\partial \mathscr{P}}{\partial y^k} + \mathscr{P} \frac{\partial^2 \mathscr{P}}{\partial y^j \partial y^k} - \frac{\lambda}{2} \frac{\partial^2 \mathscr{F}^2}{\partial y^j \partial y^k}.$$

Using the assumptions on $\mathscr{F}$ and on the projective factor $\mathscr{P}$ we can get at x_0

$$\nabla_j (\nabla_k \xi) = 3 \left(4c^2 \frac{\partial \mathscr{F}}{\partial y^j} \frac{\partial \mathscr{F}}{\partial y^k} - \frac{\lambda}{2} \frac{\partial^2 \mathscr{F}^2}{\partial y^j \partial y^k} \right) \xi$$

and hence

$$\nabla_j (\nabla_k \xi) = 3 \left(4 c^2 \frac{y^j y^k}{\|y\|^2} - \lambda \, \delta^{jk} \right) \xi,$$

where $\delta^{jk} \in \{0, 1\}$ such that $\delta^{jk} = 1$ if and only if $j = k$.

Let us introduce polar coordinates $y^1 = r \cos t$, $y^2 = r \sin t$ in the tangent space $T_{x_0} M$. We can express the curvature vector field, its first and second covariant derivatives along the indicatrix curve $\{(\cos t, \sin t); \ 0 \leqslant t < 2\pi\}$ as follows:

$$\xi = \lambda \frac{d}{dt}, \quad \nabla_1 \xi = 3c\lambda \cos t \frac{d}{dt}, \quad \nabla_2 \xi = -3c\lambda \sin t \frac{d}{dt}, \quad \nabla_1 (\nabla_2 \xi) = 12 \, c^2 \lambda \sin 2t \frac{d}{dt},$$

$$\nabla_1 (\nabla_1 \xi) = \lambda \left(12 \, c^2 \cos^2 t - \lambda \right) \frac{d}{dt}, \quad \nabla_2 (\nabla_2 \xi) = \lambda \left(12 \, c^2 \sin^2 t - \lambda \right) \frac{d}{dt}.$$

Since $c \lambda \neq 0$, the vector fields

$$\frac{d}{dt}, \quad \cos t \frac{d}{dt}, \quad \sin t \frac{d}{dt}, \quad \cos t \sin t \frac{d}{dt}, \quad \cos^2 t \frac{d}{dt}, \quad \sin^2 t \frac{d}{dt}$$

are contained in the infinitesimal holonomy algebra $\mathfrak{hol}^*_{x_0} (M)$. It follows that the generator system

$$\left\{ \frac{d}{dt}, \quad \cos t \frac{d}{dt}, \quad \sin t \frac{d}{dt}, \quad \cos 2t \frac{d}{dt}, \quad \sin 2t \frac{d}{dt} \right\}$$

of the Fourier algebra $F(\mathbb{S}^1)$ (c.f. Eq. (12.59)) is contained in the infinitesimal holonomy algebra $\mathfrak{hol}^*_{x_0} (M)$. Hence the assertion follows from Proposition 8.

We remark that the standard Funk plane and the Bryant-Shen 2-spheres are connected, projectively flat Finsler manifolds of non-zero constant curvature. Moreover, in each of them, there exists a point $x_0 \in M$ and an adapted local coordinate system centered at x_0 with the following properties: the Finsler norm $\mathscr{F}(x_0, y)$ and the projective factor $\mathscr{P}(x_0, y)$ at x_0 are given by $\mathscr{F}(x_0, y) = \|y\|$ and by $\mathscr{P}(x_0, y) = c \cdot \|y\|$ with some constant $c \in \mathbb{R}$, $c \neq 0$, where $\|y\|$ is an Euclidean norm in the tangent space at x_0. Using Theorem 1 we can obtain

Theorem 2 *The closed holonomy groups of the standard Funk plane and of the Bryant-Shen 2-spheres are maximal, that is diffeomorphic to the orientation preserving diffeomorphism group of* $\mathbb{S}^1$.

References

1. M. Ackermann, R. Hermann, *Sophus Lie's Transformation Group Paper* (Mathematical Science Press, Brookline, 1975).
2. W. Barthel, Nichtlineare Zusammenhänge und deren Holonomiegruppen. J. Reine Angew. Math. **212**, pp. 120–149 (1963)
3. A. Borel, A. Lichnerowicz, Groupes d'holonomie de varietes riemanniennes. CR Acad. Sci. Paris **234**, 1835–1837 (1952)
4. R. Bryant, Finsler structures on the 2-sphere satisfying K = 1, in *Finsler Geometry, Contemporary Mathematics*, vol. 196 (American Mathematical Society, Providence, 1996), pp. 27–42
5. R. Bryant, Projectively flat Finsler 2-spheres of constant curvature. Sel. Math. N. Ser. **3**, 161–204 (1997)
6. S.S. Chern, Z. Shen, *Riemann-Finsler Geometry*. Nankai Tracts in Mathematics, vol. 6 (World Scientific, Singapore, 2005)
7. M. Crampin, D.J. Saunders, Holonomy of a class of bundles with fibre metrics. Publ. Math. Debr. **81**, 199–234 (2012)
8. Ch. Ehresmann, Les connexions infinitésimales dans un espace fibré différentiable. Colloque de Topologie, Bruxelles, (1950), pp. 29–55
9. M.R. Herman, Sur le groupe des difféomorphismes du tore. Ann. Inst. Fourier **23**, 75–86 (1973)
10. Y. Katznelson, *An Introduction to Harmonic Analysis* (Cambridge University Press, Cambridge, 2004)
11. S. Kobayashi, *Transformation Groups in Differential Geometry*. Ergebnisse der Mathematik und ihrer Grenzgebiete, vol. 70 (Springer, Berlin, 1972)
12. I. Kolar, P.W. Michor, J. Slovak, *Natural Operations in Differential Geometry* (Springer, Berlin, 1993)
13. L. Kozma, On Landsberg spaces and holonomy of Finsler manifolds. Contemp. Math. **196**, 177–185 (1996)
14. L. Kozma, Holonomy structures in Finsler geometry. Part 5, in *Handbook of Finsler Geometry*, ed. by P.L. Antonelli (Kluwer Academic, Dordrecht, 2003), pp. 445–490
15. A. Kriegl, P.W. Michor, *The Convenient Setting for Global Analysis*. Surveys and Monographs, vol. 53 (AMS, Providence, 1997)
16. B. Li, Z. Shen, On a class of projectively flat Finsler metrics with constant flag curvature. Int. J. Math. **18**, 1–12 (2007)
17. P.W. Michor, *Gauge Theory for Fiber Bundles*. Monographs and Textbooks in Physical Sciences. Lecture Notes, vol. 19 (Bibliopolis, Napoli, 1991)
18. Z. Muzsnay, P.T. Nagy, Tangent Lie algebras to the holonomy group of a Finsler manifold. Commun. Math. **19**, 137–147 (2011)
19. Z. Muzsnay, P.T. Nagy, Finsler manifolds with non-Riemannian holonomy. Houst. J. Math. **38**, 77–92 (2012)
20. Z. Muzsnay, P.T. Nagy, Witt algebra and the curvature of the Heisenberg group. Commun. Math. **20**, 33–40 (2012)
21. Z. Muzsnay, P.T. Nagy, Characterization of projective Finsler manifolds of constant curvature having infinite dimensional holonomy group. Publ. Math. Debr. **84**, 17–28 (2014)
22. Z. Muzsnay, P.T. Nagy, Projectively flat Finsler manifolds with infinite dimensional holonomy. Forum Mathematicum **27**, 767–786 (2015)

23. Z. Muzsnay, P.T. Nagy, Finsler 2-manifolds with maximal holonomy group of infinite dimension. Differ. Geom. Appl. **39**, 1–9 (2015)
24. S. Numata, On Landsberg spaces of scalar curvature. J. Korean Math. Soc. **12**, 97–100 (1975)
25. H. Omori, *Infinite-Dimensional Lie Groups*. Translation of Mathematical Monographs, vol. 158 (American Mathematical Society, Providence, 1997)
26. G. Randers, On an asymmetrical metric in the fourspace of general relativity. Phys. Rev. (2) **59**, 195–199 (1941)
27. Z. Shen, *Differential Geometry of Spray and Finsler Spaces* (Kluwer Academic, Dordrecht, 2001)
28. Z. Shen, Two-dimensional Finsler metrics of constant flag curvature. Manuscripta Math. **109**, 349–366 (2002)
29. Z. Shen, Projectively flat Randers metrics with constant flag curvature. Math. Ann. **325**, 19–30 (2003)
30. Z. Shen, Projectively flat Finsler metrics with constant flag curvature. Trans. Am. Math. Soc. **355**, 1713–1728 (2003)
31. Z.I. Szabó, Positive definite Berwald spaces. Tensor, New Ser. **35**, 25–39 (1981)
32. J. Teichmann, Regularity of infinite-dimensional Lie groups by metric space methods. Tokyo J. Math. **24**, 29–58 (2001)
33. J.A. Wolf, Differentiable fibre spaces and mappings compatible with Riemannian metrics. Michigan Math. J. **11**, 65–70 (1964)

Chapter 13
Lepage Manifolds

Olga Rossi

Abstract Lepage manifolds represent a geometric structure arising from differential equations, and related with them as close as possible. Geometric properties of Lepage manifolds reflect geometrical, topological and dynamical properties of differential equations, and vice versa. After symplectic geometry, Lepage manifolds represent a next step in understanding interrelations between differential equations, geometry and physics. The aim of this chapter is to introduce the new concept, and to stimulate deeper studies in this direction as well as applications in differential geometry, differential equations, dynamical systems, exterior differential systems, calculus of variations, and mathematical physics.

13.1 Introduction

Lepage manifold is a fibred manifold endowed with a certain differential form—Lepage $(n + 1)$-form where n is the dimension of the base manifold. Lepage manifolds represent a geometric structure arising from differential equations, and related with them as close as possible. It turns out that geometric properties of Lepage manifolds reflect geometric, topological and dynamical properties of differential equations, and vice versa. The aim of this chapter is to introduce the new concept and to stimulate further research in this direction, as well as many potential applications.

After symplectic geometry, Lepage manifolds represent a next step in understanding interrelations between differential equations, geometry and physics. In a similar way as symplectic manifolds constitute foundations of analytical mechanics and the theory of dynamical systems, Lepage manifolds become fundamental for differential equations in general, with strong applications in the calculus of variations and geometric field theories.

O. Rossi (✉)
Department of Mathematics, Faculty of Science, The University of Ostrava, 30. dubna 22, 701 03
Ostrava, Czech Republic
e-mail: olga.rossi@osu.cz

© Springer International Publishing AG 2017   321
G. Falcone (ed.), *Lie Groups, Differential Equations, and Geometry*,
UNIPA Springer Series, DOI 10.1007/978-3-319-62181-4_13

The use of differential forms in the calculus of variations and its applications in physics goes back to É. Cartan, Poincaré, Whittaker, Carathéodory, Lepage, Dedecker and Weinstein (see, e.g., [9, 11, 16, 72, 79, 92, 94]). A great success was the development of symplectic geometry as a beautiful and powerful tool for geometrization of classical Hamiltonian mechanics (see, e.g., Abraham and Marsden [1], Arnold et al. [4], de León and Rodrigues [14], Libermann and Marle [73]), and with many applications in physics, the theory of differential equations, dynamical systems, control theory, and elsewhere. Nevertheless, it has turned out that there are many variational systems for which the use of symplectic manifolds is inappropriate. Indeed, a symplectic structure is naturally inherited only by a particular class of variational problems, namely those described by variational integrals of one independent variable with first-order regular Lagrangians, independent upon the parameter of the curves (time). The generalization of the symplectic structure arising in classical mechanics to multiple integrals is by no means trivial. It means to find out proper geometric structures for general variational problems, including degenerate Lagrangians (implicit variational equations) and higher-order Lagrangian systems, in order to generalize Lagrangian theory, Hamilton and Hamilton–Jacobi theory and Noether theory. The next challenge is then to include non-variational equations in the picture, and to understand the property "to be variational".

Despite great efforts, the attempts to include more general Lagrangians or even non-variational equations in the symplectic setting have not really been satisfactory. It has turned out that the difficulties are of geometric origin: instead of tangent and cotangent bundles, one has to consider *jet bundles*,[1] and symplectic forms have to be replaced by some other geometric objects. There is a vast literature which cannot be mentioned here; to illustrate the state of art and the diversity of approaches we refer at least to the books [7, 13, 27, 32, 34, 41, 52, 59, 66], and selected papers [8, 10, 15, 18, 20, 21, 24–26, 28–31, 33, 38–40, 42, 45, 48, 51, 60, 63, 71, 75, 76, 81, 82, 85, 91] to mention just a few.

Concerning the variational case, roughly speaking, there are two mainstream settings: one based on multisymplectic or polysymplectic forms, and the other based on the *Cartan form and its generalizations*. Within the latter approach, the central object for formulation of Lagrangian and Hamiltonian field theories is the Poincaré–Cartan form, or more generally, a *Lepage n-form* where n is the dimension of the base of the underlying fibred manifold. Following the idea of Lepage [72], examples of these forms were introduced to the calculus of variations in the seminal works of Carathéodory, Dedecker, Krupka, Betounes, Olver, Gotay, Crampin and Saunders (see e.g. [5, 9, 12, 17, 30, 45, 46, 55, 77]). The general concept of Lepage n-form, or, *Lepage equivalent of a Lagrangian*, is due to Krupka [45, 48] who used it to build, systematically, a general theory of higher order variational integrals on fibred manifolds, generalizing the Lagrange, Hamilton, Hamilton–Jacobi and Noether

---

[1]The concept introduced by Ehresmann in the 1950s [22, 23]; a standard book on this subject is [86].

theories of classical mechanics [52]. Subsequently, instead of a Lepage equivalent of a Lagrangian, we suggested to consider better a certain $(n + 1)$-form which in this context becomes a *Lepage equivalent of a system of variational equations* (see [56, 58–60, 62, 81, 83]). This step was motivated mainly by the idea to obtain a global Hamilton theory, independent upon a choice of a particular Lagrangian; in this new setting, Hamilton equations are the same for all equivalent Lagrangians. At the same time this approach avoids certain unpleasant discrepancies which appear in Hamiltonian formulation of Lagrangian theories related with particular Lagrangians (as reported, e.g., in [17, 19, 35, 59, 60, 68, 69, 87]), and allows one to treat Lagrangian systems globally in terms of the theory of exterior differential systems even in case when a global Lagrangian does not exist [59, 60, 63, 81].

It took quite a long time to find a proper generalization of Lepage $(n + 1)$-forms to non-variational equations [53, 78]. The generalization is not straightforward, and appears within the theory of variational sequences (for the introduction to the variational sequence theory we refer to [49, 50], and [52]). This progress motivated us to introduce the concept of a *Lepage manifold* as defined in this article.

Every Lepage manifold naturally carries an exterior differential system (EDS). A fundamental result on the Inverse Problem of the Calculus of Variations [48] implies that the corresponding system of (generally higher-order) differential equations (ODEs if $n = 1$ and PDEs otherwise) for integral sections of this EDS is variational, and so represents Euler–Lagrange equations of possibly locally defined Lagrangians, if and only if the generating Lepage $(n + 1)$-form is *closed* (up to a contact equivalence). This property makes Lepage manifolds very suitable structures to study *variational equations*, and at the same time a good background to study *non-variational differential equations* in the context of the calculus of variations making use of variational structures and techniques. With Lepage manifolds, many particular results known so far and fragmented in the calculus of variations and in the geometric theory of differential equations can be put together in a unified framework, linked together, and extended. Lepage manifolds naturally arise wherever differential equations of dynamical type are concerned (i.e. related with dynamical forms, vector fields, sprays, semisprays, Ehresmann connections). Through this relationship they are naturally present in Riemannian geometry, Finsler geometry, symplectic geometry, and Lagrangian geometry, as well as in analytical mechanics, classical field theories, relativity, string theories, and other physical systems. The present exposition aims to show the potential of this new framework, and to stimulate further research in this direction, as well as study of applications, namely in differential geometry, differential equations and mathematical physics.

The present article is structured as follows: In Sect. 13.2 we introduce *Lepage forms* (of any degree) in the context of the variational sequence, and we remind the solution of the Inverse Problem of the Calculus of Variations. We follow the paper [78] where the reader can find more detailed exposition, as well as other aspects and applications of the theory of variational sequences. Section 13.3 is devoted to *Lepage manifolds*. After defining the concept we show the relationship with variational equations and with non-variational equations, and we introduce the important class of *regular* Lepage manifolds. We also remind a theorem by Krupka et al. [54], which makes possible to transfer 'non-variational' Lepage manifolds to the 'variational' ones. In Sect. 13.4 we pay a particular attention to Lepage manifolds of order zero which are a natural framework for general *Hamiltonian systems*. Many results of this section are taken from the paper [36]. Then we present original results on *Legendre transformation* and *Poisson structure* for the generalized (and not necessarily variational) Hamiltonian systems, and, remarkably, we introduce a new concept of a *directional Poisson bracket* in covariant Hamiltonian field theory. The last two sections summarize some applications of Lepage manifolds in classical field theories and in mechanics. Finally we briefly show that (and how) Lepage manifolds naturally arise in *Riemann* and *Finsler geometry*.

We assume that the reader has a basic knowledge of jet bundles and related geometric structures as introduced, e.g., in [86]. As for the calculus of variations in jet bundles, the reader can be guided by the books [13, 59], and [52].

13.2 Lepage Forms

13.2.1 Jet Bundles and Contact Forms

All manifolds and mappings are smooth, and summation over repeated indices applies.

Let $\pi : Y \to X$ be a fibred manifold (surjective submersion) with fibre dimension m, and $\pi_r : J^r Y \to X$ its r-jet prolongation; usually we consider $r = 1, 2$, and for the sake of simplicity we sometimes denote Y as $J^0 Y$. We also denote by $\pi_{r,k} : J^r Y \to J^k Y$ the natural projections. $J^r Y$ is the manifold of r-jets of local sections γ of π; a point in $J^r Y$ (the equivalence class of sections of π at $x \in X$, with the same derivatives up to the order r) is denoted by $j_x^r \gamma$. In what follows we shall omit the word "local" and speak just about "sections" meaning "local sections".

Any section γ of π can be naturally prolonged to a section $J^r \gamma$ of the fibred manifold $\pi_r : J^r Y \to X$, defined by $J^r \gamma(x) = j_x^r \gamma$. Sections of π_r of this kind are called *holonomic* and have to be distinguished from (general) sections of π_r which need not be a form of a prolongation of a section of π.

Let (x^i, y^σ) denote local fibred coordinates on Y, and (x^i, y_J^σ) the associated coordinates on $J^r Y$, where J is a multiindex, $0 \leq |J| \leq r$, $J = (j_1, \ldots, j_p)$ with $1 \leq j_1 \leq j_2 \leq \cdots \leq j_p \leq n$ and $p \leq r$. In formulas where the summation convention applies we shall also use the functions y_J^σ, $J = (j_1, \ldots, j_p)$, $p \leq r$, where $1 \leq j_1, j_2, \ldots, j_r \leq n$; these functions are defined naturally with help of the above coordinate functions using the symmetry of partial derivatives. Next, we set

$$\omega_0 = dx^1 \wedge dx^2 \wedge \cdots \wedge dx^n, \quad \omega_j = i_{\partial/\partial x^j}\omega_0, \quad \omega_{Jj} = i_{\partial/\partial x^j}\omega_J. \tag{13.1}$$

Jet bundles over fibred manifolds carry a canonical *contact structure*. A differential form on $J^r Y$ is called *contact* if it vanishes on r-jet prolongations of all sections of π. It easily follows that every q-form for $q > n$ is contact, so it is convenient to introduce a refined concept: A contact q-form η is called k-contact if for every π_r-vertical vector field ξ the contraction $i_\xi \eta$ is $(k-1)$-contact, where $k = 1, 2, \ldots q$. 0-contact forms, i.e. such that $i_\xi \eta = 0$ whenever ξ is a π_r-vertical vector field, are more often referred to as *horizontal forms*.

Horizontal forms of degree n where n is the dimension of the base manifold X are essential in the calculus of variations, since they serve as *Lagrangians*. In what follows, we shall denote the sheaf of Lagrangians of order r over π by $\Lambda^r_{n,X}$.

An important structural Lemma states that *every q-form on $J^r Y$, if lifted to $J^{r+1}Y$, has an invariant decomposition*

$$\pi^*_{r+1,r}\eta = h\eta + p_1\eta + \cdots + p_q\eta \tag{13.2}$$

where $h\eta$ is horizontal, and $p_k\eta$ are k-contact forms $(1 \leqslant k \leqslant q)$ [48].

If applied to the exterior derivative of a function f, the splitting (13.2) gives the horizontal differential $d_H f = h df$ and vertical differential $d_V f = p_1 df$. The n components of the horizontal differential are called *total derivatives*; they take the form

$$d_i f = \frac{\partial f}{\partial x^i} + \sum_{|J|=0}^{r} \frac{\partial f}{\partial y^\sigma_J} y^\sigma_{Ji}. \tag{13.3}$$

In view of the canonical decomposition above we often say that a form η is *at most k-contact* if $p_i\eta = 0$ for all $i > k$, and *at least k-contact* if $p_i\eta = 0$ for all $i < k$. Note that if $q \geqslant n + 1$ then η is contact, and it is *at least $(q-n)$-contact*, i.e. the contact components $p_0, p_1, \ldots, p_{q-n-1}$ of η are equal to zero. If, moreover, $p_{q-n}\eta = 0$, we speak about a *strongly contact* form.

Contact 1-forms on $J^r Y$ annihilate a distribution called *Cartan distribution* and denoted $\mathscr{C}_r$. The annihilator $\mathscr{C}^0_r$ of $\mathscr{C}_r$ is generated by the following local contact forms:

$$\omega^\sigma_J = dy^\sigma_J - y^\sigma_{Ji} dx^i, \quad 0 \leqslant |J| \leqslant r - 1. \tag{13.4}$$

It is also worth notice that the *contact ideal* of order r is generated by contact 1-forms and exterior derivatives of contact 1-forms on $J^r Y$ [49].

Finally, we recall that a vector field ξ on $J^r Y$ is called a *contact symmetry* if it is a symmetry of the contact ideal on $J^r Y$; contact symmetries are characterized by the condition $\mathscr{L}_\xi \mathscr{C}^0_r \subset \mathscr{C}^0_r$. In particular, for every π-projectable vector field ζ on Y the r-jet prolongation $J^r\zeta$ is a contact symmetry. Conversely, if ξ on $J^r Y$ is a π_r-projectable contact symmetry, then $\xi = J^r\zeta$ for some π-projectable vector field ζ on Y.

13.2.2   The Variational Sequence

Now we are prepared to introduce the variational sequence due to Krupka [49] (see also [52] and [78] for review).[2]

Let us denote by $\mathbb{R}_Y$ the constant sheaf over $\mathbb{R}$, and by Λ_q^r the sheaf of q-forms on $J^r Y$. Next, we let $\Lambda_{0,c}^r = \{0\}$, and $\Lambda_{q,c}^r$ be the sheaf of contact q-forms, if $q \leq n$, or the sheaf of strongly contact q-forms, if $q > n$, on $J^r Y$. We set

$$\Theta_q^r = \Lambda_{q,c}^r + \left(d\Lambda_{q-1,c}^r \right), \tag{13.5}$$

where $d\Lambda_{q-1,c}^r$ is the image of $\Lambda_{q-1,c}^r$ by the exterior derivative d, and $\left(d\Lambda_{q-1,c}^r \right)$ denotes the sheaf generated by the presheaf $d\Lambda_{q-1,c}^r$. We note that $\Theta_q^r = 0$ for $q > \operatorname{corank} \mathscr{C}_r$.

The subsequence

$$0 \to \Theta_1^r \to \Theta_2^r \to \Theta_3^r \to \cdots \tag{13.6}$$

of the de Rham sequence

$$0 \to \mathbb{R}_Y \to \Lambda_0^r \to \Lambda_1^r \to \Lambda_2^r \to \Lambda_3^r \to \cdots \tag{13.7}$$

is an exact sequence of soft sheaves. The quotient sequence

$$0 \to \mathbb{R}_Y \to \Lambda_0^r \to \Lambda_1^r/\Theta_1^r \to \Lambda_2^r/\Theta_2^r \to \Lambda_3^r/\Theta_3^r \to \cdots \tag{13.8}$$

is called the *variational sequence* of order r.

Theorem 13.2.1 ([49]) *The variational sequence is an acyclic resolution of the constant sheaf $\mathbb{R}_Y$.*

The above assertion means that the variational sequence is locally exact with the exception of $\mathbb{R}_Y$. As a consequence, due to the abstract de Rham theorem, we obtain the following fundamental result:

Corollary 13.2.2 ([49]) *The cohomology groups of the cochain complex of global sections of the variational sequence are identified with the de Rham cohomology groups $H_{\mathrm{dR}}^q Y$ of the manifold Y.*

Since the variational sequence is a quotient sequence, the elements in the individual columns are elements of the quotient sheaves Λ_q^r/Θ_q^r, i.e. they are *equivalence classes of local r-th order differential q-forms*. We denote by $[\rho] \in \Lambda_q^r/\Theta_q^r$ the class of $\rho \in \Lambda_q^r$. Morphisms in the variational sequence are then by construction *quotients of the exterior derivative d*. We denote

$$\mathscr{E}_q : \Lambda_q^r/\Theta_q^r \to \Lambda_{q+1}^r/\Theta_{q+1}^r, \tag{13.9}$$

---

[2]For basics of sheaf theory needed below we refer to standard books, e.g. [6] or [93].

so that

$$\mathscr{E}_q([\rho]) = [d\rho]. \tag{13.10}$$

Now, $\mathscr{E}_q([\rho]) = [d\rho] = 0$ means that there exists $[\eta] \in \Lambda^r_{q-1}/\Theta^r_{q-1}$ such that $[\rho] = \mathscr{E}_{q-1}([\eta]) = [d\eta]$. If, moreover, $H^q_{\mathrm{dR}} Y = \{0\}$, then the class $[\eta]$ has a *global* representative.

The name "variational sequence" comes from the direct relationship with the calculus of variations. Namely, as proved in [49, 51], *the sheaf* Λ^r_n/Θ^r_n *is isomorphic with a subsheaf of the sheaf of Lagrangians* $\Lambda^{r+1}_{n,X}$, *and the quotient mapping* $\mathscr{E}_n$: $\Lambda^r_n/\Theta^r_n \to \Lambda^r_{n+1}/\Theta^r_{n+1}$ *is the Euler–Lagrange mapping.*

The kernel of $\mathscr{E}_n$ then consists of *null-Lagrangians (variationally trivial Lagrangians)*, and its image, which, by exactness is *the kernel of the next morphism*,

$$\mathscr{E}_{n+1} : \Lambda^r_{n+1}/\Theta^r_{n+1} \to \Lambda^r_{n+2}/\Theta^r_{n+2}, \tag{13.11}$$

represents *variational equations*. Therefore $\mathscr{E}_{n+1}$ is called *Helmholtz morphism*. The name refers to the famous "Helmholtz conditions", necessary and sufficient conditions for differential equations to be variational (as they stand) [3, 37, 47]. Remarkably, the Helmholtz morphism, discovered within the variational sequence theory, has not been known in the classical calculus of variations.

13.2.3   Source Forms and Lepage Forms

Classes in the variational sequence can be represented by *global differential forms* which play a fundamental role in the calculus of variations and the theory of differential equations on manifolds. To introduce them we shall need on $J^r Y$ the ideal generated by lifts of *contact 1-forms of order one*, whose elements take the form $\pi^*_{r,1}\omega \wedge \eta$ where $\omega \in \mathscr{C}^0_1$ (the annihilator of $\mathscr{C}_1$) and η is of order r. Since locally they are expressed as $\omega^\sigma \wedge \eta_\sigma$, where $\omega^\sigma = dy^\sigma - y^\sigma_i dx^i$, $1 \leq \sigma \leq m$, are the basic local contact 1-forms of order 1, we say that forms belonging to this ideal are $\{\omega^\sigma\}$-*generated*.

Definition 13.2.3 ([65])   Let $k \geq 1$. By *source form* of degree $n + k$ and order r we shall mean a $\{\omega^\sigma\}$-generated k-contact $(n + k)$-form on $J^r Y$.

Source $(n+1)$-forms are called *dynamical forms*, source $(n+2)$-forms are called *Helmholtz-like forms*.[3]

The question how to construct examples of source forms is solved by the *interior Euler operator* $\mathscr{I}$ (see, e.g., [2, 43, 44, 53]). The interior Euler operator reflects in

---

[3] As explained later, for the calculus of variations the most important examples of source forms of degree $n + 1$ are *Euler–Lagrange forms*, and of source forms of degree $n + 2$ the *Helmholtz forms*.

an intrinsic way the procedure of getting a distinguished global representative of a class $[\rho] \in \Lambda_q^r/\Theta_q^r$ for $q > n$ by applying to ρ the operator p_{q-n} and factorization by Θ_q^r.

Let $k \geqslant 1$. For $\rho \in \Lambda_{n+k}^r$ we have $\pi_{r+1,r}^*\rho = p_k\rho + p_{k+1}\rho + \cdots + p_{k+n}\rho$; we set

$$\mathscr{I}(\rho) = \frac{1}{k}\omega^\sigma \wedge \sum_{|J|=0}^{r} (-1)^{|J|} d_J \left(i_{\partial/\partial y_j^\sigma} p_k\rho \right) , \tag{13.12}$$

where $d_J = d_{j_1} \ldots d_{j_p}$ for $|J| = p$. It can be shown by means of the partition of unity arguments that this formula defines a global form $\mathscr{I}(\rho)$.

The operator $\mathscr{I} : \Lambda_{n+k}^r \to \Lambda_{n+k}^{2r+1}$ has the following properties:

1. For every $\rho \in \Lambda_{n+k}^r$, $\mathscr{I}(\rho)$ is a source form of degree $n + k$.
2. $\mathscr{I}(p_k\rho) = \mathscr{I}(\rho)$.
3. $\mathscr{I}(\rho)$ belongs to the same class as $\pi_{2r+1,r}^*\rho$, i.e., $\pi_{2r+1,r}^*\rho - \mathscr{I}(\rho) \in \Theta_{n+k}^{2r+1}$.
4. $\mathscr{I}^2 = \mathscr{I}$, up to a canonical projection; precisely, $\mathscr{I}^2(\rho) = \pi_{4r+3,2r+1}^*\mathscr{I}(\rho)$.
5. Ker $\mathscr{I} = \Theta_{n+k}^r$.
6. For $k = 1$, $\mathscr{I}(\rho)$ is the *unique source form* representing the class $[\rho]$. If $k > 1$, $\mathscr{I}(\rho)$ is no longer a unique source form equivalent with ρ (precisely, with $\pi_{2r+1,r}^*\rho$).[4] We say that $\mathscr{I}(\rho)$ is a *canonical source form* for $[\rho]$.

By construction, $\mathscr{I}(\rho)$ is a k-contact form, however, in general it is different from $p_k\rho$. The difference is given by means of the so-called *residual operator* $\mathscr{R}$ as follows:

$$p_k\rho = \mathscr{I}(\rho) + p_k dp_k \mathscr{R}(\rho) \tag{13.13}$$

(up to appropriate canonical projections—see the footnote below). It is worth notice that $\mathscr{R}(\rho)$ is a local strongly contact $(n + k - 1)$-form, and that, contrary to $\mathscr{I}(\rho)$, the form $\mathscr{R}(\rho)$ need not be unique and need not be global.

The operator $\mathscr{I}$ enables one to introduce fundamental differential forms as follows:

Definition 13.2.4 ([53]) Let $k \geqslant 0$. A $(n + k)$-form ρ on $J^r Y$ is called *Lepage form of degree k* if

$$p_{k+1}d\rho = \mathscr{I}(d\rho) . \tag{13.14}$$

From the definition we immediately see that

- *every q-form on Y, $q \geqslant n$*, is a Lepage form,
- *every closed q-form on $J^r Y$, $q \geqslant n$*, is a Lepage form,

---

[4]For the sake of simplicity of notations, we shall often omit the pull-back and identify a form with its lift or projection.

and we can obtain the structure of Lepage forms:

Theorem 13.2.5 ([78])   *Equation* (13.14) *has the solution*

$$\pi_{r+1,r}^{*}\rho = \theta_{p_k\rho} + dv + \mu = \theta_{p_k\rho} + p_{k+1}dv + \eta\,, \tag{13.15}$$

where

$$\theta_{p_k\rho} = p_k\rho - p_{k+1}\mathscr{R}(dp_k\rho)\,, \tag{13.16}$$

v is an arbitrary at least $(k + 1)$-contact $(n + k - 1)$-form, and μ (resp. η) is an arbitrary at least $(k + 2)$-contact form.

For every choice of v and μ (resp. η) the Lepage forms (13.15) belong to the same variational class $[\rho] \in \Lambda_{n+k}^{s}/\Theta_{n+k}^{s}$ (for a proper s), i.e. $\pi_{s,r}^{*}\rho - \theta_{p_k\rho} \in \Theta_{n+k}^{s}$.

In what follows, we shall denote by $\{\rho\}$ the family of (possibly local) Lepage forms containing ρ, as characterized by Theorem 13.2.5.

The form $\theta_{p_k\rho}$ is completely determined by its k-contact part $p_k\rho$. It was discovered in [78]. For $k = 0$ this is the famous Cartan form [11, 13, 25], hence we call it *Cartan form of degree k*. As seen from the defining formula (13.16), the Cartan form is generically local. However, from the formula (13.15) it follows that if $n = \dim X = 1$ and $r \geq 0$ (mechanics and higher-order mechanics), or, if $n > 1$ and $r = 1$ (first order field theory), then *for every $k \geq 0$ the Cartan form $\theta_{p_k\rho}$ is global*.

It is worth notice that an equivalent definition of a Lepage $(n + k)$-form of order r obviously is as follows:

$$\pi_{r+1,r}^{*}d\rho = \mathscr{I}(d\rho) + F, \text{ where } F \text{ is at least } (k + 2)\text{-contact.} \tag{13.17}$$

In view of (13.15) we also say that ρ is a *Lepage equivalent of $p_k\rho$*. As proved in [78], *every k-contact $(n + k)$-form has a global Lepage equivalent of degree k*. For $k = 0$ (i.e. $p_0\rho = h\rho$) the definition gives *Lepage equivalent of a Lagrangian $h\rho = \lambda$* as introduced by Krupka in [45] and [48]. The dynamical form $p_1d\rho = p_1d\theta_\lambda$ is then the *Euler–Lagrange form $E_\lambda = \mathscr{I}(d\lambda)$*, and $E_n : \lambda \to E_\lambda$ is the *Euler–Lagrange mapping*. For $k > 0$, the most important Lepage forms are *Lepage equivalents of source forms*. If $k = 1$, the definition gives *Lepage equivalent of a dynamical form ε* (first introduced in [56] for locally variational dynamical forms), and $p_2d\rho = p_2d\theta_\varepsilon$ is the *canonical Helmholtz form $H_\varepsilon = \mathscr{I}(d\varepsilon)$*. The mapping $E_{n+1} : \varepsilon \to H_\varepsilon$ is then the *Helmholtz mapping*. For $k = 2$ we get Lepage equivalents of source forms η of degree $n + 2$ which are called *Helmholtz-like forms*, and the corresponding mapping is $E_{n+2} : \eta \to \mathscr{I}(d\eta)$. This case has been studied for $n = 1$ in [64] and [74].

We also note that formula (13.15) immediately yields:

Corollary 13.2.6 *If $n = \dim X = 1$, then for every $k \geq 0$ and every $r \geq 1$, any k-contact $(n + k)$-form has a unique global Lepage equivalent.*

13.2.4 The Inverse Problem in the Variational Sequence

We start with the following definition:

Definition 13.2.7 ([48]) A dynamical form ε is called *(globally) variational* if there exists a Lagrangian λ such that $\varepsilon = E_\lambda$ (possibly up to a jet projection). ε is called *locally variational* if the domain of ε can be covered by open sets U_ι such that for every ι, $\varepsilon|_{U_\iota}$ is variational.

The famous *Inverse Problem of the Calculus of Variations* is the question under what conditions a dynamical form ε is locally/globally variational. In terms of the variational sequence this transfers to the question on the image of the Euler–Lagrange mapping, or due to the exactness of the sequence, on the kernel of the Helmholtz mapping. In this way the inverse problem in its local and global form extends to each column of the variational sequence, and is solved by the exactness of the sequence. With help of Lepage forms, the solution of the Inverse Problem is obtained in full generality and in an elegant way, transferring the whole problem to the Poincaré Lemma and de Rham cohomology.

Before presenting the theorem, recall a convenient modification of the Poincaré homotopy operator, adapted to the contact structure. Denote by $\mathscr{P}$ the Poincaré homotopy operator. Then $\mathscr{A}$ defined by

$$\mathscr{A} \, p_0 \omega = 0, \quad \mathscr{A} \, p_k \omega = p_{k-1} \mathscr{P} \omega, \quad k \geqslant 1 \tag{13.18}$$

Krupka [48], satisfies $\pi_{r+1,r}^* \omega = \mathscr{A} d(\pi_{r+1,r}^* \omega) + d\mathscr{A}(\pi_{r+1,r}^* \omega)$, and is adapted to the decomposition of forms into contact components: if (locally) $\omega = d\rho$, then

$$\pi_{r+1,r}^* \rho = \mathscr{A}(\pi_{r+1,r}^* \omega) = \sum_{k=0}^{q} \mathscr{A} \, p_{k+1} \omega \,, \tag{13.19}$$

hence $p_k \rho = \mathscr{A} \, p_{k+1} \omega$. Compared to $\mathscr{P}$, the operator $\mathscr{A}$ concerns vertical curves (curves in the fibres over X) only.

Theorem 13.2.8 ([78] General Inverse Variational Problem) *Let σ be a source form of degree $n + k$, $k \geqslant 1$. The following conditions are equivalent:*

(1) *σ belongs to $\mathrm{Ker}\, E_{n+k}$.*
(2) *σ has a closed Lepage equivalent.*
(3) *Every Lepage equivalent ρ of σ satisfies $p_{k+1} d\rho = 0$.*

If $E_{n+k}(\sigma) = 0$, then there is a local source $(n + k - 1)$-form η satisfying $E_{n+k-1}(\eta) = \sigma$, and it holds $\eta = \mathscr{A}\sigma$.

If moreover $H_{\mathrm{dR}}^{n+k} Y = \{0\}$, then there is a global source $(n + k - 1)$-form η satisfying $E_{n+k-1}(\eta) = \sigma$.

Note that :

- For $k = 1$ condition (1) means that the dynamical form $\varepsilon = p_1\sigma$ is locally variational.
- Condition (3) represents precisely the *source constraints*, i.e. constraints on a source form of degree $n + k$ belong to the kernel of the corresponding morphism in the variational sequence. It is a generalization (to any degree) of the famous *Helmholtz conditions* on local variationality of a dynamical form (i.e. necessary and sufficient conditions of existence of a local Lagrangian for differential equations represented by the dynamical form).[5]
- Condition (2) is equivalent with (3), hence also represents the source constraints. However, since (up to projection) $d\rho = \sum_{i=1}^{n+1} p_{k+i}d\rho$, condition $d\rho = 0$ includes also conditions $p_{k+2}d\rho = 0, \ldots, p_{k+n+1}d\rho = 0$ which are dependent, hence superfluous. In coordinates this gives dependent conditions in form of derivatives of the Helmholtz conditions (respectively, their higher degree generalizations if $k > 1$).
- Formula $\eta = \mathscr{A}\sigma$ for $k = 1$ (i.e. if σ is a dynamical form) gives the celebrated Vainberg–Tonti Lagrangian [89, 90].

13.3 Lepage Manifolds

13.3.1 Lepage Manifolds and Differential Equations

As above, we consider a fibred manifold $\pi : Y \to X$ and its jet prolongations.

Definition 13.3.1 Let $r \geqslant 0$. By a *Lepage manifold of order r* we shall mean a fibred manifold $\pi : Y \to X$ endowed with a Lepage $(n + 1)$-form α on J^rY such that $p_1\alpha$ is a dynamical form.

A Lepage manifold of order r will be denoted by (π_r, α).

By definition, a Lepage manifold (π_r, α) is endowed with a (uniquely determined) global *dynamical form* $\varepsilon = p_1\alpha$ of order $r + 1$. In fibred coordinates

$$\varepsilon = E_\sigma \omega^\sigma \wedge \omega_0 \tag{13.20}$$

where the coefficients E_σ are functions on a coordinate domain in $J^{r+1}Y$. The dynamical form represents globally and intrinsically *differential equations of order* $r + 1$. More precisely, the equations arise as equations for sections γ of the fibred

---

[5]Helmholtz conditions ($k = 1$) in the form (3) were first presented in [57]. The coordinate form of the Helmholtz conditions has been first obtained by Helmholtz [37] for second order ODEs, and by Anderson and Duchamp [3] and Krupka [47, 48] for higher order PDEs.

For $k = 2$ (and mechanics) the explicit coordinate form of (3) can be found in [64] and [74].

manifold π such that

$$E_\sigma \circ J^{r+1}\gamma = 0. \tag{13.21}$$

Indeed, in coordinates, this is a system of $m = \dim Y - \dim X$ differential equations (ODEs if $\dim X = 1$ and PDEs if $\dim X > 1$) of order $r + 1$ for the components of sections γ of π.

Since α is a Lepage equivalent of ε, Theorem 13.2.8 tells us that *if $p_2d\alpha = 0$ then the dynamical form ε is locally variational* (and vice versa), meaning that the Helmholtz form $H_\varepsilon = E_{n+1}(\varepsilon)$ vanishes, and around every point, ε is the Euler–Lagrange form of a local Lagrangian λ. Hence the corresponding differential equations (13.21) are Euler–Lagrange equations of λ; in such a case, we say that Eq. (13.21) are *variational (as they stand)*. Moreover by Theorem 13.2.8 (2), local variationality of ε is *equivalent* with the existence of a *closed Lepage equivalent of ε* (which, however, need not be global). In other words, on a Lepage manifold (π_r, α) such that $p_2d\alpha = 0$ but $d\alpha \neq 0$, we can locally always find a *closed* Lepage form α' such that $p_1\alpha' = p_1\alpha = \varepsilon$.

On the other hand, given a Lepage manifold (π_r, α) such that $p_2d\alpha \neq 0$, it holds $H_\varepsilon \neq 0$, the Lepage family $\{\alpha\}$ cannot be (even locally) represented by a closed form, and the corresponding *equations* (13.21) *are not variational (as they stand)*.

From this analysis we can see that the concept of Lepage manifold concerns in fact three columns in the variational sequence: In the variational case and in the non-variational case, respectively, we have

$$\lambda \xrightarrow{\ E_n\ } E_\lambda \xrightarrow{\ E_{n+1}\ } 0$$

$$\varepsilon \xrightarrow{\ E_{n+1}\ } H_\varepsilon \xrightarrow{\ E_{n+2}\ } 0$$

with help of source forms, or

$$\{\rho_\lambda\} \xrightarrow{\ d\ } \{d\rho_\lambda\} = \{\alpha_{E_\lambda}\} \xrightarrow{\ d\ } 0$$

$$\{\alpha_\varepsilon\} \xrightarrow{\ d\ } \{d\alpha_\varepsilon\} = \{\beta_{H_\varepsilon}\} \xrightarrow{\ d\ } 0$$

with help of Lepage forms.

Apart from the dynamical form ε, every Lepage manifold is canonically endowed with the *exterior differential system* $\mathscr{D}_\alpha$ generated by n-forms

$$i_\xi\alpha \quad \text{where } \xi \text{ runs over all } \pi_r\text{-vertical vector fields on } J^r Y. \tag{13.22}$$

We have the following theorem (cf. [63]):

Theorem 13.3.2 *Let (π_r, α) be a Lepage manifold, $\varepsilon = p_1\alpha$. Let γ be a section of π. The following conditions are equivalent:*

(1) γ *is a solution of differential equations* (13.21) *for ε.*
(2) $J^r\gamma$ *is an integral section of $\mathcal{D}_\alpha$, i.e.*

$$J^r\gamma^* i_\xi\alpha = 0 \quad \text{for every } \pi_r\text{-vertical vector field } \xi \text{ on } J^rY. \qquad (13.23)$$

The set of holonomic solutions of $\mathcal{D}_\alpha$ does not depend on the choice of an element of the Lepage family $\{\alpha\}$.

Proof All the assertions follow easily from

$$J^r\gamma^* i_\xi\alpha = J^{r+1}\gamma^* i_\xi(p_1\alpha) = J^{r+1}\gamma^* (E_\sigma\xi^\sigma\omega_0) = \left((E_\sigma \circ J^{r+1}\gamma)(\xi^\sigma \circ \gamma)\right)\omega_0$$

$$(13.24)$$

if we take for the ξ's the r-th, resp. $(r+1)$-prolongation of a local basis $\xi_{(v)} = \partial/\partial q^v$ (hence $\xi^\sigma_{(v)} = \delta^\sigma_v$). $\qquad\qquad\square$

Remark 13.3.3 Consider the associated system $\hat{\mathcal{D}}_\alpha$ of α, i.e. the exterior differential system locally generated by n-forms $i_\xi\alpha$ where ξ runs over *all* vector fields on J^rY. Then $\mathcal{D}_\alpha \subset \hat{\mathcal{D}}_\alpha$. However, $\mathcal{D}_\alpha$ and $\hat{\mathcal{D}}_\alpha$ *have the same holonomic integral sections*. This follows from the fact that for $1 \leq i \leq n$,

$$J^r\gamma^* i_{\partial/\partial x^i}\alpha = J^{r+1}\gamma^* i_{\partial/\partial x^i}(p_1\alpha) = -\left((E_\sigma y_i^\sigma) \circ J^{r+1}\gamma\right)\omega_0, \qquad (13.25)$$

hence equations for holonomic integral sections of $\mathcal{D}_\alpha$ and $\hat{\mathcal{D}}_\alpha$ take the same form: $E_\sigma \circ J^{r+1}\gamma = 0, 1 \leq \sigma \leq m$.

Remarkably, given a Lepage manifold (π_r, α), one can consider *all* (not only holonomic) solutions of $\mathcal{D}_\alpha$. The corresponding equations are then equations for components of sections δ of the fibred manifold $\pi_r : J^rY \to X$ and take the form

$$\delta^* i_\xi\alpha = 0 \quad \text{for every } \pi_r\text{-vertical vector field } \xi \text{ on } J^rY. \qquad (13.26)$$

Equivalently they can be written in the form

$$\delta^* i_{\partial/\partial y^\sigma_{j_1\ldots j_k}}\alpha = 0, \quad 1 \leq \sigma \leq m, \; 0 \leq k \leq r, \; 1 \leq j_1 \leq \cdots \leq j_k \leq n. \qquad (13.27)$$

We note that (13.27) are *first order differential equations* (ODEs or PDEs depending on the dimension of X), and (generally) they are *not equivalent* with (13.23), i.e., with Eq. (13.21). However, if a section γ of π is a solution of (13.21) then $J^r\gamma$ is a solution of (13.27).

Equation (13.27) can be given another geometric interpretation in terms of Lepage manifolds. Namely, given a Lepage manifold (π_r, α) of order $r \geq 1$, we can consider α as a *Lepage form of order zero* for the fibred manifold $\pi_r : J^r Y \to X$. In this way we relate with (π_r, α) the Lepage manifold $((\pi_r)_0, \alpha)$ of order zero. Let us denote the *i*-contactization operator with respect to the projection π_r by $\tilde{p}_i$. The associated dynamical form

$$\tilde{\varepsilon} = \tilde{p}_1 \alpha \tag{13.28}$$

is then defined on $J^1(J^r Y)$ and called the *Hamilton form* of ε. Note that α is, indeed, the Lepage equivalent of $\tilde{\varepsilon}$ (more precisely, the Lepage equivalent of $\tilde{\varepsilon}$ is projectable onto $J^r Y$, and its projection is α). The corresponding exterior differential system is $\tilde{\mathscr{D}}_\alpha = \mathscr{D}_\alpha$ and we can see that *integral sections of $\mathscr{D}_\alpha$ are exactly those sections of π_r which are solutions of the (first order) differential equations related with the dynamical form $\tilde{\varepsilon}$.*

Definition 13.3.4 We call $\tilde{\mathscr{D}}_\alpha = \mathscr{D}_\alpha$ the *Hamiltonian system*, and the corresponding first order differential equations the *Hamilton equations* of α.

Remark 13.3.5 As we have seen, if $p_2 d\alpha = 0$ then $H_\varepsilon = 0$ and ε is locally variational. This means that the corresponding $(r+1)$-order equations for sections γ of π are variational as they stand, and they come as Euler–Lagrange equations from local Lagrangians $\lambda = \mathscr{A}\varepsilon$. Since generally $\tilde{p}_2 d\alpha \neq 0$, $H_{\tilde{\varepsilon}} \neq 0$, and the Hamilton form $\tilde{\varepsilon}$ need not be locally variational. If, however, $\tilde{p}_2 d\alpha = 0$ then $\tilde{\varepsilon} = \tilde{p}_1 \alpha$ is locally variational, and the Hamilton equations are first order Euler–Lagrange equations of local first order Lagrangians $\tilde{\lambda} = \mathscr{A}\tilde{\varepsilon}$.

It is worth note that by Theorem 13.2.8 condition $p_2 d\alpha = 0$ guarantees that $\varepsilon = p_1 \alpha$ has (at least a local) *closed* Lepage equivalent α'. Thus, *in the variational case, we can always find (at least locally) variational Hamilton equations.* Note that $\mathscr{D}_\alpha$ and $\mathscr{D}_{\alpha'}$ are different (the Hamilton equations are different), however, the holonomic solutions of $\mathscr{D}_\alpha$ and $\mathscr{D}_{\alpha'}$ are the same (indeed, $p_1 \alpha = p_1 \alpha' = \varepsilon$).

Remark 13.3.6 (Calculus of Variations) If, in particular, one has a Lepage manifold (π_r, α) where α is *closed*, then both ε and $\tilde{\varepsilon}$ are locally variational. This means that we have an *r*-th order variational principle over the fibred manifold π for ε (Lagrange theory) and an associated zero order variational principle over the fibred manifold π_r for $\tilde{\varepsilon}$ (Hamilton theory). (If $r = 0$ they, of course, coincide.) The relationship between solutions of the Euler–Lagrange equations determined by ε and Hamilton equations determined by $\tilde{\varepsilon}$ then can be characterized as follows: *holonomic solutions of the Hamilton equations coincide with r-jet prolongations of extremals.*

The case of closed α will be discussed in more detail in Sect. 13.5.

13.3.2 Regular Lepage Manifolds

From the definition of $\mathscr{D}_\alpha$ we can see that at each point $j_x^r \gamma \in J^r Y$, the rank of $\mathscr{D}_\alpha$ is less or equal to the dimension of the bundle $V\pi_r$ of π_r-vertical vectors.

Definition 13.3.7 A Lepage manifold (π_r, α) (resp. a Lepage $(n + 1)$-form α on $J^r Y$) is called *regular* if

$$\text{rank } \mathcal{D}_\alpha = \dim V\pi_r = \text{the fibre dimension over } X. \qquad (13.29)$$

Proposition 13.3.8 (π_r, α) *is regular if and only if the map*

$$V\pi_r \ni \xi \to i_\xi \alpha \in \Lambda^n(J^r Y) \qquad (13.30)$$

mapping π_r-vertical vector fields to n-forms on $J^r Y$, is injective.

Proof If α is regular then the n-forms $\eta_\sigma^J = i_{\partial/\partial y_J^\sigma}\alpha$, where $1 \leqslant \sigma \leqslant m$ and $J = \{j_1 \ldots j_k\}$, $1 \leqslant j_1 \leqslant \cdots \leqslant j_k \leqslant n$, $0 \leqslant k \leqslant r$, generating $\mathcal{D}_\alpha$, are linearly independent at each point of $J^r Y$. Hence $i_\xi \alpha = \xi_J^\sigma \eta_\sigma^J = 0$ means that $\xi_J^\sigma = 0$ for all σ and J, proving that $\xi = 0$.

Conversely, if the mapping (13.30) is injective, then $\xi_J^\sigma \eta_\sigma^J = 0$ implies that $\xi_J^\sigma = 0$ for all σ and J, $1 \leqslant \sigma \leqslant m$ and $J = \{j_1 \ldots j_k\}$, $1 \leqslant j_1 \leqslant \cdots \leqslant j_k \leqslant n$, $0 \leqslant k \leqslant r$. This means that the n-forms η_σ^J are linearly independent, so that rank $\mathcal{D}_\alpha = \dim V\pi_r$. $\qquad \square$

It is worth compare regular Lepage manifolds with symplectic and multisymplectic manifolds[6]:

- The Cauchy distribution of a regular Lepage $(n + 1)$-form need not be (and in applications usually is not) trivial (see also Example 13.3.1 below).
- Every multisymplectic manifold is a regular Lepage manifold of order zero.
- If $\dim X = 1$, $r \geqslant 0$, and $\dim J^r Y$ is odd, then every Lepage manifold (π_r, α) with α closed and regular is fibrewise symplectic (i.e., α restricted to each fibre $\pi_r^{-1}(x)$ is a symplectic form).

Example 13.3.1 ([36]) Consider the fibred manifold $\mathbb{R}^n \times \mathbb{R}^m \to \mathbb{R}^n$ where m is even, and the $(n + 1)$-form

$$\alpha = B_\sigma dy^\sigma \wedge \omega_0 + \frac{1}{2}B_{\sigma\nu}^1 dy^\sigma \wedge dy^\nu \wedge \omega_1, \quad B_{\sigma\nu}^1 = -B_{\nu\sigma}^1, \quad \det\left(B_{\sigma\nu}^1\right) \neq 0. \qquad (13.31)$$

Since rank $\alpha = \text{rank}\left(B_\sigma, B_{\sigma\nu}^1, 0, \ldots, 0\right) = m$, (π_0, α) is a regular Lepage manifold. The corresponding dynamical form is $\varepsilon = \left(B_\sigma + B_{\sigma\nu}^1 y_1^\nu\right) dy^\sigma \wedge \omega_0$. Computing $i_\xi \alpha = 0$ for a vector field

$$\zeta = \bar{\zeta}^i \frac{\partial}{\partial x^i} + \zeta^\sigma \frac{\partial}{\partial y^\sigma}, \qquad (13.32)$$

---

[6] A differential form ω of degree $q > 2$ on a manifold M is called *multisymplectic* if it is closed and nondegenerate (meaning that the Cauchy distribution of ω is zero, i.e. the condition $i_\xi \omega = 0$ implies $\zeta = 0$) [8].

and taking into account that $\omega_{11} = 0$, we get the components of ζ,

$$\zeta^\sigma = -\bar{\zeta}^1 M^{\sigma\nu} B_\nu, \quad 1 \leqslant \sigma \leqslant m, \quad \bar{\zeta}^2 = \cdots = \bar{\zeta}^n = 0, \tag{13.33}$$

where $M^{\sigma\nu}$ is the inverse matrix to $B^1_{\sigma\nu}$, and the condition $\zeta^\sigma B_\sigma = 0$. However, the last condition with ζ^σ as above takes the form of identity, since $M^{\sigma\nu} B_\sigma B_\nu$ is identically zero. Summarizing, we have obtained that the regular form α has a one-dimensional kernel spanned by the everywhere non-zero vector field

$$\frac{\partial}{\partial x^1} - M^{\sigma\nu} B_\nu \frac{\partial}{\partial y^\sigma}. \tag{13.34}$$

13.3.3 Helmholtz Symmetries

We shall close this section with an important result which yields a relationship between *non-variational* and *variational* Lepage manifolds (and hence between the corresponding non-variational and variational equations).

Definition 13.3.9 Let (π_r, α) be a Lepage manifold, $\varepsilon = p_1\alpha$ the corresponding dynamical form. A vector field ξ is called *Helmholtz symmetry* if ξ is a contact symmetry, and the flow of ξ leaves invariant the Helmholtz form H_ε.

Immediately from the definition it follows that a contact symmetry ξ is a Helmholtz symmetry if and only if

$$\mathscr{L}_\xi H_\varepsilon = 0. \tag{13.35}$$

The meaning of Helmholtz symmetries comes out by the following theorem:

Theorem 13.3.10 ([54]) *Let ε be a dynamical form such that $H_\varepsilon \neq 0$. Let ξ be a contact symmetry such that $\mathscr{L}_\xi H_\varepsilon = 0$. Then $\mathscr{L}_\xi\varepsilon$ is a locally variational dynamical form.*

If (π_r, α) is a 'non-variational' Lepage manifold, i.e. such that $p_2 d\alpha \neq 0$, then Eq. (13.35) where $\varepsilon = p_1\alpha$ possibly provides us with nontrivial Helmholtz symmetries. By the above theorem, every such a symmetry assigns to the Lepage manifold (π_r, α) another Lepage manifold, and the new Lepage manifold is variational.

More precisely, we can state the following:

Theorem 13.3.11 *Let (π_r, α) be a Lepage manifold, and assume that $p_2 d\alpha \neq 0$. Let ζ be a π-projectable vector field on Y such that $\xi = J^{r+1}\zeta$ leaves the form $p_2 d\alpha$ invariant, i.e., let $\mathscr{L}_\xi p_2 d\alpha = 0$. Then (π_r, α') where $\alpha' = \mathscr{L}_{J^r\zeta}\alpha$, is a Lepage manifold, and $p_2 d\alpha' = 0$.*

Proof We shall use the fact that the Lie derivative along a contact symmetry preserves the decomposition of differential forms into contact components (cf.

formula (13.2)). By the preceding theorem, $\varepsilon' = \mathscr{L}_\xi p_1 \alpha$ is a locally variational dynamical form. The $(n + 1)$-form $\alpha' = \mathscr{L}_{Jr\xi}\alpha$ satisfies $p_1\alpha' = p_1\mathscr{L}_{Jr\xi}\alpha = \mathscr{L}_{Jr+1\xi}p_1\alpha = \mathscr{L}_{Jr+1\xi}\varepsilon$, which is a dynamical form. Moreover, $p_2 d\alpha' = p_2 d\mathscr{L}_{Jr\xi}\alpha = p_2\mathscr{L}_{Jr\xi}d\alpha = \mathscr{L}_{Jr+1\xi}p_2 d\alpha = 0$, proving that (π_r, α') is a 'variational' Lepage manifold.   □

13.4 Hamiltonian Systems

13.4.1 Lepage Manifolds of Order Zero

Definition 13.4.1 By a *Hamiltonian system* we shall mean a Lepage manifold of order zero.

By definition, a Hamiltonian system is determined by a $(n + 1)$-form α defined on the total space Y of a fibred manifold $\pi : Y \to X$ over an n-dimensional base X. Hamilton equations are then first order differential equations for sections of π arising from the dynamical form $\varepsilon = p_1 \alpha$ on $J^1 Y$ (called the *Hamilton $(n+1)$-form* in this context), or as integral sections of the exterior differential system $\mathscr{D}_\alpha$ on Y, locally generated by n-forms $\eta_\sigma = i_{\partial/\partial y^\sigma}\alpha$, $1 \leqslant \sigma \leqslant m$.

We note that a Hamiltonian system (π_0, α) is *regular* if rank $\mathscr{D}_\alpha = m = \dim Y - \dim X$.

The local structure of Hamiltonian systems has been clarified by Haková and Krupková in [36]:

Denote by (x^i, y^σ) local fibred coordinates on Y. Then α can be expressed in the form

$$\alpha = B_\sigma dy^\sigma \wedge \omega_0 + \frac{1}{2!}B^{i_1}_{\sigma v_1}dy^\sigma \wedge dy^{v_1} \wedge \omega_{i_1} + \frac{1}{3!}B^{i_1 i_2}_{\sigma v_1 v_2}dy^\sigma \wedge dy^{v_1} \wedge dy^{v_2} \wedge \omega_{i_1 i_2}$$

$$+ \cdots + \frac{1}{(n+1)!}B^{i_1 \ldots i_n}_{\sigma v_1 \ldots v_n}dy^\sigma \wedge dy^{v_1} \wedge \cdots \wedge dy^{v_n} \wedge \omega_{i_1 \ldots i_n}, \tag{13.36}$$

where the components are totally antisymmetric in both the lower and the upper indices; note that this implies that they are also totally symmetric with respect to the interchange of pairs of indices as follows: $B^{j_1 \ldots j_p \ldots j_q \ldots j_k}_{\sigma v_1 \ldots v_p \ldots v_q \ldots v_k} = B^{j_1 \ldots j_q \ldots j_p \ldots j_k}_{\sigma v_1 \ldots v_q \ldots v_p \ldots v_k}$. The corresponding Hamilton $(n + 1)$-form then reads:

$$\varepsilon = E_\sigma \omega^\sigma \wedge \omega_0, \tag{13.37}$$

where

$$E_\sigma = B_\sigma + B^{i_1}_{\sigma v_1}y^{v_1}_{i_1} + B^{i_1 i_2}_{\sigma v_1 v_2}y^{v_1}_{i_1}y^{v_2}_{i_2} + \cdots + B^{i_1 \ldots i_n}_{\sigma v_1 \ldots v_n}y^{v_1}_{i_1} \cdots y^{v_n}_{i_n}, \tag{13.38}$$

and we note that the functions E_σ satisfy the antisymmetry conditions

$$\frac{\partial E_\sigma}{\partial y_i^\nu} = -\frac{\partial E_\nu}{\partial y_i^\sigma}.\tag{13.39}$$

Using formula (13.38) we can express α in terms of components of ε [36]:

$$\pi_{1,0}^*\alpha = E_\sigma \omega^\sigma \wedge \omega_0 + \sum_{k=1}^{n} \frac{1}{k!(k+1)!} \frac{\partial^k E_\sigma}{\partial y_{i_1}^{\nu_1} \dots \partial y_{i_k}^{\nu_k}} \omega^\sigma \wedge \omega^{\nu_1} \wedge \cdots \wedge \omega^{\nu_k} \wedge \omega_{i_1 \dots i_k}.$$
$$\tag{13.40}$$

Formula (13.38) yields the *Hamilton equations*:

$$\left(B_\sigma + B_{\sigma\nu_1}^{i_1} y_{i_1}^{\nu_1} + B_{\sigma\nu_1\nu_2}^{i_1 i_2} y_{i_1}^{\nu_1} y_{i_2}^{\nu_2} + \cdots + B_{\sigma\nu_1\dots\nu_n}^{i_1\dots i_n} y_{i_1}^{\nu_1} \dots y_{i_n}^{\nu_n} \right) \circ J^1\gamma = 0,\tag{13.41}$$

where $1 \le \sigma \le m$, and the functions B satisfy the antisymmetry conditions mentioned above. Note that the most general Hamilton PDEs are *polynomial in the first derivatives*.

The associated exterior differential system $\mathscr{D}_\alpha$ is generated by the n-forms

$$\eta_\sigma \equiv i_{\partial/\partial y^\sigma} \alpha$$
$$= B_\sigma \omega_0 + \sum_{k=1}^{n} \frac{1}{k!} B_{\sigma\nu_1\dots\nu_k}^{i_1\dots i_k} dy^{\nu_1} \wedge \cdots \wedge dy^{\nu_k} \wedge \omega_{i_1\dots i_k}, \quad 1 \le \sigma \le m.\tag{13.42}$$

Obviously, we have the following explicit characterization of *regular Hamiltonian systems*:

Proposition 13.4.2 ([36]) *Regularity of α is equivalent with any of the following conditions:*

(1) *The n-forms η_σ, $1 \le \sigma \le m$, are linearly independent at each point of their domain of definition.*
(2) *The rank of the matrix*

$$\left(B_\sigma \quad B_{\sigma\nu_1}^{i_1} \quad B_{\sigma\nu_1\nu_2}^{i_1 i_2} \quad \cdots \quad B_{\sigma\nu_1\dots\nu_n}^{i_1\dots i_n} \right)\tag{13.43}$$

where σ labels rows and the other sets of indices label columns, is maximal and equal to $m = \dim Y - \dim X$ at each point of Y.
(3) *The rank of the matrix*

$$\left(E_\sigma \quad \frac{\partial E_\sigma}{\partial y_{i_1}^{\nu_1}} \quad \frac{\partial^2 E_\sigma}{\partial y_{i_1}^{\nu_1} \partial y_{i_2}^{\nu_2}} \quad \cdots \quad \frac{\partial^n E_\sigma}{\partial y_{i_1}^{\nu_1} \dots \partial y_{i_n}^{\nu_n}} \right)\tag{13.44}$$

where σ labels rows and the other sets of indices label columns, is maximal and equal to $m = \dim Y - \dim X$ at each point of Y.

Corollary 13.4.3 ([36]) *For α be regular any of the following conditions is sufficient:*

$$\text{rank}\left(\frac{\partial^k E_\sigma}{\partial y^{\nu_1}_{i_1} \ldots \partial y^{\nu_k}_{i_k}}\right) = m, \quad 1 \leqslant k \leqslant n, \quad 1 \leqslant i_1 \leqslant i_2 \cdots \leqslant i_k \leqslant n. \quad (13.45)$$

Also for the following theorem and its proof we refer to [36]:

Theorem 13.4.4 ([36]) *If α is regular, then the ideal generated by $\mathscr{D}_\alpha$ is closed.*

Let us now assume that a Hamiltonian system (π_0, α) satisfies $p_2 d\alpha = 0$. Then as we know, the Hamilton equations are *variational*, i.e. the Hamilton form $\varepsilon = p_1\alpha$ is locally variational: its components E_σ satisfy the first order Helmholtz conditions

$$\frac{\partial E_\sigma}{\partial y^\nu_j} + \frac{\partial E_\nu}{\partial y^\sigma_j} = 0, \quad \frac{\partial E_\sigma}{\partial y^\nu} - \frac{\partial E_\nu}{\partial y^\sigma} + d_j\frac{\partial E_\nu}{\partial y^\sigma_j} = 0. \quad (13.46)$$

Theorem 13.2.8 also guarantees that there exists a *closed* $(n + 1)$-form giving rise to ε. However, in the zero order case a stronger result holds true, saying that *every variational Hamiltonian system is closed*. More precisely:

Theorem 13.4.5 *Given a Hamiltonian system (π_0, α) on Y, the following conditions are equivalent:*

(1) $p_2 d\alpha = 0$
(2) α *is closed.*

The theorem means that if an $(n+1)$-form α on Y satisfies $p_2 d\alpha = 0$ then also all the other contact components are zero, $p_k d\alpha = 0$ for $3 \leqslant k \leqslant n + 1$. This property comes from the *uniqueness* of a projectable (onto Y) Lepage equivalent of $\varepsilon = p_1\alpha$, proved in [36]:

Theorem 13.4.6 ([36]) *Let ε be a dynamical form on J^1Y such that $\varepsilon = p_1\alpha$ for some $(n + 1)$-form α on Y. The following conditions are equivalent:*

(1) ε *is locally variational.*
(2) *the $(n + 1)$-form α on Y is unique and satisfies $d\alpha = 0$.*
(3) *the components of ε satisfy the (first order) Helmholtz conditions, and α is given by formula* (13.36), *resp.* (13.40).

By the above theorem, variational Hamiltonian systems come from local first order Lagrangians

$$\lambda = \mathscr{A}\varepsilon = h\mathscr{A}\alpha = h\rho \quad (13.47)$$

where ρ is a local n-form on Y such that $d\rho = \alpha$.

13.4.2   Legendre Coordinates and Poisson Structures

There is an important case of Hamiltonian systems represented by α that is *at most 2-contact*. Hence,

$$\pi_{1,0}^{*}\alpha = E_{\sigma}\omega^{\sigma} \wedge \omega_0 + \frac{1}{2}\frac{\partial E_{\sigma}}{\partial y_i^{\nu}}\omega^{\sigma} \wedge \omega^{\nu} \wedge \omega_i \,, \tag{13.48}$$

where the antisymmetry relation (13.39) holds, so that

$$\alpha = B_{\sigma}dy^{\sigma} \wedge \omega_0 + \frac{1}{2}B_{\sigma\nu}^{i}dy^{\sigma} \wedge dy^{\nu} \wedge \omega_i \tag{13.49}$$

where $B_{\sigma\nu}^{i} = -B_{\nu\sigma}^{i}$. Since in this case $E_{\sigma} = B_{\sigma} + B_{\sigma\nu}^{i}y_i^{\nu}$, the Hamilton equations are *quasilinear first order PDEs*:

$$\left(B_{\sigma} + B_{\sigma\nu}^{i}y_i^{\nu}\right) \circ J^1\gamma = 0 \,, \tag{13.50}$$

and $\mathscr{D}_{\alpha}$ is generated by the n-forms

$$\eta_{\sigma} = B_{\sigma}\omega_0 + B_{\sigma\nu}^{i}dy^{\nu} \wedge \omega_i, \tag{13.51}$$

where $1 \leqslant \sigma \leqslant m$, and the functions $B_{\sigma\nu}^{i}$ satisfy the antisymmetry conditions mentioned above.

Assume that

$$d = \frac{m}{n+1} \tag{13.52}$$

is an integer, i.e., assume that the base and fibre dimensions satisfy the relation $d + dn = m$. Next, assume that around each point $y \in Y$ there is a fibred chart (x^i, y^K, p_K^i), where $1 \leqslant i \leqslant n, 1 \leqslant K \leqslant d$, such that

$$p_2\alpha = \pi_K^i \wedge \omega^K \wedge \omega_i \,, \tag{13.53}$$

(summation over $K = 1, \ldots, d$) where ω^K and π_K^i are the corresponding basic contact forms on J^1Y, i.e., in the associated coordinates $\left(x^i, y^K, p_K^i, y_j^K, p_{Kj}^i\right)$ on J^1Y,

$$\omega^K = dy^K - y_j^K dx^j \,, \quad \pi_K^i = dp_K^i - p_{Kj}^i dx^j \,. \tag{13.54}$$

We shall call coordinates of this kind *Legendre coordinates* on the Lepage manifold (π_0, α).

In Legendre coordinates we have

$$\pi_{1,0}^{*}\alpha = \left(A_K\omega^K + A_i^K\pi_K^i\right) \wedge \omega_0 + \pi_K^i \wedge \omega^K \wedge \omega_i \,, \tag{13.55}$$

that is,

$$\alpha = -H_K dy^K \wedge \omega_0 - H_i^K dp_K^i \wedge \omega_0 + dp_K^i \wedge dy^K \wedge \omega_i \qquad (13.56)$$

(the choice of the minus signs is a matter of convention), where

$$H_K = -A_K - p_{Kj}^j, \quad H_i^K = -A_i^K + y_i^K. \qquad (13.57)$$

We can see that:

Proposition 13.4.7 *If α admits Legendre coordinates, then α is regular, and*

$$p_3 d\alpha = 0. \qquad (13.58)$$

Proof $p_3 d\alpha = 0$ is obvious from (13.55). If ζ is a vertical vector field then $\zeta = \zeta^K \partial/\partial y^K + \xi_K^i \partial/\partial p_K^i$, and the condition $i_\zeta \alpha = 0$ yields $\zeta^K = 0$, $\xi_K^i = 0$, i.e. $\zeta = 0$, proving that α is regular. $\qquad\qquad\qquad\qquad\qquad\qquad\qquad\qquad\qquad\qquad\square$

Computing $\mathscr{D}_\alpha$ we obtain Hamilton equations in Legendre coordinates as follows:

$$\frac{\partial y^K}{\partial x^i} = H_i^K \circ \gamma, \qquad \frac{\partial p_K^j}{\partial x^j} = -H_K \circ \gamma. \qquad (13.59)$$

It is worth note that $p_3 d\alpha = 0$ does not mean that α should be closed. Therefore Legendre coordinates are defined also for a class of *non-variational Hamiltonian systems*.

Denote $\alpha = \kappa + dp_K^i \wedge dy^K \wedge \omega_i$; the splitting on the right-hand side is generally not invariant, i.e. κ is a local $(n+1)$-form. In case that α is closed we have $d\kappa = 0$, hence $\kappa = -dH \wedge \omega_0$ for a (local) function H (called *Hamilton function*), and the Hamilton equations (13.59) take the "familiar form"

$$\frac{\partial y^K}{\partial x^i} = \frac{\partial H}{\partial p_K^i} \circ \gamma, \qquad \frac{\partial p_K^j}{\partial x^j} = -\frac{\partial H}{\partial y^K} \circ \gamma. \qquad (13.60)$$

A covering of (π_0, α) by Legendre coordinates induces a "multi-Poisson structure": precisely, on Y there arises the following tensor field, to be called the *Poisson bivector 1-form*:

$$\Pi = \frac{\partial}{\partial p_K^i} \wedge \frac{\partial}{\partial y^K} \otimes dx^i. \qquad (13.61)$$

With help of Π we can introduce a *directional bracket of* (possibly local) *functions* on Y as follows: Let v be a nowhere zero vector field on X, and denote by $\tilde{v}$ a lift of v to Y. Given $f, g \in \Lambda_0^0$ we set:

$$\{f, g\}_v = \Pi(df, dg, \tilde{v}). \qquad (13.62)$$

The bracket obviously does not depend on the lift of v and it holds

$$\begin{aligned}
\{f,g\}_v &= \left(\frac{\partial}{\partial p_K^i} \wedge \frac{\partial}{\partial y^K} \otimes dx^i\right)(df,dg,v) \\
&= \left(\frac{\partial f}{\partial p_K^i}\frac{\partial g}{\partial y^K} - \frac{\partial g}{\partial p_K^i}\frac{\partial f}{\partial y^K}\right)v^i = (i_{\tilde{v}}\Pi)(df,dg)
\end{aligned}$$

(13.63)

Theorem 13.4.8 $(\Lambda_0^0, \{.,.\}_v)$ *is a Poisson algebra.*

Proof We have to prove that $\{.,.\}_v$ satisfies the following conditions:

(1) $\{f,g\}_v = -\{g,f\}_v$ (antisymmetry),
(2) $\{f,gh\}_v = g\{f,h\}_v + h\{f,g\}_v$ (Leibniz rule),
(3) $\{\{f,g\}_v,h\}_v + \{\{h,f\}_v,g\}_v + \{\{g,h\}_v,f\}_v = 0$ (Jacobi identity),

for all $f,g,h \in \Lambda_0^0$.

The antisymmetry is obvious, and the Leibniz rule easily follows from the definition. Let us show that also the Jacobi identity holds true. We have:

$$\begin{aligned}
\{\{f,g\}_v,h\}_v &= \left(\frac{\partial\{f,g\}_v}{\partial p_L^j}\frac{\partial h}{\partial y^L} - \frac{\partial h}{\partial p_L^j}\frac{\partial\{f,g\}_v}{\partial y^L}\right)v^j \\
&= \left(\left(\frac{\partial^2 f}{\partial p_L^j \partial p_K^i}\frac{\partial g}{\partial y^K} + \frac{\partial f}{\partial p_K^i}\frac{\partial^2 g}{\partial y^K \partial p_L^j} - \frac{\partial^2 g}{\partial p_L^j \partial p_K^i}\frac{\partial f}{\partial y^K} - \frac{\partial g}{\partial p_K^i}\frac{\partial^2 f}{\partial y^K \partial p_L^j}\right)\frac{\partial h}{\partial y^L} \right. \\
&\left. - \frac{\partial h}{\partial p_L^j}\left(\frac{\partial^2 f}{\partial y^L \partial p_K^i}\frac{\partial g}{\partial y^K} + \frac{\partial f}{\partial p_K^i}\frac{\partial^2 g}{\partial y^K \partial y^L} - \frac{\partial^2 g}{\partial y^L \partial p_K^i}\frac{\partial f}{\partial y^K} - \frac{\partial g}{\partial p_K^i}\frac{\partial^2 f}{\partial y^K \partial y^L}\right)\right)v^i v^j
\end{aligned}$$

Cycling f,g,h and summing up then yields zero, as desired. $\square$

Definition 13.4.9 We shall call the bracket $\{.,.\}_v$ *Poisson bracket in the direction of* v.

Given (any) 1-form η on Y, there arises a vector field $X_{\eta,v}$, defined by

$$X_{\eta,v}(g) = \Pi(\eta,dg,v) \quad \text{for every } g \in \Lambda_0^0.$$

(13.64)

We shall call $X_{\eta,v}$ the *Hamiltonian vector field of* η *in the direction of* v. We can see that $X_{\eta,v}$ is vertical and

$$X_{\eta,v} = \left(\eta_i^K \frac{\partial}{\partial y^K} - \eta_K \frac{\partial}{\partial p_K^i}\right)v^i,$$

(13.65)

where η_i^K and η_K are components of η at dp_K^i and dy^K, respectively. In particular, for η's of the form $p_1\eta = d_V f$, we get Hamiltonian vector fields associated with functions in Λ_0^0:

$$X_{f,v} = \left(\frac{\partial f}{\partial p_K^i} \frac{\partial}{\partial y^K} - \frac{\partial f}{\partial y^K} \frac{\partial}{\partial p_K^i} \right) v^i . \tag{13.66}$$

We note that

$$\{f,g\}_v = X_{f,v}(g) . \tag{13.67}$$

13.5 Cartan–Lepage Manifolds in Field Theory

Many important examples of Lepage manifolds appear in physics, namely in mechanics and field theories. In this section we shall present the Lepage manifolds setting applicable to electromagnetism (Maxwell equations), gravity (Einstein equations), gauge theories, and others. We refer also to [81] for other aspects and further references.

13.5.1 Closed Lepage Forms

Equations of motion of classical fields and gravity are second order PDEs *affine in the second derivatives*, and *variational*. This leads us to consider *closed* Lepage manifolds of order one.

The local structure of Lepage $(n + 1)$-forms we are interested in, is as follows.

Theorem 13.5.1 ([70]) *Let α be a Lepage $(n + 1)$-form on J^1Y such that $\varepsilon = p_1\alpha$ is a dynamical form affine in the second derivatives, and $d\alpha = 0$. Then in fibred coordinates α may be written as*

$$\alpha = \alpha_\varepsilon + \eta \tag{13.68}$$

where η is a closed at least 2-contact form, and where α_ε is closed and completely determined by ε. An explicit expression for α_ε is

$$\pi_{2,1}^* \alpha_\varepsilon = E_\sigma \omega^\sigma \wedge \omega_0 + \frac{1}{2} \frac{\partial E_\sigma}{\partial y_j^\nu} \omega^\sigma \wedge \omega^\nu \wedge \omega_j + \frac{\partial E_\sigma}{\partial y_{ij}^\nu} \omega^\sigma \wedge \omega_i^\nu \wedge \omega_j + \cdots$$

$$+ \frac{1}{n!(n+1)!} \frac{\partial^n E_\sigma}{\partial y_{j_1}^{\nu_1} \cdots \partial y_{j_n}^{\nu_n}} \omega^\sigma \wedge \omega^{\nu_1} \wedge \cdots \wedge \omega^{\nu_n} \wedge \omega_{j_1 \cdots j_n}$$

$$+ \frac{1}{(n!)^2} \frac{\partial^n E_\sigma}{\partial y_{j_1}^{\nu_1} \cdots \partial y_{j_{n-1}}^{\nu_{n-1}} \partial y_{j_n P}^{\nu_n}} \omega^\sigma \wedge \omega^{\nu_1} \wedge \cdots \wedge \omega^{\nu_{n-1}} \wedge \omega_p^{\nu_n} \wedge \omega_{j_1 \cdots j_n} ,$$

$$\tag{13.69}$$

where the E_σ satisfy Helmholtz conditions

$$\frac{\partial E_\sigma}{\partial y_{ij}^\nu} - \frac{\partial E_\nu}{\partial y_{ij}^\sigma} = 0,$$

$$\frac{\partial E_\sigma}{\partial y_i^\nu} + \frac{\partial E_\nu}{\partial y_i^\sigma} - 2d_k\frac{\partial E_\nu}{\partial y_{ik}^\sigma} = 0, \tag{13.70}$$

$$\frac{\partial E_\sigma}{\partial y^\nu} - \frac{\partial E_\nu}{\partial y^\sigma} + d_j\frac{\partial E_\nu}{\partial y_j^\sigma} - d_jd_k\frac{\partial E_\nu}{\partial y_{jk}^\sigma} = 0.$$

Poincaré Lemma then yields:

Theorem 13.5.2 *For a Lepage $(n+1)$-form α as above, local Lagrangians are given by the Vainberg–Tonti formula $\lambda = h\mathscr{A}\alpha = \mathscr{A}\varepsilon$. In coordinates the set of all local Lagrangians takes the form*

$$L = y^\sigma \int_0^1 E_\sigma\left(x^i, uy^\nu, uy_j^\nu, uy_{jk}^\nu\right) du + d_j\phi^j, \tag{13.71}$$

where ϕ^j are arbitrary functions. The restriction of α to a suitable open set U satisfies

$$\alpha|_U = d\theta_\lambda + d\mu, \tag{13.72}$$

where θ_λ is the Poincaré–Cartan form of a local Lagrangian for ε, and μ is an at least 2-contact n-form.

In view of these properties we can regard our Lepage manifold as a *fibred manifold*, equipped with a *class of locally equivalent Lagrangians*. We therefore have a family $\{L_\iota\}$ of Lagrange functions, each defined on an open set $U_\iota \subset J^2Y$, such that $\bigcup_\iota U_\iota = J^2Y$, and whenever $U_\iota \cap U_\kappa \neq \emptyset$, around every point of the intersection we have $L_\iota = L_\kappa + d_j\varphi^j$ for some functions φ^j. Note that in some cases we can find a global Lagrangian in the family, but in general such a global Lagrangian (even of higher order) does not exist. The obstructions come from the topology of Y (see [3]).

13.5.2 De Donder–Hamilton Equations on Lepage Manifolds

Formula (13.72) shows that there is an important family of Lepage manifolds, namely those with α 'as short as possible', or, 'as close to the Cartan form as possible'. Since $d\theta$ is at most 2-contact, this concerns α's which are *at most*

2-*contact* (leading to the choice of μ 2-contact and such that $p_3 d\mu = 0$).[7] Requiring μ be a *first order* 2-contact form, we are led to the following definition:

Definition 13.5.3 We say that (π_1, α) is a *Cartan–Lepage manifold* if α is closed, at most 2-contact and $\{\omega^\sigma\}$-generated.[8,9]

Notice that in fibred coordinates we have

$$\pi_{2,1}^* \alpha = E_\sigma \omega^\sigma \wedge \omega_0 + \frac{1}{2}\left(\frac{\partial E_\sigma}{\partial y_j^\nu} - d_k f_{\sigma\nu}^{j,k}\right)\omega^\sigma \wedge \omega^\nu \wedge \omega_j$$
$$+ \left(\frac{\partial E_\sigma}{\partial y_{ij}^\nu} - f_{\sigma\nu}^{i,j}\right)\omega^\sigma \wedge \omega_i^\nu \wedge \omega_j, \tag{13.73}$$

where E_σ satisfy the Helmholtz conditions, $f_{\sigma\nu}^{i,j} = -f_{\sigma\nu}^{j,i} = f_{\nu\sigma}^{j,i}$ are some first order functions such that $d\alpha = 0$, and the condition

$$\frac{\partial}{\partial y_{pj}^\rho}\left(\frac{\partial E_\sigma}{\partial y_i^\nu} - \frac{\partial E_\nu}{\partial y_i^\sigma}\right) = \frac{\partial f_{\sigma\nu}^{i,j}}{\partial y_p^\rho} + \frac{\partial f_{\sigma\nu}^{i,p}}{\partial y_j^\rho} \tag{13.74}$$

(meaning that α is projectable onto $J^1 Y$) holds true.

Keeping notations as in the previous sections, we shall consider the exterior differential system $\mathcal{D}_\alpha$, and equations for its holonomic integral sections (Euler–Lagrange equations) as well as equations for all its integral sections (Hamilton equations). We recall that in general Hamilton and Euler–Lagrange equations are not equivalent, as there might exist 'Hamilton extremals' that are not prolongations of extremals.

It is worth notice that *on a Lepage manifold both the Euler–Lagrange equations and the Hamilton equations are independent of a choice of a concrete Lagrangian for E.*

We can see that α is regular if rank $\alpha = m + nm$, and that in coordinates the regularity condition now reads as follows:

Theorem 13.5.4 ([60]) α *is regular if and only if at each point of* $J^1 Y$

$$\det\left(\frac{\partial E_\sigma}{\partial y_{ij}^\nu} - f_{\sigma\nu}^{i,j}\right) \neq 0. \tag{13.75}$$

---

[7] In general we cannot take simply $d\theta_\lambda$ since θ_λ need not be global.

[8] As we shall see, in this case the Hamilton equations become of De Donder type.

[9] This choice of α in the Lepage class $\{\alpha\}$ is the most simple. As we know, more generally we can add to α (any) at least 2-contact form, and/or the derivative of a 1-contact form, and the dynamical form ε, hence the Euler–Lagrange equations, remain the same. However, the Hamilton equations then may no longer be of De Donder type.

We have the following important theorem on *canonical form of a Cartan–Lepage* $(n+1)$*-form.*

Theorem 13.5.5 *Let* (π_1, α) *be a Cartan–Lepage manifold. Then around every point in* J^1Y *there is a neighbourhood* U *and functions* H *and* p^j_σ *defined on* U *such that*

$$\alpha|_U = -dH \wedge \omega_0 + dp^j_\sigma \wedge dy^\sigma \wedge \omega_j. \tag{13.76}$$

If, moreover, α *is regular, then the functions* p^j_σ *are independent, meaning that*

$$\text{rank}\left(\frac{\partial p^j_\sigma}{\partial y^\nu_k}\right) = \max = mn. \tag{13.77}$$

Proof The theorem is proved with help of the Poincaré Lemma. Since $d\alpha = 0$ we have locally $\alpha = d\rho$, and we can set $\rho = \mathscr{A}\alpha$ where $\mathscr{A}$ is the contact homotopy operator. Then $\lambda = h\mathscr{A}\alpha$ is the Vainberg–Tonti Lagrangian for the locally variational dynamical form $\varepsilon = p_1\alpha$. Explicitly, if we denote the components of ε by

$$E_\sigma = A_\sigma + B^{ij}_{\sigma\nu}y^\nu_{ij}, \quad B^{ij}_{\sigma\nu} = B^{ji}_{\sigma\nu} = B^{ij}_{\nu\sigma}, \tag{13.78}$$

we have by (13.73)

$$\alpha = \left(A_\sigma - 2F^i_{\sigma\nu}y^\nu_i\right)dy^\sigma \wedge \omega_0 + F^{ij}_{\sigma\nu}y^\sigma_i \, dy^\nu_j \wedge \omega_0$$
$$+ F^i_{\sigma\nu}dy^\sigma \wedge dy^\nu \wedge \omega_i + F^{ij}_{\sigma\nu}dy^\sigma \wedge dy^\nu_j \wedge \omega_i, \tag{13.79}$$

where

$$F^i_{\sigma\nu} = -F^i_{\nu\sigma} = \frac{1}{4}\left(\frac{\partial A_\sigma}{\partial y^\nu_i} - \frac{\partial A_\nu}{\partial y^\sigma_i} - 2\bar{d}_j f^{ij}_{\sigma\nu}\right), \tag{13.80}$$

$$F^{ij}_{\sigma\nu} = F^{ji}_{\nu\sigma} = B^{ij}_{\sigma\nu} - f^{ij}_{\sigma\nu}, \tag{13.81}$$

and $\bar{d}_i$ denotes the 'cut total derivative' $\bar{d}_i = d_i - y^\sigma_{ij}\frac{\partial}{\partial y^\sigma_j}$. Later we shall need the following identities yielded by $d\alpha = 0$:

$$\frac{\partial F^i_{\sigma\nu}}{\partial y^\rho_k} + \frac{1}{2}\left(\frac{\partial F^{ik}_{\nu\rho}}{\partial y^\sigma} - \frac{\partial F^{ik}_{\sigma\rho}}{\partial y^\nu}\right) = 0, \quad \frac{\partial F^{ij}_{\sigma\nu}}{\partial y^\rho_k} = \frac{\partial F^{ik}_{\sigma\rho}}{\partial y^\nu_j}. \tag{13.82}$$

Let $j_x^1 \gamma \in J^1 Y$. Using the definition of $\mathscr{A}$, we obtain in a neighbourhood U of $j_x^1 \gamma$

$$\rho = \mathscr{A}\alpha = \left(y^\sigma \int_0^1 (A_\sigma \circ \chi) du - 2y^\sigma y_i^\nu \int_0^1 (F_{\sigma\nu}^i \circ \chi) u\, du + y_i^\sigma y_j^\nu \int_0^1 (F_{\sigma\nu}^{ij} \circ \chi) u\, du \right) \omega_0$$

$$- \left(2y^\nu \int_0^1 (F_{\sigma\nu}^i \circ \chi) u\, du + y_j^\nu \int_0^1 (F_{\sigma\nu}^{ij} \circ \chi) u\, du \right) dy^\sigma \wedge \omega_i$$

$$+ \left(y^\sigma \int_0^1 (F_{\sigma\nu}^{ij} \circ \chi) u\, du \right) dy_j^\nu \wedge \omega_i,$$

$$\tag{13.83}$$

where $\chi : \left(u, \left(x^i, y^\sigma, y_j^\sigma \right) \right) \to \left(x^i, uy^\sigma, uy_j^\sigma \right)$ for $u \in [0, 1]$.

We shall show that one can express ρ in the form

$$\rho = -H\omega_0 + p_\sigma^i dy^\sigma \wedge \omega_i + d\phi \tag{13.84}$$

for some functions H, p_σ^i, and $(n-1)$-form ϕ. Shrinking U if necessary, we consider the map

$$\bar{\chi} : [0, 1] \times U \ni \left(v, \left(x^i, y^\sigma, y_j^\sigma \right) \right) \to \left(x^i, y^\sigma, vy_j^\sigma \right) \in U \subset J^1 Y, \tag{13.85}$$

and put

$$f^i = y_j^\nu y^\sigma \int_0^1 \left(\int_0^1 (F_{\sigma\nu}^{ij} \circ \chi) u\, du \right) \circ \bar{\chi}\, dv + \varphi^i(x^p, y^\sigma), \tag{13.86}$$

where φ^i are arbitrary functions. Using (13.82) we obtain that the f^i satisfy

$$\frac{\partial f^i}{\partial y_j^\nu} = y^\sigma \int_0^1 (F_{\sigma\nu}^{ij} \circ \chi) u\, du. \tag{13.87}$$

In this way we obtain:

$$\rho_0 = \mathscr{A}\alpha - d(f^i \omega_i) = -H\omega_0 + p_\sigma^i dy^\sigma \wedge \omega_i \tag{13.88}$$

where

$$p_\sigma^i = -2y^\nu \int_0^1 (F_{\sigma\nu}^i \circ \chi) u\, du - y_j^\nu \int_0^1 (F_{\sigma\nu}^{ij} \circ \chi) u\, du - \frac{\partial f^i}{\partial y^\sigma}$$

$$H = -y^\sigma \int_0^1 (A_\sigma \circ \chi) du + 2y^\sigma y_i^\nu \int_0^1 (F_{\sigma\nu}^i \circ \chi) u\, du \tag{13.89}$$

$$- y_i^\sigma y_j^\nu \int_0^1 (F_{\sigma\nu}^{ij} \circ \chi) u\, du + \frac{\partial f^i}{\partial x^i}.$$

The n-form ρ_0 satisfies

$$d\rho_0 = -dH \wedge \omega_0 + dp_\sigma^i \wedge dy^\sigma \wedge \omega_i = d\mathscr{A}\alpha = \alpha, \tag{13.90}$$

as desired.   □

Definition 13.5.6 We call (13.76) *canonical form of* α, and the functions H and p_σ^i Hamiltonian and momenta of α, respectively.

Note that

$$h\rho_0 = h\mathscr{A}\alpha - hd(f^i\omega_i) \tag{13.91}$$

is a family of local *first order* Lagrangians for ε (i.e. equivalent with the second order Vainberg–Tonti Lagrangian $h\mathscr{A}\alpha = \mathscr{A}\varepsilon$). The Lagrangians are determined up to $hd(\varphi^i\omega_i) = d_i\varphi^i\omega_0$.

For $\lambda_0 = h\rho_0$ we have $\lambda_0 = L_0\omega_0$ where

$$L_0 = -H + p_\sigma^i y_i^\sigma, \tag{13.92}$$

and

$$p_\sigma^i = \frac{\partial L_0}{\partial y_i^\sigma}; \tag{13.93}$$

the latter follows from formulas for H and p_σ^i (13.89) if we use (13.82). Moreover, since

$$-\frac{\partial p_\sigma^i}{\partial y_j^\nu} = F_{\sigma\nu}^{ij} = B_{\sigma\nu}^{ij} - f_{\sigma\nu}^{i,j} = \frac{\partial E_\sigma}{\partial y_{ij}^\nu} - f_{\sigma\nu}^{i,j}, \tag{13.94}$$

the regularity condition (13.75) takes the form

$$\det\left(\frac{\partial p_\sigma^i}{\partial y_j^\nu}\right) = \det\left(\frac{\partial^2 L_0}{\partial y_i^\sigma \partial y_j^\nu}\right) \neq 0. \tag{13.95}$$

We can see that if λ_0 and $\bar{\lambda}_0$ are two Lagrangians defined by the canonical form of α (13.76) then locally $\bar{\lambda}_0 = \lambda_0 + hd\varphi$, where $\lambda_0 = h\rho_0$ for some ρ_0 as above, and φ is a local horizontal $(n-1)$-form on Y. We note that if one of these Lagrangians is regular (in the sense of the regularity condition (13.95)) then all the Lagrangians in the family are regular (and this is if and only if α is regular). Thus the functions φ^i in (13.86) play the role of *admissible gauge functions*.

Remark 13.5.7 It is worth notice that for $n = \dim X > 1$, not every first order Lagrangian for E has the pleasant properties as have the Lagrangians λ_0 related with the canonical form of α. Namely, it may happen that for a local first order

Lagrangian $\lambda = L\omega_0$ the momenta $(p_L)^i_\sigma = \partial L/\partial y^\sigma_i$ and Hamiltonian $H_L = -L + (p_L)^i_\sigma y^\sigma_i$ for no φ are equal to p^i_σ and H, defined by (13.89). Moreover, even if α is regular, L need not satisfy the condition

$$\det\left(\frac{\partial^2 L}{\partial y^\sigma_i \partial y^\nu_j}\right) \neq 0. \tag{13.96}$$

For discussion on equivalent first order Lagrangians such that one satisfies (13.96) and the other does not we refer to [87].

We have seen that for regular α, the map $\left(x^i, y^\sigma, y^\sigma_j\right) \rightarrow \left(x^i, y^\sigma, p^j_\sigma\right)$ is a local coordinate transformation on J^1Y. In view of Sect. 13.4.2 the new coordinates are Legendre coordinates.

Summarizing, we have obtained:

Theorem 13.5.8 *Let (π_1, α) be a Cartan–Lepage manifold. Then in a neighbourhood of every point in J^1Y there exists a first order Lagrangian for $\varepsilon = p_1\alpha$. Moreover, if H and p^i_σ are a Hamiltonian and momenta of α defined by the canonical form of α then locally there exists a first order Lagrangian for ε such that $H = -L + p^\sigma_\sigma y^\sigma_i$ and $p^\sigma_i = \frac{\partial L}{\partial y^\sigma_i}$. Any such a Lagrangian is of the form $\lambda = \lambda_0 + hd\varphi$, where $\lambda_0 = h\rho_0$ and φ is a local horizontal $(n-1)$-form on Y.*

If, moreover, α is regular, then there exists a covering of J^1Y by Legendre coordinates, and a Poisson tensor on J^1Y; they are defined by the canonical form (13.76) of α. Moreover, every Lagrangian $\lambda = h\rho_0 + hd\varphi$ as above satisfies the regularity condition (13.95).

If we now compute the generators of the exterior differential system $\mathscr{D}_\alpha$ in Legendre coordinates, we obtain the Hamilton equations of α in the form

$$\frac{\partial(y^\sigma \circ \delta)}{\partial x^i} = \frac{\partial H}{\partial p^i_\sigma} \circ \delta, \qquad \frac{\partial(p^i_\sigma \circ \delta)}{\partial x^i} = -\frac{\partial H}{\partial y^\sigma} \circ \delta. \tag{13.97}$$

These equations are known also as *De Donder–Hamilton equations*.

The meaning of the regularity condition is as follows.

Theorem 13.5.9 *On a Lepage manifold (π_1, α) where α is closed, at most 2-contact, $\{\omega^\sigma\}$-generated and regular, the Euler–Lagrange and Hamilton equations are equivalent. Explicitly, if γ is an extremal then $J^1\gamma$ is a Hamilton extremal; conversely, every Hamilton extremal is of the form $J^1\gamma$ where γ is an extremal.*

An easy proof of this theorem comes from analysing the solutions of the exterior differential system $\mathscr{D}_\alpha$ for α (13.73) satisfying (13.75).

Note that, by construction, explicit formulas for the Hamiltonian and momenta come from an *integration procedure* using the Poincaré Lemma and *are determined by α rather than by a particular Lagrangian*.

13.5.3 Dual Jet Bundles

Our aim now is to provide a dual picture, analogously to symplectic mechanics where one considers the cotangent bundle T^*M equipped with the canonical symplectic form. The present setting is due to [10] (see also [83], and [67] for higher order mechanics).

Given a fibred manifold $\pi : Y \to X$ we shall explore that $\pi_{1,0} : J^1Y \to Y$ is an affine bundle, modelled on the vector bundle $V_X Y \otimes \pi^* T^* X \to Y$, where $V_X Y \to Y$ is the bundle of vectors tangent to Y and vertical over X. For every fixed $y \in Y$, the fibre $J_y^1 Y$ of $\pi_{1,0}$ over y is an affine space of dimension nm, hence it is associated with a vector space of dimension $(nm + 1)$, its *extended dual*, whose elements are the real-valued affine maps on $J_y^1 Y$. Collecting the extended duals and taking the union over $y \in Y$ we obtain a manifold of dimension $n + m + nm + 1$, denoted by $J^\dagger Y$, and called the *extended dual of $J^1 Y$*.

If $\phi \in J^\dagger Y$ and $\phi : J_y^1 Y \to \mathbb{R}$, we put $\pi_{1,0}^\dagger(\phi) = y$. Then $\pi_{1,0}^\dagger : J^\dagger Y \to Y$ becomes a vector bundle.

Each fibre $J_y^\dagger Y$ is a vector space containing a distinguished subspace of constant maps. The quotient space will be denoted by $J_y^* Y$, and the union over y will be denoted by $J^* Y$ and called the *reduced dual* of $J^1 Y$. We have $\dim J^* Y = \dim J^\dagger Y - 1 = \dim J^1 Y$.

Next we denote by $\rho : J^\dagger Y \to J^* Y$ the quotient map, and by $\pi_{1,0}^* : J^* Y \to Y$ the induced vector bundle projection.

Let $f_{(y)}$ denote the restriction of a function to the fibre over y. Given a chart $(U, (x^i, y^\sigma))$ on Y and $y \in U$, the affine functions $\left(1_{(y)}, y_{i(y)}^\sigma \right)$ form a basis for the vector space $J_y^\dagger Y$. Denote by $\left(P_{(y)}, P_{\sigma(y)}^i \right)$ the dual basis of $\left(J_y^\dagger Y \right)^*$; in this way, we get smooth functions P, P_σ^i locally defined on $J^\dagger Y$ such that $(x^i, y^\sigma, P, P_\sigma^i)$ are local coordinates on the extended dual. Then $(x^i, y^\sigma, P_\sigma^i)$ are coordinates on the reduced dual $J^* Y$.

It can be shown that with a choice of a volume element on X, $J^\dagger Y$ is diffeomorphic to the bundle of n-forms on $J^1 Y$, locally generated by $(\omega_0, dy^\sigma \wedge \omega_i)$ (see [81] for the proof). The global $(n + 1)$-form

$$\Omega = dP \wedge \omega_0 + dP_\sigma^i \wedge dy^\sigma \wedge \omega_i \tag{13.98}$$

then can be regarded as a *canonical multisymplectic form on $J^\dagger Y$*. Now, any local section h of the projection $\rho : J^\dagger Y \to J^* Y$ gives rise to a local closed $(n + 1)$-form

$$\Omega_h = h^* \Omega = -d\mathcal{H} \wedge \omega_0 + dP_\sigma^i \wedge dy^\sigma \wedge \omega_i, \tag{13.99}$$

on $J^* Y$, where we have introduced the function $\mathcal{H} = -(P \circ h)$ defined on $\operatorname{dom} h \subset J^* Y$ (an abstract Hamiltonian). With Ω_h the manifold $J^* Y$ becomes a *regular Cartan–Lepage manifold of order zero*.

We can consider the exterior differential system $\mathscr{D}_h = \{i_\xi \Omega_h\}$ where ξ runs over vector fields on J^*Y, vertical over X, so that integral sections $\psi : X \to J^*Y$ of $\mathscr{D}_h$ satisfy the De Donder–Hamilton equations

$$\frac{\partial (y^\sigma \circ \psi)}{\partial x^i} = \frac{\partial \mathscr{H}}{\partial P^i_\sigma} \circ \psi, \qquad \frac{\partial \left(P^j_\sigma \circ \psi \right)}{\partial x^j} = -\frac{\partial \mathscr{H}}{\partial y^\sigma} \circ \psi. \qquad (13.100)$$

From (13.99) it is clear that Ω_h induces a Poisson structure on J^*Y defined by the Poisson tensor

$$\Pi = \frac{\partial}{\partial P^i_\sigma} \wedge \frac{\partial}{\partial y^\sigma} \otimes dx^i. \qquad (13.101)$$

So far we have constructed a universal (abstract) Hamiltonian bundle, canonically associated with any jet bundle $\pi_1 : J^1Y \to X$. Now we shall establish a connection between an abstract Hamiltonian system and a concrete variational system on a Lepage manifold.

So, assume that our jet bundle π_1 is a Cartan–Lepage manifold, equipped with a Lepage $(n + 1)$-form α. As above, $\varepsilon = p_1\alpha$ will be the corresponding dynamical form (Euler–Lagrange form). On the extended dual $J^\dagger Y$ we have the canonical symplectic form Ω corresponding to a choice of volume form on X. Both the sides of the picture can be related by means of a map $\mathrm{Leg}_\alpha : J^1Y \to J^\dagger Y$ fibred over the identity of Y defined by

$$\mathrm{Leg}^*_\alpha \, \Omega = \alpha. \qquad (13.102)$$

We call Leg_α an *extended Legendre map of* α and Eq. (13.102) a *duality equation*. We may also construct a composite map $\mathrm{leg}_\alpha : J^1Y \to J^*Y$ by setting $\mathrm{leg}_\alpha = \rho \circ \mathrm{Leg}_\alpha$. This will be called the *reduced Legendre map*.

The map Leg_α is not unique; all such maps are parametrized by "gauge functions" $\varphi(x^i, y^\sigma)$. A similar remark applies to leg_α.

In the regular case we can choose Legendre coordinates on J^1Y. Using these Legendre coordinates, and the canonical coordinates on J^*Y, the map leg_α is represented by the identity mapping. We immediately obtain the following result.

Theorem 13.5.10 ([83]) *If α is regular, then every extended Legendre map is an immersion and every corresponding reduced Legendre map is a local diffeomorphism.*

We say that α is *hyper-regular* if there is an extended Legendre map Leg_α which is defined globally and which has the property that the corresponding reduced Legendre map leg_α is a diffeomorphism. Any such map gives rise to a global Hamiltonian section h by setting $h = \mathrm{Leg}_\alpha \circ \mathrm{leg}_\alpha^{-1}$. If α is regular rather than hyper-regular, then we may construct local sections h in this way. The negative coordinate representative of such a section h will be the function $\mathscr{H} = -(P \circ h)$, a *Hamiltonian function* corresponding to α. Corresponding to a choice of the extended Legendre map we get a family of Hamiltonian functions parametrized by φ.

With help of the definition of the Legendre maps, and using Legendre coordinates, we easily obtain the following result.

Theorem 13.5.11 ([83]) *For a hyper-regular α on J^1Y the following assertions are true:*

(1) $\mathrm{Leg}_\alpha^* \Omega = \mathrm{leg}_\alpha^* h^* \Omega = \alpha$.
(2) $\mathrm{rank}\, h^* \Omega = \mathrm{rank}\, \mathscr{D}_h = \mathrm{rank}\, \alpha = \mathrm{rank}\, \mathscr{D}_\alpha = m + nm$.
(3) $\mathrm{leg}_\alpha^* \mathscr{D}_h = \mathscr{D}_\alpha$.
(4) *If $\psi : X \to J^*Y$ is an integral section of $\mathscr{D}_h$, then $\mathrm{leg}_\alpha^{-1} \circ \psi = J^1\gamma$ where γ is a section of $\pi : Y \to X$, and it is an integral section of $\mathscr{D}_\alpha$.*
(5) *Every integral section of $\mathscr{D}_\alpha$ is of the form $J^1\gamma$, and $\psi = \mathrm{leg}_\alpha \circ J^1\gamma$ is an integral section of $\mathscr{D}_h$.*

13.6 Lepage Manifolds in Mechanics: The Metric, Symplectic and Poisson Structures

In the last section we shall consider a fibred manifold $\pi : Y \to X$ over a base of dimension 1. In this case equations for sections of π are ordinary differential equations, and Lepage $(n + 1)$-forms are 2-forms. We emphasize that for $n = 1$ there occurs a substantial simplification: *every dynamical form has a unique (global) Lepage equivalent.* This means that we have a *one-to-one correspondence between dynamical forms and Lepage manifolds.* Moreover, there is one-to-one correspondence between *Lepage manifolds defined by closed forms* and *variational equations* (precisely speaking, locally variational dynamical forms):

Theorem 13.6.1 ([56]) *Let (π_r, α) be a Lepage manifold and assume that α is closed. Then $\varepsilon = p_1\alpha$ is locally variational. Moreover, α is a unique Lepage equivalent of ε.*

Conversely, let ε be a locally variational dynamical form on J^sY, $s \geqslant 0$. Then there exists a unique Lepage manifold (π_r, α) such that $\varepsilon = p_1\alpha$. It holds $r = s - 1$ and $d\alpha = 0$.

Explicit formula for α can be found in [56] (see also [59]).

In what follows we shall be interested in Lepage manifolds which are a background for *second order ordinary differential equations,* and in particular, for *geometric mechanics.* As usual in this situation, we shall take $Y = \mathbb{R} \times M$, where M is a smooth manifold, $\dim M = m$, and $X = \mathbb{R}$. Then $J^1Y = \mathbb{R} \times TM$, $J^2Y = \mathbb{R} \times T^2M$, etc. We shall use fibred coordinates adapted to the product structure of Y, and a global coordinate on $\mathbb{R}$. This means that the transformation rules between (t, q^σ) and $(\bar{t}, \bar{q}^\sigma)$ satisfy $\bar{t} = t$ and $\partial \bar{q}^\sigma / \partial t = 0$. The associated coordinates on J^1Y and J^2Y will be denoted by $(t, q^\sigma, \dot{q}^\sigma)$ and $(t, q^\sigma, \dot{q}^\sigma, \ddot{q}^\sigma)$, respectively. We set $\omega^\sigma = dq^\sigma - \dot{q}^\sigma dt$, $\dot{\omega}^\sigma = d\dot{q}^\sigma - \ddot{q}^\sigma dt$.

Apart from the general case, we shall be interested also in *time-independent* mechanical systems. In order to define this concept we note that the global vector field $\partial/\partial t$ on $\mathbb{R}$ trivially lifts to $\mathbb{R} \times TM$ and its prolongations. We call a differential form η time-independent, if $\mathscr{L}_{\partial/\partial t}\eta = 0$. We obtain that α is time independent if and only if ε is time-independent, and this means that $\partial E_\sigma/\partial t = 0$ for all σ.

Due to uniqueness of the Lepage equivalent in mechanics, Lepage manifolds for second order ordinary differential equations are completely characterized by the following theorem due to Krupková and Prince:

Theorem 13.6.2 ([65])

(1) *Every dynamical form ε on J^2Y has a unique Lepage equivalent. It is defined on J^3Y, and takes the coordinate form*

$$\alpha = E_\sigma \omega^\sigma \wedge dt + \frac{1}{4}\left(\frac{\partial E_\sigma}{\partial \dot{q}^\nu} - \frac{\partial E_\nu}{\partial \dot{q}^\sigma} - \frac{d}{dt}\left(\frac{\partial E_\sigma}{\partial \ddot{q}^\nu} - \frac{\partial E_\nu}{\partial \ddot{q}^\sigma}\right)\right)\omega^\sigma \wedge \omega^\nu$$

$$+ \frac{1}{2}\left(\frac{\partial E_\sigma}{\partial \ddot{q}^\nu} + \frac{\partial E_\nu}{\partial \ddot{q}^\sigma}\right)\omega^\sigma \wedge \dot{\omega}^\nu. \tag{13.103}$$

(2) *If the second order ODEs are affine in the second derivatives, i.e. such that*

$$E_\sigma = A_\sigma(t, q^\rho, \dot{q}^\rho) + g_{\sigma\nu}(t, q^\rho, \dot{q}^\rho)\ddot{q}^\nu \tag{13.104}$$

then the Lepage equivalent is projectable onto J^2Y and reads

$$\alpha = A_\sigma\omega^\sigma \wedge dt + \frac{1}{4}\left(\frac{\partial A_\sigma}{\partial \dot{q}^\nu} - \frac{\partial A_\nu}{\partial \dot{q}^\sigma} - 2\frac{\partial g_{[\sigma\nu]}}{\partial t} - 2\frac{\partial g_{[\sigma\nu]}}{\partial q^\rho}\dot{q}^\rho\right)\omega^\sigma \wedge \omega^\nu$$

$$+ g_{(\sigma\nu)}\omega^\sigma \wedge d\dot{q}^\nu + g_{[\sigma\nu]}\ddot{q}^\nu\omega^\sigma \wedge dt$$

$$+ \frac{1}{4}\left(\frac{\partial g_{\sigma\rho}}{\partial \dot{q}^\nu} - \frac{\partial g_{\nu\rho}}{\partial \dot{q}^\sigma} - 2\frac{\partial g_{[\sigma\nu]}}{\partial \dot{q}^\rho}\right)\ddot{q}^\rho\omega^\sigma \wedge \omega^\nu \tag{13.105}$$

where $(\sigma\nu)$, resp. $[\sigma\nu]$ denotes symmetrization, resp. antisymmetrization in the indicated indices.

(3) *The Lepage equivalent of (13.104) is projectable onto J^1Y if and only if*

$$g_{\sigma\nu} = g_{\nu\sigma}, \quad \frac{\partial g_{\sigma\nu}}{\partial \dot{q}^\rho} = \frac{\partial g_{\sigma\rho}}{\partial \dot{q}^\nu}. \tag{13.106}$$

In this case

$$\alpha = A_\sigma\omega^\sigma \wedge dt + \frac{1}{4}\left(\frac{\partial A_\sigma}{\partial \dot{q}^\nu} - \frac{\partial A_\nu}{\partial \dot{q}^\sigma}\right)\omega^\sigma \wedge \omega^\nu + g_{\sigma\nu}\omega^\sigma \wedge d\dot{q}^\nu. \tag{13.107}$$

Definition 13.6.3 Dynamical forms characterized by property (3) of the above theorem are called *semi-variational*.

Relations (13.106) are integrability conditions for a function $f(t, q^\rho, \dot{q}^\rho)$ such that

$$g_{\sigma v} = \frac{\partial^2 f}{\partial \dot{q}^\sigma \partial \dot{q}^v}, \tag{13.108}$$

and which is determined up to a linear term. Moreover, it can be shown [59, 61] that every regular semi-variational dynamical form ε splits canonically into a sum of a variational dynamical form ε_g, coming from a Lagrangian $\tau = T\,dt$ determined by the $g_{\sigma v}$, and a first order dynamical form ϕ (a "force"), so that the semi-variational equations take the form

$$\frac{\partial T}{\partial q^\sigma} - \frac{d}{dt}\frac{\partial T}{\partial \dot{q}^\sigma} = \phi_\sigma. \tag{13.109}$$

This means that α splits to

$$\alpha = \alpha_g + \alpha_\phi \tag{13.110}$$

where α_g is the Lepage equivalent of the Euler–Lagrange form of T, and α_ϕ is the Lepage equivalent of ϕ. Note that (due to variationality) $d\alpha_g = 0$.

If the matrix $(g_{\sigma v})$ is regular and time and velocity independent, it determines a *metric* g on M. Then (up to the sign) T is the kinetic energy of the metric g, and Eq. (13.109) take the (contravariant) form

$$\ddot{q}^\sigma + \Gamma^\sigma_{v\rho}\dot{q}^v\dot{q}^\rho = g^{\sigma v}\phi_v, \tag{13.111}$$

where Γ is the Levi-Civita connection of g. If g is regular, time independent and 0-homogeneous in the velocities, it is a Finsler metric, and the corresponding equations with $\phi = 0$ are equations for geodesics on the Finsler manifold M. In a general case, we call a regular morphism $g(t, q^\rho, \dot{q}^\rho)$ satisfying the integrability conditions (13.106) a *variational metric* (more about this see [61] and [80]).

We remind that ε is variational if and only if $p_2 d\alpha \equiv H_\varepsilon = 0$ (Helmholtz conditions), or, equivalently, if α is closed. Then the force $\phi = \phi_\sigma \omega^\sigma \wedge dt$ appearing in Eq. (13.109) is a first-order *locally variational form* (i.e. satisfies the first-order Helmholtz conditions). It can be shown that this is if and only if ϕ takes the form of *Lorentz force* (see, e.g., [61] or [59] for details).

As we have seen above, second order ordinary differential equations are modelled on Lepage manifolds of order 2, or, if they are semi-variational or variational, on Lepage manifolds of order 1.

In the sequel we shall consider a Lepage manifold (π_1, α) over a fibred manifold $\pi : \mathbb{R} \times M \to \mathbb{R}$. Let us summarize what we already know: (π_1, α) is endowed with

- the *dynamical form* $\varepsilon = p_1\alpha$ which is *semi-variational or variational* depending on whether $d\alpha \neq 0$ or $d\alpha = 0$,
- the *exterior differential system* $\mathcal{D}_\alpha$, generated by the *co-distribution* Δ^0 on $\mathbb{R} \times TM$ spanned by 1-forms $i_\xi\alpha$ where ξ runs over all π_1-vertical vector fields on J^1Y, and having the same holonomic solutions as ε.

Since α is a 2-form, we also have

- the characteristic distribution χ_α (and its annihilator χ_α^0) of α on $\mathbb{R} \times TM$. Recall that

$$\chi_\alpha = \text{span}\{\zeta \mid i_\zeta\alpha = 0\}, \tag{13.112}$$

and its annihilator consists of 1-forms $i_\xi\alpha$ where ξ runs over all vector fields on $\mathbb{R} \times TM$. Thus, $\chi_\alpha \subset \Delta_\alpha$, and both the distributions have the same holonomic solutions which coincide with the solutions of the ODEs defined by ε.

- Moreover, if α is *closed* (ε is locally variational), then α has a local canonical form

$$\alpha = -dH \wedge dt + dp_\sigma \wedge dq^\sigma \tag{13.113}$$

where the functions H, p_σ are defined as components in the canonical basis of the Cartan form θ_λ of a local Lagrangian $\lambda = h\mathscr{A}\alpha$, i.e. $\theta_\lambda = \mathscr{A}\alpha = Hdt + p_\sigma dq^\sigma$.

From now on, let us assume that (π_1, α) is a *regular* Lepage manifold, i.e. $\text{rank}\,\alpha = \dim V\pi_1 = 2m$. Then the distribution Δ_α has *rank one* and coincides with the characteristic distribution χ_α. Moreover, we get the following additional structures:

- *Variational metric g* [61] defined by

$$g_{\sigma\nu} = \frac{\partial E_\sigma}{\partial \ddot{q}^\nu}. \tag{13.114}$$

Indeed, by the regularity assumption on α this matrix is regular. Consequently, *on every fibre of* $\mathbb{R} \times M$, *g induces a velocity dependent metric*. In case that α is time independent, g becomes time-independent and projects onto a velocity dependent metric on M. This may be, for example, a Finsler metric if g is 0-homogeneous in velocities, or a Riemannian metric on M if g does not depend on velocities.
- A first order dynamical form ϕ (a *force*), defined by the canonical splitting $\varepsilon = \varepsilon_g + \phi$; the force is variational or non-variational, depending on whether α is closed or not.
- Global vector field Γ on $\mathbb{R} \times TM$, spanning the distribution $\Delta_\alpha = \chi_\alpha$. It is defined by $i_\Gamma\alpha = 0$ and $dt(\Gamma) = 1$. Γ is a semispray and its integral sections identify with solutions of the equations determined by ε. In coordinates,

$$\Gamma = \frac{\partial}{\partial t} + \dot{q}^\sigma \frac{\partial}{\partial q^\sigma} - g^{\sigma\nu}A_\nu \frac{\partial}{\partial \dot{q}^\sigma}. \tag{13.115}$$

- A splitting of the tangent bundle to $\mathbb{R} \times TM$ associated with Γ (e.g. [84, 85, 88])

$$T(\mathbb{R} \times TM) = \text{span}\{\Gamma\} \oplus H_\Gamma \oplus V\pi_{1,0} \qquad (13.116)$$

where $V\pi_{1,0}$ is the bundle of the 'very vertical vectors' (vertical over $\mathbb{R} \times M$) and

$$H_\Gamma = \text{span} \left\{ \frac{\partial}{\partial y^\sigma} + \frac{1}{2} \frac{\partial \Gamma^\nu}{\partial \dot{q}^\sigma} \frac{\partial}{\partial \dot{q}^\nu} \right\} \qquad (13.117)$$

where $\Gamma^\sigma = -g^{\sigma\nu} A_\nu$. This splitting induces a splitting of the cotangent bundle which gives an adapted basis of 1-forms (dual basis to $(\Gamma, H_\sigma, \partial/\partial \dot{q}^\sigma)$) as follows: $(dt, \omega^\sigma, \psi^\sigma)$, where

$$\psi^\sigma = d\dot{q}^\sigma - \frac{1}{2} \frac{\partial \Gamma^\nu}{\partial \dot{q}^\sigma} \omega^\nu - \Gamma^\sigma dt . \qquad (13.118)$$

In the adapted basis the Lepage 2-form takes a simple expression

$$\alpha = g_{\sigma\nu} \omega^\sigma \wedge \psi^\nu . \qquad (13.119)$$

- *If α is closed, time independent and regular*: The splitting of α (13.113) is invariant and defines the *symplectic form* $\omega = dp_\sigma \wedge dq^\sigma$, and hence also the *Poisson structure* $\Pi = \frac{\partial}{\partial p_\sigma} \wedge \frac{\partial}{\partial q^\sigma}$ on TM. Obviously, Γ then splits into a vector field on $\mathbb{R}$ and a time independent vector field on TM which is the *Hamilton vector field* for ω.
- More generally, on a *regular* Lepage manifold with *closed* α (locally variational dynamical form ε, corresponding to variational equations), apart from a (generalized) metric structure g, we have also *fibrewise symplectic and Poisson structures*.
- Considering the *dual jet bundle*, we get in the same way as in the case of field theory in the previous section, the *duality* between the Lepage manifold (π_1, α) and the bundle (J^*Y, Ω_h). For a detailed exposition we refer to [67].

Lepage manifolds arise wherever differential equations of dynamical type (i.e. equations defined by dynamical forms) are concerned. In particular, they are behind any system of first-order ODEs (a vector field, dynamical system) and regular higher-order ODEs (a semispray, a spray), a symplectic structure, or behind any extremal principle (e.g. kinetic energy, Finsler function, etc.). Let us discuss the latter two cases in more detail:

Lepage Manifolds in Riemann and Finsler Geometry Let (M, g) be a Riemannian manifold (actually, the signature of the metric is inessential). Denote by (x^i), $1 \leq i \leq m$, local coordinates on M. We have on TM the induced kinetic energy, $T = \frac{1}{2} g_{ij} \dot{x}^i \dot{x}^j$, which is a first-order Lagrangian for the fibred manifold $\pi : \mathbb{R} \times M \to \mathbb{R}$. The Euler–Lagrange equations, that are equations for the geodesics on M, are represented by a second-order vector field (semispray) Γ on $\mathbb{R} \times TM$. If we denote by (dt, ω^i, ψ^i) the basis of one forms on $\mathbb{R} \times TM$, dual to the natural splitting of $\mathbb{R} \times TM$ induced by Γ, we get the Lepage 2-form $\alpha = g_{ij}\omega^i \wedge \psi^j$, which is *closed*. We easily check that Γ spans the kernel of α, so that (π_1, α), where $\pi_1 : \mathbb{R} \times TM \to \mathbb{R}$, is the Lepage manifold naturally related with (M, g).

A similar situation arises in Finsler geometry. Denote by $T^o M$ the slit tangent bundle of M (i.e. let $T^o M$ be TM with the zero section excluded). If F is a Finsler function on $T^o M$, and g is the corresponding Finsler metric tensor

$$g_{ij} = \frac{\partial^2 F^2}{\partial \dot{x}^i \partial \dot{x}^j}, \tag{13.120}$$

the equations for Finsler geodesics are Euler–Lagrange equations of the Lagrangian F^2. Let Γ be the semispray on $\mathbb{R} \times T^o M$ defined by the Euler–Lagrange equations, and (dt, ω^i, ψ^i) the basis of one forms on $\mathbb{R} \times T^o M$, dual to the natural splitting of $\mathbb{R} \times T^o M$ induced by Γ. The Lepage manifold associated with (M, F) is (π_1, α) where $\pi_1 : \mathbb{R} \times T^o M \to \mathbb{R}$, and $\alpha = g_{ij}\omega^i \wedge \psi^j$. Again, α is closed.

Notice that, in both the cases, α invariantly splits to a sum $-dH \wedge dt + \omega$, where ω is the related symplectic form.

Acknowledgements Research supported by grant No. 14-02476S Variations, Geometry and Physics of the Czech Science Foundation, and by the IRSES project LIE-DIFF-GEOM (EU FP7, nr 317721).

References

1. R. Abraham, J.E. Marsden, *Foundations of Mechanics*, 2nd edn. (The Benjamin/Cummings Publishing Company, Reading, 1978)
2. I. Anderson, The Variational Bicomplex, Utah State University, Technical Report (1989)
3. I. Anderson, T. Duchamp, On the existence of global variational principles. Am. J. Math. **102**, 781–867 (1980)
4. V.I. Arnold, V.V. Kozlov, A.I. Neishtadt, *Mathematical Aspects of Classical and Celestial Mechanics* (Springer, Berlin, 2006)
5. D.E. Betounes, Extension of the classical Cartan form. Phys. Rev. D **29**, 599–606 (1984)
6. G.E. Bredon, *Sheaf Theory* (McGraw-Hill, New York, 1967)
7. R. Bryant, P. Griffiths, D. Grossmann, *Exterior Differential Systems and Euler-Lagrange Partial Differential Equations*. Chicago Lectures in Mathematics (University of Chicago Press, Chicago, 2003)
8. F. Cantrijn, L.A. Ibort, M. de León, Hamiltonian structures on multisymplectic manifolds. Rend. Sem. Mat. Univ. Pol. Torino **54**, 225–236 (1996)
9. C. Carathéodory, Über die Variationsrechnung bei mehrfachen Integralen. Acta Szeged Sect. Scient. Mathem. **4**, 193–216 (1929)
10. J.F. Cariñena, M. Crampin, L.A. Ibort, On the multisymplectic formalism for first order field theories. Diff. Geom. Appl. **1**, 345–374 (1991)
11. É. Cartan, *Lecons Sur Les Invariants IntÉgraux* (Hermann, Paris, 1922)
12. M. Crampin, D.J. Saunders, The Hilbert-Carathéodory and Poincaré–Cartan forms for higher-order multiple-integral variational problems. Houston J. Math. **30**(3), 657–689 (2004)
13. M. de León, P.R. Rodrigues, *Generalized Classical Mechanics and Field Theory* (North-Holland, Amsterdam, 1985)

14. M. de León, P.R. Rodrigues, *Methods of Differential Geometry in Analytical Mechanics* (North-Holland, Amsterdam, 1989)
15. M. de León, J.C. Marrero, D.M. de Diego, A new geometric setting for classical field theories. Banach Cent. Publ. **59**, 189–209 (2003)
16. P. Dedecker, Calcul des variations, formes différentielles et champs géodésiques, in *Coll. internat. du C.N.R.S.*, Strasbourg, 1953 (C.N.R.S., Paris, 1954), pp. 17–34
17. P. Dedecker, On the generalization of symplectic geometry to multiple integrals in the calculus of variations, in *Lecture Notes in Mathematics*, vol. 570 (Springer, Berlin, 1977), pp. 395–456
18. P. Dedecker, Sur le formalisme de Hamilton–Jacobi–E. Cartan pour une intègrale multiple d'ordre supérieur. C. R. Acad. Sci. Paris Sér. I **299**, 363–366 (1984)
19. P. Dedecker, Existe-t-il, en calcul des variations, un formalisme de Hamilton-Jacobi-É. Cartan pour les intégrales multiples d'ordre supérieur? C. R. Acad. Sci. Paris Sér. 1 **298**, 397–400 (1984)
20. A. Echeverria-Enriquez, M.C. Muñoz-Lecanda, N. Román-Roy, Multivector field formulation of hamiltonian field theories: equations and symmetries. J. Phys. A Math. Gen. **32**, 8461–8484 (1999)
21. A. Echeverria-Enriquez, M.C. Muñoz-Lecanda, N. Román-Roy, On the multimomentum bundles and the Legendre maps in field theories. Rep. Math. Phys. **45**, 85–105 (2000)
22. C. Ehresmann, Les prolongements d'une variété différentiable: 1. calcul des jets, prolongement principal. C. R. Acad. Sci. Paris **233**, 598–600 (1951)
23. C. Ehresmann, Les prolongements d'une space fibré différentiable. C. R. Acad. Sci. Paris **240**, 1755–1757 (1955)
24. M. Ferraris, M. Francaviglia, On the global structure of the Lagrangian and Hamiltonian formalisms in higher order calculus of variations, in *Proceedings of the International Meeting on Geometry and Physics*, Florence, 1982, ed. by M. Modugno (Pitagora, Bologna, 1983), pp. 43–70
25. P.L. Garcia, The Poincaré-Cartan invariant in the calculus of variations. Symp. Math. **14**, 219–246 (1974)
26. P.L. Garcia, J. Muñoz, On the geometrical structure of higher order variational calculus, in *Modern Developments in Analytical Mechanics I: Geometrical Dynamics*, ed. by S. Benenti, M. Francaviglia, A. Lichnerowicz. Proceedings of the IUTAM-ISIMM Symposium, Torino, 1982 (Accad. delle Scienze di Torino, Torino, 1983), pp. 127–147
27. G. Giachetta, L. Mangiarotti, G. Sardanashvily, *New Lagrangian and Hamiltonian Methods in Field Theory* (World Scientific, Singapore, 1997)
28. G. Giachetta, L. Mangiarotti, G. Sardanashvily, Covariant Hamilton equations for field theory. J. Phys. A Math. Gen. **32**, 6629–6642 (1999)
29. H. Goldschmidt, S. Sternberg, The Hamilton-Cartan formalism in the calculus of variations. Ann. Inst. Fourier **23**, 203–267 (1973)
30. M.J. Gotay, A multisymplectic framework for classical field theory and the calculus of variations, I. Covariant Hamiltonian formalism, in *Mechanics, Analysis and Geometry: 200 Years After Lagrange*, ed. by M. Francaviglia, D.D. Holm (North Holland, Amsterdam, 1990), pp. 203–235
31. M.J. Gotay, J.M. Nester, G. Hinds, Presymplectic manifolds and the Dirac–Bergmann theory of constraints. J. Math. Phys. **19**, 2388–2399 (1978)
32. M.J. Gotay, J.A. Isenberg, J.E. Marsden, R. Montgomery, *Momentum Mappings and the Hamiltonian Structure of Classical Field Theories with Constraints* (Springer, New York, 1992)
33. K. Grabowska, Lagrangian and Hamiltonian formalism in field theory: a simple model. J. Geom. Mech. **2**, 375–395 (2010)
34. P.A. Griffiths, *Exterior Differential Systems and the Calculus of Variations*. Progress in Mathematics, vol. 25 (Birkhäuser, Boston, 1983)
35. D.R. Grigore, On a generalization of the Poincaré–Cartan form in higher-order field theory, in *Variations, Geometry and Physics* (Nova Science Publishers, New York, 2008), pp. 57–76

36. A. Haková, O. Krupková, Variational first-order partial differential equations. J. Differ. Equ. **191**, 67–89 (2003)
37. H. Helmholtz, Ueber die physikalische Bedeutung des Prinzips der kleinsten Wirkung. J. für die reine u. angewandte Math. **100**, 137–166 (1887)
38. H.A. Kastrup, Canonical theories of Lagrangian dynamical systems in physics. Phys. Rep. **101**, 1–167 (1983)
39. J. Kijowski, A finite-dimensional canonical formalism in the classical field theory. Commun. Math. Phys. **30**, 99–128 (1973)
40. J. Kijowski, W. Szczyrba, Multisymplectic manifolds and the geometrical construction of the Poisson brackets in the classical field theory, in *Geometrie symplectique et physique mathematique*. Colloques internationaux du Centre national de la recherche scientifique, vol. 237 (C.N.R.S., Paris, 1975), pp. 347–378
41. J. Kijowski, W.M. Tulczyjew, *A Symplectic Framework for Field Theories*. Lecture Notes in Physics, vol. 107 (Springer, Berlin, 1979)
42. I. Kolář, Some geometric aspects of the higher order variational calculus, in *Geometrical Methods in Physics*, ed. by D. Krupka. Proceedings of the Conference on Differential Geometry and its Applications. Nové Město na Moravě, Sept. 1983, vol. 2 (J. E. Purkyně University, Brno, 1984), pp. 155–166
43. M. Krbek, J. Musilová, Representation of the variational sequence by differential forms. Rep. Math. Phys. **51**(2–3), 251–258 (2003)
44. M. Krbek, J. Musilová, Representation of the variational sequence by differential forms. Acta Appl. Math. **88**(2), 177–199 (2005)
45. D. Krupka, Some geometric aspects of variational problems in fibred manifolds. Folia Fac. Sci. Nat. Univ. Purk. Brunensis, Physica, Brno (Czechoslovakia) **14**, 65 pp. (1973). ArXiv:math-ph/0110005
46. D. Krupka, A map associated to the Lepagean forms of the calculus of variations in fibred manifolds. Czechoslov. Math. J. **27**, 114–118 (1977)
47. D. Krupka, On the local structure of the Euler-Lagrange mapping of the calculus of variations, in *Proceedings of the Conference on Differential Geometry and its Applications, Nové Město na Moravě* (Czechoslovakia), 1980 (Charles University, Prague, 1982), pp. 181–188. ArXiv:math-ph/0203034
48. D. Krupka, Lepagean forms in higher order variational theory, in *Modern Developments in Analytical Mechanics I: Geometrical Dynamics*, ed. by S. Benenti, M. Francaviglia, A. Lichnerowicz. Proceedings of the IUTAM-ISIMM Symposium, Torino, 1982 (Accad. Sci. Torino, Torino, 1983), pp. 197–238
49. D. Krupka, Variational sequences on finite order jet spaces, in *Differential Geometry and Its Applications*, ed. by J. Janyška, D. Krupka. Conference Proceeding, Brno, 1989 (World Scientific, Singapore, 1990), pp. 236–254
50. D. Krupka, Variational sequences in mechanics. Calc. Var. **5**, 557–583 (1997)
51. D. Krupka, Global variational theory in fibred spaces, in *Handbook of Global Analysis* (Elsevier Science B. V., Amsterdam, 2008), pp.773–836
52. D. Krupka, *Introduction to Global Variational Geometry* (Atlantis Press, Amsterdam, 2015)
53. D. Krupka, J. Šeděnková, Variational sequences and Lepage forms, in *Differential Geometry and Its Applications*, ed. by J. Bureš, O. Kowalski, D. Krupka, J. Slovák. Conference Proceeding, Prague, August 2004 (Charles University, Prague, 2005), pp. 617–627
54. D. Krupka, O. Krupková, G. Prince, W. Sarlet, Contact symmetries of the Helmholtz form. Differ. Geom. Appl. **25**, 518–542 (2007)
55. D. Krupka, O, Krupková, D. Saunders, The Cartan form and its generalizations in the calculus of variations. Int. J. Geom. Meth. Mod. Phys. **7**, 631–654 (2010)
56. O. Krupková, Lepagean 2-forms in higher order Hamiltonian mechanics, I. Regularity. Arch. Math. (Brno) **22**, 97–120 (1986)
57. O. Krupková, On the inverse problem of the calculus of variations for ordinary differential equations. Math. Bohem. **118**, 261–276 (1993)

58. O. Krupková, A geometric setting for higher-order Dirac–Bergmann theory of constraints. J. Math. Phys. **35**, 6557–6576 (1994)
59. O. Krupková, *The Geometry of Ordinary Variational Equations*. Lecture Notes in Mathematics, vol. 1678 (Springer, Berlin, 1997)
60. O. Krupková, Hamiltonian field theory. J. Geom. Phys. **43**, 93–132 (2002)
61. O. Krupková, Variational metric structures. Publ. Math. Debr. **62**, 461–495 (2003)
62. O. Krupková, Lepage forms in the calculus of variations, in *Variations, Geometry and Physics* (Nova Science Publishers, New York, 2008), pp. 29–56
63. O. Krupková, Variational equations on manifolds, in *Advances in Mathematics Research*, vol. 9 (Nova Science Publishers, New York, 2009), pp. 201–274
64. O. Krupková, R. Malíková, Helmholtz conditions and their generalizations. Balkan J. Geom. Appl. **15**(1), 80–89 (2010)
65. O. Krupková, G.E. Prince, Lepage forms, closed 2-forms and second-order ordinary differential equations. Russ. Math. (Iz. VUZ) **51**(12), 1–16 (2007)
66. O. Krupková, D.J. Saunders (eds.), *Variations, Geometry and Physics* (Nova Science Publishers, New York, 2009), p. 360
67. O. Krupková, D.J. Saunders, Affine duality, and Lagrangian and Hamiltonian systems. Int. J. Geom. Methods Mod. Phys. **8**, 669–697 (2011)
68. O. Krupková, D. Smetanová, On regularization of variational problems in first-order field theory, in *Proceedings of the 20th Winter School "Geometry and Physics" (Srní, 2000)*. Rend. Circ. Mat. Palermo (2), Suppl. No. 66 (2001), pp. 133–140
69. O. Krupková, D. Smetanová, Legendre transformation for regularizable Lagrangians in field theory. Lett. Math. Phys. **58**, 189–204 (2001)
70. O. Krupková, D. Smetanová, Lepage equivalents of second order Euler–Lagrange forms and the inverse problem of the calculus of variations. J. Nonlin. Math. Phys. **16**, 235–250 (2009)
71. B. Kupershmidt, Geometry of jet bundles and the structure of Lagrangian and Hamiltonian formalisms, in *Lecture Notes in Mathematics*, vol. 775 (Springer, Berlin, 1980), pp. 162–217
72. T. Lepage, Sur les champs géodésiques du calcul des variations. Bull. Acad. Roy. Belg. Cl. des Sci. **22**, 716–729 (1936)
73. P. Libermann, C.-M. Marle, in *Symplectic Geometry and Analytical Mechanics*. Mathematics and Its Applications (D. Reidel, Dordrecht, 1987)
74. R. Malíková, On a generalization of Helmholtz conditions. Acta Math. Univ. Ostrav. **17**, 11–21 (2009)
75. L. Mangiarotti, M. Modugno, Fibred spaces, jet spaces and connections for field theories, in *Proceedings of the Meeting 'Geometry and Physics', 1982, Florence* (Pitagora, Bologna, 1982), pp. 135–165
76. J.E. Marsden, S. Pekarsky, S. Shkoller, M. West, Variational methods, multisymplectic geometry and continuum mechanics. J. Geom. Phys. **38**, 253–284 (2001)
77. P.J. Olver, Equivalence and the Cartan form. Acta Appl. Math. **31**, 99–136 (1993)
78. M. Palese, O. Rossi, E. Winterroth, J. Musilová, Variational sequences, representation sequences and applications in physics, 61 pp. ArXiv:1508.01752v1 [math-ph]; SIGMA **12**, 045 45 pp. (2016)
79. H. Poincaré, *Les méthodes nouvelles de la Mécanique céleste*, vol. 3 (Gauthier-Villars, Paris, 1899)
80. O. Rossi, Homogeneous differential equations and the inverse problem of the calculus of variations. Publ. Math. Debr. **84**, 165–188 (2014)
81. O. Rossi, Geometry of variational partial differential equations and Hamiltonian systems, in *Geometry of Jets and Fields, Banach Center Publications*, vol. 110 (Inst. of Math., Polish of Academy of Sciences, Warszawa, 2016), pp. 219–237
82. O. Rossi, D.J. Saunders, Dual jet bundles, Hamiltonian systems and connections. Diff. Geom. Appl. **35**, 178–198 (2014)
83. O. Rossi, D. Saunders, Lagrangian and Hamiltonian duality. J. Math. Sci. **218**, 813–816 (2016)

84. W. Sarlet, Geometrical structures related to second-order equations, in *Differential Geometry and Its Applications*, ed. by D. Krupka, A. Švec. Conference Proceedings, Brno, 1986 (D. Reidel, Dordrecht, 1986), pp. 279–288
85. D.J. Saunders, Jet fields, connections and second-order differential equations. J. Phys. A Math. Gen. **20**, 3261–3270 (1987)
86. D.J. Saunders, *The Geometry of Jet Bundles*. London Mathematical Society Lecture Note Series, vol. 142 (Cambridge University Press, Cambridge, 1989)
87. D.J. Saunders, The regularity of variational problems. Contemp. Math. **132**, 573–593 (1992)
88. D.J. Saunders, A new approach to the nonlinear connection associated with second order (and higher-order) differential equation fields. J. Phys. A Math. Gen. **30**, 1739–1743 (1997)
89. E. Tonti, Variational formulation of nonlinear differential equations I, II. Bull. Acad. Roy. Belg. Cl. Sci. **55**, 137–165, 262–278 (1969)
90. M.M. Vainberg, *Variational Methods in the Theory of Nonlinear Operators* (GITL, Moscow, 1959) (in Russian)
91. A.M. Vinogradov, A spectral sequence associated with a non-linear differential equation, and algebro-geometric foundations of Lagrangian field theory with constraints. Soviet Math. Dokl. **19**, 144–148 (1978)
92. A. Weinstein, *Lectures on Symplectic Manifolds*. CBMS Regional Conference Series in Mathematics, vol. 29 (American Mathematical Society, Providence, 1977), 48 pp.
93. R.O. Wells, *Differential Analysis on Complex Manifolds* (Springer, Berlin, 1980)
94. E.T. Whittaker, *A Treatise on Analytical Dynamics of Particles and Rigid Bodies* (The University Press, Cambridge, 1917)

Printed in the United States
By Bookmasters